# Dynamical Systems and Control

# Stability and Control: Theory, Methods and Applications

A series of books and monographs on the theory of stability and control

Edited by A.A. Martynyuk
*Institute of Mechanics, Kiev, Ukraine*
and V. Lakshmikantham
*Florida Institute of Technology, USA*

*Stability and Control: Theory, Methods and Applications*
*Volume 22*

# Dynamical Systems and Control

EDITED BY

## Firdaus E. Udwadia
University of Southern California
USA

## H. I. Weber
Pontifical Catholic University of Rio de Janeiro
Brazil

## George Leitmann
University of California, Berkeley
USA

**CHAPMAN & HALL/CRC**

A CRC Press Company
Boca Raton London New York Washington, D.C.

### Library of Congress Cataloging-in-Publication Data

Dynamical systems and control / edited by F.E. Udwadia, G. Leitmann, and H.I. Weber
      p. cm. (Stability and control ; v.22)
      Includes bibliographical references and index.
      ISBN 0-415-30997-2 (alk. paper)
      1. Dynamics. 2. Differentiable dynamical systems. 3. Control theory. I. Udwadia, F.E. II.
Leitmann, George. III. Weber, H. (Hans) IV. International Workshop on Dynamics and Control
(11th : 2000 : Rio de Janeiro, Brazil) V. Title. VI. Series.

QA845.D93 2004
003'.85—dc22
                                                     2004043591

**Visit the CRC Press Web site at www.crcpress.com**

# Contents

# Part III

# Contributors

**N.U. Ahmed,** School of Information Technology and Engineering, Department of Mathematics, University of Ottawa, Ottawa, Ontario

**José Manoel Balthazar,** Instituto de Geociências e Ciências Exatas – UNESP – Rio Claro, Caixa Postal 178, CEP 13500-230, Rio Claro, SP, Brasil

**Débora Belato,** DPM – Faculdade de Engenharia Mecânica – UNICAMP, Caixa Postal 6122, CEP 13083-970, Campinas, SP, Brasil

**Luiz Bevilacqua,** Laboratório Nacional de Computação Científica – LNCC, Av. Getúlio Vargas 333, Rio de Janeiro, RJ 25651-070, Brasil

**D. Blank,** Institut für Neuroinformatik, ETHZ/UNIZH, Winterthurerstraße 190, CH-8057 Zürich

**R.M.L.R.F. Brasil,** Dept. of Structural and Foundations Engineering, Polytechnic School, University of São Paulo, P.O. Box 61548, 05424-930, SP, Brazil

**Edson Cataldo,** Universidade Federal Fluminense (UFF), Departamento de Matemática Aplicada, PGMEC-Programa de Pós-Graduação em Engenharia Mecânica, Rua Mário Santos Braga, S/No-24020, Centro, Niterói, RJ, Brasil

**Ye-Hwa Chen,** The George W. Woodruff School of Mechanical Engineering, Georgia Institute of Technology, Atlanta, Georgia 30332, USA

**E. Crück,** Laboratoire de Recherches Balistiques et Aérodynamiques, BP 914, 27207 Vernon Cedex, France

**Seyyed Said Dana,** Graduate Studies in Mechanical Engineering, Mechanical Engineering Department, Federal University of Paraiba, Campus I, 58059-900 Joao Pessoa, Paraiba, Brazil

**Christophe Deissenberg,** CEFI, UMR CNRS 6126, Université de la Méditerranée (Aix-Marseille II), Château La Farge, Route des Milles, 13290 Les Milles, France

**José João de Espíndola,** Department of Mechanical Engineering, Federal University of Santa Catarina, Brazil

**Gustav Feichtinger,** Institute for Econometrics, OR and Systems Theory, University of Technology, Argentinierstrasse 8, A-1040 Vienna, Austria

**J.L.P. Felix,** School of Mechanical Engineering, UNICAMP, P.O. Box 6122, 13800-970, Campinas, SP, Brazil

**A. Fenili,** School of Mechanical Engineering, UNICAMP, P.O. Box 6122, 13800-970, Campinas, SP, Brazil

**Henryk Flashner,** Department of Aerospace and Mechanical Engineering, University of Southern California, Los Angeles, CA 90089-1453

**Agenor de Toledo Fleury,** Control Systems Group/Mechanical & Electrical Engineering Division, IPT/ São Paulo State Institute for Technological Research, P.O. Box 0141, 01064-970, São Paulo, SP, Brazil

**F.J. Garzelli,** Dept. of Structural and Foundations Engineering, Polytechnic School, University of São Paulo, P.O. Box 61548, 05424-930, SP, Brazil

**Michael Golat,** Department of Aerospace and Mechanical Engineering, University of Southern California, Los Angeles, CA 90089-1453

**Francisco Alvarez Gonzalez,** Dpto. Economia Cuantitativa, Universidad Complutense, Madrid, Spain

**Richard F. Hartl,** Institute of Management, University of Vienna, Vienna, Austria

**Daniel J. Inman,** Center for Intelligent Material Systems and Structures, Virginia Polytechnic Institute and State University, Blacksburg, VA 24061-0261, USA

**Fernando Kokubun,** Department of Physics, Federal University of Rio Grande, Rio Grande, RS, Brazil

**Peter Kort,** Department of Econometrics and Operations Research and CentER, Tilburg University, Tilburg, The Netherlands

**G. Leitmann,** College of Engineering, University of California, Berkeley CA 94720, USA

**Vicente Lopes, Jr.,** Department of Mechanical Engineering – UNESP-Ilha Solteira, 15385-000 Ilha Solteira, SP, Brazil

**S. Mancuso,** Rice University, Houston, Texas, USA

**Naor Moraes Melo,** Graduate Studies in Mechanical Engineering, Mechanical Engineering Department, Federal University of Paraiba, Campus I, 58059-900 Joao Pessoa, Paraiba, Brazil

**Luciano Luporini Menegaldo,** São Paulo State Institute for Technological Research, Control System Group / Mechanical and Electrical Engineering Division, P.O. Box 0141, CEP 01604-970, São Paulo-SP, Brazil

**A. Miele,** Rice University, Houston, Texas, USA

**Helio Mitio Morishita,** University of São Paulo, Department of Naval Architecture and Ocean Engineering, Av. Prof. Mello Moraes, 2231, Cidade Universitária 05508-900, São Paulo, SP, Brazil

**Frederico Ricardo Ferreira de Oliveira,** Mechanical Engineering Department/ Escola Politécnica, USP – University of São Paulo, P.O. Box 61548, 05508-900, São Paulo, SP, Brazil

**Humberto Piccoli,** Department of Materials Science, Federal University of Rio Grande, Rio Grande, RS, Brazil

**Eduard Reithmeier,** Institut für Meß- und Regelungstechnik, Universität Hannover, 30167 Hannover, Germany

**Rubens Sampaio,** Pontifícia Universidade Católica do Rio de Janeiro (PUC-Rio), Departamento de Engenharia Mecânica, Rua Marquês de São Vicente, 225, 22453-900, Gávea, Rio de Janeiro, Brasil

**N. Seube,** Ecole Nationale Supérieure des Ingénieurs des Etudes et Techniques d'Armement, 29806 BREST Cedex, France

**Marianna A. Shubov,** Department of Mathematics and Statistics, Texas Tech University, Lubbock, TX, 79409, USA

**João Morais da Silva Neto,** Department of Mechanical Engineering, Federal University of Santa Catarina, Brazil

**Simplicio Arnaud da Silva,** Graduate Studies in Mechanical Engineering, Mechanical Engineering Department, Federal University of Paraiba, Campus I, 58059-900 Joao Pessoa, Paraiba, Brazil

**Jessé Rebello de Souza Junior,** University of São Paulo, Department of Naval Architecture and Ocean Engineering, Av. Prof. Mello Moraes, 2231, Cidade Universitária 05508-900, São Paulo, SP, Brazil

**Valder Steffen, Jr.,** School of Mechanical Engineering Federal University of Uberlândia, 38400-902 Uberlândia, MG, Brazil

**R. Stoop,** Institut für Neuroinformatik, ETHZ/UNIZH, Winterthurerstraße 190, CH-8057 Zürich

**F.E. Udwadia,** Department of Aerospace and Mechanical Engineering, Civil Engineering, Mathematics, and Operations and Information Management, 430K Olin Hall, University of Southern California, Los Angeles, CA 90089-1453

**Vladimir Veliov,** Institute for Econometrics, OR and Systems Theory, University of Technology, Argentinierstrasse 8, A-1040 Vienna, Austria

**T. Wang,** Rice University, Houston, Texas, USA

**Hans Ingo Weber,** DEM - Pontifícia Universidade Católica – PUC – RJ, CEP 22453-900, Rio de Janeiro, RJ, Brasil

# Preface

This book contains some of the papers that were presented at the 11th International Workshop on Dynamics and Control in Rio de Janeiro, October 9–11, 2000. The workshop brought together scientists and engineers in various diverse fields of dynamics and control and offered a venue for the understanding of this core discipline to numerous areas of engineering and science, as well as economics and biology. It offered researchers the opportunity to gain advantage of specialized techniques and ideas that are well developed in areas different from their own fields of expertise. This cross-pollination among seemingly disparate fields was a major outcome of this workshop.

The remarkable reach of the discipline of dynamics and control is clearly substantiated by the range and diversity of papers in this volume. And yet, all the papers share a strong central core and shed understanding on the multiplicity of physical, biological and economic phenomena through lines of reasoning that originate and grow from this discipline.

I have separated the papers, for convenience, into three main groups, and the book is divided into three parts. The first group deals with fundamental advances in dynamics, dynamical systems, and control. These papers represent new ideas that could be applied to several areas of interest. The second deals with new and innovative techniques and their applications to a variety of interesting problems that range across a broad horizon: from the control of cars and robots, to the dynamics of ships and suspension bridges, to the determination of optimal spacecraft trajectories to Mars. The last group of papers relates to social, economic, and biological issues. These papers show the wealth of understanding that can be obtained through a dynamics and control approach when dealing with drug consumption, economic games, epidemics, neo-cortical synchronization, and human posture control.

This workshop was funded in part by the US National Science Foundation and CPNq. The organizers are grateful for the support of these agencies.

Firdaus E. Udwadia

# PART I

# A Geometric Approach to the Mechanics of Densely Folded Media

Luiz Bevilacqua

*Laboratório Nacional de Computação Científica – LNCC*
*Av. Getúlio Vargas 333, Rio de Janeiro, RJ 25651-070, Brasil*
*Tel: 024-233.6024, Fax: 024-233.6167, E-mail: bevi@lncc.br*

To date, the analysis of densely folded media has received little attention. The stress and strain analysis of these types of structures involves considerable difficulties because of strong nonlinear effects. This paper presents a theory that could be classified as a geometric theory of folded media, in the sense that it ultimately leads to a kind of geometric constitutive law. In other words, a law that establishes the relationship between the geometry and other variables such as the stored energy, the apparent density and the mechanical properties of the material. More specifically, the theory presented here leads to fundamental governing equations for the geometry of densely folded media, namely, wires, plates and shells, as functions of the respective slenderness ratios. With the help of these fundamental equations other relationships involving the apparent density and the energy are obtained. The structure of folded media according to the theory has a fractal representation and the fractal dimension is a function of the material ductility. Although at present we have no experiments to test the conjectures that arise from our analytical developments, the theory developed here is internally consistent and therefore provides a good basis for designing meaningful experiments.

## 1   Introduction

Crush a sheet of paper till it becomes a small ball. This is an example of what we will call a folded medium. That is, we have in mind strongly folded media. The mechanical behavior of these kinds of structures could be analyzed as a very dense set of interconnected structural pieces in such a way as to form a continuum. The initial difficulty of using the classical solid mechanics approach, in this case, lies in the definition of the proper geometry. A strongly folded medium, except when folded following very strict rules, doesn't present a regular pattern. When we crush a piece of paper the simple elements that compose the final complex configuration are distributed at random and in different sizes. So, a preliminary problem to be solved is gaining an understanding of both the local and the global geometry.

Let us think, for instance, of a paper or metal sheet densely folded to take the shape of a ball. Classical structural analysis presents serious difficulties in determining the final configuration, for this involves a complex combination of buckling, post-buckling, nonlinearities – both geometric and material – large displacements, just to mention a few. If we are basically interested only in the geometry, is it possible to establish a simple "global law" that would correlate some appropriate variables leading to the characterization of the final shape? The aim of this paper is to answer this question. A simple law is proposed as a kind of geometric constitutive equation that is different in nature from the classical concept of constitutive laws in mechanics. Some consequences are drawn from this basic law concerning mass distribution, work and energy used in the packing process.

We believe that the results are plausible, that is, there are no violations of basic principles, and there are no contradictions concerning the expected behavior of a real material. But, to be recognized as scientifically valid, the theory need to be tested against experimental results. Despite the fact that rigorous experiments are missing, the development of a coherent theory is important, both from the viewpoint of obtaining comments and suggestions on it, and from the viewpoint of developing experimental methodologies.

In the next sections we will examine densely folded wires and densely folded plates and shells. To the best of our knowledge, the current technical literature does not include references on this subject. We have exposed the basic ideas of this theory in [1] and [2]. This paper, however, is self-contained, it is a kind of closure where the concepts are presented more clearly and precisely. Except for the dynamic behavior, which is not included here, the other references are not necessary to understand this paper. The dynamics of folded media still need further analysis. The ideas advanced in [2] are at an exploratory stage and need several corrections.

## 2   Densely Folded Wires

Let us assume that a thin wire is pushed into a box with two predominant dimensions, length ($L$) and width ($h$), while the depth is approximately equal to the wire diameter, much smaller than $L$ and $h$. It is difficult to make a prediction about the geometry of the wire inside the box. The amount of wire packed in a box depends on the wire diameter, the energy expended in the process and the mechanical properties of the wire material, particularly its ductility. We will explain later what is understood as ductility in the present context.

How all those variables correlate with each other will be discussed in the sequel. It is possible, however, to anticipate some dependence relationships by appealing to common sense. It is expected, for instance, that by decreasing the wire diameter, while keeping all other variables constant, the length of the wire packed inside the box will increase. Also, by decreasing the ductility, that is, the capacity to

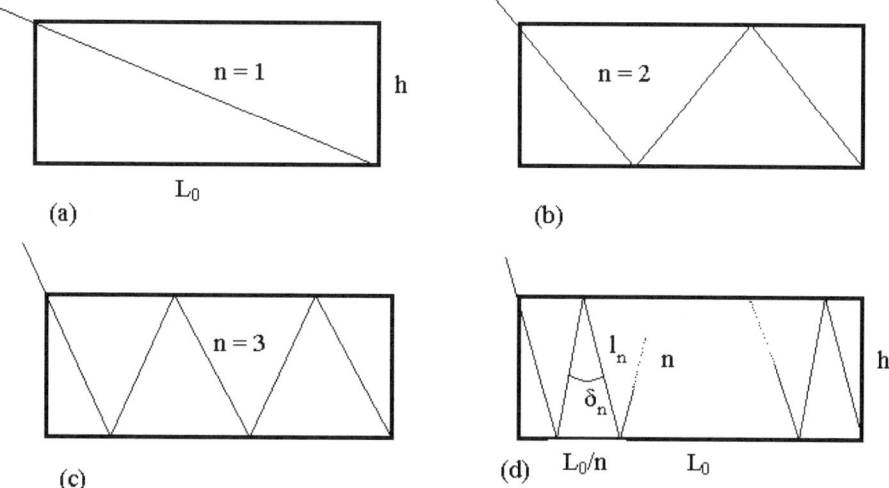

Figure 1 An ideal packing of a wire inside the box $[L_0 \times H]$.

accumulate plastic deformation, while keeping all the other variables constant, it is intuitively acceptable that the wire length in the box will increase. Other interpretations are not so straightforward and will be discussed in the proper section.

Consider a thin box $[L_0 \times h]$ and assume that a wire with diameter $\phi$ is pushed into the box. In order to simplify the problem, it is assumed in the sequel that plastic hinges will appear in the process, such that, after reaching the final stable configuration, the wire geometry can be reduced to a sequence of straight segments linked together, through plastic hinges, in the shape of a broken random line. It will be assumed throughout this paper that we are dealing with a perfectly plastic material.

Let us start with an ideal case, consisting of the configuration sequence following the pattern shown in Figure 1.

It can be easily shown that for the $n$-th term in the sequence:

$$l_n = L_0 \left( \frac{1}{(2n-1)^2} + \beta^2 \right)^{1/2} \tag{1a}$$

and

$$L_n = \sum_{m=1}^{M} l_n^{(m)} = L_0 \left( 1 + (2n-1)^2 \beta^2 \right)^{1/2} , \tag{2a}$$

where $\beta = h/L_0$.

Clearly $L_n$ is the total length of the wire inside the box corresponding to the $n$-th term. For very thin wires relative to the box dimensions, i.e., $\phi \ll L_0$ and $\phi \ll h$, the number of folds is large, $n \gg 1$. The segment $l_n$ and the total length $L_n$ can then be approximated by:

$$l_{n_L} \cong L_0 \beta = h , \tag{1b}$$

$$L_{n_L} \cong 2n_L \beta L_0 = 2n_L h . \tag{2b}$$

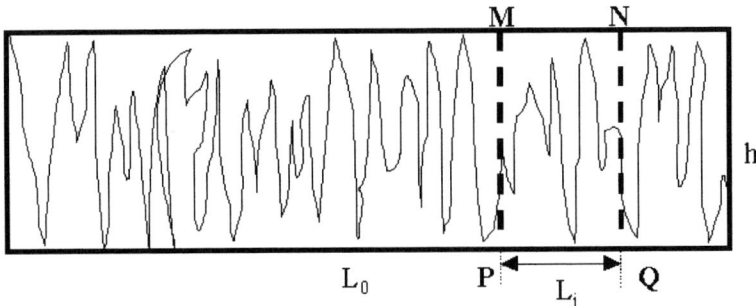

Figure 2  Appropriate geometry of a wire densely packed in a box.

For this limit case, where $n$ is very large we set $n = n_L$. It follows from (1b) and (2b) that, in this limit case, the wire occupies the total area of the box. Assuming that the material is incompressible, the mass conservation principle requires:

$$L_n \phi \leq L_0 h \,, \tag{3}$$

where the inequality sign holds when the confined wire fills up the box. Therefore for $n \to n_L$ we may write:

$$\frac{L_n}{L_0} \cong \frac{h}{\phi} \cong \left(\frac{\phi}{h}\right)^{-1} . \tag{4}$$

Combining (4) and (2b) the limit value $n_L$ can be estimated by:

$$n_L \cong \frac{1}{2} \frac{L_0}{\phi} . \tag{5}$$

This is a limit value and clearly corresponds to the case where the wire fills up the box. In general we might expect the wire geometry inside the box to be similar to the line depicted in Figure 2, only partially covering the box in a random way.

It is not likely that the wire will be so densely packed as to fill up the box. For the general case the expression (4) can then be written in the following form:

$$\frac{L_n}{L_0} \cong e_0 \left(\frac{h}{\phi}\right)^{P} \cong e_0 \left(\frac{\phi}{h}\right)^{-p} , \tag{6}$$

where the exponent $p$ is less than or equal to 1 ($p \leq 1$). If $p = 1$ the line representing the wire will cover the entire region $[L_0 \times h]$, if it is less than 1 it will only partially cover this same region. The constant $e_0$ may be adjusted to fit experimental results. Expression (6) can be put into a more convenient form for the current notation of the fractionary geometry:

$$\Gamma = e_0 \rho^{1-D} \tag{7a}$$

or

$$\log \Gamma = \log e_0 + (1 - D) \log \rho \,, \tag{7b}$$

where $\Gamma = L/L_0$, $\rho = \phi/h$, and $D$ is the fractal dimension. We have dropped the subscript $n$ in $L_n$ for the sake of simplicity.

If the fractal dimension of the line representing the wire equals two ($D = 2$), that is, the line fills up the plane, expression (7a) reduces to (4), as it should be. For the other limit, $D = 1$, the one-dimensional Euclidean geometry is preserved, $\Gamma = e_0$. In particular, for $e_0 = 1$ we get $L_n = L$, that is, there is no folding at all. The extreme cases have no practical interest, but they provide a good assessment of the theory, showing that there is no contradiction, and the conjecture is plausible.

The lower bound $D = 1$, corresponding to folding-free configurations, arises from two distinct origins. The first appears as a consequence of geometric constraints. Indeed, if $h = \phi$, the wire fits the box perfectly. There is no room for bending, the stress distribution on the wire cross-section is uniform. We are in the presence of a pure axial force and the wire collapses under simple compression. The material properties do not play any particular role in this limit case. The second possibility has to do with the material properties and is independent of the geometry. Indeed, if the material is perfectly stiff, that is, does not admit any plastic strain, the collapse occurs without any permanent deformation. In other words, there is no stable folding configuration, which is contrary to one of the fundamental requirements of this theory.

In real cases, however, we have an intermediate situation. Taking into account the discussion above, we may use the following criteria to establish the range of validity of the theory:

1. The ratio $\rho = \phi/h$ must remain inferior to 0.1: $\rho < 0.1$.
2. The contribution to the total dissipated plastic work ($W_t$) due to pure axial strain state ($W_a$) should be much smaller than the contribution due to bending ($W_b$): $W_a \leq 0.1W_t$.

The above conditions may be very strict, but only properly conducted experiments can give a conclusive answer.

Figure 3 depicts the expected variation of $\log \Gamma$ against $\log \rho$ for different values of $D$. We have assumed that $e_0$ is constant for all values of $D$, which could not be strictly true. More generally, put $e_0 = e_0(D)$, in which case the lines corresponding to $D_1$, $D_2$ and $D_3$ would not converge to the same point on the vertical axis.

From Figure 3 it is clear that the packing capacity for a given value of $\rho$ depends on the fractal dimension. Increasing values of $D$ correspond to increasing packing capacities $\Gamma$. This means that $D$ measures the propensity to incorporate plastic deformation. An experiment leading to points on the line with slope $(1 - D_1)$ in Figure 3 indicates geometric and material conditions much more favorable to incorporating permanent plastic deformation than an experiment that follows the line with slope $(1 - D_3)$.

In order to define more precisely this behavior we will introduce the notion of apparent ductility. This notion will be better understood along with the determination of the dissipated plastic work.

As mentioned before, only perfect plastic materials are considered here and the final configuration is stable. This means that if the box is removed after the packing process, the final geometry will be preserved. The energy considered here is therefore the net energy necessary to introduce permanent plastic deformation.

Let us start again with the ideal case. According to the fundamental assumptions plastic hinges will form in the vertices of the line representing the folded wire. The

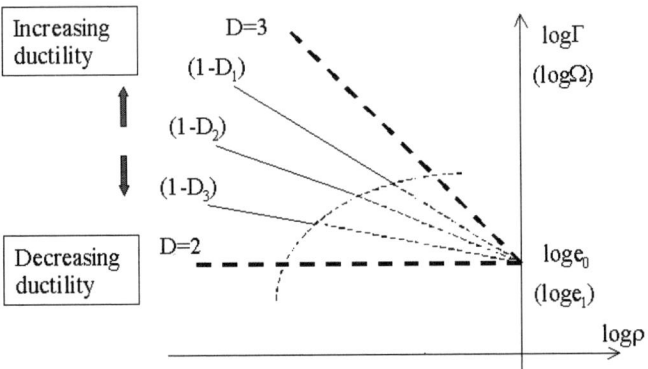

Figure 3 Expected variations of the packing capacity with the wire rigidity. Increasing fractal dimensions $D_1 > D_2 > D_3$ correspond to decreasing rigidity.

net work necessary to produce a rotation equal to $\theta_n$ in a typical plastic hinge as shown in Figure 1d is approximately equal to:

$$\tau_n = \hat{k}\sigma_Y \phi^3 \theta_n \,,$$

where $\hat{k}$ is a constant, and $\sigma_Y$ is the yield stress of the perfect plastic material.

Now using the notation shown in Figure 2, the rotation can be written as:

$$\theta_n = \pi - \delta_n \,.$$

But $\delta_n$ is of the order of $\dfrac{L_0}{nh}$ and for very large $n$, $\delta_n$ is very small. Therefore we may write:

$$\theta_n \cong g(U)\pi \,,$$

where $g(U)$ is a correction factor to take into account the material hardening, that is, the maximum rotation capacity of a typical hinge. The function $g(U)$ expresses the material capacity to accumulate plastic deformation. If $g(U) = 1$ the material is extremely ductile and if $g(U) = 0$ there is no possible bending without failure; it is an extreme case of a brittle material. We are defining $g(U)$ as a function of the material ductility $U$ that will be discussed below.

The total dissipated plastic work is given therefore by:

$$W_n = n\tau_n \cong n\hat{k}\sigma_Y g(U)\pi\phi^3 \,. \tag{8}$$

Introducing the value of $n$ given by (2b) we obtain:

$$W_n = \frac{\pi}{2}\hat{k}\sigma_Y \frac{L_{n_L}}{h} g(U)\phi^3 \,.$$

Or with equation (7a):

$$W_n = \frac{\pi}{2}\hat{k}e_0\sigma_Y L_0 h^2 g(U)\rho^{4-D} \,. \tag{9}$$

Now defining the reference plastic work as

$$W_R = \frac{\pi}{2} \hat{k} \sigma_Y h^3 G(U, \beta) ,$$

where

$$G(U, \beta) = \frac{1}{\beta} g(U) \quad \text{and} \quad \beta = \frac{h}{L_0} ,$$

we obtain:

$$W_n = W_R \rho^{4-D} . \tag{10}$$

Call $W_R$ the reference dissipated plastic work. $G(U, \beta)$ is the apparent ductility that involves both material properties $g(u)$ and the geometry $\beta$.

For the purpose of the present paper the ductility may be defined as:

$$U = 1 - \frac{\sigma_Y}{E \varepsilon_u} ,$$

where $\varepsilon_u$ is the ultimate strain at fracture, and $E$ is the Young modulus.[1]

$U$ varies from zero, when $\varepsilon_u = \varepsilon_Y$, and in this case the material is very stiff, and the failure is characterized by brittle fracture with no plastic deformation, to 1, for the ideal case of unlimited elongation at fracture $\varepsilon_u \to \infty$.

The packing capacity increases with the material ductility $g(u)$ and with the ratio $\beta = h/L_0$. The parameter $\beta$ can be interpreted as the geometric ductility. We may assert therefore that, for a fixed value of $h$, the fractal dimension will be an increasing function of the apparent ductility $G(U, \beta)$.

Dropping the subscript $n$ in (10) for the sake of simplicity, we finally obtain:

$$\tau = \frac{W}{W_R} = \rho^{4-D} . \tag{11}$$

We will call $\tau$ the dissipated plastic work density.

To illustrate the variation of dissipated plastic work with ratio $\rho$ and packing capacity $\Gamma$ consider the three points M, N and P shown in Figure 4. The points P and N correspond to the same value of $\rho$, and also to the same wire diameter, provided that $h$ is fixed. We assume that the material properties are constant for all wires. Now clearly $\Gamma_a < \Gamma_b$ and $D_2 < D_1$. Now from (11):

$$\frac{W_N}{W_P} = \rho^{-D_1 + D_2} . \tag{12}$$

But since $D_2 < D_1$ and $\rho < 1$ the right hand term in (12) is greater than one. Then from (12) we may write:

$$\frac{W_N}{W_P} > 1 .$$

---

[1] For materials displaying a stress-strain curve that can be approximated by a bi-linear law, with $\sigma_u$ as the ultimate stress at fracture, the ductility reads:

$$U = \frac{1}{2} \left( 1 + \frac{\sigma_Y}{\sigma_u} \right) \left( 1 - \frac{\sigma_Y}{E \sigma_u} \right) .$$

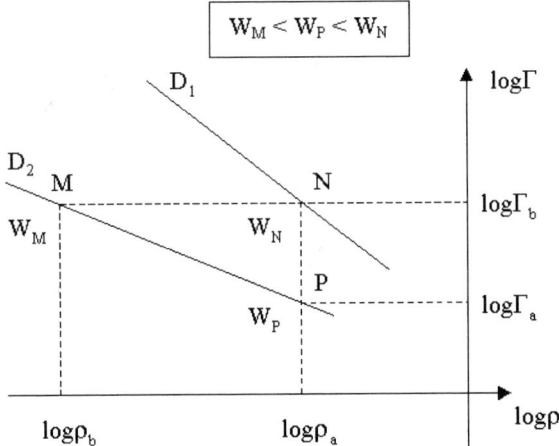

Figure 4   Packing capacities for different combinations of the wire diameter, packing capacity and apparent ductility.

Finally we may conclude that $W_P < W_N$.

Consider now the points M and N. For these two points the packing capacity is the same and $\rho_b < \rho_a$. Therefore from (11) and (7a):

$$\frac{W_M}{W_N} = \left(\frac{\rho_b}{\rho_a}\right)^3 . \tag{13}$$

But since $\rho_b < \rho_a$ the right hand term in (13) is less than one, therefore

$$\frac{W_M}{W_N} < 1 .$$

Then $W_M < W_N$.

Finally consider the points P and M, on the line corresponding to the same fractal dimension. From (11), given that $\rho_a > \rho_b$ and since the two points belong to a line with the same fractal dimension, that is, the same apparent ductility we have immediately:

$$W_M < W_P .$$

From equations (7b) and (11) it is possible to find an explicit expression for the fractal dimension as a function of the folding capacity $\Gamma$ and the plastic work density $\tau$:

$$D = 1 - 3 \frac{\log\left(\dfrac{\Gamma}{e_0}\right)}{\log \tau - \log\left(\dfrac{\Gamma}{e_0}\right)} . \tag{14}$$

Define now the packing density $\Omega$, as the ratio between the apparent specific mass $\mu$, per unit box length, and the specific mass of the wire $\mu_0$, per unit wire length. The apparent specific mass is defined as the total weight of the wire packed inside the box divided by the box length $L_0$. This definition implies the homogenization of the specific mass, making it uniformly distributed along the box in what, in general, is a good approximation. After some simple calculations it is then possible to write:

$$\mu = \frac{\mu_0 \sum_{m=1}^{M} l_n^{(m)}}{L} \qquad n \to n_L, \tag{15a}$$

$$\Omega = e_0 \rho^{1-D},$$

or

$$\log \Omega = \log e_0 + (1 - D) \log \rho, \tag{15b}$$

where $\Omega = \mu/\mu_0$.

Therefore the same law governing the packing capacity $\Gamma$ also applies to the packing density $\rho$. The representation in Figure 3 is equally valid for this case.

Analytically the parameter $e_0$ in (15a,b) is the same as in (7a,b). Only experimental evidence can confirm this result.

The limit cases have the same interpretation as for the packing capacity. Putting $e_0 = 1$, the limit case $D = 1$ is satisfied, for the virtual specific mass will coincide with the wire specific mass. But there is no strong reason to abandon from the very beginning the hypothesis of having $e_0 = e_0(D)$. This might well be the case, and at the present stage only experimental data can provide answers on this subject.

Now, consider the region MNPQ shown in Figure 2. It is plausible to assume that the plastic energy stored in the wire inside the box is uniformly distributed along the length $L_0$. Therefore the energy stored in MNPQ $W_j$, is $L_j/L_0$ times the total energy necessary to pack the wire inside the box $[L_0 \times h]$. But also the reference energy $W_R$ is proportional to $L_0$ by definition, therefore the reference energy corresponding to the box $[L_j \times h]$ $W_{Rj}$ is also $L_j/L_0$ times $W_R$ corresponding to the box $[L_0 \times h]$. Since the plastic work density $\tau$ is the ratio $W_j/W_{Rj}$ it is easily seen that $\tau$ is invariant for any sub-region $[L_j \times h]$ of $[L_0 \times h]$. Combining (7a) and (11) we get:

$$\Gamma_j = e_0 E_j^{1-D} = e_0 E^{1-D} = \Gamma.$$

That is the packing capacity is the same for the box $[L_0 \times h]$ and for any of its sub-regions $[L_j \times h]$. This result is coherent with the hypothesis of the uniform distribution of the specific mass introduced before. It is also a confirmation of the intrinsic self-similarity property required by the structure of the fractal geometry.

The above discussions lead to some conclusions that can be summarized as follows:

**Proposition 1**     *1. The geometry of densely folded wires packed in a two-dimensional box – [L × h], L > h – has a random structure characterized by a fractal dimension $1 < D < 2$, provided that:*

    *i. The slenderness ratio defined by $\rho = \phi/h$, where $\phi$ is the wire diameter, is sufficiently small. That is, $\rho \ll 1$.*

    *ii. The material is perfectly plastic.*

    *iii. The final configuration is stable.*

2. *The packing capacity* $\Gamma = L/L_0$, *and the packing density* $\Omega = \mu/\mu_0$, *vary with the slenderness ratio according to the power law:*

$$\Gamma = e_0 \rho^{1-D},$$

*and*

$$\Omega = e_0 \rho^{1-D},$$

*where $L$ is the wire length packed in the box; $L_0$ is the box length; $\mu$ is the apparent specific mass; $\mu_0$ is the wire specific mass per unit length. The parameter $e_0$ is to be determined experimentally.*

3. *The fractal dimension $D$ depends on the apparent ductility. Wires folding in a configuration with a high fractal dimension $D$ will have a corresponding high apparent ductility. For a given value of the packing capacity $\Gamma$, the fractal dimension is a function of the dissipated plastic work density $\tau$:*

$$D = 1 - 3 \frac{\log\left(\dfrac{\Gamma}{e_0}\right)}{\log \tau - \log\left(\dfrac{\Gamma}{e_0}\right)}.$$

4. *If the packing capacity is governed by a power law as given in item 2 above, then the geometry representing the wire final configuration is self-similar. Conversely, if the geometry is self-similar the packing configuration has a fractal structure as indicated in item 2 above.*

# 3   Densely Folded Shells

Let us move to a more complex case. Consider a uniform spherical thin shell with radius equal to $R$ under uniform external pressure $p$ as shown in Figure 5.

Here, just as in the previous case, the shell is made of a perfect plastic material. When $p$ reaches a critical value the shell collapses to form a complex surface composed of small tiles, in general of arbitrary shape, disposed around the rigid sphere of radius $R_0$. The pressure continues to act till the entire shell is confined within the "spherical crust" bounded by two spheres, $R_0$ and $R_0 + h$. That is, the original shell is packed inside the "spherical crust" of thickness equal to $h$.

Let us start with an ideal configuration as in the case of folded wires. Assume a regular folding such that the tiles have the shape of isosceles triangles and the fundamental element of 3-D geometry, that is, the surface generator consists of a pyramid whose basis is an equilateral triangle, and the height is equal to $h$ as shown in Figure 6. A pineapple shell provides a good approximation to visualize this type of surface, which is a concave polyhedron.

The edges of this surface, that is, the common lines of adjacent tiles are the rupture lines. The tiles rotate about these lines, developing a relatively complex mechanism and accumulating permanent plastic strain till the final configuration is reached. The final configuration is stable. The original shell remains folded within the "spherical crust" without any external or internal restrain.

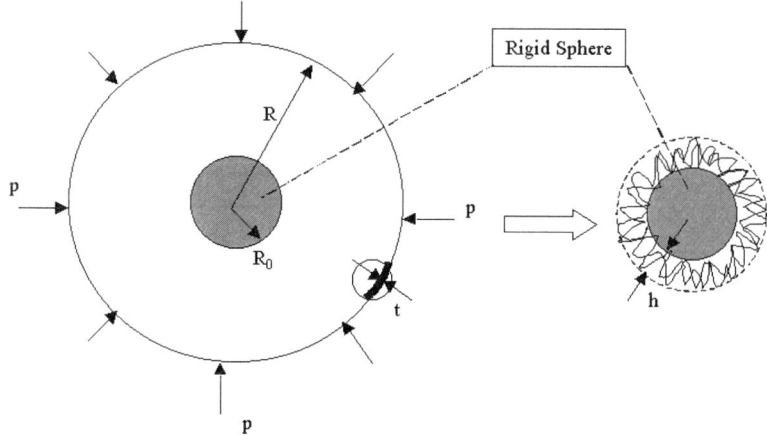

Figure 5  Collapse of a spherical thin shell under the action of an external pressure $p$. The shell is made of a perfect plastic material.

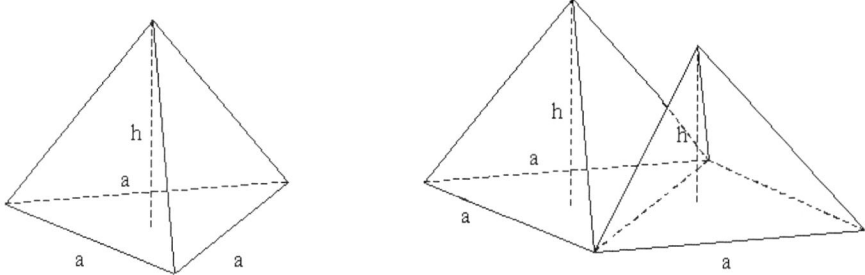

Figure 6  Elements of the ideal geometry of a folded surface. (a) The basic generator. (b) Composition of two basic elements.

The principle of mass conservation for incompressible materials gives:

$$nS_L t \leq 3\pi R^2 t,$$

where $S_L$ is the lateral surface of the generator, $n$ is the number of elements in the polyhedron, and $t$ is the shell thickness. The lateral surface can be easily calculated:

$$S_L = \frac{3}{2}a^2 \left[ \left(\frac{h}{a}\right)^2 + \frac{1}{12} \right]^{1/2}. \tag{16a}$$

For $h/a \gg 1$ we may write:

$$S_L \cong \frac{3}{2}ah. \tag{16b}$$

Introducing this value of $S_L$ in the above mass conservation equation we get:

$$n \cong k^* \frac{R^2}{ah}.$$

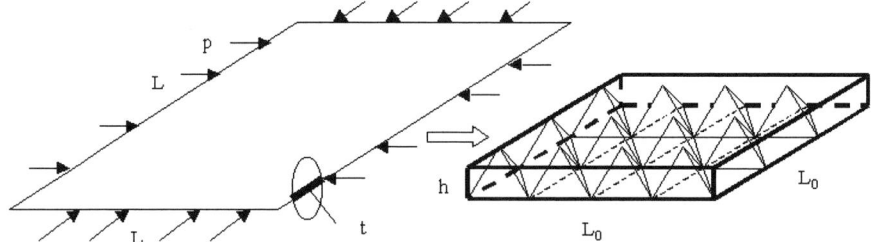

Figure 7 Schematic representation of a thin plate forced into a box $L_0 \times L_0 \times h$.

In the limit, $a \to t$, the packing capacity for the given geometric characteristics, $R$, $R_0$, $t$ and $h$ reaches the upper limit. The 2-D original medium – the shell with radius $R$ has been compacted to a 3-D medium – a solid inside the "crust." In the limit, the number of elements is:

$$n \to n_L \cong k^* \frac{R^2}{th} \, . \tag{17}$$

On the other hand, if the shell is reduced to a 3-D solid packed inside the "crust" and since $h \ll R_0$ we may also write:

$$4\pi R^2 t \cong \hat{k}^* \, 4\pi R_0^2 h \, ,$$

from which follows

$$\left( \frac{R}{R_0} \right)^2 \cong e_0^* \frac{h}{t} \tag{18}$$

with

$$\left( \frac{R^2}{R_0^2} \right) = \Gamma^* \quad \text{and} \quad \frac{t}{h} = \rho^*$$

the equation above can be rewritten:

$$\Gamma^* = e_0^* \rho^{*2-D^*} \, , \tag{19a}$$

or

$$\log \Gamma^* = \log e_0^* + (2 - D^*) \log \rho^* \, . \tag{19b}$$

This is the generalized expression for packed thin shells under external pressure crushed against a rigid sphere. The extension of this expression to the case of packing thin plates under the action of in-plane compression forces in a box $[L_0 \times L_0 \times h]$ is not difficult. The confinement scheme is shown in Figure 7.

The variable $\Gamma^*$ for this case reads:

$$\Gamma^* = \frac{L^2}{L_0^2} \, .$$

Clearly the power law governing the variation of the packing capacity $\Gamma^*$ with the ratio $t/h$ is coherent with the two limiting cases. For $D^* = 3$ the final configuration has the Euclidean dimension of a solid. This hypothesis representing an ideal case was used to obtain equation (18). Clearly for $D^* = 3$ the generalized expressions (19a,b) reproduce the expected result. The second limit case occurs for $D^* = 2$. This means that the folded shell does not change its basic geometric characteristic, remaining with the same Euclidean dimension $D^* = 2$ proper to the original shell before the collapse. Introducing this value in the generalized equation it is easily seen that $\Gamma^* \to e_0^*$. The conservation of the Euclidean dimension prevails in this case, and there is no distinction between the original and the final geometry. The parameter $e_0^*$ can be a function of $D^*$; we leave this hypothesis open depending on experimental confirmation.

There are two distinct explanations for a folding-free configuration. One of them is of a geometric nature, and the other is related to the material properties.

Suppose first that $R \to R_0$, that is, the thin shell is adherent to the rigid sphere from the very beginning. As a matter of fact there is no possible packing for this initial geometry; it follows immediately from (19a) that $\Gamma^* \to 1$ consequently $h \to t$. The shell thickness is equal to the crust thickness, and the basic assumptions of the theory are violated in this case.

The second possibility arises from the material properties. If the material is perfectly stiff, that is, does not admit plastic deformation, there will be no dissipated plastic work and the theory fails. As mentioned for the case of thin wires, in practice we face intermediate situations. Similarly, the following criteria can be useful to establish the limits of applicability of the theory:

1. The ratio $\rho^* = t/h$ must be smaller than 0.1: $\rho^* < 0.1$.

2. The contribution to the total dissipated plastic work $(W_t^*)$ due to the pure membrane component of the stress state $(W_m^*)$ should be much smaller than the contribution due to the bending component of the stress state $(W_b^*)$: $W_m^* \leq 0.1 W_t^*$.

Just as in the case of wires, it is reasonable to admit that larger values of $D$ denote very dense packing configurations. We may say that the apparent ductility, that is, the capacity to accumulate plastic deformation in the folding process, increases with $D$. The notion of apparent ductility is similar to that introduced for folded wires. It is better understood when presented together with the discussion on the dissipated plastic work.

Consider a shell made of an ideal plastic material. The edges of the final polyhedral surface produced in the process of confinement are rupture lines that accumulate plastic strain. The plastic deformation developed at each edge is proportional to the relative rotation $\theta_n$ of the adjacent tiles and to the edge length $l_n$. We may write therefore:

$$\tau_n^* = \hat{k}^* \sigma_Y t^2 l_n \theta_n \,.$$

The total dissipated plastic work can be written as:

$$\tau_n^* = \hat{k}^* \sigma_Y t^2 \pi g^*(U) L_n \,, \tag{20}$$

where $L_n$ stays for the length of all edges added together, and the rotation $\theta_n$ has been substituted by the average value:

$$\theta_n = \pi g^*(U)$$

to take into account the material ductility as explained before.

Now, let $n$ be the number of generators. The set of triangles corresponding to the basis of the generators are the faces of a convex polyhedron circumscribing the sphere $R_0$, in the ideal case. The Euler theorem states that:

$$F + V - E = 2,$$

where $F$ is the number of faces, $V$ is the number of vertices and $E$ the number of edges. For a polyhedron composed of a very large number of triangular faces, the number of edges is approximately triple the number of vertices. So from Euler's formula and with $A \cong 3V$ we get:

$$V = \frac{F}{2} - 1,$$

from which follows:

$$A \cong \frac{3F}{2} - 3.$$

Now, with $F = n$ and for very large $n$, the total number of edges of the polyhedron circumscribed to the sphere $R_0$ can be estimated by:

$$A \cong \frac{3n}{2}.$$

The total length of the edges connecting the basis of the fundamental elements or generators is therefore $^3/_2 na$. Now adding the edges at the lateral surfaces of each pyramid – see Figure 6 – with length approximately equal to $h$ each edge, we get:

$$L_n \cong n \left( 3h + \frac{3}{2}a \right).$$

In the limit as $a \to t$ and since $h \gg t$ the second term within brackets in the above expression can be disregarded in the presence of $h$. Therefore in the limit for very large $n$, $n \to n_L$, we get:

$$L_{n_L} \cong \frac{R^2}{t}. \tag{21}$$

The plastic work dissipated in the packing process is therefore:

$$W_{n_L}^* \cong \hat{k}^* \pi \sigma_p g^*(U) t R^2.$$

Now with the help of (19a) we obtain:

$$W_{n_L}^* \cong \hat{k}^* e_0^* \pi \sigma_Y g^*(U) h R_0^2 \rho^{3-D^*}.$$

Now with $W_R$ as the reference plastic work defined as

$$W_R^* \cong \hat{k}^* e_0^* \pi \sigma_Y h^3 G^*(U),$$

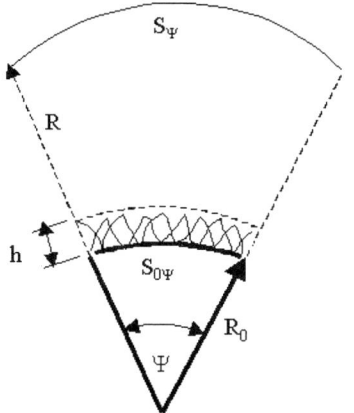

Figure 8  Confinement within a solid angle $\Psi$.

where $G(U, \beta^*) = g(U)/(\beta^*)^2$ and $\beta^* = h/R_0$. Finally, calling $\tau^*$ the plastic work density we obtain:

$$\tau^* = \frac{W^*}{W_R^*} = \rho^{3-D^*} ,\tag{23}$$

where we have dropped the subscript $n_L$ for the sake of simplicity.

Now from (19a) and (23) we arrive at an expression that gives the fractal dimension as a function of the packing capacity and the plastic work density:

$$D = 2 - \frac{\log\left(\dfrac{\Gamma^*}{e_0^*}\right)}{\log \tau^* - \log\left(\dfrac{\Gamma^*}{e_0^*}\right)} .\tag{24}$$

Just as in the previous case and following a similar reasoning, the equation governing the packing density is readily obtained:

$$\Omega^* = e_0^*(\rho^*)^{2-D^*} ,\tag{25a}$$

or

$$\log \Omega^* = \log e_0^* + (2 - D^*) \log \rho^* ,\tag{25b}$$

where $\Omega^*$ is the ratio between the apparent specific mass of the shell $\mu^*$ and the specific mass of the $\mu_0^*$. As in the previous comments concerning the packed wires, the theoretical development leads to the same parameter $e_0^*$ for both the packing capacity and the packing density. We would like to point out once more that experimental confirmation is essential to verify this analytical result. It is not impossible to have the parameter $e_0^*$ a function of $D^*$, that is, $e_0^* = e_0^*(D^*)$.

Examining the equations (19a,b) it is possible to conclude that the self-similarity requirement of a fractal structure is satisfied. Given the ratio $t/h$, any portion of the confined shell comprised by a solid angle $\Psi$ will have the same packing capacity. Indeed, the packing capacity is the ratio defined by the area of the original shell over the area of the "crust" containing the packed shell. Now if we take the portion encompassed by the solid angle $\Psi$, the ratio between the areas $S_\Psi$ and $S_{0\Psi}$ will obey the same relation valid for the complete shell $R^2/R_0^2$ given that the ratio $t/h$, and the mechanical properties of the material are preserved. Therefore the portion of the shell confined inside the volume $[h \times S_{0\Psi}]$ is self-similar to the entire shell packed inside the "crust."

The main results of the above discussion can be formalized as follows:

**Proposition 2**     1. *The geometry of densely folded plates and shells packed in a 3-D container with volume $V$ – a box $V = L^2 h$, $L \gg h$ or a crust $V = 4\pi R^2 h$, $R \gg h$ – has a fractal structure characterized by a fractal dimension $2 < D^* < 3$, provided that:*

   *i. The slenderness ratio defined by $\rho^* = t/h$ is sufficiently small where $t$ is the plate or shell thickness. That is, $\rho^* \ll 1$.*

   *ii. The material is perfectly plastic.*

   *iii. The final configuration is stable.*

2. *The packing capacity $\Gamma^* = S/S_0$ and the packing density $\Omega^* = \mu^*/\mu_0^*$ vary with the slenderness ratio according to the power law:*

$$\Gamma^* = e_0^*(\rho^*)^{2-D^*},$$

*and*

$$\Omega^* = e_0^*(\rho^*)^{2-D^*},$$

*where $S$ is the total area of the plate or shell packed inside the container; $S_0$ is the reference area $S_0 = V/h$; $\mu^*$ is the apparent specific mass; $\mu_0^*$ is the specific mass of the plate or shell per unit area. The value of $e_0^*$ has to be determined experimentally.*

3. *The fractal dimension $D^*$ depends on the apparent ductility. Plates and shells folding in a configuration with a high fractal dimension $D^*$ will have a corresponding high apparent ductility. For a given value of the packing capacity $\Gamma^*$, the fractal dimension is a function of the dissipated plastic work density $\tau^*$:*

$$D = 2 - \frac{\log\left(\dfrac{\Gamma^*}{e_0^*}\right)}{\log \tau^* - \log\left(\dfrac{\Gamma^*}{e_0^*}\right)}.$$

4. *If the packing capacity is governed by a power law as given in item 2 above, then the geometry representing the plate or shell final configuration is self-similar. Conversely, if the geometry is self-similar the packing configuration has a fractal structure as proposed in item 2 above.*

# 4 Conclusions

The fundamental ideas of this paper are formalized in propositions 1 and 2. We can summarize the key conjectures as follows:

1. The geometry of perfect plastic, densely packed thin wires, thin plates and shells has a fractal structure. The packing capacity varies with the slenderness ratio $\phi/h$ for wires or $t/h$ for plates and shells, according to a power law.

2. The fractal dimension depends on the material ductility. A corresponding high fractal dimension characterizes highly ductile materials. The fractal dimension can therefore be considered as a global material property.

3. Moreover, the packing density varies with the slenderness ratio according to the same fractal dimension as the packing capacity.

4. If the previous statements are true, the ratio between the packing capacity and the plastic work density is independent of the fractal packing dimension. The ratio between these two variables is a function of $\rho$ only.

5. The geometry of the confined solid is self-similar. Any part is similar to the whole.

6. The governing equations for the packing capacity or packing density as functions of the slenderness ratio or the reduced plastic work density involve constants that must be determined from experimental data.

The consideration of additional conditions like locking effects, more realistic material properties like elastic–plastic behavior and the influence of the packing procedure as well, would introduce difficulties that are only worthwhile dealing with if the simplified theory works well. These improvements would not change the fundamental conclusions advanced here.

It is very important to make it clear that the theory, as presented here, is only valid for very thin wires, plates or shells. That is, the slenderness ratio must be small, otherwise the main assumptions supporting the theory fail to apply. This is apparent particularly for the expressions involving the reduced packing plastic work.

Finally, as was stated in the introduction, we have not found references dealing with this subject other than those listed below. On the other hand, the current technical literature is very rich in papers and books dealing with the fractal representation of nature and physical phenomena. We end by listing several references that have no connection with this paper. Possible applications to geophysics, to natural and artificial membranes can be further investigated [3–5]. Although from a different point of view, where the fractal character of the geometry is determined by biological processes of growth rather than from the action of external forces, the lungs can be examined with a similar technique [6]. The mathematical foundations of fractal geometry can be found in [7, 8].

# Acknowledgements

This research has been partially supported by the Carlos Chagas Filho Foundation – FAPERJ – through the program "Cientista do nosso Estado" and by the National Council for Scientific and Technological Research – CNPq.

# References

1. Bevilacqua, L. (2000) Constitutive laws for strong geometric non-linearity. *Journal of the Brazilian Society for Mechanical Sciences*, **XIII** (2), 217–229.
2. Bevilacqua, L. (1999) Dynamic Characterization of Fractal Surfaces. In *Proceedings EURODINAME-99, Dynamic Problems in Mechanics and Mechatronics*, Universität Ulm, pp. 285–292.
3. Turcotte, D.L. (1997) *Fractal and Chaos in Geology and Geophysics*. Cambridge University Press.
4. Barabasi, A.-L. and Stanley, H.E. (1997) *Fractal Concepts in Surface Growth*. Cambridge University Press.
5. Feder, J. (1988) *Fractals*. Plenum Press, New York and London.
6. Khoo, M.C.K. (ed.) (1996) *Bioengineering Approaches to Pulmonary Physiology and Medicine*. Plenum Press, New York.
7. Falconer K. (1990) *Fractal Geometry: Mathematical Foundations and Applications*. John Wiley & Sons, Chichester and New York.
8. Tricot, C. (1995) *Curves and Fractal Dimensions*. Springer Verlag, New York.

# On a General Principle of Mechanics and Its Application to General Non-Ideal Nonholonomic Constraints

F.E. Udwadia

*Department of Aerospace and Mechanical Engineering, Civil Engineering, Mathematics, and Operations and Information Management*
*430K Olin Hall, University of Southern California, Los Angeles, CA 90089-1453*
*E-mail: fudwadia@usc.edu*

In this paper we prove a general minimum principle of analytical dynamics that encompasses nonideal constraints. We show that this principle reduces to Gauss's Principle when the constraints are ideal. We use this new principle to obtain the general, explicit, equation of motion for systems with nonideal constraints. An example for a nonholonomically constrained system where the constraints are nonideal is presented.

## 1 Introduction

The motion of complex mechanical systems is often mathematically modeled by what we call their equations of motion. The current formalisms (Lagrange's equations [1], Gibbs–Appell equations [2, 3], generalized inverse equations [4]) have as their foundation D'Alembert's Principle which states that, at each instant of time during the motion of the mechanical system, the sum total of the work done by the forces of constraint under virtual displacements is zero. Such constraint forces are often referred to as being ideal. D'Alembert's Principle is equivalent to a principle that was first stated by Gauss [5] and is referred to nowadays as Gauss's Principle. Recently, Gauss's Principle was used by Udwadia and Kalaba [4, 6], in conjunction with the concept of the Penrose inverse of a matrix, to obtain a simple and general set of equations for holonomically and nonholonomically constrained mechanical systems.

Though these two fundamental principles of mechanics are often useful to adequately model mechanical systems, there are, however, numerous situations where they are not valid. Such systems have, to date, been left outside the preview of the Lagrangian framework. As stated by Goldstein (1981, page 14), "This [total work done by forces of constraint equal to zero] is no longer true if sliding friction is present, and we must exclude such systems from our [Lagrangian] formulation." [7] And Pars (1979) in his treatise [8] on analytical dynamics writes, "There are in fact systems for which the principle enunciated [D'Alembert's Principle] ... does not hold. But such systems will not be considered in this book." Newtonian approaches are usually used to deal with the problem of sliding friction. [7] For general systems with nonholonomic constraints, [7] the inclusion into the framework of Lagrangian dynamics of constraint forces that do work has remained to date an open problem in analytical dynamics, for neither D'Alembert's Principle nor Gauss's Principle is then applicable.

In this paper we obtain a general principle of analytical dynamics that encompasses nonideal constraints. It generalizes Gauss's Principle to situations where the forces of constraint *do* work under virtual displacements. It therefore brings nonideal constraints within the scope of Lagrangian mechanics. The power of the principle is exhibited by the simple and straightforward manner in which it yields the general, explicit equation of motion for constrained mechanical systems where the constraints may not be ideal. We provide an illustrative example in which we obtain the explicit equations of motion for a nonholonomic mechanical system first proposed by Appell.

Since the principle obtained in this paper is a generalization of Gauss's Principle, and since most texts today do not have an adequate description of it, we start with a short description of Gauss's work. In 1829, Gauss [5] published a landmark paper entitled, "On a New Universal Principle of Mechanics," in what is today the *Journal fur reine und angewandte Mathematik*. In it, he presented a universal principle of mechanics under the assumption that the constraints acting on the mechanical system under consideration are ideal. Several features of this paper are worth noting: (1) The paper is only 3 pages long; (2) Gauss shows us a totally new line of thinking by considering the deviations of the motion of the constrained system from what they might have been were there no constraints acting on it; (3) The mathematics involved is trivial, essentially the use of the cosine rule for a triangle, and the use of D'Alembert's Principle; the result is proved in a single paragraph; (4) Gauss, though well aware of the usefulness of Lagrangian coordinates, purposively uses Cartesian inertial coordinates to state and derive his results; he does not bother with generalized coordinates, or Lagrange's equations; and (5) Gauss develops the only general global minimum principle in analytical dynamics.

The reason for the detailed exposition above is because in this paper we shall follow the spirit of Gauss's line of reasoning. In Section 2 we present a statement of the problem, and in Section 3 we present our new minimum principle applicable to nonideal constraints. Section 4 applies this new minimum principle to obtain the general, explicit equations of motion where the constraints may be nonideal, and in Section 5 we present an illustrative example showing the simplicity with which this general equation yields results for nonholonomic, nonideal constraints. Section 6 gives the conclusions.

# 2 Statement of the Problem of Constrained Motion with Nonideal Constraints

Consider a mechanical system comprised of $n$ particles, of mass $m_i$, $i = 1, 2, 3, \ldots, n$. We shall consider an inertial Cartesian coordinate frame of reference and describe the position of the $j$-th particle in this frame by its three coordinates $x_j$, $y_j$ and $z_j$. Let the impressed forces on the $j$-th mass in the X-, Y- and Z- directions be given, $F_{x_j}$, $F_{y_j}$, $F_{z_j}$ respectively. Then the equation of motion the of 'unconstrained' mechanical system can be written as

$$M\ddot{x} = F(x, \dot{x}, t), \qquad x(0) = x_0, \qquad \dot{x}(0) = \dot{x}_0. \tag{1}$$

Here the matrix $M$ is a $3n$ by $3n$ diagonal matrix with the masses of the particles in sets of three along the diagonal, $x = [x_1, y_1, z_1, \ldots, x_n, y_n, z_n]^T$, and the dots refer to differentiation with respect to time. Similarly, the $3n$-vector ($3n$ by 1 column vector) $F = [F_{x_1}, F_{y_1}, F_{z_1}, \ldots, F_{x_n}, F_{y_n}, F_{z_n}]^T$. By 'unconstrained' we mean that the components of the $3n$-vector of velocity, $\dot{x}(0)$, can be independently prescribed. We note that the acceleration, $a(t)$, of the unconstrained system is then simply given by

$$a(t) = M^{-1} F(x, \dot{x}, t). \tag{2}$$

The impressed force $3n$-vector $F(x, \dot{x}, t)$ is a known vector, i.e., it is a known function of its arguments.

Let this system now to be subjected to a further set of $m = h + s$ constraints of the form

$$\varphi(x, t) = 0 \tag{3}$$

and

$$\psi(x, \dot{x}, t) = 0, \tag{4}$$

where $\varphi$ is an $h$-vector and $\psi$ an $s$-vector. We shall assume that the initial conditions satisfy these constraint equations at time $t = 0$, i.e., $\varphi(x_0, 0) = 0$, $\dot{\varphi}(x_0, 0) = 0$, and $\psi(x_0, \dot{x}_0, 0) = 0$.

Assuming that equations (3) and (4) are sufficiently smooth, we differentiate equation (3) twice with respect to time, and equation (4) once with respect to time, to obtain the equation

$$A(x, \dot{x}, t)\ddot{x} = b(x, \dot{x}, t), \tag{5}$$

where the matrix $A$ is $m$ by $3n$, and $b$ is the $m$-vector that results from carrying out the differentiations. We place no restrictions on the rank of the matrix $A$.

This set of constraint equations include among others, the usual holonomic, nonholonomic, scleronomic, rheonomic, catastatic and acatastatic varieties of constraints; combinations of such constraints may also be permitted in equation (5). It is important to note that equation (5), together with the initial conditions ($\varphi(x_0, 0) = 0$, and $\psi(x_0, \dot{x}_0, 0) = 0$), is equivalent to equations (3) and (4). In what follows we shall, for brevity, drop the arguments of the various quantities, unless needed for clarification.

Consider the mechanical system at any instant of time $t$. Let us say we know its position, $x(t)$, and its velocity, $\dot{x}(t)$, at that instant. The presence of the constraints whose kinematic description is given by equations (3) and (4) causes the acceleration, $\ddot{x}(t)$, of the constrained system to differ from its unconstrained acceleration, $a(t)$, so that the acceleration of the constrained system can be written as

$$\ddot{x}(t) = a(t) + \ddot{x}^c(t)\,, \tag{6}$$

where $\ddot{x}^c$ is the *deviation* of the acceleration of the constrained system from what it would have been had there been no constraints imposed on it at the instant of time $t$. Alternatively, upon premultiplication of equation (6) by $M$, we see that at the instant of time $t$,

$$M\ddot{x}(t) = Ma(t) + M\ddot{x}^c(t) = F + F^c \tag{7}$$

and so a force of constraint, $F^c$, is brought into play that causes the deviation in the acceleration of the constrained system from what it might have been (i.e., $a$) in the absence of the constraints.

Thus the constrained mechanical system is described so far by the matrices $M$ and $A$, and the vectors $F$ and $b$. The determination from equations (7) and (5) of the acceleration $3n$-vector, $\ddot{x}$, of the constrained system, and of the constraint force $3n$-vector, $F^c$, constitutes an under-determined problem and cannot, in general, be solved. Additional information related to the nature of the force of constraint $F^c$ is required and is situation-specific. Thus, to obtain an equation of motion for a given mechanical system under consideration, additional information – beyond that contained in the four quantities $M$, $A$, $F$ and $b$ – needs to be provided by the mechanician who is modeling the motion of the specific system.

Let us assume that we have this additional information (for some specific mechanical system) regarding the constraint force $3n$-vector, $F^c$, at each instant of time $t$ in the form of the work done by this force under virtual displacements of the mechanical system at time $t$. A virtual displacement of the system at time $t$ is defined as *any* $3n$-vector, $\nu(t)$, that satisfies the relation

$$A(x, \dot{x}, t)\nu(t) = 0\,. \tag{9}$$

The mechanician modeling the motion of the system then provides the work done, $W^c(t)$, under virtual displacements by $F^c$ through knowledge of a vector at each instant of time $t$, so that

$$\nu^T F^c \equiv W^c(t) = \nu^T C(x, \dot{x}, t)\,. \tag{10}$$

At any given instant of time $t$, $W^c$ may be *positive*, *zero* or *negative*. The determination of $C$ is left to the mechanician, and is most often done by inspection of, and/or experimentation with, the mechanical system that he is attempting to mathematically model. We note that from a dimensional analysis standpoint, the units of $C$ are those of force. Yet, we will show that, in general, this vector $C$ cannot be treated as a 'given' or 'impressed force'; and, it does *not* equal to the additional constraint force acting on the mechanical system. In Ref. [12] we explain the general nature of the specification of the nonideal constraint force $F^c$ given by equation (10).

For example, upon examination of a given mechanical system, the mechanician could decide that $C \equiv 0$ (for all time $t$) is a good enough approximation to the behavior of the actual force of constraint $F^c$ in the system under consideration. In that situation, equation (10) reduces to

$$\nu^T F^c \equiv W^c(t) = 0,\qquad(11)$$

which is, of course, D'Alembert's Principle, and the constraints are now referred to as being ideal. Though this approximation is a useful one in many practical situations, it is most often, still only an approximation, at best. More generally, the mechanician would be required to provide the $3n$-vector $C(x, \dot{x}, t)$, and when $C \neq 0$, the constraints are called nonideal.

Hence the specification of constrained motion of a mechanical system where the constraints are nonideal requires *in addition* to the knowledge of the four quantities, $M$, $A$, $F$ and $b$, also knowledge of the vector $C$. By "knowledge" we mean, as before, knowledge of these quantities as known functions of their respective arguments.

# 3  A General Principle of Mechanics

We again consider the mechanical system at time $t$, and assume that we know $x(t)$ and $\dot{x}(t)$ at that time. Since by equation (7), $F^c = M\ddot{x} - F$, equation (10) can be rewritten at time $t$, as

$$\nu^T(M\ddot{x} - F - C) = 0,\qquad(12)$$

where $\nu$ is any virtual displacement at time $t$, and $\ddot{x}$ is the acceleration of the constrained system.

Now we consider a *possible* acceleration, $\hat{\ddot{x}}(t)$ of the mechanical system at time $t$; that is, *any* $3n$-vector that satisfies the constraint equation $A\hat{\ddot{x}} - b$ at that time. Since the acceleration of the constrained system, $\ddot{x}$, at time $t$ must also satisfy the same equation, we must have

$$A(\hat{\ddot{x}} - \ddot{x}) = Ad = 0\qquad(13)$$

and so by virtue of relation (9), the $3n$-vector $d = \hat{\ddot{x}} - \ddot{x}$ at the time $t$, then qualifies as a virtual displacement! Hence by (12) at time $t$, we must have

$$d^T(M\ddot{x} - F - C) = 0.\qquad(14)$$

We now present two Lemmas.

**Lemma 1** *For any symmetric $k$ by $k$ matrix $Y$, and any set of $k$-vectors $e$, $f$ and $g$,*

$$(e - g, e - g)_Y - (e - f, e - f)_Y = (g - f, g - f)_Y - 2(e - f, g - f)_Y,\qquad(15)$$

*where we define $(a, b)_Y \equiv a^T Y b$ for the two $k$-vectors $a$ and $b$.*

**Proof:** This identity can be verified directly. □

For short, in what follows, we shall call $a^T Y a$ the $Y$-norm of the vector $a$ (actually it is the square of the $Y$-norm).

**Lemma 2** *Any vector $d = \hat{\ddot{x}} - \ddot{x}$ satisfies at time $t$, the relation*

$$(M\hat{\ddot{x}} - (F+C), (M\hat{\ddot{x}} - (F+C))_{M^{-1}} - (M\ddot{x} - (F+C), M\ddot{x} - (F+C))_{M^{-1}} =$$

$$= (d, d)_M + 2(M\ddot{x} - (F+C), d). \tag{16}$$

**Proof:** Set $k = 3n$, $Y = M$, $e = M^{-1}(F+C)$, $f = \ddot{x}$, and $g = \hat{\ddot{x}}$ in relation (15). The result follows. $\qquad\square$

We are now ready to state the general minimum principle of analytical dynamics.

  *A constrained mechanical system subjected to nonideal constraints evolves in time in such a manner that its acceleration $3n$-vector, $\ddot{x}$ at each instant of time minimizes the quadratic form*

$$G_{ni}(\hat{\ddot{x}}) = (M\hat{\ddot{x}} - (F+C), M\hat{\ddot{x}} - (F+C))_{M^{-1}} \tag{17}$$

*over all 'possible' acceleration $3n$-vectors, $\hat{\ddot{x}}$, at that instant of time.*

**Proof:** For the constrained mechanical system described by equations (1)–(3) and (10), the $3n$-vector $d$ satisfies relation (14); hence the last member on the right-hand side of equation (16) becomes zero. Since $M$ is positive definite, the scalar $(d, d)_M$ on the right-hand side of (16) is always positive for $d \neq 0$. By virtue of (16), the minimum of (17) must therefore occur when $\hat{\ddot{x}} = \ddot{x}$. $\qquad\square$

**Remark 1:** We note that the units of $C$ are those of force; furthermore $C$ needs to be prescribed (at each instant of time) by the mechanician, based upon examination of the given specific mechanical system whose equations of motion (s)he wants to write. From equation (7), we have $F^c = M\ddot{x} - F$ where $\ddot{x}$ is the acceleration of the constrained system. Were we to replace $\ddot{x}$ on the right-hand side of this relation by any particular *possible* acceleration $\hat{\ddot{x}}$, we would obtain the corresponding *possible* force of constraint relevant to this *possible* acceleration as $\hat{F}^c = M\hat{\ddot{x}} - F$. The quadratic form (17) can be rewritten as

$$G_{ni}(\hat{\ddot{x}}) = (\hat{F}^c - C, \hat{F}^c - C)_{M^{-1}} \tag{18}$$

and hence the minimization of $G_{ni}(\hat{\ddot{x}})$ over all possible vectors $\hat{\ddot{x}}$, leads to the following alternative statement.

  *A constrained mechanical system evolves in time so that the $M^{-1}$-norm of the force of constraint that is generated less the prescribed vector $C$ is minimized, at each instant of time, over all possible forces of constraint $\hat{F}^c$ at that time.*

**Remark 2:** It should be noted that, in general, $F^c \neq C$. In fact as seen from equation (10) the quantity $(F^c - C)$ at time $t$ is *that part of the total force of constraint* that does *no* work under virtual displacements $\nu$ at time $t$.

**Remark 3:** Thinking of the vector $C$ as a force that is prescribed by the mechanician at time $t$, $c = M^{-1}C$ is the acceleration that this force would engender in the unconstrained system at that time; similarly $a = M^{-1}F$ is the acceleration that the impressed force $F$ would engender in the *unconstrained* system at the time $t$. Denoting at time $t$,

$$a_{ni} = a + c \tag{19}$$

we can rewrite (17) as

$$G_{ni}(\hat{\ddot{x}}) = (\hat{\ddot{x}} - a_{ni}, \hat{\ddot{x}} - a_{ni})_M . \tag{20}$$

Hence we have the following alternative understanding of constrained motion: *a constrained mechanical system evolves in time in such a way that at each instant of time the M-norm of the <u>deviation</u> of its acceleration from $a_{ni}$ is a minimum. The acceleration $a_{ni}$ at any time t is the acceleration of the unconstrained system under the combined action of the impressed forced F (acting on the system at time t) and the prescribed force C (that describes the nature of the nonideal constraints acting at time t).*

**Remark 4:** We note from the proof that at each instant of time $t$ the minima in (17)–(19) are *global*, since we do not restrict the *possible* accelerations in magnitude, as long as they satisfy the relation $A\hat{\ddot{x}} = b$.

**Remark 5:** Comparing this principle with other fundamental principles of analytical dynamics (like Hamilton's Principle [7, 8], which is an extremal principle), this then appears to be the *only* general global minimum principle in analytical dynamics.

**Remark 6:** The general principle stated above reduces to Gauss's Principle [5] when $C \equiv 0$, and all the constraints are ideal.

We next show how this new principle can be used to obtain the equation of motion of constrained systems when the forces of constraint are not ideal.

## 4   Explicit Equations of Motion for Systems with Nonideal Constraints

Let us denote $r = M^{1/2}(\ddot{x} - a - c)$, so that

$$\ddot{x} = M^{-1/2} + a + c \tag{21}$$

and the relation $A\ddot{x} = b$ becomes,

$$Br = (AM^{-1/2})r = b - Aa - Ac, \tag{22}$$

where $B = AM^{-1/2}$. Then the general principle (20) reduces to minimizing $\|r\|^2$, subject to the condition $Br = b - Aa - Ac$. But the solution of this problem is simply [9] [1]

---

[1]Actually, instead of the Moore–Penrose inverse we could use any so-called 1–4 generalized inverse, see [9].

$$r = B^+(b - Aa - Ac), \tag{23}$$

where $B^+$ stands for the Moore–Penrose inverse [9] of the matrix $B$. Substituting for $r$ in equation (21) yields the explicit equation of motion for the system as

$$\ddot{x} = M^{-1/2}B^+(b - Aa - Ac) + a + c =$$

$$= a + M^{-1/2}B^+(b - Aa) - M^{-1/2}B^+BM^{1/2}M^{-1}C + M^{-1}C =$$

$$= a + M^{-1/2}B^+(b - Aa) + M^{-1/2}(I - B^+B)M^{-1/2}C. \tag{24}$$

Premultiplying equation (24) by $M$, one obtains the general form for the equation of motion of a constrained system with nonideal constraints as

$$M\ddot{x} = F + M^{1/2}B^+(b - Aa) + M^{1/2}(I - B^+B)M^{-1/2}C = F + F_i^c + F_{ni}^c. \tag{25}$$

When $C \equiv 0$, all the constraints are ideal, and we obtain the results obtain in Refs. [4, 8].

From the right-hand side of equation (25) we notice that the total force acting on the constrained system is made up of 3 members: (1) the impressed force $F$; (2) the force $F_i^c = M^{1/2}B^+(b - Aa)$ *which would exist were all the constraints ideal,* $C \equiv 0$; and (3) the force $F_{ni}^c = M^{1/2}(I - B^+B)M^{-1/2}C$ that is brought into play only when the constraints are nonideal so that the mechanician needs to provide the vector $C(x, \dot{x}, t)$ that describes the nonideal nature of the constraints at each instant of time $t$.

One last point, what does the general principle of mechanics look like in generalized Lagrangian coordinates, $q$? As will be seen from its proof, to obtain the general minimum principle of mechanics and the equation of motion for constrained systems with nonideal constraints in Lagrangian coordinates all one has to do is make the substitutions: $x \to q$, $\dot{x} \to \dot{q}$, $\ddot{x} \to \ddot{q}$, $M \to M(q, t)$, $F \to Q$, and $\hat{\ddot{x}} \to \hat{\ddot{q}}$ in equations (1)–(25).[2] Equation (1) now becomes Lagrange's equation of motion for the unconstrained system. Note that, in general, $q$, would now be a vector of dimension $k$, so that the positive definite matrix $M(q, t)$ will be $k$ by $k$, and the matrix $A(q, \dot{q}, t)$ will be accordingly $m$ by $k$.

The general principle of analytical dynamics in Lagrangian coordinates then becomes:

*A constrained mechanical system subjected to nonideal constraints evolves in time in such a manner that its acceleration, $\ddot{q}$, at each instant of time minimizes the quadratic form*

$$G_{ni}(\hat{\ddot{q}}) = (M(q,t)\hat{\ddot{q}} - Q(q,\dot{q},t) - C(q,\dot{q},t), M(q,t)\hat{\ddot{q}} - Q(q,\dot{q},t) - C(q,\dot{q},t))_{M^{-1}} \tag{26}$$

*over all 'possible' accelerations, $\hat{\ddot{q}}$, that satisfy the equation $A(q,\dot{q},t)\hat{\ddot{q}} = b(q,\dot{q},t)$, at that instant of time. For clarity, we have shown the arguments of the various quantities explicitly.*

---

[2]We assume that the constraints do not alter the rank of the matrix $M(q, t)$. This is usually the case in analytical dynamics.

And the general equation of constrained motion with nonideal constraints becomes:

$$M\ddot{q} = Q + M^{1/2}(b - Aa) + M^{1/2}(I - B^+B)M^{-1/2}C = Q + Q_i^c + Q_{ni}^c, \quad (27)$$

where, for clarity, we have suppressed the arguments of the various quantities. The vector $Q$ is the impressed or given force. The force $Q_i^c$ is caused by the presence of the constraints and as though they are ideal; the force $Q_{ni}^c$ is brought into play because of the nonideal nature of the constraints. This nonideal character of the constraints is prescribed by the mechanician through the work done under virtual displacements by the constraint forces as $W(t) = \nu^T(t)C(q, \dot{q}, t)$. Equation (27), which we obtained here by a very different approach, is identical to that obtained in Refs. [13] and [14].

Though not in any detail, our approach has been inspired by the central idea used by Gauss [5] in developing his results. One now sees why Gauss, in his original paper, did not bother to use Lagrangian coordinates, despite the fact that he used angle coordinates all the time for his astronomical measurements of comet motions.

# 5 Illustrative Example

To illustrate the simplicity with which we can write out the equations of motion for nonholonomic systems with nonideal constraints we consider here a generalization of a well-known problem that was first introduced by Appell [10].

Consider a particle of unit mass moving in a Cartesian inertial frame subjected to the known impressed (given) forces $F_x(x, y, z, t)$, $F_y(x, y, z, t)$ and $F_z(x, y, z, t)$ acting in the X-, Y- and Z-directions. Let the particle be subjected to the nonholonomic constraint $\dot{x}^2 + \dot{y}^2 - \dot{z}^2 = 2\alpha g(t)$, where $\alpha$ is a given scalar constant and the $g$ is a given, known function of its arguments. Appell [10] takes $\alpha = 0$, and he describes a physical mechanism that would yield his constraint.

Furthermore, Appell [10] assumed that the constraint is ideal. Let us generalize his example and say that the mechanician (who has supposedly examined the physical mechanism which is being modeled here) ascertains that this nonholonomic constraint subjects the particle to a force that is proportional to the square of its velocity and opposes its motion, so that the virtual work done by the force of constraint (under a virtual displacement $\nu$) on the particle is prescribed (by the mechanician) as

$$W^c(t) = -a_0\nu^T(t)\begin{bmatrix} \dot{x} \\ \dot{y} \\ \dot{z} \end{bmatrix}\frac{u^2}{|u|} = -a_0\nu^T(t)\begin{bmatrix} \dot{x} \\ \dot{y} \\ \dot{z} \end{bmatrix}|u|, \quad (28)$$

where $u(t)$ is the speed of the particle, and $a_0$ is a given constant. It should be pointed out that this force arises *because of the presence of the constraint*, $\dot{x}^2 + \dot{y}^2 - \dot{z}^2 = 2\alpha g(t)$. Were this constraint to be removed, this force (whose nature is described by relation (28)) would disappear. Thus the constraint is nonholonomic and nonideal. We shall obtain the explicit equations of motion of this system.

On differentiating the constraint equation with respect to time, we obtain

$$[\dot{x}\ \dot{y}\ -\dot{z}]\begin{bmatrix} \dot{x} \\ \dot{y} \\ \dot{z} \end{bmatrix} = \alpha\dot{g}, \tag{29}$$

where $\dot{g} = \dfrac{dg}{dt}$. Thus we have $A = [\dot{x}\ \dot{y}\ -\dot{z}]$, and the scalar $b = \alpha\dot{g}$. Since $M = I_3$, $B = A$. Hence $B^+ = \dfrac{1}{u^2}[\dot{x}\ \dot{y}\ -\dot{z}]$, and the vector $C = -\dfrac{a_0 u^2}{|u|}[\dot{x}\ \dot{y}\ \dot{z}]$. The unconstrained acceleration $a = [F_x\ F_y\ F_z]$. The equation of motion for this constrained system can now be written down directly by using equation (25). It is given by

$$\begin{bmatrix} \ddot{x} \\ \ddot{y} \\ \ddot{z} \end{bmatrix} = \begin{bmatrix} F_x \\ F_y \\ F_z \end{bmatrix} + \frac{\alpha\dot{g} - \dot{x}F_x - \dot{y}F_y + \dot{z}F_z}{u^2}\begin{bmatrix} \dot{x} \\ \dot{y} \\ -\dot{z} \end{bmatrix} -$$

$$-\frac{a_0}{|u|}\begin{bmatrix} \dot{y}^2 + \dot{z}^2 & -\dot{x}\dot{y} & \dot{x}\dot{z} \\ -\dot{y}\dot{x} & \dot{x}^2 + \dot{z}^2 & \dot{y}\dot{z} \\ \dot{x}\dot{z} & -\dot{y}\dot{z} & \dot{x}^2 + \dot{y}^2 \end{bmatrix}\begin{bmatrix} \dot{x} \\ \dot{y} \\ \dot{z} \end{bmatrix}, \tag{30}$$

which simplifies to

$$\begin{bmatrix} \ddot{x} \\ \ddot{y} \\ \ddot{z} \end{bmatrix} = \begin{bmatrix} F_x \\ F_y \\ F_z \end{bmatrix} + \frac{\alpha\dot{g} - \dot{x}F_x - \dot{y}F_y + \dot{z}F_z}{u^2}\begin{bmatrix} \dot{x} \\ \dot{y} \\ -\dot{z} \end{bmatrix} - \frac{a_0}{|u|}\begin{bmatrix} 2\dot{x}\dot{z}^2 \\ 2\dot{y}\dot{z}^2 \\ 2\dot{z}(\dot{x}^2 + \dot{y}^2) \end{bmatrix}. \tag{31}$$

The first term on the right-hand side is the impressed force. The second term on the right-hand side is the constraint force $F_i^c$ that would prevail were all the constraints ideal so that they did no work under virtual displacements. The third term on the right-hand side of equation (31) is the contribution, $F_{ni}^c$ to the total constraint force generated by virtue of the fact that the constraint force is not ideal, and its nature in the given physical situation is specified by the vector $C$, which gives the work done by this constraint force under virtual displacements. Note that $F_{ni}^c \neq C$.

We observe that it is because we do not eliminate any of the $x$'s or the $\dot{x}$'s (as is customarily done in the development of the equations of motion for constrained systems) that we can explicitly assess the effect of the 'given' force, and of the components $F_i^c$ and $F_{ni}^c$ on the motion of the constrained system.

Equations (25) and (27) can also be used to directly obtain the equations of motion for holonomic nonideal constraints, like when sliding friction may be present. Examples of such systems and others may be found in Refs. [11] and [12] which obtain the same general equation (27), but by very different routes.

# 6 Conclusions

This paper develops new fundamental insights into the evolution of constrained motion where the forces of constraint are general and may do work under virtual displacements.

Its main contributions are the following.

1. We obtain a minimum principle of mechanics that is general enough to include constraints forces that may do *positive, zero* or *negative* work under virtual displacements.

2. Unlike other fundamental principles of analytical dynamics (e.g., Hamilton's Principle), which are, strictly speaking, extremal principles, the principle obtained here is a global minimum principle. Furthermore, unlike principles like Hamilton's Principle that involve integrals over time, this principle is satisfied at *each* instant of time as the dynamics of the constrained system evolves.

3. The principle reduces, as it must, to Gauss's Principle when the constraint forces are ideal, and they do not work under virtual displacements.

4. The power of this new principle is illustrated by the simple way it allows us to obtain the general equation of motion for holonomically and nonholonomically constrained mechanical systems subjected to nonideal constraints.

# References

1. Lagrange, J.L. (1787) *Mecanique Analytique*. Mme Ve Courcier, Paris.
2. Gibbs, J.W. (1879) On the Fundamental Formulae of Dynamics. *Am. J. Math.*, **2**, 49–64.
3. Appell, P. (1899) Sur une Forme Generale des Equations de la Dynamique. *C. R. Acad. Sci.*, Paris, **129**, 459–460.
4. Udwadia, F.E. and Kalaba, R.E. (1992) A New Perspective on Constrained Motion. *Proc. Roy. Soc. Lon.*, **439**, 407–410.
5. Gauss, C.F. (1829) Über ein neues allgemeines Grundgesetz der Mechanik. *Journ. für reine und angewandte Mathematik*, **4**, 232–235.
6. Kalaba, R.E. and Udwadia, F.E. (1993) Equations of Motion for Nonholonomic, Constrained Dynamical Systems via Gauss's Principle. *Journal of Applied Mechanics*, **60**, 662–668.
7. Goldstein, H. (1981) *Classical Mechanics*. Addison-Wesley, Reading, MA.
8. Pars, L.A. (1979) *A Treatise on Analytical Dynamics*. Oxbow Press, Woodbridge, Connecticut.
9. Udwadia, F.E. and Kalaba, R.E. (1996) *Analytical Dynamics: A New Approach*. Cambridge University Press, England.
10. Appell, P. (1911) Example de Mouvement d'un Point Assujeti a une Liason Exprimee par une Relation Non Lineaire entre les Composantes de la Vitesse. *Comptes Rendus*, 48–50.

11. Udwadia, F.E. and Kalaba, R.E. (2000) Lagrangian Dynamics and Non-Ideal Constraints. *Journal of Aerospace Engineering*, **13**, 17–24.

12. Udwadia, F.E. and Kalaba, R.E. (2001) Explicit Equations of Motion for Constrained Systems with Non-Ideal Constraints. *Journal of Applied Mechanics*, **68**, 462–467.

13. Udwadia, F.E. and Kalaba, R.E. (2002) On the Foundations of Analytical Dynamics. *Journal of Nonlinear Mechanics*, **37**, 1079–1090.

14. Udwadia, F.E. and Kalaba, R.E. (2002) What is the General Form of the Explicit Equation for Constrained Mechanical Systems? *Journal of Applied Mechanics*, **69**, 335–339.

# Mathematical Analysis of Vibrations of Nonhomogeneous Filament with One End Load

Marianna A. Shubov

*Department of Mathematics and Statistics, Texas Tech University*
*Lubbock, TX, 79409, USA*
*Tel: (806) 742-2336, E-mail: marianna.shubov@ttu.edu*

We consider a class of nonselfadjoint operators generated by the equation and the boundary conditions, which govern small vibrations of an ideal filament with homogeneous Dirichlet boundary condition at one end and a heavy load at the other end. The filament has a nonconstant density and is subject to a viscous damping with a nonconstant damping coefficient. The boundary conditions contain an arbitrary complex parameter. We present the spectral asymptotics for the aforementioned one-parameter family of nonselfadjoint operators and investigate such properties of the root vectors of each operator as completeness and minimality in the state space of the system. In the forthcoming papers, based on the results of the present paper, we will prove the Riesz basis property of the eigenfunctions. The spectral results obtained in the aforementioned papers will allow us to solve boundary and/or distributed controllability problems for the filament using the spectral decomposition method.

## 1   Introduction and Statement of Problem

The present paper is devoted to the mathematical investigation of the initial-boundary value problem that describes small vibrations of an ideal filament which is fixed at one end and carries a heavy load at the other end. In current mathematical literature, there exists many works in which the motions of different types of filaments (both linear and nonlinear) have been investigated from purely theoretical and computational points of view (see [5, 7–15, 20–22] and references therein). In this paper, we study the problem which can be considered as a generalization of the results discussed in the monograph [13]. The author of [13], D.R. Merkin, has started from the general ideas of mechanics and then has derived static and dynamic equations of filaments, has formulated numerous boundary-value problems, and has presented solutions for some of them. Many technological applications have been given in [13] and several computer algorithms have been discussed. In our work, we consider a more general case of a filament with spatially nonhomogeneous parameters and a distributed viscous damping, and we answer several questions that have not been raised in [13] or other sources. For example, we present here a precise asymptotics of the spectrum for a loaded filament and give a rigorous description of such properties of the eigenstates as completeness, minimality, and in the forthcoming paper, we will prove the Riesz basis property of eigenstates in an appropriate state space of the system. The latter results will be instrumental for the solutions of different problems on boundary and distributed controllability of a filament. Such questions are beyond the scope of monograph [13]. In connection with our future applications of the spectral results to controllability problems, we would like to mention paper [6]. An important engineering model of an "overhead crane," i.e., a motorized platform moving along a horizontal bench with a flexible cable attached to the platform is considered in [6]. The cable is supposed to be homogeneous and to have no damping. A boundary control is proposed that guarantees uniform exponential stability. In their proofs, the authors of [6] have introduced the Lyapunov function (the generalized energy) and have derived the estimate from above in the exponential form. Our goal is different; namely, we would like not only to prove an exact controllability result for the system, but also to give an explicit expression for the control law in the terms of the spectral characteristics of the problem. Finally, we would like to mention two interesting papers [16, 17] by V. Pivovarchik, in which the author has addressed the question of the reconstruction of the string parameters based on the spectral characteristics of the appropriate Sturm–Liouville problem.

The model considered in the present work is well known in mechanics and before we formulate the problem, we recall some notions from mechanics. A filament is known as a one-dimensional system that under the action of external forces can make up a form of any geometrical curve. A filament that does not show any resistance to bending and torsion will be called an ideal or absolutely flexible filament. In what follows, under the term "filament," we will understand an ideal inextensible one-dimensional system. "One-dimensional" means that the sizes of the cross-section are negligibly small in comparison with the length of the system. In spite of the fact that an ideal filament is an abstraction, in many cases threads, ropes, cables, and chains clamped in appropriate ways correspond to this model quite well. That is why mathematical theory of an ideal filament might be of great practical importance.

We consider a nonhomogeneous inextensible filament of the length $l$ which has been hung up at one end with a load of the weight $M$ at the other end. In the state of equilibrium, the filament is hanging along a certain vertical line which we will consider as the $x$-axis. Disregarding the sizes of the load, we will study small vibrations of a filament near equilibrium assuming that these vibrations take place in one plane. The fact that the filament is nonhomogeneous means that the linear density $\rho$ depends on the chosen point, i.e., $\rho = \rho(x)$, $x \in [0, l]$. Note that the inhomogeneity may be caused either by inhomogeneity of the material or by the differences in the areas of the cross-sections. In what follows, we will use the following notations: $S(x)$ – the area of the cross-section of the filament at the point $x$; $\rho(x)$ – the density of the filament (mass per unit length) at the point $x$; $M$ – the mass of the point load at the end $x = l$; $g$ – the gravitational constant; $T(x)$ – the tension of the filament at the point $x$. The equation of small transversal vibrations of a filament can be given in the following form:

$$\rho(x)S(x)u_{tt}(x,t) = (T(x)u_x(x,t))_x . \tag{1.1}$$

As is known (see [13] and references therein), the tension of the filament $T(x)$ at the point $x$ is the sum of the gravity force of the part of the filament which is located below $x$, the gravity force of the point mass $M$, and the vertical component of the exterior force applied to this mass.

We will consider small oscillations of the filament. By small oscillations we mean that the angle between the $x$-axis and the tangent line to the filament is sufficiently small (see Fig. 1). The latter assumption will allow us to develop a linear theory of vibrations of the damped nonhomogeneous filament which is subject to nonconservative boundary conditions.

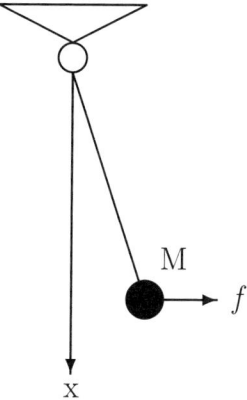

Figure 1

The tension at the point $x$ can be given in the following form:

$$T(x) = Mg + \int_x^l \rho(\tau)S(\tau)d\tau . \tag{1.2}$$

Combining Eqs. (1.1) and (1.2), we arrive at the equation describing small transversal vibrations of the filament

$$\rho(x)S(x)u_{tt}(x,t) = \left( \left( Mg + g \int_x^l \rho(\tau)S(\tau)d\tau \right) u_x(x,t) \right)_x . \tag{1.3}$$

Assume that there exists a viscous damping with the damping coefficient **2d**. This assumption leads to the modification of Eq. (1.3)

$$\rho(x)S(x)u_{tt}(x,t) = \left(\left(Mg + g\int_x^l \rho(\tau)S(\tau)d\tau\right)u_x(x,t)\right)_x - 2\mathbf{d}(x)u_t(x,t). \quad (1.4)$$

At the end $x = 0$, we will consider the homogeneous Dirichlet boundary condition

$$u(0,t) = 0. \quad (1.5)$$

The boundary condition at the end $x = l$ can be derived if one takes into account the equation of motion of the end mass. The following forces are acting on the point mass $M$: a) the gravity force $Mg$ is acting downward in vertical direction; b) the tension force $T$ at the point $x = l$ is acting upward along the tangent line; c) the exterior force $f(t)$ which is assumed to act in the horizontal direction (see Fig. 1). Thus, we obtain the following equation which describes the transversal motion of the point mass $M$:

$$Mu_{tt}(l,t) = -T(l)u_x(l,t) + f(t). \quad (1.6)$$

Substituting $T$ from Eq. (1.2) into Eq. (1.6), we obtain

$$Mu_{tt}(l,t) = -Mgu_x(l,t) + f(t). \quad (1.7)$$

We will study even more general boundary conditions at $x = l$ than (1.7). Namely, assume that we have the following one-parameter family of boundary conditions:

$$Mu_{tt}(l,t) + Mg\,u_x(l,t) = f(t) + hu_t(l,t), \ h \in \mathbb{C} \cup \{\infty\}. \quad (1.8)$$

Introducing in a standard fashion the initial conditions, we arrive at the following initial-boundary value problem:

$$\rho(x)S(x)u_{tt}(x,t) = \left(\left((Mg + g\int_x^l \rho(\tau)S(\tau)d\tau\right)u_x(x,t)\right)_x - 2\mathbf{d}(x)u_t(x,t), \quad (1.9)$$

$$u(0,t) = 0, \quad (1.10)$$

$$Mu_{tt}(l,t) + Mgu_x(l,t) = f(t) + hu_t(l,t), \qquad h \in \mathbb{C} \cup \{\infty\}, \quad (1.11)$$

$$u(x,0) = g_1(x), \qquad u_t(x,0) = g_2(x). \quad (1.12)$$

Now we describe the properties of the coefficients. Assume that the coefficients are real-valued functions, and they are subject to the following conditions:

$$\rho,\ S \in W_1^2(0,l), \qquad \mathbf{d} \in W_1^1(0,l), \qquad (\rho S)(x) \geq C > 0 \quad \text{for} \quad x \in [0,l]. \quad (1.13)$$

(We use the standard notation $W_m^l$ for the Sobolev space [1] of functions having weak derivatives up to the order $l$ in the space $L^m$.)

**Remark 1.1:** We do not impose on $\mathbf{d}$ the restriction $\mathbf{d} \geq 0$ since all our results are valid without this assumption. However, we will call this coefficient "viscous damping." Certainly, this phrase is some abuse of terminology because, rigorously speaking, only a nonnegative coefficient $\mathbf{d}$ can be called a viscous damping.

From now on, the problem given by (1.9)–(1.12) is our main object of interest.

In the conclusion of this section, we briefly outline the content of the paper. In Section 2, we give an operator setting of the problem in the so-called energy space (the state space of the system). We introduce the main matrix differential operator; asymptotic and spectral properties of this operator are our interest. In this section, we also reproduce the main result from our paper [20] (see Theorem 2.1 below), in which we present the precise asymptotics of the spectrum of the problem. In Section 3, we formulate one of the main results of the present paper – Theorem 3.1. To make the exposition self-contained, we recall all necessary definitions. In Section 4, we prove that the set of eigenmodes is complete in the state space of the system.

# 2 Operator Reformulation of the Problem

In this section, we reformulate the initial-boundary value problem given by (1.9)–(1.12). Namely, we represent it as the first order in time evolution equation in the state space which is a Hilbert space equipped with the so-called energy norm.

It is convenient to introduce the notations

$$p(x) = \rho(x)S(x), \qquad r(x) = Mg + g \int_x^l p(\tau)d\tau. \tag{2.1}$$

Using the properties of the functions $\rho$ and $S$ from (1.13), we can easily see that the functions $p$ and $r$ enjoy the following properties:

$$p \in W_1^2(0,l), \quad p \geq C > 0, \quad r \text{ is an absolutely continuous function.} \tag{2.2}$$

In terms of (2.1), Eq. (1.9) has the form

$$p(x)u_{tt}(x,t) = (r(x)u_x(x,t))_x - 2\mathbf{d}(x)u_t(x,t). \tag{2.3}$$

## 2.1 Energy balance

Let $\mathcal{E}$ be the energy of the system, $\mathcal{D}$ be the loss of energy due to the dampings, and $\mathcal{W}$ be the power of the external force. Let the quantities $\mathcal{E}, \mathcal{D}$, and $\mathcal{W}$ be defined for a smooth function $u(x,t)$, $x \in [0,l]$, $t \geq 0$, by the following formulas:

$$\mathcal{E}(t) = \frac{1}{2} \int_0^l \left( r(x)|u_x(x,t)|^2 + p(x)|u_t(x,t)|^2 \right) dx + M|u_t(l,t)|^2, \tag{2.4}$$

$$\mathcal{D}(t) = -2 \int_0^l \mathbf{d}(x)|u_t(x,t)|^2 dx + \operatorname{Re} h \, |u_t(l,t)|^2, \tag{2.5}$$

$$\mathcal{W}(t) = f(t)\,\mathrm{Re}\,u_t(l,t)\,. \qquad (2.6)$$

The following statement holds.

**Lemma 2.1** *If $u$ satisfies Eq. (1.9) and the boundary conditions given by (1.10) and (1.11), then*

$$\frac{d\mathcal{E}}{dt} = \mathcal{D} + \mathcal{W}\,. \qquad (2.7)$$

**Proof:** Let us multiply Eq. (2.3) by $\overline{u}_t(x,t)$ and integrate over the interval $[0,l]$. We have

$$\int\limits_0^l p(x)u_{tt}(x,t)\overline{u}_t(x)dx = \int\limits_0^l (r(x)u_x(x,t))_x\overline{u}_t(x,t)dx - 2\int\limits_0^l \mathbf{d}(x)|u_t(x,t)|^2dx\,. \qquad (2.8)$$

Integrating by parts in Eq. (2.8), we reduce it to the form

$$\int\limits_0^l p(x)u_{tt}(x,t)\overline{u}_t(x)dx = -\int\limits_0^l r(x)u_x(x,t)\overline{u}_{xt}(x,t)dx -$$

$$-2\int\limits_0^l \mathbf{d}(x)|u_t(x,t)|^2dx + r(x)u_x(x,t)\overline{u}_t(x,t)\,\bigg|_0^l\,. \qquad (2.9)$$

Consider the out of integral term in (2.9). Due to boundary condition (1.10) and definitions (2.1), we obtain

$$r(x)u_x(x,t)\overline{u}_t(x,t)\big|_0^l = r(l)u_x(l,t)\overline{u}_t(l,t) - r(0)u_x(0,t)\overline{u}_t(0,t) = Mgu_x(l,t)\overline{u}_t(l,t)\,. \qquad (2.10)$$

From boundary condition (1.11), we have

$$Mgu_x(l,t)\overline{u}_t(l,t) = f(t)\overline{u}_t(l,t) + h|u_t(l,t)|^2 - Mu_{tt}(l,l)\overline{u}_t(l,t)\,. \qquad (2.11)$$

Combining (2.8) with (2.10) and (2.11), we obtain

$$\int\limits_0^l p(x)u_{tt}(x,t)\overline{u}_t(x)dx = -\int\limits_0^l r(x)u_x(x,t)\overline{u}_{x,t}(x,t)dx -$$

$$-2\int\limits_0^l \mathbf{d}(x)|u_t(x,t)|^2dx - Mu_{tt}(l,t)\overline{u}_t(l,t) + h|u_t(l,t)|^2 + f(t)\overline{u}_t(l,t)\,. \qquad (2.12)$$

In the next step, let us take the equation which is complex conjugated to Eq. (2.3), multiply it by $u_t(x, t)$, and integrate over the interval $[0, l]$. If we add together the latter equation and Eq. (2.12), we arrive at the following result:

$$\int\limits_0^l p(x) \frac{d}{dt} |u_t|^2 dx = -\int\limits_0^l r(x) \frac{d}{dt} |u_x(x, t)|^2 dx - 4 \int\limits_0^l \mathbf{d}(x) |u_t(x, t)|^2 dx -$$

$$- M \frac{d}{dt} |u_t(l, t)|^2 + 2 \operatorname{Re} h |u_t(l, t)|^2 + 2f(t) \operatorname{Re} u_t(l, t). \qquad (2.13)$$

It is obvious that Eq. (2.13) implies (2.7). The lemma is completely shown. $\qquad \square$

## 2.2 Evolution equation

Let $\mathcal{H}$ be a Hilbert space of three-component vector-valued functions $U = (u_0, u_1, u_2)^T$ (here and below the script '$T$' means the transposition) obtained as a closure of smooth functions satisfying the condition $u_0(0) = 0$ with respect to the following energy norm:

$$\|U\|_{\mathcal{H}}^2 = \frac{1}{2} \int\limits_0^l \left[ r(x) |u_0'(x)|^2 + p(x) |u_1(x)|^2 \right] dx + M |u_2|^2. \qquad (2.14)$$

As follows from (2.14), the energy space $\mathcal{H}$ can be represented in the form

$$\mathcal{H} = W_{2,r}^1(0, l) \oplus L_p^2(0, l) \oplus \mathbb{C}, \qquad (2.15)$$

where $W_{2,r}^1(0, l)$ is a weighted Sobolev space, and $L_p^2(0, l)$ is a weighted $L^2$ space. Namely, if $f \in W_{2,r}^1(0, l)$ and $g \in L_p^2(0, l)$, then

$$\|f\|^2 = \int\limits_0^l r(x) [|f'(x)|^2 + |f(x)|^2] dx < \infty \quad \text{and} \quad \|g\|^2 = \int\limits_0^l p(x) |g(x)|^2 dx < \infty.$$

$$(2.16)$$

Since both $r$ and $p$ are positive continuous functions which stay away from zero, the spaces $W_{2,r}^1(0, l)$ and $L_p^2(0, l)$ are metrically equivalent to the standard $W_2^1(0, l)$ and $L^2(0, l)$ spaces. If $h = \infty$, then the function $u_0$ has to satisfy an additional condition $u_0(l) = 0$. In $\mathcal{H}$ we consider a matrix differential operator given by the differential expression

$$\mathcal{L}_h = -i \begin{pmatrix} 0 & 1 & 0 \\ \dfrac{1}{p(x)} \dfrac{d}{dx} \left( r(x) \dfrac{d}{dx} \right) & -\dfrac{2\mathbf{d}(x)}{p(x)} & 0 \\ -g \dfrac{d}{dx} \cdot \bigg|_{x=l} & 0 & \dfrac{h}{M} \end{pmatrix}, \qquad (2.17)$$

defined on the following domain:

$$\mathcal{D}\left(\mathcal{L}_h\right) = \left\{ U \in \mathcal{H} : u_0 \in W_2^2(0,l),\ u_1 \in W_2^1(0,l),\ u_2 = u_1(l),\ u_1(0) = 0 \right\}. \quad (2.18)$$

The notation $A \equiv -g\dfrac{d}{dx}\cdot\bigg|_{x=l}$ means that for a differentiable function $f$, the following relation holds:

$$Af = -gf'(l).$$

**Remark 2.1:** In what follows, we are looking for a solution $u(x,t)$ with finite energy (2.14). It means that the vector $U = (u(x,t), u_t(x,t), u_t(l,t))^T$ has to belong to the energy space $\mathcal{H}$ in the sense that for fixed $t$, the function $U(\cdot,t)$ is an element from $\mathcal{H}$.

By a direct computation, one can show that the initial-boundary value problem given by (1.9)–(1.12) without a forcing term ($f = 0$) is equivalent to the following evolution equation in $\mathcal{H}$

$$U_t(x,t) = i\left(\mathcal{L}_h U\right)(x,t), \qquad U(x,t)\big|_{t=0} = G(x),$$

$$\tag{2.19}$$

$$G(x) = (g_1(x),\ g_2(x),\ g_3)^T, \qquad x \in (0,l).$$

The dynamics generator $\mathcal{L}_h$ of the evolution problem defined by (2.19) is our main object of interest. In a series of forthcoming papers [20–22], we will show that $\mathcal{L}_h$ is a nonselfadjoint operator which belongs to a class of Riesz spectral operators in the sense of N. Dunford [2]. The latter statement means that this nonselfadjoint operator admits the spectral decomposition which is similar to the spectral decomposition of a selfadjoint operator. In fact, in the aforementioned series of our works, we do not need the most general definition of a spectral operator. The operators we consider here are Riesz spectral operators with discrete spectra. To formulate the definition that will be adopted for a Riesz spectral operator, we recall an important notion of a Riesz basis.

**Definition 2.1** *Let $R(H)$ be the class of bounded linear operators in $H$. A sequence of vectors $\{\psi_n\}_{n=1}^\infty \subset H$ is called a Riesz basis if there exists an operator $A \in R(H)$ such that $A^{-1} \in R(H)$, and the system of vectors $\{A\psi_n\}_{n=1}^\infty$ forms an orthonormal basis in $H$. $A$ is called an orthogonalizer of $\{\psi_n\}_{n=1}^\infty$. If $A$ is an orthogonalizer, then all other orthogonalizers have the form $UA$, where $U$ is an arbitrary unitary operator. The norm $\|A\|$ is uniquely defined by the basis. For any Riesz basis $\{\psi_n\}_{n=1}^\infty \subset H$, there is a unique biorthogonal basis $\{\psi_n^*\}_{n=1}^\infty$ defined by the relations: $(\psi_n, \psi_m^*) = \delta_{nm}$.*

**Definition 2.2** *An operator $\mathcal{L}$ in a complex separable Hilbert space $H$ is called Riesz spectral if it has the following properties.*

   (i) *$\mathcal{L}$ is an either bounded or closed unbounded operator which is defined on a dense domain $D(\mathcal{L}) \subset H$;*

   (ii) *$\mathcal{L}$ has a discrete spectrum;*

*(iii) only a finite number of the eigenvectors have finite chains of associate vectors;*

*(iv) the system of root vectors (eigenvectors and associate vectors together) forms a Riesz basis (a linear isomorphic image of an orthonormal basis) in $H$.*

Definition 2.2 can be reformulated as follows. $\mathcal{L}$ is an operator whose matrix in an appropriate basis has a Jordan normal form with a finite number of nontrivial Jordan cells of a finite length each. Definition 2.2 is also equivalent to the following one.

**Definition 2.3** *a) Let $\{\psi_n\}_{n=1}^{\infty}$ be a Riesz basis in $H$, and let $\{\lambda_n\}_{n=1}^{\infty}$ be a sequence of complex numbers. Define an operator $S$ in $H$ by the formula:*

$$S\varphi = \sum_{n=1}^{\infty} \lambda_n(\varphi, \psi_n^*)\psi_n \qquad (2.20)$$

*on the dense domain*

$$D(S) = \left\{ \varphi \in \mathcal{H} : \sum_{n=1}^{\infty} |\lambda_n|^2 |(\varphi, \psi_n^*)|^2 < \infty \right\}. \qquad (2.21)$$

*The operators of type (2.20), (2.21) are called scalar operators.*

*b) An operator $\mathcal{L}$ in $H$ is called a spectral operator if it can be represented in the form:*

$$\mathcal{L} = S + N, \qquad (2.22)$$

*where $S$ is a scalar operator, and $N$ is a bounded finite rank nilpotent operator (i.e., there exists a positive integer $k$ such that $N^k = 0$), which commutes with $S$.*

(In a more general definition of a spectral operator [2], spectral representation (2.20) may be continuous, i.e., it involves integration with respect to a spectral measure, and $N$ may be a quasinilpotent operator, i.e., it is bounded and its spectrum $\sigma(N) = \{0\}$).

In the present paper, we address the properties (i) and (ii) from Definition 2.2. The property (iii) has been proven in our work [20], where we have derived a precise spectral asymptotics of the dynamics generator $\mathcal{L}_h$. To present a whole picture, we will reproduce the main result of [20] as Theorem 2.1 below. In the present paper and in [21], we discuss the properties of the set of root vectors of the operator $\mathcal{L}_h$. To recall the notion of root vectors, we point out that generally $\mathcal{L}_h$ is a nonselfadjoint operator in the state space $\mathcal{H}$. As we will show (see Theorem 3.1 below), this operator may have multiple eigenvalues of a finite multiplicity each. In addition to an eigenvector, corresponding to a given eigenvalue $\lambda$, it may be a finite chain of *associate vectors* ($\psi$ is an associate vector of a linear operator $A$ of the order $m$ corresponding to the eigenvalue $\lambda$ if $(A - \lambda I)^m \psi \neq 0$ and $(A - \lambda I)^{m+1}\psi = 0$). The entire collection of eigenvectors and associate vectors is called the set of *root vectors*. Returning to the operator $\mathcal{L}_h$, in the paper [21], we will discuss the Riesz basis property of the root vectors in the state space $\mathcal{H}$. We will recall all necessary definitions and provide some background from harmonic analysis. To obtain the aforementioned results about the root vectors, we need very detailed information on the spectrum and eigenfunctions of the operator $\mathcal{L}_h$. We point out here that the problem of the Riesz basis property of root vectors has been addressed in a number of our works [23, 25–27, 29, 30] devoted to the mathematical analysis of the problems arising in the study of vibrations of different flexible structures. In our paper [22], we will prove the exact boundary and distributed controllability results for the filament model using the spectral decomposition method introduced by D.L. Russell [18, 19] in the late sixties. The method used in [22] is the generalization of the method which has been utilized in our control-theoretical papers [24, 28, 31]. In the conclusion of this section, we reproduce our results on the spectral asymptotics from our work [20].

**Theorem 2.1** *The operator $\mathcal{L}_h$ has a countable set of eigenvalues $\{\lambda_n^h\}_{n\in\mathbb{Z}}$. This set is located in a strip parallel to the real axis and has only two points of accumulation: $+\infty$ and $-\infty$ in the sense that $\operatorname{Re}\lambda_n^h \to \pm\infty$ as $n \to \pm\infty$ and $\operatorname{Im}\lambda_n^h \to const$ as $n \to \pm\infty$ (see (2.23) below).*

1. *There exists $n_0 > 0$ such that for $|n| \geq n_0$, the following asymptotic formula is valid:*

$$\lambda_n^h = \mathcal{M}^{-1}\left(n + \frac{1}{2}\operatorname{sgn}n\right)\pi + i\mathcal{M}^{-1} + O\left(|n|^{-1}\right), \quad |n| \longrightarrow \infty. \quad (2.23)$$

*Two constants $\mathcal{M}$ and $\mathcal{N}$ in (2.23) are given by the formulas*

$$\mathcal{M} = \int_0^l \sqrt{\frac{p(\eta)}{r(\eta)}}\, d\eta, \qquad \mathcal{N} = \int_0^l \frac{\mathbf{d}(\eta)}{\sqrt{p(\eta)r(\eta)}}\, d\eta. \quad (2.24)$$

2. *The corresponding eigenvector $\Phi_n^h$ of the operator $\mathcal{L}_h$ can be represented in the form*

$$\Phi_n^h(x) = \left((i\lambda_n^h)^{-1}F_n^h(x), F_n^h(x), F_n^h(l)\right)^T, \quad (2.25)$$

*where the following asymptotic representation for the function $F_n^h$ is valid as $|n| \to \infty$:*

$$F_n^h(x) = \left(\frac{r(x)}{p(x)}\right)^{1/4}\exp\left\{\int_0^x \frac{gp(\eta)}{2r(\eta)}\, d\eta\right\}\cos\left[\lambda_n^h\int_0^x \sqrt{\frac{p(\eta)}{r(\eta)}}\, d\eta - \right.$$

$$\left. -i\int_0^x \frac{\mathbf{d}(\eta)}{\sqrt{p(\eta)r(\eta)}}\, d\eta + \omega\right] + O(|n|^{-1}), \quad |n| \to \infty, \quad (2.26)$$

*with $\omega$ being defined as*

$$\omega = \frac{\pi}{2} + \frac{i}{4}\ln\left(\frac{r(0)}{p(0)}\right). \quad (2.27)$$

*The estimate $O(|n|^{-1})$ is uniform with respect to $x \in [0, l]$.*

# 3   Properties of the Dynamics Generator $\mathcal{L}_h$

In this section, we prove several results on the properties of the operator $\mathcal{L}_h$. Before we formulate these results, we recall some definitions [3, 4].

**Definition 3.1**   *A closed linear operator $A$, acting on a Hilbert space $H$ and defined on the domain $D(A)$, which is dense in $H$, is said to be dissipative if $\operatorname{Im}(AF, F)_H \geq 0$ for $F \in D(A)$.*

**Definition 3.2**   *A dissipative operator $A$, defined on a domain $D(A)$ which is dense in $H$, is said to be simple if $A$ and $A^*$ do not have a common invariant subspace on which they coincide.*

**Definition 3.3**   *A dissipative operator $A$ in $H$ is said to be maximal if it does not admit any extensions.*

**Definition 3.4**   *An eigenvalue $\lambda$ of a bounded operator $A$ in a Hilbert space $H$ is said to be normal if and only if*

*a) $\lambda$ is an isolated point of the spectrum of $A$,*

*b) the algebraic multiplicity of $\lambda$ is finite, and*

*c) the range $(A - \lambda I)H$ of the operator $A - \lambda I$ is closed.*

Now we are in a position to formulate the first theorem in this section.

**Theorem 3.1**   *a) $\mathcal{L}_h$ is a closed nonselfadjoint operator in the energy space $\mathcal{H}$.*

*b) If $\mathbf{d} \geq 0$ is not identically equal to zero and $\operatorname{Re} h \leq 0$, or if $\mathbf{d} = 0$ and $\operatorname{Re} h < 0$, then $\mathcal{L}_h$ is a dissipative operator in $\mathcal{H}$.*

*c) If $\mathbf{d} > 0$, then $\mathcal{L}_h$ is a simple operator in $\mathcal{H}$.*

*d) When $\mathcal{L}_h$ is a dissipative operator, then it is maximal.*

*e) $\mathcal{L}_h^{-1}$ exists and is a compact operator in $\mathcal{H}$, i.e., the operator $\mathcal{L}_h$ has a purely discrete spectrum consisting of normal eigenvalues.*

**Proof:** a) We start with the formula for the adjoint operator $\mathcal{L}_h^*$. To derive this formula, we use the standard definition of an adjoint operator according to which

$$(\mathcal{L}_h U, V)_{\mathcal{H}} = (U, \mathcal{L}_h^* V)_{\mathcal{H}}, \quad U \in D(\mathcal{L}_h), \quad V \in D(\mathcal{L}_h^*), \qquad (3.1)$$

where $(\cdot, \cdot)_{\mathcal{H}}$ is the inner product generated by the energy norm (2.14). After straightforward calculations, we obtain that the adjoint operator $\mathcal{L}_h^*$ is given by the differential expression

$$
\mathcal{L}_h^* = -i \begin{pmatrix} 0 & 1 & 0 \\ \dfrac{1}{p(x)}\dfrac{d}{dx}\left(r(x)\dfrac{d}{dx}\right) & \dfrac{2\mathbf{d}(x)}{p(x)} & 0 \\ -g\dfrac{d}{dx}\cdot\bigg|_{x=l} & 0 & -\dfrac{\bar{h}}{M} \end{pmatrix}, \tag{3.2}
$$

defined on the following domain:

$$
\mathcal{D}\left(\mathcal{L}_h^*\right) = \left\{ U \in \mathcal{H} : u_0 \in W_2^2(0,l),\ u_1 \in W_2^1(0,l),\ u_2 = u_1(l),\ u_1(0) = 0 \right\}. \tag{3.3}
$$

Note that formulas (3.2) and (3.3), describing the adjoint operator $\mathcal{L}_h^*$, can be obtained from formulas (2.17) and (2.18) in which $\mathbf{d}$ and $h$ have been replaced with $(-\mathbf{d})$ and $(-\bar{h})$ respectively. The fact that $\mathcal{L}_h$ is closed will be shown at the end of the proof of statement e).

b) To prove that $\mathcal{L}_h$ is dissipative, let us consider the inner product $(\mathcal{L}_h\Phi, \Phi)_\mathcal{H}$ with $\Phi = (\varphi_0, \varphi_1, \varphi_2)^T \in D(\mathcal{L}_h)$. We have

$$
2(\mathcal{L}_h\Phi, \Phi)_\mathcal{H} = -i \left\{ \int_0^l \left[ r(x)\varphi_1'(x)\bar{\varphi}_0'(x) + (r(x)\varphi_0'(x))'\bar{\varphi}_1(x) - \right. \right.
$$

$$
\left. - 2\mathbf{d}(x)|\varphi_1(x)|^2 \right] dx - Mg\varphi_0'(l)\bar{\varphi}_2 + h|\varphi_2|^2 \bigg\} =
$$

$$
= -i \left\{ \int_0^l [r(x)\varphi_1'(x)\bar{\varphi}_0'(x) + r(x)\varphi_0'(x)\varphi_1'(x)]\, dx - \right.
$$

$$
\left. - 2\int_0^l \mathbf{d}(x)|\varphi_1(x)|^2 dx - Mg\varphi_0'(l)\bar{\varphi}_2 + h|\varphi_2|^2 + r(x)\varphi_0'(x)\bar{\varphi}_1(x)\bigg|_0^l \right\}. \tag{3.4}
$$

Taking into account the boundary conditions $\varphi_1(0) = 0$, $\varphi_1(l) = \varphi_2$, and the relation $r(l) = Mg$, we obtain from (3.4)

$$
2\,\mathrm{Im}\,(\mathcal{L}_h\Phi, \Phi)_\mathcal{H} = 2\int_0^l \mathbf{d}(x)|\varphi_1(x)|^2 dx - \mathrm{Re}\,h\,|\varphi_2|^2. \tag{3.5}
$$

This completes the proof of statement b).

c) To prove that $\mathcal{L}_h$ is a simple operator in $\mathcal{H}$, we use a contradiction argument, i.e., we assume that there exists a nontrivial vector $F = (f_0, f_1, f_2)^T$ such that $F \in D(\mathcal{L}_h)$, $F \in D(\mathcal{L}_h^*)$, and the following equation holds:

$$
\mathcal{L}_h F = \mathcal{L}_h^* F. \tag{3.6}
$$

Taking into account formulae (2.17), (2.18), (3.2), and (3.3), we rewrite Eq. (3.6) component-wise and have

$$\frac{\mathbf{d}(x)}{p(x)}f_1(x) = 0\,, \qquad hf_2 + \bar{h}f_2 = 0\,. \tag{3.7}$$

Since $\mathbf{d} > 0$ and $p > 0$, we immediately get $f_1 = 0$. Due to the fact that $f_2 = f_1'(l)$, we obtain that $f_2 = 0$. Thus, to satisfy Eq. (3.7), the vector $F$ must have the form $(f_0, 0, 0)^T$, $f_0 \in W_2^2(0, l)$. It can be easily verified that such a vector belongs to an invariant subspace of $\mathcal{L}_h$ if and only if $F = (f_0, 0, 0)^T \in Ker(\mathcal{L}_h)$. Indeed, the fact that $F$ belongs to the invariant subspace of $\mathcal{L}_h$ means that the image $\mathcal{L}_h F = G$ must be of the form $G = (g_0, 0, 0)^T$. Taking into account formula (2.17) for the operator $\mathcal{L}_h$, we conclude that $F \in Ker(\mathcal{L}_h)$. The inverse implication is obvious. As is known, for a dissipative operator, the following relation is valid:

$$\mathcal{H} = Ker(\mathcal{L}_h) \oplus \overline{\mathcal{R}(\mathcal{L}_h)}\,, \tag{3.8}$$

where by $\oplus$, we have denoted the orthogonal sum of the subspaces, and by $\mathcal{R}(\mathcal{L}_h)$ we have denoted the range of the operator $\mathcal{L}_h$. We also know, that for any linear operator

$$\mathcal{H} = \mathcal{N}(\mathcal{L}_h^*) \oplus \overline{\mathcal{R}(\mathcal{L}_h)}\,, \tag{3.9}$$

where $\mathcal{N}(\mathcal{L}_h^*)$ is the null-space of the operator $\mathcal{L}_h^*$. Comparing (3.8) and (3.9), we obtain that

$$Ker(\mathcal{L}_h) = Ker(\mathcal{L}_h^*)\,. \tag{3.10}$$

Let us show that $Ker(\mathcal{L}_h) = \{0\}$. Indeed, using formula for $\mathcal{L}_h$, we can see that $Ker(\mathcal{L}_h)$ is a subspace of $\mathcal{H}$ spanned by the vector $F = (f_0, 0, 0)^T$, where $f_0$ satisfies the following boundary-value problem:

$$\frac{d}{dx}\left(r(x)\frac{df_0}{dx}\right) = 0\,, \qquad f_0(0) = f_0'(l) = 0\,. \tag{3.11}$$

Integrating (3.11), we obtain that

$$\frac{df_0(x)}{dx} = \frac{C}{r(x)}\,. \tag{3.12}$$

Taking into account that $f_0'(l) = 0$ and $r(l) > 0$, we obtain from (3.12) that $f_0'(x) = 0$, $x \in [0, l]$. Combining the latter equation for $f_0$ with the condition $f_0(0) = 0$, we obtain that $f_0 = 0$. Statement c) is also shown.

d) To prove that $\mathcal{L}_h$ is a maximal operator, it suffices to show that [3]

$$(\mathcal{L}_h + iI)D(\mathcal{L}_h) = \mathcal{H}\,. \tag{3.13}$$

As will be shown in the proof of statement e), the operator $\mathcal{L}_h^{-1}$ exists and is compact. Thus, the following relation is valid:

$$(\mathcal{L}_h + iI)D(\mathcal{L}_h) = (I + i\mathcal{L}_h^{-1})\mathcal{L}_h D(\mathcal{L}_h)\,. \tag{3.14}$$

Since zero belongs to the resolvent set of the operator $\mathcal{L}_h$, we have that $\mathcal{L}_h D(\mathcal{L}_h) = \mathcal{H}$. Note that the operator $(I + i\mathcal{L}_h^{-1})$ is a linear isomorphism of $\mathcal{H}$ onto $\mathcal{H}$. Indeed, if the previous statement is not valid, then we would have that $(-i)$ is an eigenvalue of $\mathcal{L}_h$, which contradicts statement c). This completes the proof of statement d).

e) Finally, we prove that $\mathcal{L}_h^{-1}$ exists and is a compact operator. Let us consider the equation

$$\mathcal{L}_h \Psi = Y, \qquad \Psi \in D(\mathcal{L}_h), \qquad Y \in \mathcal{H}. \tag{3.15}$$

We will show that Eq. (3.15) has a unique solution for any $Y \in \mathcal{H}$ and give a precise formula for this solution, i.e., we will give a formula for the resolvent operator. Rewriting Eq. (3.15) component-wise and assuming that $\Psi = (\psi_0, \psi_1, \psi_2)^T$ and $Y = (y_0, y_1, y_2)^T$, we obtain the following system:

$$\psi_1 = iy_0, \tag{3.16}$$

$$\frac{1}{p(x)} \frac{d}{dx} \left( r(x) \frac{d\psi_0(x)}{dx} \right) - \frac{2\mathbf{d}(x)}{p(x)} \psi_1 = iy_1, \tag{3.17}$$

$$-g\psi_0'(l) + \frac{h}{M} \psi_2 = iy_2. \tag{3.18}$$

Substituting (3.16) into (3.17), we obtain

$$\frac{d}{dx} \left( r(x) \frac{d\psi_0(x)}{dx} \right) = 2i\mathbf{d}(x) y_0(x) + iy_1(x). \tag{3.19}$$

Denoting

$$y_3(x) = 2i\mathbf{d}(x) y_0(x) + iy_1(x), \tag{3.20}$$

and then integrating Eq. (3.19), we obtain

$$\frac{d\psi_0(x)}{dx} = \frac{1}{r(x)} \int_x^l y_3(\tau) d\tau + \frac{C_1}{r(x)}. \tag{3.21}$$

Integrating Eq. (3.21) once again, we obtain

$$\psi_0(x) = \int_0^x \frac{dt}{r(t)} \int_t^l y_3(\tau) d\tau + C_1 \int_0^x \frac{dt}{r(t)} + C_2. \tag{3.22}$$

Since $\psi_0(0) = 0$, we have $C_2 = 0$. Now let us choose $C_1$ in such a way that $\psi_2 = \psi_1(l)$. To this end by combining Eqs. (3.18) and (3.21), we get

$$\psi_2 = \psi_0'(l) \frac{Mg}{h} + iy_2 \frac{M}{h} = \frac{Mg}{h} \frac{C_1}{r(l)} + iy_2 \frac{M}{h}. \tag{3.23}$$

Recalling that $Mg = r(l)$ and taking into account (3.16), we obtain the equation for $C_1$

$$\frac{C_1}{h} + \frac{iM}{h} y_2 = iy_0(l),$$

which gives us

$$C_1 = ih \left( y_0(l) - My_2 \right). \tag{3.24}$$

Thus, given $Y \in \mathcal{H}$, we have found the precise formula for $\Psi$

$$\Psi(x) = \left( \int_0^x \frac{dt}{r(t)} \int_t^l y_3(\tau) d\tau + C_1 \int_0^x \frac{dt}{r(t)}, \quad iy_0(x), \quad iy_0(l) \right)^T. \qquad (3.25)$$

Now we have to check that $\Psi$ given by (3.25) belongs to $D(\mathcal{L}_h)$, which can be done in a straightforward manner.

At last, we can easily show that $\mathcal{L}_h^{-1}$ is compact. Note that $D(\mathcal{L}_h)$, defined by (2.18), is a closed subspace of the space $\mathcal{H}_1 = W_2^2(0,l) \times W_2^1(0,l) \times \mathbb{C}$, while the energy space $\mathcal{H}$ is linearly isomorphic to a closed subspace of the space $\mathcal{H}_2 = W_2^1(0,l) \times L_2(0,l) \times \mathbb{C}$. It follows from the above proof that $\mathcal{L}_h^{-1}$ is a bounded operator from $\mathcal{H}$ onto $D(\mathcal{L}_h)$ if $D(\mathcal{L}_h)$ is equipped with the norm of $\mathcal{H}_1$. Therefore, $\mathcal{L}_h^{-1}$ is a compact operator since the embedding $\mathcal{H}_1 \hookrightarrow \mathcal{H}_2$ is compact [1].

To complete the proof, it remains to be seen that $\mathcal{L}_h$ is a closed operator because this operator is inverse to a compact operator $\mathcal{L}_h^{-1}$ defined on the entire energy space $\mathcal{H}$. The theorem is completely shown. $\qquad \square$

# 4   Completeness of the Root Vectors in the Energy Space

In this section, we prove an important property of the set of root vectors of the operator $\mathcal{L}_h$, i.e., we prove that this set is complete in the energy space. As a consequence of the main result of the previous section, we already have that the set of root vectors is minimal in $\mathcal{H}$. (Recall that an infinite set of vectors in a Hilbert space is said to be *minimal* if any vector from this set does not belong to a closed linear span of the remaining vectors.) The fact of minimality of the root vectors results from the following two properties: (i) the operators $\mathcal{L}_h$ and $\mathcal{L}_h^{-1}$ have the same sets of root vectors, and (ii) the operator $\mathcal{L}_h$ is compact. However, the fact that an infinite set of vectors is complete and minimal in $\mathcal{H}$ is not enough to claim that this set forms a Riesz basis. In our next paper we will prove the Riesz basis property of the root vectors of $\mathcal{L}_h$ in $\mathcal{H}$.

Our first result in this section is concerned with the relationship between the operator $\mathcal{L}_h$ with $h \neq 0$ and the operator $\mathcal{L}_0$ ($\mathcal{L}_0$ coincides with $\mathcal{L}_h$ when $h = 0$).

**Theorem 4.1** *For any $h_1$, $h_2 \in \mathbb{C} \cup \{\infty\}$, $(\operatorname{Re} h_1, \operatorname{Re} h_2 \geq 0)$, the difference $\mathcal{L}_{h_1} - \mathcal{L}_{h_2}$ is a rank-one operator. In particular, each operator $\mathcal{L}_h$ is a rank-one perturbation of the operator $\mathcal{L}_0$.*

**Proof:** Since we know that zero does not belong to the spectrum of the operator $\mathcal{L}_h$ for any $h$ $(\operatorname{Re} h \geq 0)$, it suffices to prove that the difference

$$T_{h_1 h_2} = \mathcal{L}_{h_1}^{-1} - \mathcal{L}_{h_2}^{-1} \qquad (4.1)$$

is a one-dimensional operator in $\mathcal{H}$. As follows from formula (3.25), for any vector $Y = (y_0, y_1, y_2)^T \in \mathcal{H}$, we have

$$\left( \mathcal{L}_{h_1}^{-1} - \mathcal{L}_{h_2}^{-1} \right) Y = \left( i(h_1 - h_2)(y_0(l) - My_2) \int_0^x \frac{dt}{r(t)}, 0, 0 \right)^T. \qquad (4.2)$$

Formula (4.2) means that the operator $T_{h_1 h_2}$ maps any vector $Y \in \mathcal{H}$ into the multiple of one and the same standard vector

$$V = \left( \int\limits_0^x (r(t))^{-1} dt, 0, 0 \right)^T .$$

This completes the proof of the theorem.                                                                    □

**Corollary 4.1** We mention that the relation similar to (4.1) is valid for the adjoint operator as well. Namely, for any $h_1, h_2 \in \mathbb{C} \cup \{\infty\}$, $(\text{Re}\, h_1, \text{Re}\, h_2 \geq 0)$, there exists a rank-one operator $\hat{T}_{h_1 h_2}$ such that

$$\left( \mathcal{L}_{h_2}^* \right)^{-1} = \left( \mathcal{L}_{h_1}^* \right)^{-1} + \hat{T}_{h_1 h_2} . \tag{4.3}$$

We will use the result of this theorem to prove that the set of root vectors of the operator $\mathcal{L}_h$ is complete in the energy space $\mathcal{H}$. This proof is based on the following Keldysh theorem [4].

**Theorem 4.2** (M. Keldysh). *Let $A = B(I + S)$, where $B = B^* \in \Sigma_p$, $p < \infty$, and $S \in \Sigma_\infty$. (Recall that a compact operator $K$ in a Hilbert space $H$ belongs to the class $\Sigma_p$, $p < \infty$, if the sequence of the eigenvalues of the operator $(K^* K)^{1/2}$ belongs to $l^p$; $\Sigma_\infty$ is the class of all compact operators.) If the operator $A$ is injective, (i.e., $\text{Ker}\, A = \{0\}$), then the system of its root vectors is complete in $H$.*

Now we are in a position to formulate the following statement.

**Theorem 4.3** *The set of root vectors of the operator $\mathcal{L}_h$ is complete in the energy space $\mathcal{H}$.*

**Proof:** First of all, we note that for the operator $\mathcal{L}_0^*$ the following relation is valid:

$$\mathcal{L}_0^* = \mathcal{L} + \mathcal{D}, \qquad \mathcal{L} = \mathcal{L}^*, \qquad \mathcal{D} \in R(\mathcal{H}), \tag{4.4}$$

where $\mathcal{L}$ is a selfadjoint operator in $\mathcal{H}$ defined by the formula

$$\mathcal{L} = -i \begin{pmatrix} 0 & 1 & 0 \\ \dfrac{1}{p(x)} \dfrac{d}{dx}\left( r(x) \dfrac{d}{dx} \right) & 0 & 0 \\ \left. -g \dfrac{d}{dx} \cdot \right|_{x=l} & 0 & 0 \end{pmatrix}, \tag{4.5}$$

on the domain (2.18); $\mathcal{D}$ is a bounded operator in $\mathcal{H}$ defined by the formula

$$\mathcal{D} = -i \begin{pmatrix} 0 & 0 & 0 \\ 0 & \dfrac{2\mathbf{d}(x)}{p(x)} & 0 \\ 0 & 0 & 0 \end{pmatrix}, \tag{4.6}$$

and $R(\mathcal{H})$ is a class of all bounded operators on $\mathcal{H}$. Let us rewrite (4.3) assuming that $h_1 = 0$, $h_2$ is replaced with $h$ and $\hat{T}_h$ is a rank-one operator, i.e., $\hat{T}_h = \hat{T}_{0h_2}$. We have

$$(\mathcal{L}_h^*)^{-1} = (\mathcal{L}_0^*)^{-1} + \hat{T}_h \,. \tag{4.7}$$

Using representation (4.4), we can write

$$(\mathcal{L}_0^*)^{-1} = (\mathcal{L} + \mathcal{D})^{-1} + \hat{T}_h = [(I + \mathcal{D}\mathcal{L}^{-1})^{-1} + \hat{T}_h\mathcal{L}]\mathcal{L}^{-1} \,. \tag{4.8}$$

As we know, $(\mathcal{L}_h^*)^{-1} \in \Sigma_\infty$. The proof that $\mathcal{L}^{-1} \in \Sigma_\infty$ can be obtained from the proof of Theorem 3.1 by repeating all steps and assuming that $h = \mathbf{d} = 0$. As follows from the asymptotics of the spectrum of the selfadjoint operator $\mathcal{L}$ (see Theorem 2.1), the set of its eigenvalues $\{\lambda_n^\circ\}_{n\in\mathbb{Z}}$ has the following property: $\{(\lambda_n^\circ)^{-1}\}_{n\in\mathbb{Z}} \in l^p$, $p > 1$. Since for any selfadjoint compact operator $K$, its s-numbers (the eigenvalues of the operator $(K^*K)^{1/2}$) coincide with the moduli of the eigenvalues, we obtain that $\mathcal{L} \in \Sigma_p$. It is also clear that in formula (4.8), we have that $\mathcal{D}\mathcal{L}^{-1} \in \Sigma_p$, $\hat{T}_h\mathcal{L}$ is a rank-one operator in $\mathcal{H}$, and also the operator $(I + \mathcal{D}\mathcal{L}^{-1})^{-1}$ exists and can be represented in the form

$$(I + \mathcal{D}\mathcal{L}^{-1})^{-1} = I + V \,, \qquad V \in \Sigma_p \,. \tag{4.9}$$

Indeed, the existence of the operator $(I + \mathcal{D}\mathcal{L}^{-1})^{-1}$ follows from the fact that the operator $(\mathcal{L}_0^*)^{-1}$ exists. To find the expression for the operator $V$ from (4.9), let us apply the operator $(I + \mathcal{D}\mathcal{L}^{-1})$ to the equation from (4.9). We have

$$I = (I + \mathcal{D}\mathcal{L}^{-1}) + V(I + \mathcal{D}\mathcal{L}^{-1}) \,, \tag{4.10}$$

and thus we obtain that

$$V = -\mathcal{D}\mathcal{L}^{-1}(I + \mathcal{D}\mathcal{L})^{-1} \,. \tag{4.11}$$

Since $(I + \mathcal{D}\mathcal{L}^{-1}) \in R(\mathcal{H})$ and $\mathcal{D}\mathcal{L}^{-1} \in \Sigma_p$, we obtain the desired result (4.9). Substituting (4.9) into (4.8), we have

$$(\mathcal{L}_h^*)^{-1} = (I + V + \hat{T}_h\mathcal{L})\mathcal{L}^{-1} = (I + \hat{V})\mathcal{L}^{-1} \,, \qquad \hat{V} \in \Sigma_p \,. \tag{4.12}$$

From (4.12), we have the following representation:

$$\mathcal{L}_h^{-1} = \mathcal{L}^{-1}(I + \hat{V}^*) \,. \tag{4.13}$$

If we identify $\mathcal{L}_h^{-1}$, $\mathcal{L}^{-1}$, and $\hat{V}^*$ from (4.13) with the operators $A$, $B$, and $S$ from Theorem 4.2, we immediately obtain the fact of completeness of the root vectors of the operator $\mathcal{L}_h^{-1}$ (and, therefore, $\mathcal{L}_h$).

The theorem is shown. $\qquad\qquad\qquad\qquad\qquad\qquad\qquad\qquad\qquad\Box$

# Acknowledgment

Partial support by the National Science Foundation Grants DMS-0072247 and ECS-0080441 and Advanced Research Program of Texas-01 Grant 0036-44-045 is highly appreciated.

# References

1. Adams, R.A. (1975) *Sobolev Spaces*. Academic Press, New York.

2. Dunford, N. and Schwatz, J.T. (1971) *Linear Operators, Part III: Spectral Operators*. Wiley Interscience Inc., New York and London.

3. Foias, C. and Sz.-Nagy, B. (1970) *Harmonic Analysis of Operators on Hilbert Space*. Elsevier, New York.

4. Gohberg, I.Ts. and Krein, M.G. (1969) Introduction to Theory of Linear Nonselfadjoint Operators. In *Trans. of Math. Monogr.*, vol. 18, AMS, Providence, RI.

5. Bechtel, S., Bollinger, K.D., Cao, J.Z. and Forest, M.G. (1995) Torsional Effects in High-Order Viscoelastic Thin-Filament Models. *SIAM J. Appl. Math.*, **55** (1), 58–99.

6. Coron, J.M. and Andrea-Norel, B. (2000) Exponential Stabilization of an Overhead Crane with Flexible Cable via a Back-Stepping Approach. *Automatica*, **36**, 587–593.

7. Gomilko, A. and Pivovarchik, V. (1999) On Bases of Eigenfunctions of Boundary Problem Associated with Small Vibrations of Damped Nonsmooth Inhomogeneous String. *Asymp. Anal.*, **20** (3-4), 301–315.

8. Goriely, A. and Tabor, M. (1997) Nonlinear Dynamics of Filaments, I and II. Dynamical Instabilities, *Phys. D*, **105** (3-4), 20–44, 45–61.

9. Ispolov, Y.G. (1997) Oscialltions of an Extendable Flexible Filament with Small Sag. *Prikl. Mat. Mekh.*, **61** (5), 766–773 [in Russian, Transl. in *J. Appl. Math. Mech.*, **61** (5), 743–750].

10. Klein, R. and Majda, A.J. (1991) Self-Stretching of a Perturbed Vortex Filament, I. The Asymptotic Equation for Derivations from a Straight Line. *Phys. D*, **49** (3), 323–352.

11. Klein, R. and Ting, L. (1995) Theoretical and Experimental Studies of Slender Vortex Filaments. *Appl. Math. Lett.*, **8** (2), 45–50.

12. Langer, J. and Perline, R. (1996) The Planar Filament Equation. In *Mechanics Day (Waterloo, ON, 1992)*, Fields Inst. Commun., **7**, 171–180, Amer. Math. Soc., Providence, RI.

13. Merkin, D.R. (1980) *Introduction to the Mechanics of a Flexible Filament*. "Nauka," Moscow (in Russian).

14. Nishiyama, T. and Tani, A. (1996) Initial and Initial-Boundary Value Problems for a Vortex Filament with or without Axial Flow. *SIAM J. of Math. Anal.*, **27** (4), 1015–1023.

15. Nizette, M. and Goriely, A. (1999) Towards a Classification of Euler–Kirchhoff Filaments. *J. Math. Phys.*, **40** (6), 2830–2866.

16. Pivovarchik, V.N. (1997) Inverse Problem for a Smooth String with Damping at One End, *Journ. Operator Theory*, **38**, 243–263.

17. Pivovarchik, V.N. (1999) Direct and Inverse Problems for a Damped String. *Journ. Operator Theory*, **42**, 189–220.

18. Russell, D.L. (1978) Conrollability, Stabilizability Theory for Linear Partial Differential Equations. Recent Progress and Open Questions. *SIAM J. Contr. Opt.*, **20**, 639–739.

19. Russell, D.L. (1967) Nonharmonic Fourier Series in the Control Theory of Distributed Parameter Systems. *J. Math. Anal. Appl.*, **18**, 542–559.

20. Shubov, M.A., Belinskiy, B.P. and Martin, C.F. (2001) Asymptotics of Eigenfrequencies and Eigenmodes of Nonhomogeneous Inextensible Filament with One End Load. *Math. Methods in Applied Sciences*, **24**, 1139–1167.

21. Shubov, M.A. (2002) Riesz Basis Property of the Generalized Eigenfunctions for the Problem of Small Vibrations of Nonhomogeneous Filament with One End Load. Preprint of Texas Tech University.

22. Shubov, M.A. and Belinskiy, B.P. (2002) Boundary and Distributed Controllability of Nonhomogeneous Filament with One End Load. Preprint of Texas Tech University.

23. Shubov, M.A. (1996) Basis Property of Eigenfunctions of Nonselfadjoint Operator Pencils Generated by Equation of Nonhomogeneous Damped String. *Integr. Eqs. Oper. Theory*, **25**, 289–328.

24. Shubov, M.A. (1998) Spectral Operators Generated by 3-Dimensional Damped Wave Equation and Application to Control Theory. In *Spectral and Scattering Theory*, A.G. Ramm (ed.), Plenum Press, New York, pp. 177–188.

25. Shubov, M.A. (1997) Spectral Operators Generated by Damped Hyperbolic Equations. *Integ. Eqs. Oper. Theory*, **28**, 358–372.

26. Shubov, M.A. (1999) Spectral Operators Generated by Timoshenko Beam Model. *Systems and Control Letters*, **38** (4-5), 249–258.

27. Shubov, M.A. (1999) The Riesz Basis Property of the System of the Root Vectors of a Nonhomogeneous Damped String. Transformation Operators Method. *Methods of Analysis and Applications*, **6** (4), 571–592.

28. Shubov, M.A. (2000) Exact Controllability of Timoshenko Beam. *IMA J. Contr. Inform.*, **17**, 375–395.

29. Shubov, M.A. (1995) Asymptotics of Resonances and Geometry of Resonance States in the Problem of Scattering of Acoustical Waves by a Spherically Symmetric Inhomogeneity of the Density. *Diff. Int. Eqs.*, **8** (5), 1073–1115.

30. Shubov, M. A. (2000) Riesz Basis Property of Root Vectors of Nonselfadjoint Operators Generated by Radial Damped Wave Equations. *Adv. Diff. Eqs.*, **5** (4-6), 623–656.

31. Shubov, M.A. (1998) Exact Boundary and Distributed Controllability of Radial Damped Wave Equation. *J. Mathem. Pures et Appl.*, **77**, 415–437.

# Expanded Point Mapping Analysis of Periodic Systems

Henryk Flashner[†] and Michael Golat[‡]

Department of Aerospace and Mechanical Engineering
University of Southern California, Los Angeles, CA 90089-1453
Tel: (213) 740-0489, Fax: (213) 740-8071, E-mail: hflashne@usc.edu

A new numerical-analytical method for the analysis of nonlinear periodic systems referred to as an Expanded Point Mapping (EPM) is presented. The Expanded Point Mapping approach combines the *cell to cell mapping* and *point mapping* methods to form a comprehensive methodology for the study of the global behavior of nonlinear periodic systems. The proposed method is applicable to multi-degree of freedom systems, multi-parameter systems, and allows analytical studies of local stability characteristics of steady state solutions. In the paper, the theoretical basis for the EPM method is formulated and a procedure for the analysis of nonlinear dynamical systems is presented. Analysis of a pendulum with a periodically excited support in the plane is used to illustrate the method. The results demonstrate the efficiency and accuracy of the proposed approach in analyzing nonlinear periodic systems.

## 1 Introduction

Many important problems of practical interest are modeled by periodic differential equations that are both highly nonlinear and contain periodically varying parameters. These problems are of great mathematical challenge and have attracted a great deal of theoretical research (see Cesari [1], Hale [2], Urabe [3], and Yakubovich and Starzhinskii [4]). Moreover, they have many applications in various fields of science and engineering (see Bolotin [5], Kauderer [6], Lichtenberg and Lieberman [7], and Hayashi [8]).

---

[†]Professor, Corresponding Author
[‡]Graduate Student

The majority of analytical methods for this class of problems can be divided into two categories: perturbation analysis using asymptotic expansion in terms of small parameters and numerical simulation (see Bogoliubov and Mitropolsky [9], Nayfeh and Mook [10], Nayfeh [11], Moon [12], and Thompson and Stewart [13]). Perturbation methods allow for an analytical treatment of the problem but, being local in nature, are valid only for small parameter variations and in a small neighborhood of an equilibrium or periodic solution. Consequently, global domains of attraction can only be qualitatively extrapolated from the localized behavior around fixed points and periodic solutions. Moreover, perturbation methods are in general difficult to implement for systems of order greater than two. Numerical simulation studies are well suited for investigating a system's behavior in the entire state space. It is straightforward to perform digital simulations on a nonlinear problem. However, as is the case with all numerical solutions, qualitative behavior of the system is often difficult to deduce, especially for high-order systems. In the case of systems with more than one degree-of-freedom, or with possibly more than one independent parameter, the amount of computational effort required is prohibitive.

It is from these aforementioned deficiencies of direct numerical techniques that a number of combined numerical-analytical methods have been developed. An attractive way to analyze periodic systems is to formulate their discrete-time dynamics by defining a *point mapping* (see Bernussou [14]). By doing this, the dynamics of the original system is given in terms of difference (algebraic) equations, rather than in terms of time varying differential equations. Point mapping techniques are widely used today not only to study theoretical aspects of discrete-time systems but also to provide a computational basis for understanding their global dynamics (see Guckenheimer and Holmes [15] and Poincaré [16]). Similarly, the method of *cell to cell mapping* has been extensively employed for studying the global behavior of nonlinear systems. A significant amount of research involving cell to cell mapping techniques has been performed by Hsu and his coworkers [17–19] and Guttalu *et al.* [20]. For a detailed treatment of *cell to cell mapping*, the reader is referred to the research monograph by Hsu [21].

In this paper, a new numerical-analytical method referred to as an Expanded Point Mapping (EPM) is introduced. The Expanded Point Mapping approach combines key elements of the *cell to cell mapping* and *point mapping* methods to form a comprehensive methodology to study the global behavior of nonlinear periodic systems in conjunction with analytical investigations of stability characteristics of equilibrium points and periodic solutions. The proposed method is applicable to multi-degree of freedom systems, and the analytical procedure is formulated for any order of approximation. Consequently, the approach allows convenient analysis of multi-dimensional and multi-parameter systems. The global analysis performed by the EPM formulation consists of obtaining all possible equilibrium and periodic solutions and determining their domains of attraction. Moreover, the EPM technique generates an analytical expression representing each periodic solution allowing local stability analysis of the equilibrium and periodic solutions computed in the global analysis part of the approach. In addition, the EPM approach provides the tools to express the solutions as a function of system parameters which allows the study of stability characteristics as function of these parameters and subsequently obtaining bifurcation conditions.

The paper is organized as follows. In Section 2 the point mapping method and an algorithm to obtain its approximation are summarized. Section 3 briefly reviews the *cell to cell mapping* method. In Section 4 the Expanded Point Mapping method is formulated, and a procedure for its application to analysis of nonlinear periodic systems is described. In Section 5 the proposed approach is demonstrated in a study of an inverted pendulum with a support periodically excited in the plane. Finally, concluding remarks are provided in Section 6.

## 2 Point Mapping Analysis

Consider a dynamical system described by a set of $N$ ordinary differential equations

$$\dot{\mathbf{x}}(t) = \mathbf{f}(t, \mathbf{x}(t), \mathbf{s}),\tag{1}$$

where $t \in \mathbb{R}^+$ denotes time, $\mathbf{x}(t) \in \mathbb{R}^N$ is the state vector, $\mathbf{s} \in \mathbb{R}^L$ is the parameter vector, and $\mathbf{f} : \mathbb{R}^+ \times \mathbb{R}^N \times \mathbb{R}^L \to \mathbb{R}^N$ is analytic for every $\mathbf{x}$ for a given value of the parameter vector $\mathbf{s}$. Moreover, $\mathbf{f}$ is assumed to be periodic in $t$ with period $T$, i.e., $\mathbf{f}(t, \mathbf{x}, \mathbf{s}) = \mathbf{f}(t+T, \mathbf{x}, \mathbf{s})$. To analyze the equations in (1), we observe the state of the system at discrete instances every $T$ units of time. A *point mapping* (or *Poincaré map*) is the dynamic relationship between the state of the system at $t = nT$ and the state at $t = (n+1)T$ which can be expressed as set of difference (algebraic) equations

$$\mathbf{x}(n+1) = \mathbf{G}(\mathbf{x}(n), \mathbf{s}), \qquad n = 0, 1, 2, \ldots,\tag{2}$$

where $\mathbf{x}(n)$ and $\mathbf{x}(n+1)$ are the states of the system at $t = nT$ and $t = (n+1)T$, respectively. Note that since $\mathbf{f}$ is analytic, the differential equations (1) satisfy Lipschitz conditions, and therefore $\mathbf{G}$ is one-to-one for a fixed value of $\mathbf{s}$.

The point mapping formulation (2) allows for convenient determination of periodic solutions and their stability characteristics. A periodic solution of period $KT$, or in short a *P-K* solution of the point mapping (2) for some $\mathbf{s} = \mathbf{s}^*$, consists of $K$ distinct points $\mathbf{x}^*(j)$, $j = 1, 2, \ldots, K$ such that

$$\mathbf{x}^*(m+1) = \mathbf{G}^m(\mathbf{x}^*(1), \mathbf{s}^*), \qquad m = 1, 2, \ldots, K-1,$$
$$\mathbf{x}^*(1) = \mathbf{G}^K(\mathbf{x}^*(1), \mathbf{s}^*),\tag{3}$$

where $\mathbf{G}^m$ is the point mapping $\mathbf{G}$ applied $m$ times. From this definition, an equilibrium state of (2) can be viewed as a P-1 solution. Also, an equilibrium point $\mathbf{x}^*$ of (1) which satisfies $\mathbf{f}(t, \mathbf{x}^*, \mathbf{s}^*) = 0$ $\forall t$ is clearly a solution of the corresponding point mapping (2). Finding a *P-K* solution of the point mapping $\mathbf{G}$ is equivalent to finding a periodic solution of the original continuous time system. Having found a *P-K* solution by solving the algebraic system of equations (2), its local stability properties (in the sense of Liapunov) are determined by the eigenvalues of the matrix $\mathbf{H}$ given by

$$\mathbf{H} = \prod_{j=1}^{K} \mathbf{H}^*(j), \quad \text{where} \quad \mathbf{H}^*(j) = \left.\frac{\partial \mathbf{G}}{\partial \mathbf{x}}\right|_{\mathbf{x}=\mathbf{x}^*(j),\, \mathbf{s}=\mathbf{s}^*}.\tag{4}$$

The eigenvalues of $\mathbf{H}$ are denoted by $\lambda_i(\mathbf{H})$, $i = 1, 2, \ldots, N$. Using the theory of linear difference equations (see Miller [22]), the local stability of a $P$-$K$ solution can be summarized as follows:

1. A $P$-$K$ solution of (2) is locally asymptotically stable if and only if $|\lambda_i(\mathbf{H})| < 1$ $\forall i = 1, 2, \ldots, N$.

2. A $P$-$K$ solution of (2) is unstable if $|\lambda_i(\mathbf{H})| > 1$ for some $i = 1, 2, \ldots, N$.

3. If $|\lambda_i(\mathbf{H})| = 1$ for some $i = j$ and $|\lambda_i(\mathbf{H})| < 1$ $\forall i = 1, 2, \ldots, N$ $i \neq j$, then the linearization process is inconclusive. Moreover, if this condition occurs while $\mathbf{s}$ is varied, bifurcation conditions are established.

The *domain* or *basin of attraction* $B_{\mathbf{x}^*}$ of an asymptotically stable $P$-$K$ point $\mathbf{x}^*$ is defined as

$$B_{\mathbf{x}}^* = \left\{ \mathbf{x} \in \mathbb{R}^N \mid \mathbf{G}^{jK}(\mathbf{x}) \to \mathbf{x}^*, \; j \to \infty \right\}. \tag{5}$$

## 2.1   Truncated point mappings

The algorithm for obtaining an approximation, in analytic form, for the point mapping $\mathbf{G}(\mathbf{x}(n), \mathbf{s})$ of (2) was first introduced by Flashner [23] and Flashner and Hsu [24]. The underlying philosophy and the main steps of the algorithm are presented in the following. For more details about this procedure see Guttalu and Flashner [25, 26].

Since the function $\mathbf{f}(t, \mathbf{x}(t), \mathbf{s})$ in equation (1) is analytic, it can be expressed by a series of the form

$$\mathbf{f}(t, \mathbf{x}, \mathbf{s}) = \mathbf{p}(t, \mathbf{x}, \mathbf{s}) = \sum_r^P b_{j_1 \cdots j_N}^{(i)f}(t, \mathbf{s}) x_1^{j_1} \ldots x_N^{j_N}, \qquad i = 1, 2, \ldots, N, \tag{6}$$

where $\mathbf{p}(t, \mathbf{x}, \mathbf{s})$ denotes a vector homogeneous multinomial of degree $P$ in the state variables $x_i$, $i = 1, 2, \ldots, N$. The symbol $\sum_r^P$ denotes summation over all sequences $j_1, \ldots, j_r$ such that $\sum_{k-1}^N j_k = r$ and $r = 1, 2, \ldots, P$, i.e., a vector homogeneous multinomial of degree $P$ contains terms where the sum of the state variable exponents is less than or equal to $P$. To abbreviate notation denote the set of indices $j_1, \ldots, j_r$ by $E_r$. Then we denote the coefficients of the vector homogeneous multinomial in (6) by the symbol $b_{E_r}^{(i)f}(t, \mathbf{s})$. These coefficients, in general, can be functions of the parameter vector $\mathbf{s}$. The algorithm for determining point mappings is based on multinomial functions by truncating at some degree $k$. Let $\mathbf{p}(t, \mathbf{x}, \mathbf{s})$ be a multinomial of degree greater than $k$. Define the truncation operation at degree $k$ by

$$\mathbf{p}_{(k)}(t, \mathbf{x}, \mathbf{s}) = \overline{\mathbf{p}(t, \mathbf{x}, \mathbf{s})}^k = \sum_r^k b_{E_r}^{(i)f}(t, \mathbf{s}) x_1^{j_1} \ldots x_N^{j_N}. \tag{7}$$

Note that in our definition of the multinomial $\mathbf{p}(t, \mathbf{x}, \mathbf{s})$, we assume no terms of order zero in the components of $\mathbf{x}$. There is no hard forcing in the system (1). For the case of $P \to \infty$ in equation (6), $\mathbf{f}(t, \mathbf{x}, \mathbf{s})$ is approximated by a $k$-jet denoted by $j^k \mathbf{f}$, i.e., by a Taylor series expansion. Let $\mathbf{p}(t, \mathbf{x}, \mathbf{s})$ and $\mathbf{q}(t, \mathbf{x}, \mathbf{s})$ be two vector homogeneous multinomials. It can be shown (see Poston and Stewart [27]) that the $k$-jets obey the rules in equations (8–10). Namely

$$j^k(\mathbf{p} + \mathbf{q}) = j^k\mathbf{p} + j^k\mathbf{q}\,, \tag{8}$$

$$j^k(\mathbf{p} \cdot \mathbf{q}) = \overline{j^k\mathbf{p} \cdot j^k\mathbf{q}}\,, \tag{9}$$

$$j^k(\mathbf{p} \circ \mathbf{q}) = \overline{j^k\mathbf{p} \circ j^k\mathbf{q}}\,, \tag{10}$$

The computational algorithm for evaluating a point mapping is based on the fact that the operations of multinomial addition, multiplication, and telescoping (composition) can be interchanged with the truncation operation. Moreover, the Runge–Kutta method of integration can be expressed as a sequence of multinomial telescoping and truncation operations which finally results in a truncated multinomial expression for the point mapping $\mathbf{G}(\mathbf{x}(n), \mathbf{s})$ of (2). Using the Runge–Kutta method, the state of the system at time $t = t_p + h$ can be formulated as follows (see Henrici [28]):

$$\mathbf{x}(t_p + h) = \mathbf{x}(t_p) + h \sum_{m=1}^{M} d_m \mathbf{k}_m(t, \mathbf{x}, \mathbf{s})\,, \tag{11}$$

where $M$ is the order of the Runge–Kutta method, $h$ is a fixed time step, and $d_m$ are certain constants determined by the Runge–Kutta method. The vectors $\mathbf{k}_m$ are given by

$$\mathbf{k}_m(t, \mathbf{x}, \mathbf{s}) = \mathbf{f}(t_p + ha_m, \mathbf{x}(t_p) + hc_m\mathbf{k}_{m-1}, \mathbf{s})\,, \qquad m = 1, 2, \ldots, M\,, \tag{12}$$

where the constants $a_m$ and $c_m$ are defined by the particular Runge–Kutta method implemented, with $a_1 = 0$ and $c_1 = 0$.

To derive the algorithm, the period $T$ is divided into $N_t$ subintervals of length $h = T/N_t$. We are interested in computing a $k$-jet of the point mapping $\mathbf{G}$ in (2). The point mapping algorithm takes advantage of the structure of the dynamic equation for the system in (11). It can be shown that each function $\mathbf{k}_m$ can be expressed by telescoping (composition) and addition of truncated multinomials, resulting in $j^k\mathbf{k}_m(i; \mathbf{x}, \mathbf{s}) = \mathbf{p}_{(k)}^m(i; \mathbf{x}, \mathbf{s})$, $m = 1, 2, \ldots, M$. Similarly, equation (11) consists of additions of truncated multinomials yielding

$$j^k\mathbf{x}(t_p + ih) = \mathbf{x} + h \sum_{m=1}^{M} d_m \mathbf{p}_{(k)}^m(i; \mathbf{x}, \mathbf{s}) = \mathbf{Q}_{(k)}(i; \mathbf{x}, \mathbf{s})\,.$$

Repeating this procedure $N_t - 1$ times yields

$$\mathbf{x}((n+1)T) = \overline{\mathbf{Q}_{(k)}(N_t; (\mathbf{Q}_{(k)}(N_t - 1; (\mathbf{Q}_{(k)}(N_t - 2; (\cdots (\mathbf{Q}_{(k)}(1; \mathbf{Q}_{(k)}(0, \mathbf{x}, \mathbf{s}), \mathbf{s})), \mathbf{s}))), \mathbf{s})}^k$$

$$= j^k\mathbf{G}(\mathbf{x}(nT), \mathbf{s}) \triangleq \mathbf{G}_{(k)}(\mathbf{x})\,, \tag{13}$$

where $\mathbf{Q}_{(k)}(0; \mathbf{x}, \mathbf{s}) = \mathbf{x}$. It is extremely important to note that the *truncated point mapping* for $\mathbf{G}(\mathbf{x}(n), \mathbf{s})$ is *exact* up to the $k$-th order multinomial term. Note, that because of the characteristics of the truncation operation (see equations 8–10) $\mathbf{G}_{(k)}$ can be computed by manipulating multinomials of degree not larger than $k$.

While obtaining the mapping in (13), a truncation of the parameter vector $\mathbf{s}$ can be performed simultaneously by a scheme similar to the one used for the truncation of the state vector $\mathbf{x}$. The algorithm devised here separately keeps track of the powers of the components of the parameter vector $\mathbf{s}$ in addition to the state vector $\mathbf{x}$. Assume that the coefficients of the elements of the parameter vector $\mathbf{s}$ of $\mathbf{f}$ given in equation (6) are analytic, i.e., they consist of powers of the components of $\mathbf{s}$. Therefore, we can take the $p$-jet of the coefficients with respect to the powers of $s_j$, $j = 1, 2, \ldots, L$ and obtain $b_{Er}^{(i)f}(t, \mathbf{s}) = \sum_q^p \sigma_{Er,Eq}^{(i)}(t) s_1^{k_1} \cdots s_L^{k_L}$. The symbol $\sum_q^p$ denotes summation over all sequences $k_1, \ldots, k_q$ such that $\sum_{m=1}^L k_m = q$ and $q = 1, 2, \ldots, p$. Defining the coefficients of $\mathbf{f}(t, \cdot, \cdot)$ as $\sigma_{Er,Eq}^{(i)}(t)$ yields

$$j_{\mathbf{x}}^k j_{\mathbf{s}}^p \mathbf{f}(t, \mathbf{x}, \mathbf{s}) = \sum_r^k \sum_q^p \sigma_{Er,Eq}^{(i)}(t) x_1^{j_1} \cdots x_N^{j_N} s_1^{k_1} \cdots s_L^{k_L} .$$

Assume that the truncation on the elements of $\mathbf{s}$ was done up to the power $p$. Then, the corresponding truncated point mapping is denoted by

$$x((n+1)T) = \overline{\mathbf{G}(\mathbf{x}(nT), \mathbf{s})}^{k\,p} = \mathbf{G}_{(k,p)}(\mathbf{x}(nT), \mathbf{s}) . \tag{14}$$

## 3 Cell Mapping Analysis

We now briefly discuss a general computational technique based on discretization of the state space which can be employed for analyzing the global behavior of a given dynamical system of the type (1). We use here the derivation and characteristic developments provided in Hsu [21]. In this formulation, the state space is thought of not as a continuum but instead as a collection of *cells*. Each cell occupies a prescribed region of state space as defined by (15). Every cell in the discretized state space or *cell state space* is to be treated as a state entity. The cell state space is constructed by simply dividing each state variable component $x_i$, $i = 1, 2, \ldots, N$ into a large number of intervals of uniform size $h_i$. The new cell variables $z_i$ are defined to contain the state variables $x_i$ such that

$$\left(z_i - \frac{1}{2}\right) h_i \leq x_i \leq \left(z_i + \frac{1}{2}\right) h_i . \tag{15}$$

By definition, $z_i$ in (15) is an integer. A *cell vector* $\mathbf{z}$ is defined to be an $N$-tuple $z_i$, $i = 1, 2, \ldots, N$. Clearly, a point $\mathbf{x} \in \mathbb{R}^N$ with components $x_i$ belongs to a cell $\mathbf{z} \in S$ with components $z_i$ if and only if $x_i$ and $z_i$ satisfy (15) for all $i$. The space $S$ consisting of elements which are $N$-tuples of integers is referred to as an $N$-dimensional *cell space*. If $N_{c_i}$ denotes the number of cells along the $x_i$-axis, then the cell space $S$ contains a total of $N_{c_1} \times N_{c_2} \times \ldots \times N_{c_N}$ cells.

In the cell state space $S$, one can define a cell-to-cell mapping dynamical system in the form

$$\mathbf{z}(n+1) = \mathbf{C}(\mathbf{z}(n)), \quad \mathbf{C} : S \to S, \quad n \in \mathbb{Z}^+ , \tag{16}$$

where $\mathbf{C}$ is referred to as a *simple cell mapping* (SCM) and is perceived as a mapping from a set of integers to a set of integers. Equation (16) then describes the evolution of a cell dynamical system in $N$-dimensional cell state space $S$. The dynamics of the cell mapping system (16) is characterized by singular cells consisting of equilibrium cells and periodic cells. A *periodic cell* of period $K \in \mathbb{Z}^+$ (or a *P-K* cell) is a set of $K$ distinct cells $\{\mathbf{z}^*(k) \mid k = 1, 2, \ldots, K\}$ such that

$$\mathbf{z}^*(m + 1) = \mathbf{C}^m(\mathbf{z}^*(1)), \qquad m = 1, 2, \ldots, K - 1,$$
$$\mathbf{z}^*(1) = \mathbf{C}^K(\mathbf{z}^*(1)), \tag{17}$$

where $\mathbf{C}^m$ means the mapping $\mathbf{C}$ applied $m$-times. This set is said to constitute a *periodic motion* of period $K$ or simply a *P-K motion*.

To obtain a cell mapping associated with the periodic dynamical system (1), the following procedure is usually implemented. First, construct a cell space structure in the state space region with cell size $h_i$ in the $x_i$-direction. Let $\mathbf{x}^{(d)}(t_p)$ denote the center point of the cell $\mathbf{z}(t_p)$ so that $x_i^{(d)}(t_p) = h_i z_i(t_p)$ $i = 1, 2, \ldots, N$. Integrate the state equations (1) over one period of time from time $t = t_p$ to time $t = t_p + T$. Suppose that the trajectory starting from $\mathbf{x}^{(d)}(t_p)$ terminates at $\mathbf{x}^{(d)}(t_p + T)$. The cell in which $\mathbf{x}^{(d)}(t_p + T)$ lies is taken to be $\mathbf{z}(t_p + T)$, the image cell of $\mathbf{z}(t_p)$. Specifically,

$$z_i(t_p + T) = C_i(\mathbf{z}(t_p)) = Int\left[\frac{x_i^{(d)}(t_p + T)}{h_i} + \frac{1}{2}\right], \qquad i = 1, 2, \ldots, N, \tag{18}$$

where $Int(y)$, for any real number $y$, represents the largest integer which is less than or equal to $y$, i.e., $Int(y) \leq y$. This process of finding image cells is repeated for every cell in the cell state space $S$. The mapping $\mathbf{C}$ so determined is a cell mapping associated with the dynamical system (1). Figure 1 depicts a two-dimensional simple cell mapping given by $\mathbf{C}(\mathbf{z}_1) \to \mathbf{z}_2$, $\mathbf{C}(\mathbf{z}_2) \to \mathbf{z}_2$, $\mathbf{C}(\mathbf{z}_3) \to \mathbf{z}_4$, etc.

A trajectory of the cell-to-cell dynamical system (16) starting from an initial cell state $\mathbf{z}(0)$ is referred to as a cell sequence of (16) which is the set of integers given by $\{\mathbf{z}(k), \ k = 1, 2, \ldots\}$. In practical applications, one is usually interested in a bounded region of the state space which contains a finite number of cells. Once the state variable goes beyond a certain positive or negative magnitude, we are no longer interested in further evolution of the system (1). Taking advantage of this, all the cells outside this finite region can be lumped together into a single cell, a *sink cell*, which will be assumed to map into itself in the mapping scheme, i.e., $\mathbf{z}_{Sink}(n + 1) = \mathbf{C}(\mathbf{z}_{Sink}(n))$. Due to the finite number of cells in state space, all cell sequences must terminate with a finite number of cell mappings into one of the steady cell states, namely an equilibrium cell, group of periodic cells, or the sink cell. This property is of crucial importance in developing the global cell mapping algorithm by Hsu and Guttalu [19]. This algorithm that unravels the global behavior of the system (1), provides the following characteristics of the system:

1. Location of the equilibrium states and periodic solutions in a given region of the state space.

2. Domains of attraction associated with the asymptotically stable equilibrium states and periodic solutions.

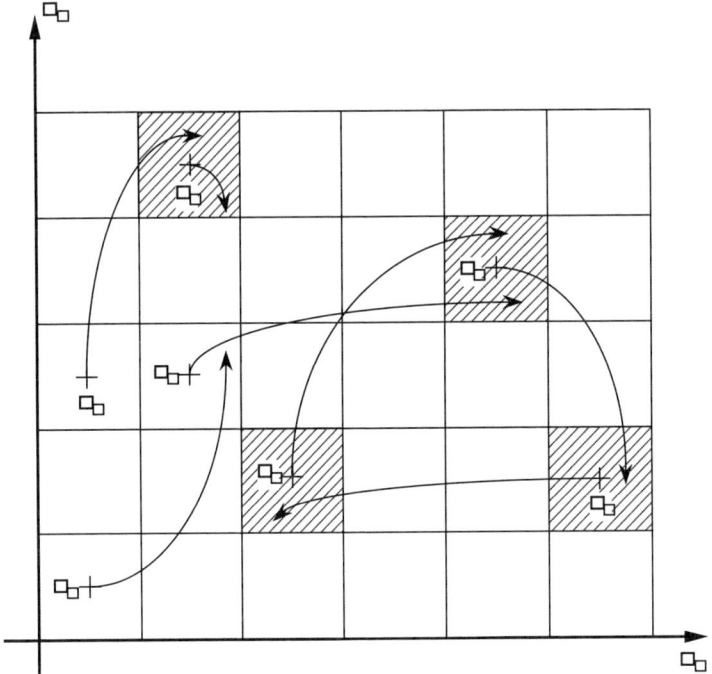

Figure 1 Example of a simple cell mapping with periodic (shaded) and transient cells. $\mathbf{z}_2$ is a $P$-$K$ solution of period 1. $\mathbf{z}_3$, $\mathbf{z}_4$, and $\mathbf{z}_5$ constitute a $P$-$K$ solution of period 3.

3. Step-by-step evolution of the global behavior of the system starting from any initial state within the cell state space.

Figure 1 illustrates several arbitrary trajectories for a hypothetical two-dimensional simple cell mapping. For this case $\mathbf{z}_2$ is a P-1 cell and $\mathbf{z}_3 \to \mathbf{z}_4 \to \mathbf{z}_5$ are P-3 cells. Cells $\mathbf{z}_1$, $\mathbf{z}_6$, and $\mathbf{z}_7$ are transient cells lying within the domains of attraction for the respective $P$-$K$ solutions.

# 4  Expanded Point Mapping

In this section, the Expanded Point Mapping (EPM) is introduced. Section 4 is divided into two parts. First, the Expanded Point Mapping method is formulated and its theoretical justification discussed. Then a procedure for EPM-based global analysis and local analytical study of periodic solutions of nonlinear periodic systems is described.

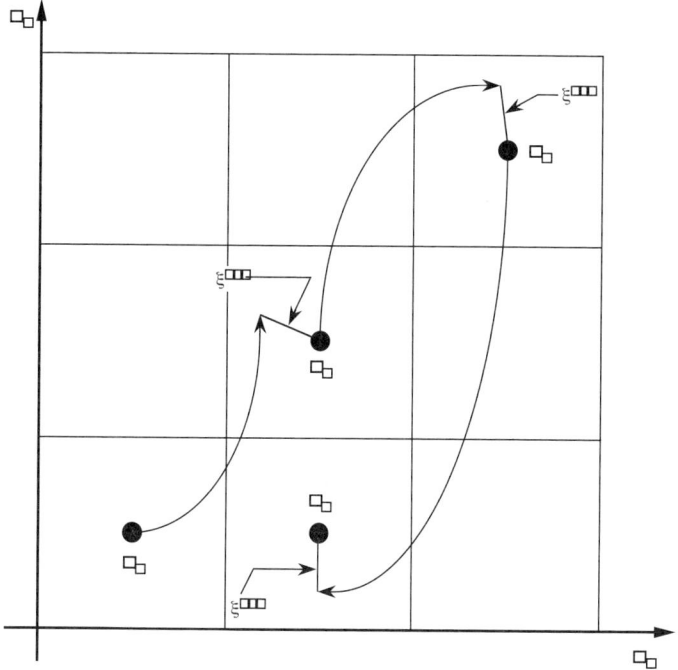

Figure 2  Simple cell mapping spatial discretization error.

## 4.1  EPM formulation

### 4.1.1  State space discretization

The region of interest in the state space is discretized into a finite number of cells $N_C + 1$ in the manner described in Section 3. Thus forming a cell state space of dimension $N_C = N_{c_1} \times N_{c_2} \times \ldots \times N_{c_N}$ and a corresponding sink cell.

### 4.1.2  Cell space analysis

Define $\phi^{*(i)}(t) \in \mathbb{R}^N$ to be a trajectory starting at the center point of each cell $\mathbf{z}_i$ for $i = 1, 2, \ldots, N_C$. As in the simple cell mapping approach the center point of each cell is integrated from $t = t_p$ to $t = t_p + T$ with $t_p = 0$ for simplicity. Each cell center point is represented by $\phi^{*(i)}(0)$, and the corresponding cell center point trajectory termination is $\phi^{*(i)}(T)$.

### 4.1.3  Analytical study of trajectory evolution

The truncated point mapping approach is used to precisely locate trajectories inside each individual cell. The perturbation (error) about $\phi^{*(i)}(t)$ is denoted as $\boldsymbol{\xi}^{(i)}(t)$ (see Figure 2), where $\mathbf{x}(t) = \phi^{*(i)}(t) + \boldsymbol{\xi}^{(i)}(t)$. Substitution of $\mathbf{x}(t)$ into equation (1) yields

$$\dot{\boldsymbol{\phi}}^{*(i)}(t) + \dot{\boldsymbol{\xi}}^{(i)}(t) = \mathbf{f}(t, \boldsymbol{\phi}^{*(i)}(t) + \boldsymbol{\xi}^{(i)}(t), \mathbf{s}). \tag{19}$$

Assuming that $\mathbf{f}$ is analytic, we apply a Taylor series expansion about $\boldsymbol{\phi}^{*(i)}(t)$ to the right-hand side of (19). Recalling that $\boldsymbol{\phi}^{*(i)}(t)$ is a solution of equation (1), the perturbed trajectory satisfies

$$\dot{\boldsymbol{\xi}}^{(i)}(t) = \mathbf{p}_{(m)}^{(i)}(t, \boldsymbol{\xi}^{(i)}(t), \mathbf{s}) + \mathbb{R}_{(m+1)}^{(i)}(t, \boldsymbol{\xi}^{(i)}(t), \mathbf{s}), \tag{20}$$

where $\mathbf{p}_{(m)}^{(i)}(t, \boldsymbol{\xi}^{(i)}(t), \mathbf{s})$ is a vector homogeneous multinomial of degree $m$ in $\boldsymbol{\xi}^{(i)}(t)$. Notice that (20) is written as a perturbation about a trajectory. Consequently, it does not possess terms of order zero, and satisfies the hard forcing criteria of Section 2.1. The remainder term $\mathbb{R}_{(m+1)}^{(i)}(t, \boldsymbol{\xi}^{(i)}(t), \mathbf{s})$ is a higher order multinomial of $O(\|\boldsymbol{\xi}^{(i)}(t)\|^{m+1})$. In some neighborhood of $\boldsymbol{\phi}^{*(i)}(t)$ for which $\|\boldsymbol{\xi}^{(i)}(t)\| < \rho$, the truncated system

$$\dot{\boldsymbol{\xi}}^{(i)}(t) = \mathbf{p}_{(m)}^{(i)}(t, \boldsymbol{\xi}^{(i)}(t), \mathbf{s}), \qquad i = 1, 2, \ldots, N_C, \tag{21}$$

provides a good approximation to the dynamics of the perturbed system. A truncated point mapping for each individual cell $\mathbf{z}_i$ is obtained by integrating (21) from $t = t_p$ to $t = t_p + T$ where again $t_p = 0$ for simplicity.

$$\boldsymbol{\xi}^{(i)}(T) = \mathbf{G}_{(m)}^{(i)}(\boldsymbol{\xi}^{(i)}(0), \mathbf{s}), \qquad i = 1, 2, \ldots, N_C. \tag{22}$$

Hence, the analytical $m$-th order approximation of the trajectory is given by:

$$\mathbf{x}(T) = \boldsymbol{\phi}^{*(i)}(T) + \mathbf{G}_{(m)}^{(i)}(\boldsymbol{\xi}^{(i)}(0), \mathbf{s}), \qquad i = 1, 2, \ldots, N_C. \tag{23}$$

Note that equation (22) defines the mapping of all points within the cell $\mathbf{z}_i$ and is accurate to degree $m$. This is the fundamental tenet of the Expanded Point Mapping approach. Defining the norm $\| \cdot \|$ to be the infinity norm, i.e., $\|\boldsymbol{\xi}^{(i)}(t)\| = \max_j |\xi_j^{(i)}(t)|$ $j = 1, 2, \ldots, N$, with an appropriate scaling, then $\|\boldsymbol{\xi}^{(i)}(0)\|_\infty < \sigma_i$ defines a rectangular region. This rectangular region is considered by Hsu [21] as a definition of a cell occupying a region of state space in the same spirit as previously developed in Section 3. Therefore, $\boldsymbol{\xi}^{(i)}(T)$ defines a map of any point within the entire cell $\mathbf{z}_i$. It is critical to reiterate the fact the map $\boldsymbol{\xi}^{(i)}(T)$ can be evaluated to any order of multinomial approximation $m$ and can be developed for systems of arbitrary dimension $N$. By definition, a periodic solution $(\mathbf{x}(0), \mathbf{x}(T), \ldots, \mathbf{x}(kT))$ of period $KT$ satisfies:

$$
\begin{aligned}
\mathbf{x}(0) \quad &\rightarrow \quad \mathbf{x}(T), \\
\mathbf{x}(T) \quad &\rightarrow \quad \mathbf{x}(2T), \\
&\vdots \\
\mathbf{x}((K-1)T) \quad &\rightarrow \quad \mathbf{x}(KT), \\
\mathbf{x}(KT) \quad &\rightarrow \quad \mathbf{x}(0).
\end{aligned}
\tag{24}
$$

### 4.1.4 Local stability

The analytical expression for the truncated point mapping (22) permits us to study the stability properties of the Expanded Point Mapping. The local stability characteristics of periodic solutions can be obtained by using the point mapping theory developed in Section 2. Consider $\mathbf{G}^{(i)}_{(m)}(\boldsymbol{\xi}^{(i)}(k), \mathbf{s})$, it will consist of a linear part and subsequent higher order terms up to degree $m$

$$\mathbf{G}^{(i)}_{(m)}(\boldsymbol{\xi}^{(i)}(k), \mathbf{s}) = \mathbf{H}^{(i)}(\boldsymbol{\xi}^{(i)}(k), \mathbf{s})\boldsymbol{\xi}^{(i)}(k) + \cdots .$$

Applying Liapunov's indirect method and its extension (see Jury [29] and Vidyasagar [30]) to the $P$-$K$ solution $\mathbf{x}(j)$, $j = 1, 2, \ldots, K$ yields

$$\mathbf{H}(\mathbf{s}) = \prod_{j=1}^{K} \mathbf{H}^{(j)}(\boldsymbol{\xi}^{(j)}(k), \mathbf{s}) .$$

The local stability of $\mathbf{x}(j)$ is determined by the eigenvalues of $\mathbf{H}$. There exists a general relationship between linearized flows and Poincaré maps (see Guckenheimer and Holmes [15]) using Floquet theory. For the nonlinear system $\dot{\mathbf{x}} = \mathbf{f}(t, \mathbf{x}, \mathbf{s})$ the characteristic multipliers for the linearized flow are identical to the eigenvalues of the linearized Poincaré map. Consequently, the characteristic multipliers of the linearized system

$$\dot{\boldsymbol{\xi}}^{(i)}(t) = \left.\frac{\partial \mathbf{f}}{\partial \mathbf{x}}\right|_{\mathbf{x}=\phi^{*(i)}(t)} \boldsymbol{\xi}^{(i)}(t)$$

will be identical to the eigenvalues of the linear part of the cell truncated point mapping

$$\boldsymbol{\xi}^{(i)}(T) = \mathbf{G}^{(i)}_{(1)}(\boldsymbol{\xi}^{(i)}(0), \mathbf{s}) = \mathbf{H}^{(i)}(\boldsymbol{\xi}^{(i)}(0), \mathbf{s})\boldsymbol{\xi}^{(i)}(0) .$$

### 4.1.5 Bifurcation analysis

The truncated point mapping analytical expression (22) also contains functional relations between system parameters. Therefore, the Expanded Point Mapping approach can be used to perform bifurcation studies by writing the functional, $\lambda_i[\mathbf{H}(\mathbf{s})]$ $i = 1, 2, \ldots, N$. If for some $\mathbf{s} = \mathbf{s}_0$, $|\lambda_i[\mathbf{H}(\mathbf{s})]| = 1$, the periodic solution at hand may lose stability and new solutions may come into existence. The new solutions can then be analytically studied by writing an $m$-th order approximation of the periodic solution for $\mathbf{s} = \mathbf{s}_0 + \boldsymbol{\epsilon}$ where $\|\boldsymbol{\epsilon}\|$ is small.

## 4.2 EPM procedure

Based on Section 4.1 a trajectory $\mathbf{x}(k)$ with $t = kT$ is computed for any initial condition of (1) in the region of interest as follows:

1. Discretize the state space region of interest into $N_C$ cells.

2. Find the mapping of each cell center point $\phi^{*(i)}(T)$.

3. Compute the $m$-th order approximation $\mathbf{G}^{(i)}_{(m)}(\boldsymbol{\xi}^{(i)}(k), \mathbf{s})$ of the mapping inside each cell.

4. Set $t_p = 0 \rightarrow k = 0$.

5. Calculate the perturbation $\boldsymbol{\xi}^{(i)}(k) = \mathbf{x}(k) - \boldsymbol{\phi}^{*(i)}(0)$.

6. Apply the perturbation $\boldsymbol{\xi}^{(i)}(k)$ to the truncated point mapping (22) and obtain $\boldsymbol{\xi}^{(i)}(k+1) = \mathbf{G}_{(m)}^{(i)}(\boldsymbol{\xi}^{(i)}(k), \mathbf{s})$.

7. Calculate where $\mathbf{x}(k)$ maps to by using $\mathbf{x}(k+1) = \boldsymbol{\phi}^{*(i)}(T) + \boldsymbol{\xi}^{(i)}(k+1)$.

8. Increment $k \rightarrow k+1$ and repeat steps 5 thru 8 until the trajectory $\mathbf{x}(k)$ satisfies one of three possible outcomes.

   (a) The trajectory $\mathbf{x}(k)$ maps to the sink cell.
   (b) The trajectory $\mathbf{x}(k)$ evolves into a $P$-$K$ solution.
   (c) The trajectory $\mathbf{x}(k)$ remains within the region under study and exhibits chaotic behavior, i.e., it is aperiodic.

9. Trajectory evaluation is performed for initial conditions of interest using steps 4–8 above.

10. Local stability analysis is performed by evaluating $\mathbf{H}(\mathbf{s})$ along the trajectory for $\mathbf{s} = \mathbf{s}_0$.

11. Bifurcation analysis is performed by studying the eigenvalues of $\mathbf{H}(\mathbf{s})$ as a function of $\mathbf{s}$.

Figure 3 illustrates a cell truncated point mapping $\mathbf{G}_{(m)}^{(1)}(\boldsymbol{\xi}^{(1)}(k), \mathbf{s})$ resulting from the EPM procedure. All of the points within cell $\mathbf{z}_1$ (dark shaded region) map to the lightly shaded EPM mapping region. Note that the EPM map partially covers several cells, $\mathbf{z}_2, \ldots, \mathbf{z}_9$, and is a curvilinear region. For comparison the simple cell mapping $\boldsymbol{\phi}^{*(1)}(T)$ is shown, where the SCM map is the crosshatched area within cell $\mathbf{z}_3$.

# 5 Pendulum with Periodically Excited Support in the Plane

Consider a pendulum subjected to simultaneous horizontal and vertical planar excitations as depicted in Figure 4.

## 5.1 System dynamic equations

The system possesses a viscous damping torque proportional to the angular velocity, $T_{\text{viscous}} = c_{mL} \cdot \dot{\theta} = m \cdot L \cdot c \cdot \dot{\theta}$. The governing differential equation of motion for the pendulum is

$$\frac{d^2\theta}{dt^2} + c\frac{d\theta}{dt} + \frac{g}{L}\sin\theta + \frac{1}{L}\frac{d^2X}{dt^2}\cos\theta + \frac{1}{L}\frac{d^2Y}{dt^2}\sin\theta = 0, \qquad (25)$$

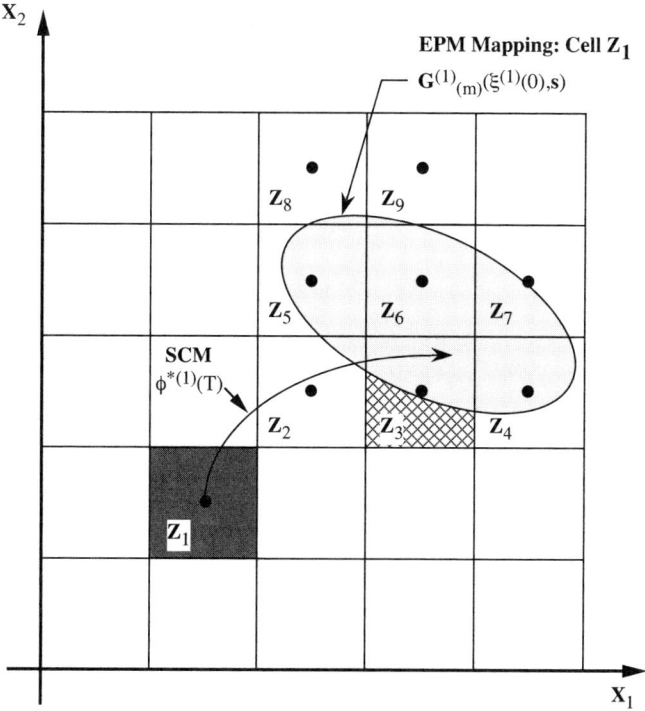

Figure 3 EPM mapping for cell $\mathbf{z}_1$.

where $m =$ mass of the pendulum bob, $L =$ length of the pendulum, $c =$ rotational damping coefficient, $g =$ acceleration of gravity, $\theta =$ angular displacement, $X(t) =$ $x$-axis excitation, and $Y(t) = y$-axis excitation. Consider a periodic support motion of the form

$$X(t) = LX_0 \cos \omega t, \qquad Y(t) = LY_0 \cos(\omega t + \phi). \qquad (26)$$

Let $\tau = \omega t$, $\omega_\theta^2 = g/L$, $c = \mu \omega_\theta$, then the equation of motion can be written as follows

$$\frac{d^2\theta}{d\tau^2} + \frac{\mu}{r}\frac{d\theta}{d\tau} + \frac{1}{r^2}\sin\theta - X_0\cos\theta\cos\tau - Y_0\sin\theta\cos(\tau+\phi) = 0, \qquad (27)$$

where $r = \omega/\omega_\theta$ is the ratio of the forcing frequency $\omega$ to the natural frequency $\omega_\theta$ of the pendulum. Define the state variables $x_1 = \theta$, $x_2 = d\theta/d\tau = \dot\theta$, and the parameter variable $x_3 = r$, and get the state representation of the system as follows

$$\dot{x}_1 = x_2,$$
$$\dot{x}_2 = -\frac{\mu}{x_3}x_2 - \frac{1}{x_3^2}\sin x_1 + X_0\cos x_1\cos\tau + Y_0\sin x_1\cos(\tau+\phi). \qquad (28)$$

The values of system parameters selected were $\mu = 0.1$, $\phi = \pi/2$, $X_0 = 0.5$, $Y_0 = 0.5$, and $x_3 = 0.75$. Equation (28) was used for the cell center point integration process, $\phi^{*(i)}(0) \rightarrow \phi^{*(i)}(T)$, $i = 1, 2, \ldots, N_C$, with $T = 2\pi$. The $m$-jet

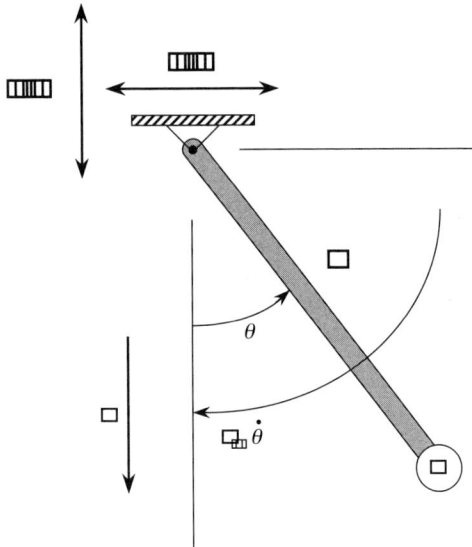

Figure 4  Parametrically forced pendulum with viscous damping.

approximation of system dynamic equations (see equation (21)) for $m = 4$ is given by:

$$\dot{\xi}_1^{(i)}(\tau) = \xi_2^{(i)},$$

$$\dot{\xi}_2^{(i)}(\tau) = \left\{ -X_0 \sin\left[\phi_1^{*(i)}(\tau) + \tau\right] - \frac{1}{x_3^2}\cos\left[\phi_1^{*(i)}(\tau)\right] \right\} \xi_1^{(i)} - \frac{\mu}{x_3}\dot{\xi}_2^{(i)} +$$

$$+ \frac{1}{2!}\left\{ -X_0 \cos\left[\phi_1^{*(i)}(\tau) + \tau\right] + \frac{1}{x_3^2}\sin\left[\phi_1^{*(i)}(\tau)\right] \right\}\left(\xi_1^{(i)}\right)^2 + \quad (29)$$

$$+ \frac{1}{3!}\left\{ X_0 \sin\left[\phi_1^{*(i)}(\tau) + \tau\right] + \frac{1}{x_3^2}\cos\left[\phi_1^{*(i)}(\tau)\right] \right\}\left(\xi_1^{(i)}\right)^3 +$$

$$+ \frac{1}{4!}\left\{ X_0 \cos\left[\phi_1^{*(i)}(\tau) + \tau\right] - \frac{1}{x_3^2}\sin\left[\phi_1^{*(i)}(\tau)\right] \right\}\left(\xi_1^{(i)}\right)^4 .$$

The truncated point mapping algorithm is applied directly to (29) in order to calculate each mapping of points within the cell, i.e., the cell's truncated point mapping. This completes the preliminary EPM problem definition, computational parameter specifications, etc.

## 5.2   Results

To demonstrate the EPM procedure the global behavior of (28) will be presented. For the EPM analysis the region of state space studied in $\mathbb{R}^2$ was $-\pi \leq x_1 < \pi$ and $-\pi \leq x_2 < \pi$. This system was previously analyzed using simple cell mapping (see Guttalu and Flashner [31]).

### 5.2.1 *Periodic solutions and basins of attraction*

We proceed with a fourth-order EPM analysis of the system by taking $51 \times 51$ intervals in the phase plane (equivalent to 2,601 cells in cell state space $S$) and determine each individual cell's truncated point mapping, $\mathbf{G}_{(4)}^{(i)}(\boldsymbol{\xi}^{(i)}(0), \mathbf{s})$ $i = 1, 2, \ldots, 2601$. EPM trajectory evolution was then performed for an initial condition resolution of $451 \times 451$. The results indicated there were three periodic orbits (near $x_1 \approx -0.6\pi$, $x_1 \approx -0.2\pi$, $x_1 \approx 0.2\pi$) as shown in Figure 5 where the location of the $P$-$K$ points found by the EPM method were indicated by the symbol "+" enclosed in small boxes. The periodic orbits were obtained by direct integration of (28) with the initial states being the location of the $P$-$K$ points. Two of the orbits were found to be asymptotically stable (indicated by solid curves) while the third is unstable (indicated by a dashed curve). In Section 5.2.2 we demonstrate how the EPM $P$-$K$ solution stability character was determined. Figure 6 shows the corresponding EPM domains of attraction. The initial conditions in the phase space were divided into four categories. Specifically, trajectories that map to the two asymptotically stable orbits with the darker region corresponding to the smaller Group A orbit and the lighter region to the larger Group C orbit, those that map to the unstable Group B orbit (indicated by "x"), and those that map to the sink cell (blank regions).

In order to assess the accuracy of the EPM method an "exact" periodic solution response and corresponding domains of attraction was constructed by continuously integrating each initial condition until the response either went to a periodic solution or left the region of state space $S$ under study. The Direct Integration periodic solutions and basins of attraction are depicted in Figures 7–8 respectively. First, we observe that the locations of the $P$-$K$ points determined by the EPM method were remarkably accurate to reproduce the "exact" orbits without requiring additional numerical calculations. Table 1 documents a representative sample of EPM $P$-$K$ points. Groups A and C were virtually identical to the "exact" results while Group C differs by no more than 0.5%. Furthermore, it is important to note a fundamental difference between the EPM and Direct Integration periodic solutions displayed in Figures 5 and 7, respectively. The EPM method accurately located the unstable solution while Direct Integration did not. An additional numerical technique would be required to ascertain the existence of unstable solutions if one utilizes Direct Integration.

With regard to the EPM and Direct Integration basins of attraction for the two asymptotically stable solutions (Groups A and C), they match extremely well as depicted in Figures 6 and 8. Within the limits of the initial condition discretization no discernable differences exist between the two basins of attraction. In addition, the EPM technique determines the unstable manifold boundary between the two asymptotically stable solutions in the form of initial conditions that map to sink cell whereas Direct Integration does not. Again, additional alternative methods would be required if one was relying solely on Direct Integration.

Finally, Figures 9–10 display the results of applying a simple cell mapping analysis for the same level of state space discretization utilized for the EPM study, i.e., a cell array of dimension $51 \times 51$. For this case the simple cell mapping method did not detect the unstable solution and the two asymptotically stable $P$-$K$ cell locations (Groups A and C) were grossly inaccurate. Moreover, only the most coarse features

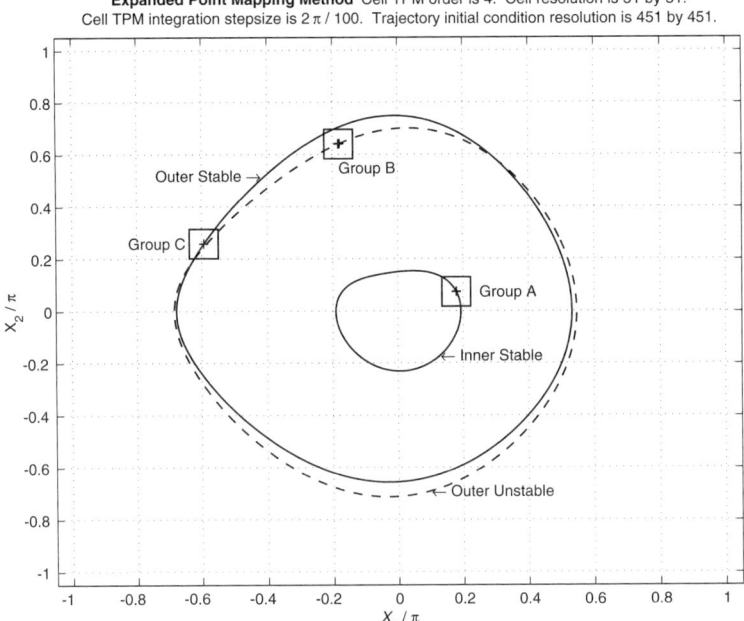

Figure 5  EPM $P$-$K$ points.

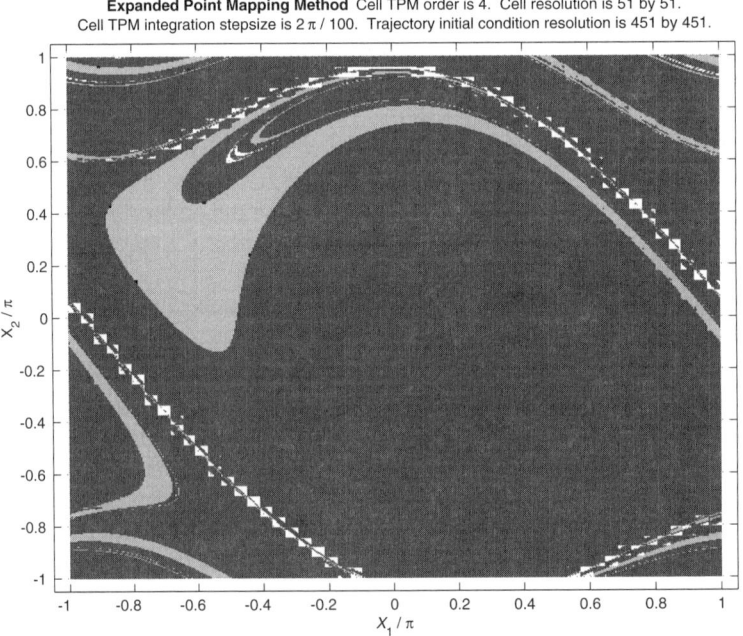

Figure 6  EPM basins of attraction.

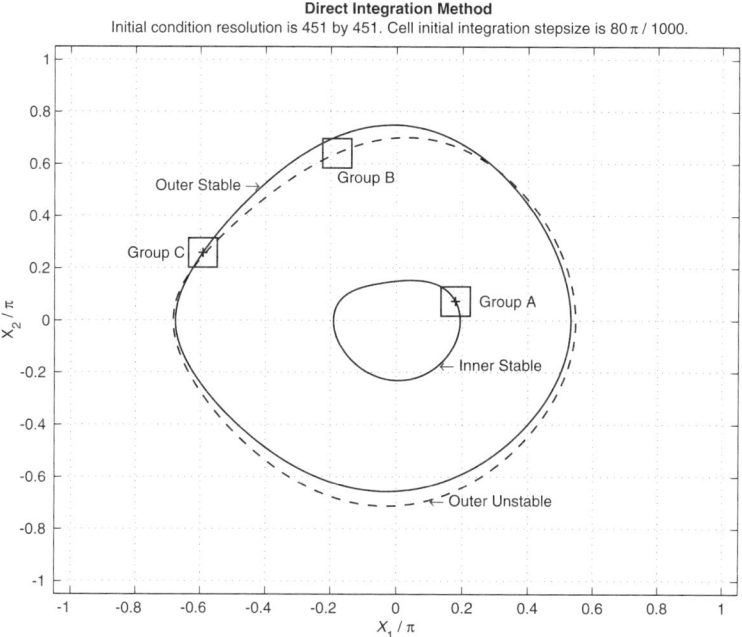

Figure 7  Direct integration periodic solutions.

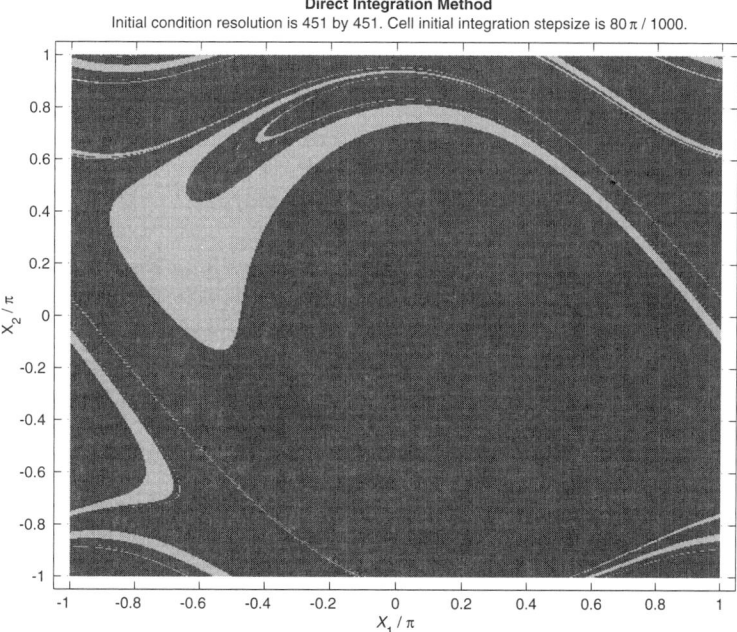

Figure 8  Direct integration basins of attraction.

Table 1   EPM $P$-$K$ points vs. direct integration P-1 point.

|  |  | Percent Deviation from Direct Integration (%) |
|---|---|---|
| | Group A P-1 points | |
| 1 | $(0.56325, 0.23582)$ | $(+0.01, +0.08)$ |
| 2 | $(0.56335, 0.23573)$ | $(+0.03, +0.04)$ |
| | Direct Integration P-1 point | |
| | $(0.56319, 0.23564)$ | |
| | Group B P-1 points | |
| 1 | $(-0.57128, 2.0179)$ | $(+0.30, +0.030)$ |
| 2 | $(-0.57243, 2.0174)$ | $(+0.50, +0.005)$ |
| | Direct Integration P-1 point | |
| | $(-0.5695, 2.0173)$ | |
| | Group C P-1 points | |
| 1 | $(-1.8631, 0.81491)$ | $(+0.04, -0.08)$ |
| 2 | $(-1.8618, 0.81707)$ | $(-0.03, +0.20)$ |
| | Direct Integration P-1 point | |
| | $(-1.8624, 0.81553)$ | |

of their corresponding basins of attraction were uncovered with accompanying distortions in the basin boundary positions. Clearly, Figures 9–10 illustrate the SCM state space discretization error introduced when utilizing a relatively large cell size $h_i$ for the system (28). In order to obtain the same level of EPM $P$-$K$ solution accuracy it would be necessary to increase the SCM cellular resolution to $451 \times 451$ (equivalent to 203,401 cells in cell state space $S$). This observation points out a tremendous advantage of the EPM method in comparison with SCM, namely a 78 fold reduction in total cells required. It is worthwhile repeating the fact that EPM results for the $51 \times 51$ cellular space very closely approximate the "exact" results obtained by direct numerical integration.

### 5.2.2   Stability analysis

For this specific example problem the linearized flow of (28) is

$$\dot{\boldsymbol{\xi}}^{(i)} = \begin{bmatrix} 0 & 1 \\ \left\{ -X_0 \sin\left[\phi_1^{*(i)}(\tau) + \tau\right] - \dfrac{1}{x_3^2}\cos\left[\phi_1^{*(i)}(\tau))\right] \right\} & \left(\dfrac{-\mu}{x_3}\right) \end{bmatrix} \boldsymbol{\xi}^{(i)}(\tau). \quad (30)$$

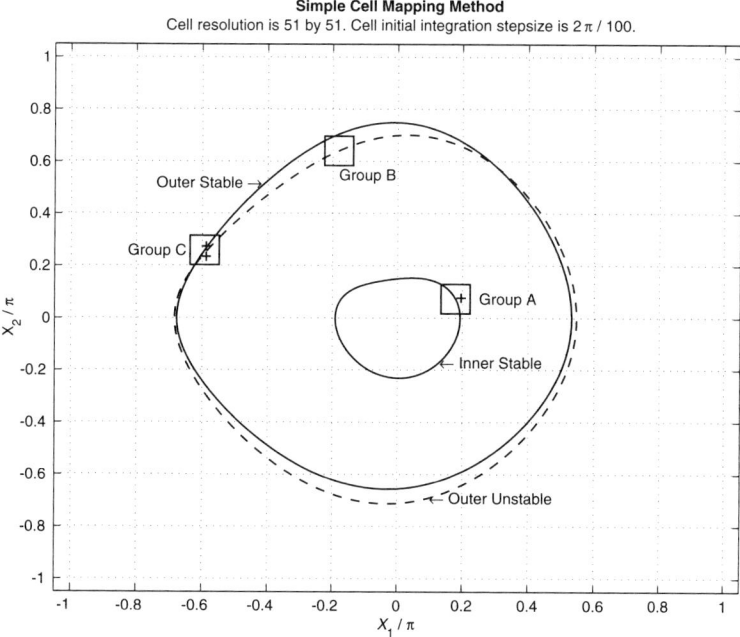

Figure 9  SCM *P-K* cells.

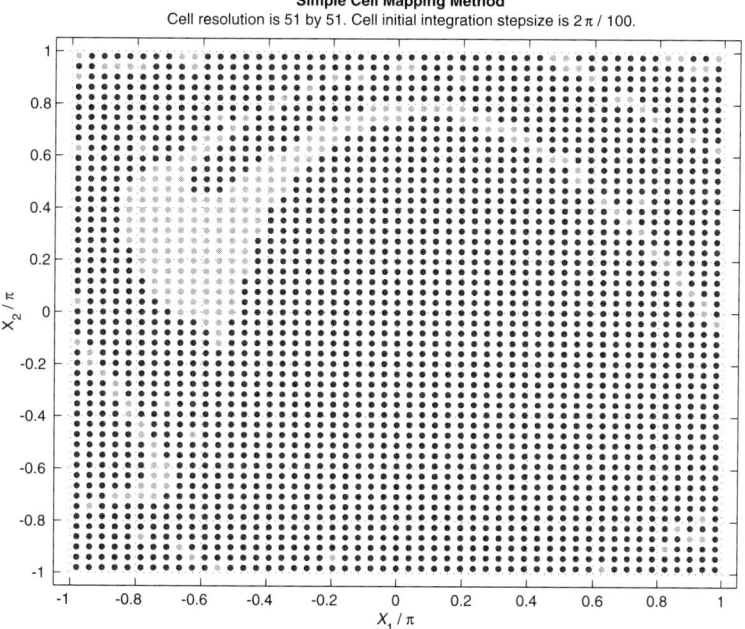

Figure 10  SCM basins of attraction.

Applying Liouville's Formula directly to (30) we obtain

$$\prod_{i=1}^{2} \lambda_i = \exp\left[\int_0^{T=2\pi} \left(\frac{-\mu}{x_3}\right) ds\right] .$$

Substituting $\mu$ and $x_3$ yields

$$\lambda_1 \cdot \lambda_2 = \exp\left[2\pi\left(\frac{-0.1}{0.75}\right)\right] = 0.432679 \tag{31}$$

to six significant digits. Since $\lambda_1 \cdot \lambda_2$ is determined analytically, it is possible to verify the multiplication of the calculated eigenvalues.

As explained in Section 4.1.4 the local $P$-$K$ solution stability is determined by the eigenvalues of the matrix $\mathbf{H} = \prod_{j=1}^{K} \mathbf{H}^{(j)}(\boldsymbol{\xi}^{(j)}(k), \mathbf{s})$. For the P-1 solutions represented by Group A in Figure 5

$$H^{(\text{Group A})} = \begin{bmatrix} -0.165032 & 0.465830 \\ -0.803010 & -0.355164 \end{bmatrix} \tag{32}$$

where the eigenvalues of $\mathbf{H}$ were $\lambda_{1,2} = -0.260098 \pm 0.604176i$ with $i = \sqrt{-1}$. The corresponding magnitudes were $|\lambda_1| = |\lambda_2| = 0.657784$, implying asymptotic stability. Note that $\lambda_1 \cdot \lambda_2 = 0.432680$, which is identical to (31) to five significant digits. Turning our attention to the P-1 points represented by Group C in Figure 5 we have

$$H^{(\text{Group C})} = \begin{bmatrix} 0.855133 & -0.822730 \\ 1.29839 & -0.743216 \end{bmatrix} \tag{33}$$

with eigenvalues $\lambda_{1,2} = 0.0559584 \pm 0.655399i$, magnitudes $|\lambda_1| = |\lambda_2| = 0.657784$, and $\lambda_1 \cdot \lambda_2 = 0.432680$, implying asymptotic stability of Group C. Again $\lambda_1 \cdot \lambda_2$ is identical to (31) to five significant digits. For Group B in Figure 5 we have

$$H^{(\text{Group B})} = \begin{bmatrix} 3.54822 & -3.85470 \\ 1.17039 & -1.14955 \end{bmatrix} \tag{34}$$

with eigenvalues $\lambda_1 = 2.20219$, $\lambda_2 = 0.196477$, and again $\lambda_1 \cdot \lambda_2 = 0.432680$. Hence, the Group B P-1 solution in Figure 5 represents a saddle point. The white space in Figure 6 shows the location of points mapping to the sink cell (corresponding to the unstable manifold of the saddle point), this is in agreement with the result that the stable and unstable manifolds of a saddle point separate various domains of attraction.

### 5.2.3  Trajectory evolution accuracy

Using the previous results from Section 5.2.1 where the cell resolution utilized was $51 \times 51$ a SCM and EPM trajectory generated from the same initial condition were compared separately with an "exact" integration solution. The results are

Table 2  CPU time requirements for basins of attraction and periodic solutions.

| Method | CPU Seconds | Initial Condition Resolution | Cellular Resolution |
|---|---|---|---|
| EPM | 3,649 | $451 \times 451$ | $51 \times 51$ |
| SCM | 14,193 | $451 \times 451$ | $451 \times 451$ |
| Direct Integration | 283,424 | $451 \times 451$ | N/A |

shown in Figures 11–12. In this instance, the SCM trajectory diverged rapidly from the "exact" integration and ultimately went to the incorrect periodic response as shown in Figure 11. This illustrates the potential pitfalls of a coarse SCM cellular resolution and the numerical error introduced due to cell discretization effects. In contrast, the EPM response closely matches the "exact" response as depicted in Figure 12. Within the limits of the plot resolution the EPM method has reproduced a trajectory virtually identical to continuous integration.

The computational time requirements for the EPM, SCM, and Direct Integration methods were also compared. All computations were performed on a Motorola StarMax 3000. In Section 5.2.1 we examined $451 \times 451$ initial conditions in the region of interest and determined both an "exact" basin of attraction and EPM basin of attraction. To make a meaningful comparison a SCM basin of attraction plot was constructed with a cellular resolution of $451 \times 451$ and each cell center point matching the $451 \times 451$ initial conditions used for both the "exact" and EPM basins of attraction. The cpu times are listed in Table 2. The cost for Direct Integration was approximately 78 times as high as needed by the EPM technique. Similarly, the SCM cpu cost was approximately 4 times greater than the EPM cpu cost. Clearly, the EPM method possesses a significant computational advantage over both the "exact" and SCM methods. Furthermore, it was interesting to note the SCM *P-K* cell locations were still not as accurate as the EPM *P-K* points. Again, this points out SCM cell discretization error, albeit small, exists even at *fine* cellular resolutions.

Finally, an additional EPM study was performed whereby the cellular resolution of the example problem was systematically lowered and the effects on both basins of attraction and *P-K* solution accuracy were examined. It was found for a fourth-order Expanded Point Mapping the cellular resolution could be reduced to $17 \times 17$ and still accurately locate the two stable and one unstable *P-K* solutions. The only inaccuracies introduced were in the boundaries separating the various basins of attraction. Thus, the EPM approach was extremely robust with respect to *large* cell sizes.

# 6  Concluding Remarks

A new numerical-analytical method called Expanded Point Mapping (EPM) was formulated. This methodology combines the *cell to cell mapping* and *point mapping* methods to investigate the global behavior and stability characteristics of equilib-

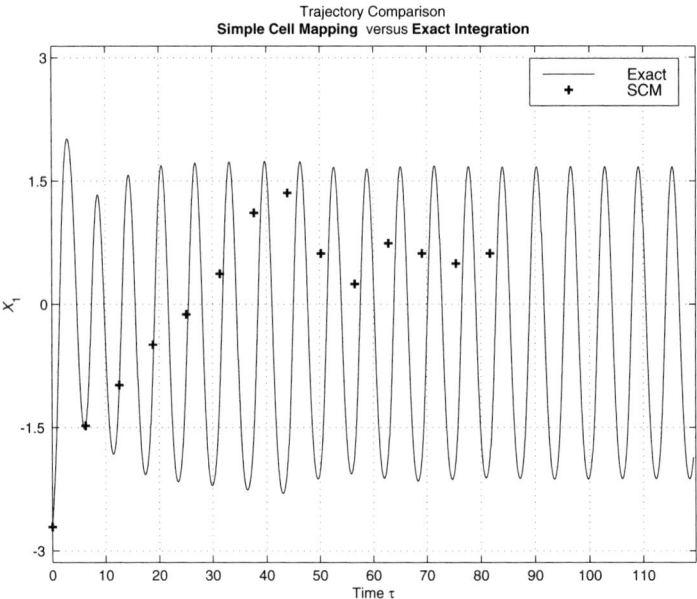

Figure 11   Trajectory comparison. SCM versus exact integration.

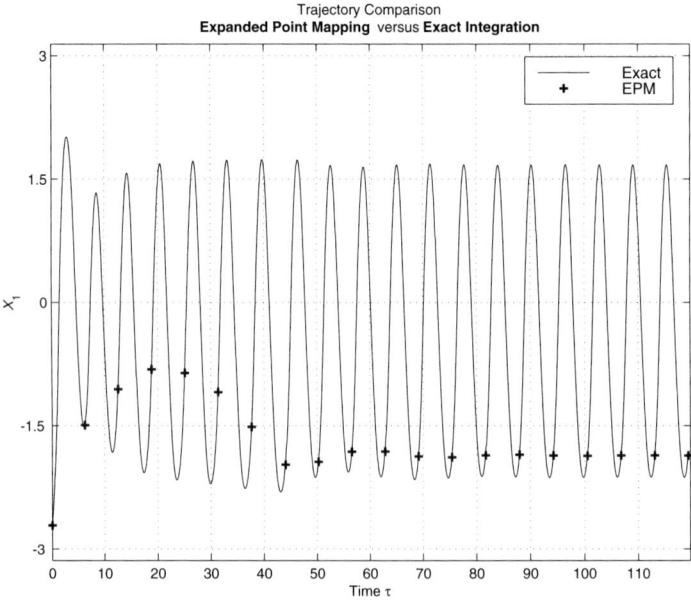

Figure 12   Trajectory comparison. EPM versus exact integration.

rium points and periodic solutions of nonlinear time-varying systems. The proposed method has the following features:

- The EPM method has a well defined theoretical basis stemming from cell mapping and point mapping theories. The analytic expression for the mapping of points within a cell is exact up to any desired degree of approximation $m$.

- The EPM method is formulated for multi-dimensional systems. This feature is an important advantage over other analytic approaches for which analysis of systems of order greater than two is very cumbersome, and sometimes even impossible.

- Local stability studies of periodic solutions can be performed analytically on the $m$-th order analytic approximation of the point mapping within a given cell.

- The trajectory evolution for EPM is based upon a continuum in state space. Hence, the EPM approach has the ability to generate any number of trajectories. This feature allows an accurate global analysis of the system and is an advantage over cell mapping-based global analysis.

## Acknowledgment

The research reported here was partially supported by the National Science Foundation grant CMS-9700467.

## References

1. Cesari, L. (1971) *Asymptotic Behavior and Stability Problems in Ordinary Differential Equations.* Academic Press, New York, 3rd edition.
2. Hale, J.K. (1963) *Oscillations in Non-linear Systems.* McGraw-Hill, New York.
3. Urabe, M. (1967) *Non-linear Autonomous Oscillations.* Academic Press, New York.
4. Yakubovich, V.A. and Starzhinskii, V.M. (1975) *Linear Differential Equations with Periodic Coefficients.* Vols 1 and 2, John Wiley and Sons, New York.
5. Bolotin, V.V. (1964) *The Dynamic Stability of Elastic Systems.* Holden-Day, San Francisco.
6. Kauderer, H. (1958) *Nichtlineare Mechanik.* Springer-Verlag, Berlin.
7. Lichtenberg, A.J. and Lieberman, M.A. (1083) *Regular and Stochastic Motions.* Springer-Verlag, New York.
8. Hayashi, C. (1964) *Nonlinear Oscillations in Physical Systems.* McGraw-Hill, New York.
9. Bogoliubov, N.N. and Mitropolsky, Y.A. (1961) *Asymptotic Methods in the Theory of Nonlinear Oscillations.* Hindustan Publishing Company, Delhi.
10. Nayfeh, A.H. and Mook, D.T. (1979) *Nonlinear Oscillations.* John Wiley and Sons, New York.

11. Ali Hasan Nayfeh (1973) *Perturbation Methods*. John Wiley and Sons, New York.

12. Moon, F.C. (1987) *Chaotic Vibrations*. John Wiley and Sons, New York.

13. Thompson, J.M.T. and Stewart, H.B. (1986) *Nonlinear Dynamics and Chaos*. John Wiley and Sons, Chichester.

14. Bernussou, J. (1977) *Point Mapping Stability*. Oxford, Pergamon.

15. Guckenheimer, J. and Holmes, P. (1983) *Nonlinear Oscillations, Dynamical Systems, and Bifurcations of Vector Fields*. Springer-Verlag, New York.

16. Poincaré, H. (1899) *Les Methodes Nouvelles de la Mecanique Celeste*. 3 Volumes, Gauthier-Villar, Paris.

17. Hsu, C.S. (1980) A Theory of Cell-to-Cell Mapping Dynamical Systems. *ASME Journal of Applied Mechanics*, **47**, 931–939.

18. Hsu, C.S. and Chiu, H.M. (1986) A Cell Mapping Method for Nonlinear Deterministic and Stochastic Systems, Part I. The method of analysis. *ASME Journal of Applied Mechanics*, **53**, 695–701.

19. Hsu, C.S. and Guttalu, R.S. (1986) An Unravelling Algorithm for Global Analysis of Dynamical Systems: An Application of Cell-to-Cell Mappings. *ASME Journal of Applied Mechanics*, **47**, 940–948.

20. Guttalu, R.S. (1981) On Point Mapping Methods for Studying Nonlinear Dynamical Systems. Ph.D. dissertation, University of California, Berkeley.

21. Hsu, C.S. (1987) *Cell-to-Cell Mapping: A method of global analysis for nonlinear systems*. Springer-Verlag, New York.

22. Miller, K.S. (1968) *Linear Difference Equations*. Benjamin, New York.

23. Flashner, H. (1979) A Point Mapping Study of Dynamical Systems. Ph.D. dissertation, University of California, Berkeley.

24. Flashner, H. and Hsu, C.S. (1983) A Study of Nonlinear Periodic Systems via the Point Mapping Method. *International Journal for Numerical Methods in Engineering*, **19**, 185–215.

25. Guttalu, R.S. and Flashner, H. (1989) Periodic Solutions of Nonlinear Autonomous Systems by Approximate Point Mappings. *Journal of Sound and Vibration*, **129**, 291–311.

26. Guttalu, R.S. and Flashner, H. (1955) Analysis of Bifurcation and Stability of Periodic Systems using Truncated Point Mappings. In *Nonlinear Dynamics: New Theoretical and Applied Results*, pp. 230–255, Akademie Verlag, Berlin.

27. Poston, T. and Stewart, I. (1978) *Catastrophe Theory and its Applications*. Pitman Publishing, London.

28. Henrici, P. (1962) *Discrete Variable Methods in Ordinary Differential Equations*. John Wiley and Sons, New York.

29. Jury, E.I. (1974) *Inners and Stability of Dynamic Systems*. John Wiley and Sons, New York.

30. Vidyasagar, M. (1978) *Nonlinear Systems Analysis*. Prentice-Hall, Englewood Cliffs, New Jersey.

31. Guttalu, R.S. and Flashner, H. (1994) A Numerical Study of the Dynamics of a Pendulum with a Moving Support. *ASME Journal of Applied Mechanics*, **192**, 293–311.

# A Preliminary Analysis of the Phase Portrait's Structure of a Nonlinear Pendulum-Mechanical System Using the Perturbed Hamiltonian Formulation

Débora Belato,[1] Hans Ingo Weber[2]
and José Manoel Balthazar[3]

[1] *DPM – Faculdade de Engenharia Mecânica – UNICAMP*
*Caixa Postal 6122, CEP 13083-970, Campinas, SP, Brasil*
*E-mail: belato@fem.unicamp.br,*
[2] *DEM - Pontifícia Universidade Católica – PUC – RJ*
*CEP 22453-900, Rio de Janeiro, RJ, Brasil*
*E-mail: hans@mec.puc-rio.br*
[3] *Instituto de Geociências e Ciências Exatas – UNESP – Rio Claro*
*Caixa Postal 178, CEP 13500-230, Rio Claro, SP, Brasil*
*E-mail: jmbaltha@rc.unesp.br*

The majority of the features of a nonlinear system can be better understood through the evolution of their phase portrait within the domain of variation of a chosen control parameter. Generally, this analysis is done using the basin of attraction with its respective attractors, but this investigation can also be done using its transients' trajectories since they are able to reveal the geometric structure of the phase portrait. Thus, using the characteristic of transient solution, this work describes the topology of the phase portrait of a particular pendulum-mechanical system to provide a global understanding of its behavior when a chosen control parameter is varied. This is done using the perturbed Hamiltonian formulation in order to obtain a mathematical model that describes the pendulum's behavior under the action of a more complex excitation force.

# 1   Introduction

A nonlinear dynamic system can be investigated through the evolution of its phase portrait when a chosen control parameter is varied, because it indicates the way that a system loses its stability providing a picture of the behavior of their solutions. However, when this analysis is done considering just the calculation of the basins of attraction of their respective attractors, some initial geometric characteristics of the phase portrait can be lost since the transient trajectories are totally neglected. As a matter of fact, in a nonlinear system the transient solution reveals the geometric structure of its solution surface in an $n$-dimensional space, where this surface represents the total solution of a differential equations' set and the phase portrait represents only the projection of the former on the plane. Depending on the type of intersection among the transient functions we can detect both the minimum and maximum points of the solution surface (stable and unstable solutions), and the narrow geometrical structure existing among them which is given by the existence of fractals, drawing, in this way, the shape of the solution surface which is projected on the plane [2]. Thus, a preliminary study of a nonlinear system must contain a detailed investigation of its solution's set, including the behavior of its transient solutions, in order to find out the topologic structure of its phase portrait when the variation of a chosen control parameter is considered.

   In this work, we will investigate the evolution of the phase portrait of a particular nonlinear system slightly similar to that of [1] which consists in a simple pendulum whose point of support is vibrated by a two-bar linkage along a horizontal guide (Figure 1). This paper is a continuation of the work reported in [5], where the Hamiltonian formulation was used to detect and identify the nonlinear phenomena that this particular system presents with the variation of a chosen control parameter. Here, the analysis will be done adding a small damping force in the differential equations of the system in order to identify the main characteristics of the obtained solutions. A complete study of the electromotor-pendulum's behavior near the fundamental resonance region can be found in [3].

   This work is organized as follows. In Section 2 we describe the perturbed Hamiltonian differential equations. The numerical results and some comments are presented in Section 3, and in Section 4, we present our conclusions.

# 2   Mathematical Model

A scheme of the mechanical system to be investigated here is given in Figure 1. It is composed of a simple pendulum whose point of support is horizontally vibrated by a crank mechanism. The complete mathematical formulation of this problem is described in [5], where the obtained Hamilton equations are:

$$\alpha' = P + \sigma_2 F \cos \alpha \,, \tag{1}$$

$$P' = \sigma_2 PF \sin \alpha + \sigma_2^2 F^2 \cos \alpha \sin \alpha - \gamma \sin \alpha \,, \tag{2}$$

with $F = \left[1 + \dfrac{\sigma_1 \cos \tau}{(1 - \sigma_1^2 \sin^2 \tau)^{1/2}}\right] \sin \tau$, $\sigma_1 = a/b$, $P = p_\alpha/ml^2\omega$, $\sigma_2 = a/l$,

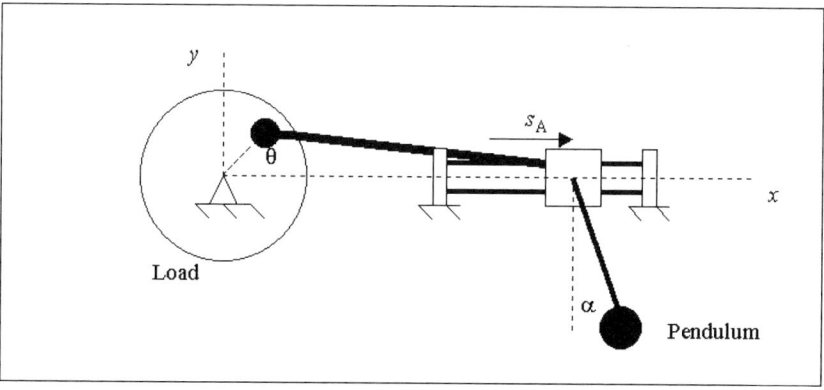

Figure 1  Schematic of the crank-pendulum mechanism.

$\gamma = g/l\omega^2$, and the primes denote derivatives concerning $\tau = \omega t$. The Hamiltonian function is defined as

$$\overline{H}^*(\alpha, P; t) = \frac{P^2}{2} + \sigma_2 PF \cos\alpha - \sigma_2^2 \frac{F^2}{2} \sin^2\alpha - \gamma\cos\alpha. \tag{3}$$

In this work, we consider a small perturbation in equations (1) and (2) which is given by a viscous damping at the pendulum's point of support equal to $c_P\alpha'$, obtaining the system equations:

$$\alpha' = P + \sigma_2 F \cos\alpha, \tag{4}$$

$$P' = \sigma_2 PF \sin\alpha + \sigma_2^2 F^2 \cos\alpha\sin\alpha - \gamma\sin\alpha - \delta(P + \sigma_2 F \cos\alpha), \tag{5}$$

with $F = \left[1 + \dfrac{\sigma_1 \cos\tau}{(1 - \sigma_1^2 \sin^2\tau)^{1/2}}\right]\sin\tau$, $\sigma_1 = a/b$, $P = p_\alpha/ml^2\omega$, $\sigma_2 = a/l$, $\gamma = g/l\omega^2$, $\delta = c_P/ml^2\omega^2$ and the primes denoting derivatives concerning $\tau$.

## 3   Numerical Results

The numerical simulations were done using the Simulink toolbox in Matlab$^{TM}$, through the integration of the equations (4) and (5), using the parameters: $a = 0.07$ m, $b = l = 0.3$ m, $g = 9.81$ m/s$^2$, i.e., $\sigma_1 = 0.233$, $\sigma_2 = 0.233$, and $\gamma$ is the chosen control parameter. The graphs below are Poincaré maps calculated every time that the trajectory of the system crosses the plane $\tau = (n+1)\pi$, $n = 1, 2, \ldots$, defining the phase portrait of the system for a given value of the control parameter.

In Figures 3 and 5, for each value of the control parameter there are two corresponding graphs: the first contains only the transient solution reaching the respective attractor for determined initial conditions; the second also contains some functions (in black) calculated using just the initial points of the transient trajectory which draws the phase portrait's structure. In reality, this last picture is

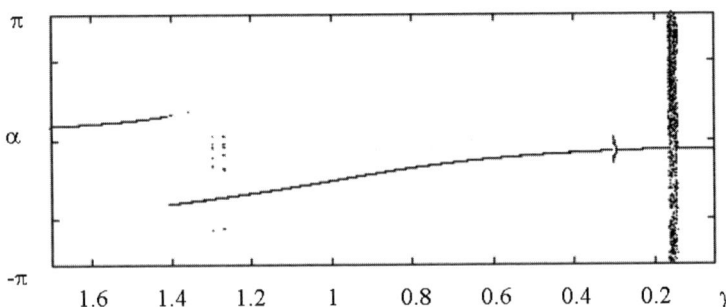

Figure 2 Bifurcation diagram of the perturbed Hamiltonian equations obtained with the variation of the control parameter ($\gamma$) and using the initial condition $(\alpha(0), P(0)) = (0, 0)$. We identify four bifurcations' values given by: $\gamma \approx 1.4$; 0.3; 0.16; 0.025.

obtained using each of the first six points of the transient solution varying the initial condition in the following way: $-\pi \leq \alpha(0) \leq \pi$ and $P(0) = 0$, which generates six independent functions inside the phase portrait. These lines indicate the borders of motion, where the solution travels before tending towards its attractor for a given initial condition, representing thus the "skeleton" of the system's general motion.

There are intersections among these six independent functions (represented by different colors and identified as "transient functions"), but not at the same point, creating a limited region which contains the attractors (for one, Figure 8c), whereas the juxtaposition of these lines means the existence of a saddle (for one, Figure 4a and 9e). As a matter of fact, the non-intersection of the transient functions at the same point, i.e., the non-existence of a fixed point for the transient functions, is due to the inclusion of the damping force in the Hamiltonian equations breaking the symmetry of the phase portrait in relation to the axis $P = 0$. Thus, the evolution of the transient functions with variation of the control parameter indicates the geometrical structure of the phase portrait (Figures 4 and 6), and their fitting by using known mathematical functions may lead to an analytical description of their behavior that will be investigated in a future work. Here, we are only interested in showing the geometrical form of the phase portrait in order to identify its irregularities and changes during variation of a control parameter.

To begin our analysis, it is important to observe that there are three main types of solutions that appear inside the phase portrait, when existence and evolution is made possible by the change in the system's stability: the first is a non-resonant stable solution, located on the bottom of the potential well, and represents the global minimum of the solution surface for the non-perturbed Hamiltonian system (its evolution in relation to the variation of the control parameter is given in Figure 2); the second is a resonant stable solution, located on the left of the phase portrait and the third is a resonant unstable solution (saddle) located on the right of the phase portrait and cannot be easily detected by numerical integration, Figure 3a. We can observe that the two resonant solutions (stable and unstable) remain in an intermediate position inside the phase portrait whereas the non-resonant solution moves towards the resonant unstable solution, and the distance between both

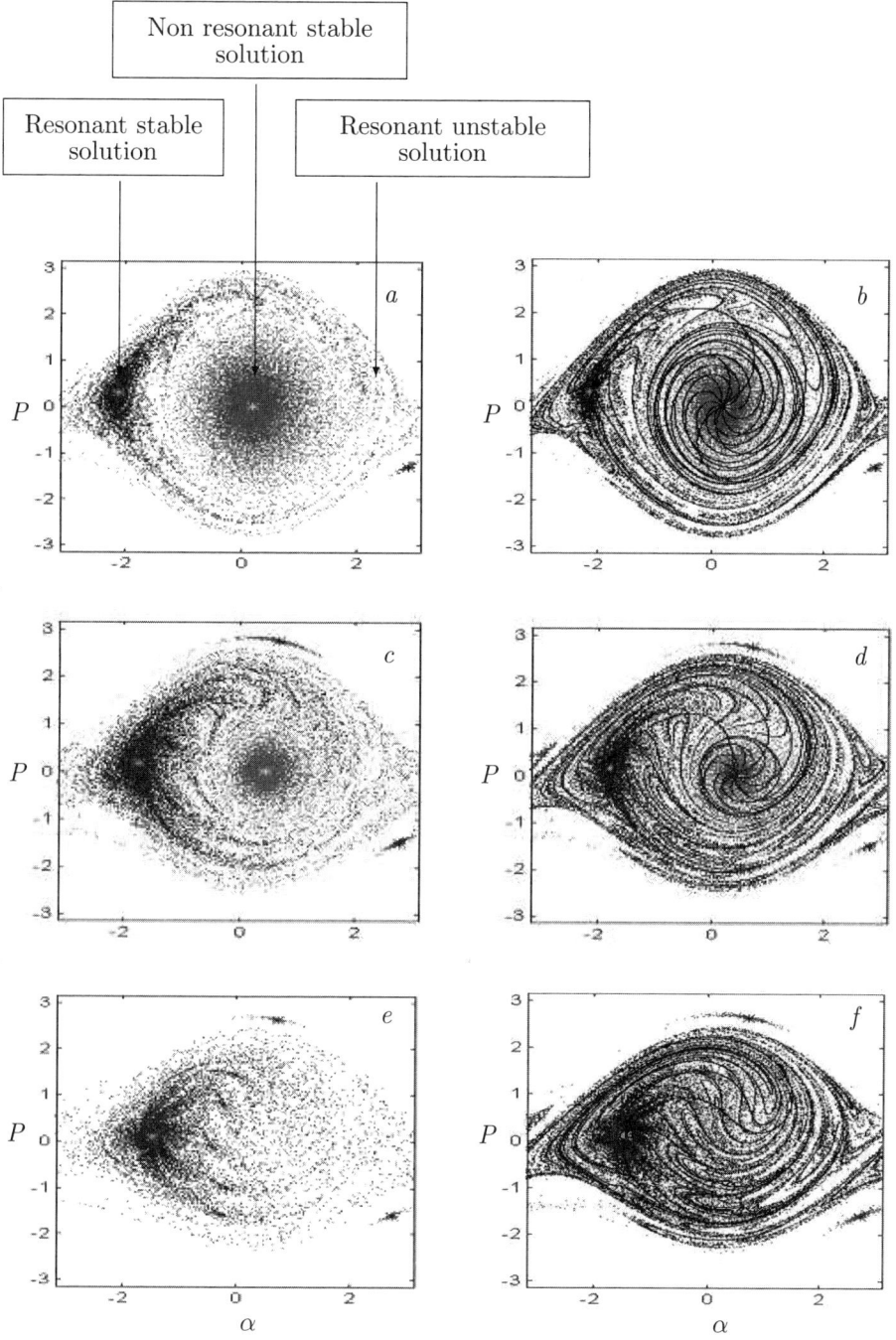

Figure 3  Phase portrait characteristics when (a,b) $\gamma = 2$; (c,d) $\gamma = 1.5$; (e,f) $\gamma = 1.25$, with the respective behavior of the transient functions in (b,d,f).

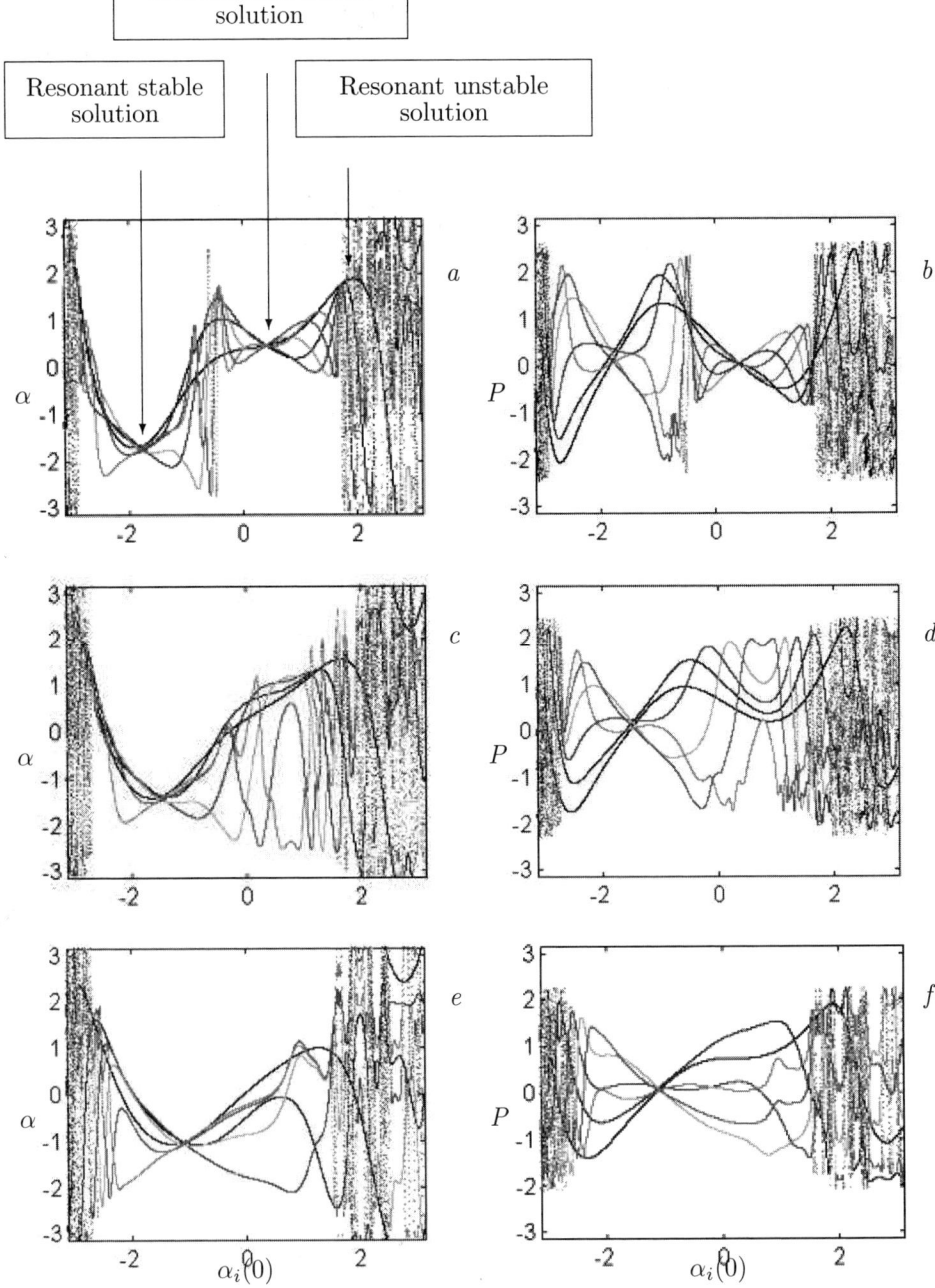

Figure 4 Transient functions characteristics for (a,b) $\gamma = 1.5$; (c,d) $\gamma = 1.25$; (e,f) $\gamma = 1.0$. (a,c,e) describe the behavior of $\alpha \times \alpha_i(0)$ and (b,d,f) describe the behavior of $P \times \alpha_i(0)$.

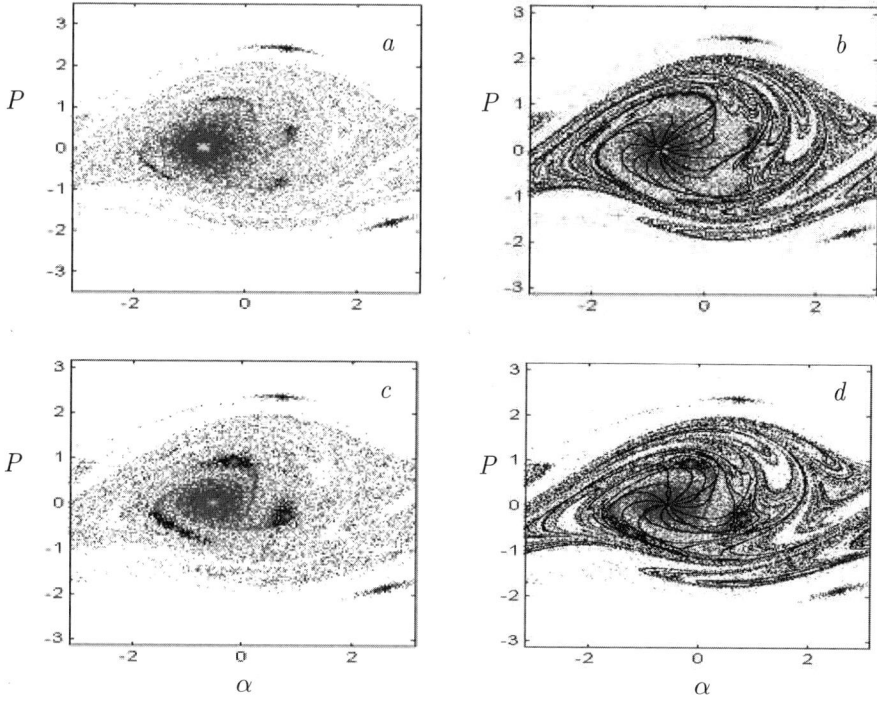

Figure 5  Phase portrait characteristics when (a,b) $\gamma = 0.8$; (c,d) $\gamma = 0.6$.

determines the loss of stability of the system at a critical value, where the saddle collides with the periodic attractor, [2]. Thus, with variation in the value of the control parameter, the basin of attraction of the resonant attractor increases its size whereas the non-resonant one diminishes, disappearing totally in this last critical value. The system undergoes a saddle-node bifurcation, resulting in the appearance of an irregular transient which leads to a new basin of a 7-period attractor, Figure 3e,f. Note, the system enters a domain where a sub-harmonic solution of several orders can be seen, culminating in a 3-period sub-harmonic, Figure 5c,d.

For $\gamma \approx 0.3$, the solution duplicates its period, where the only attractor bifurcates into two. However, there is not a complete cascade in this domain but only a discontinuous increase in the solutions amplitude, [6]. Afterwards, the only basin of attraction in the center of the phase portrait tends to diminish again, disappearing totally when $\gamma \approx 0.16$ where the system undergoes a global bifurcation and characteristics of the non-hyperbolicity of the minimum point of the solution surface appears. As a matter of fact, in this domain there occurs a complete domination of the irregular solutions in the interior of the phase portrait, which strongly diminishes the basin with a limited attractor in the center of the phase portrait, Figure 7c. At this moment, the solution undergoes a discontinuous jump to new remote attractors given by the rotations of the pendulum, [2]. We also can observe that the 1-period rotational attractors bifurcate into 3-period ones, but a detailed study of these solutions will be done in a future work.

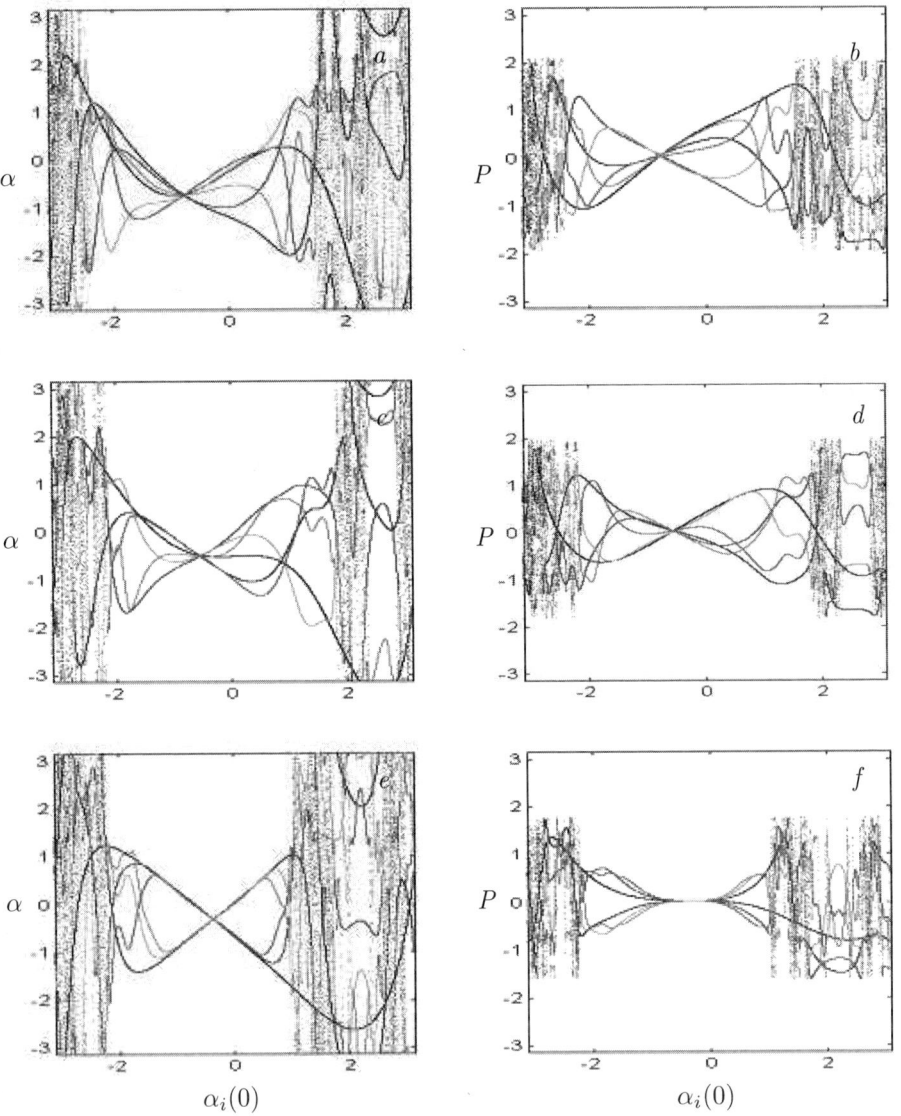

Figure 6 Transient functions for (a,b) $\gamma = 0.8$; (c,d) $\gamma = 0.6$; (e,f) $\gamma = 0.3$, with (a,c,e) reflecting $\alpha \times \alpha_i(0)$ and (b,d,f) reflecting $P \times \alpha_i(0)$.

Again, with variation of the control parameter, the basin of attraction in the center of the phase portrait (limited attractors) appears, with its motion direction inverted, and the basin gradually increases its size. Also, new small basins around the principal one are created, Figure 7d,e. In this condition, three prolongations of the transient functions occur towards the saddles $\alpha = \pm\pi$, connecting them till the region near to the boundary of the basin of attraction in the center of the phase portrait, leading all their unstable characteristics to appear near the attractor. This last event constitutes the phenomenon of multistability reported in [7] and a complete study of these observations is provided by [4].

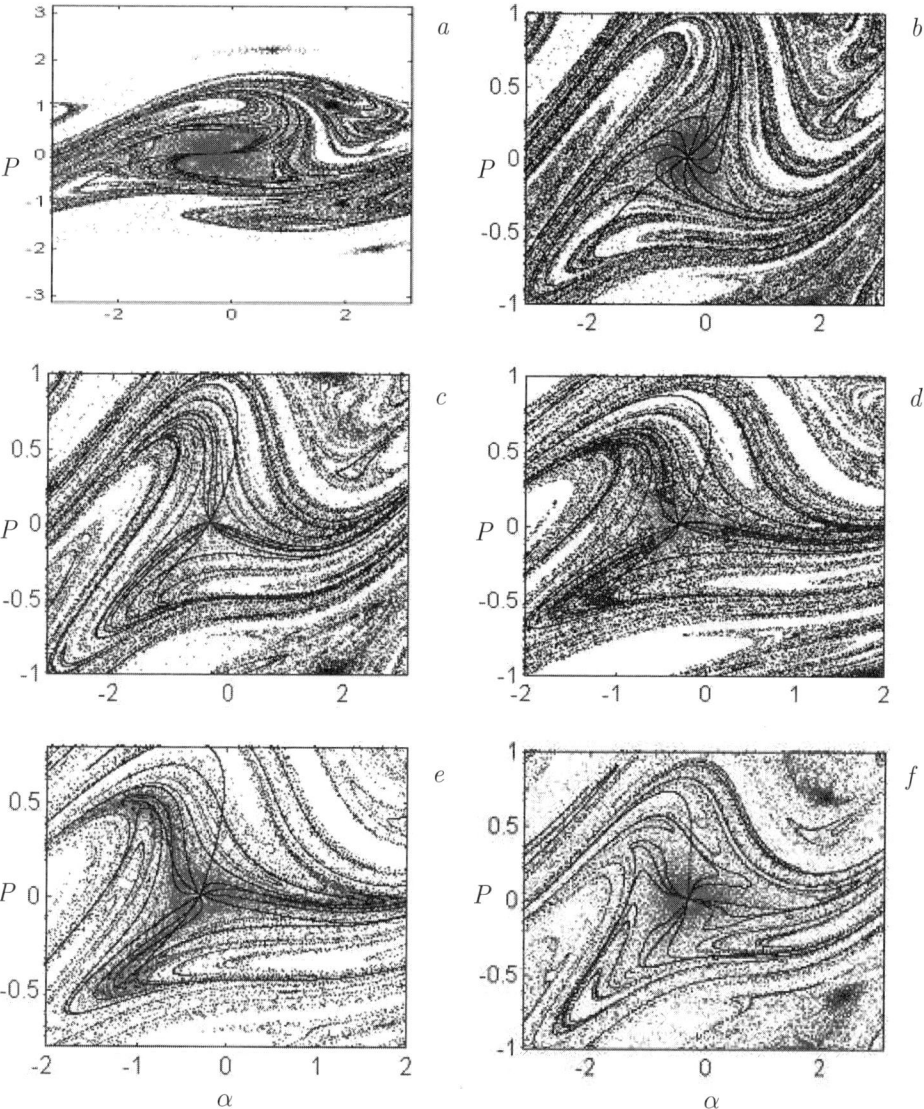

Figure 7 Phase portrait characteristics with the respective transient functions for (a) $\gamma = 0.3$; (b) $\gamma = 0.2$; (c) $\gamma = 0.144$; (d) $\gamma = 0.13$; (e) $\gamma = 0.122$; (f) $\gamma = 0.1$.

When $\gamma \approx 0.025$, the system creates another basin of attraction in the center of the phase portrait which appears under the action of an inverse saddle-node bifurcation, where the saddle and the periodic attractor reappear with their positions inverted inside the phase portrait (Figure 8). This time the saddle solution separates the two periodic attractors, and this separation process goes on until the attractors reach the basin of the two unstable solutions located near the point $\alpha = \pm\pi/2$ with $P$ tending to zero, Figure 8d. This occurs because the system works at a high excitation frequency where the stabilization of the pendulum oscillation around an

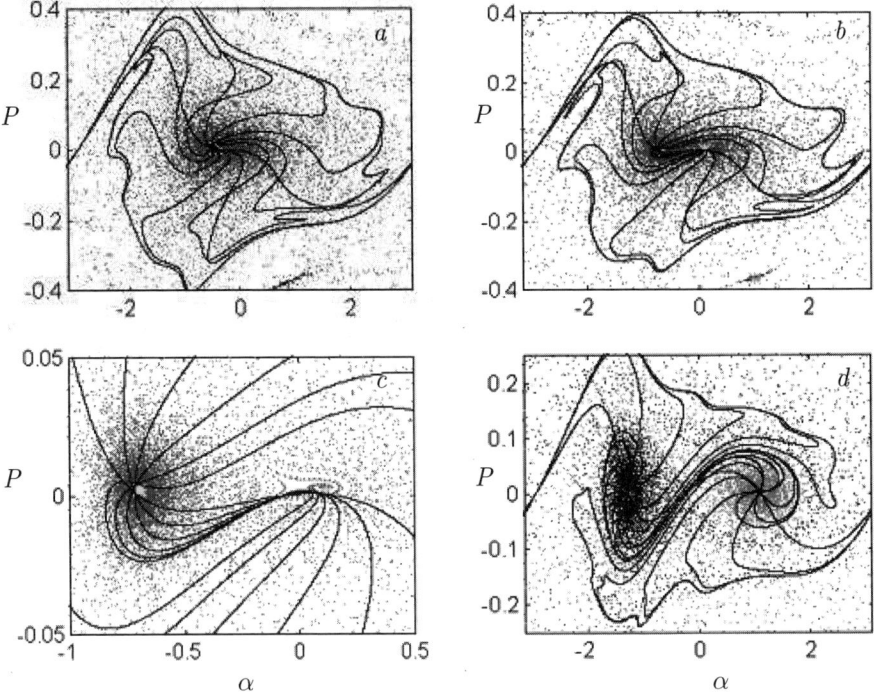

Figure 8   Phase portrait characteristics with the respective transient functions for
(a) $\gamma = 0.03$; (b) $\gamma = 0.025$; (c) $\gamma = 0.025$; (d) $\gamma = 0.01$.

unstable point can be done. We observe that in this domain of the control parameter the features of the solutions in the phase portrait have a "more stable behavior"
than the ones verified at a low excitation frequency, since in the former case the calculated transient functions are totally continuous, and they can be expressed using
Euclidian geometry, which did not happen before due to the presence of a fractal
structure surrounding the basin of the limited attractors. Thus, in this stage the
geometrical complexity of the phase portrait's structure diminishes, and the transient functions are clearly defined as a continuous line without any discontinuity
near the center of the phase portrait, Figure 9d,e,f.

A summary of all the motions of this system is presented in Figure 10, and it
reflects the behavior of the phase portrait with the variation of the chosen control
parameter. We can observe that the transient trajectory draws the structure of the
phase portrait, providing a good tool for the study of a nonlinear system, but the
comprehension of the arrangement of these lines inside the phase portrait is a topic
that must be better explored.

# 4   Conclusion

In this work, a particular perturbed Hamiltonian system is analyzed through evolution of its phase portrait in order to obtain all the bifurcational changes of the
system. Thus we can identify the three different bifurcational behaviors responsible

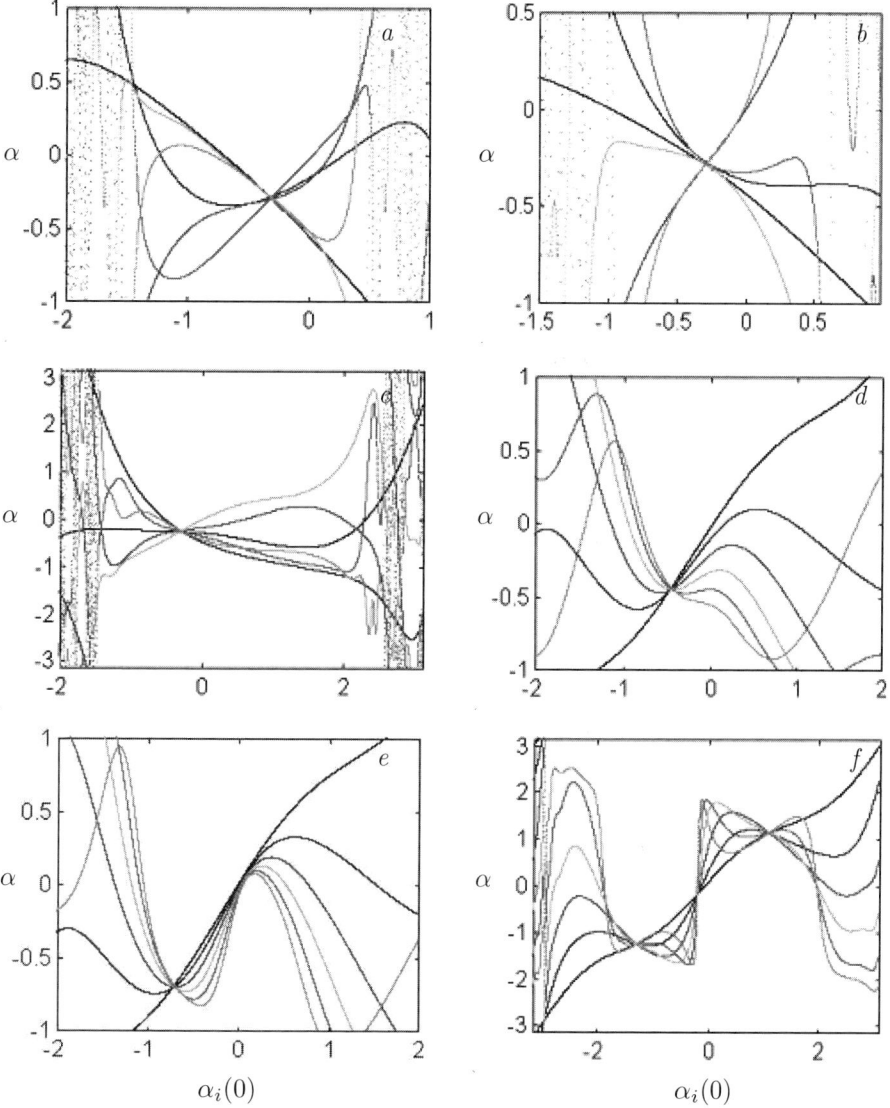

Figure 9  Graphs $\alpha \times \alpha_i(0)$ representing the transient functions for (a) $\gamma = 0.2$; (b) $\gamma = 0.144$; (c) $\gamma = 0.1$, (d) $\gamma = 0.03$; (e) $\gamma = 0.025$; (f) $\gamma = 0.01$.

for: (a) the destruction of a periodic attractor by a saddle-node bifurcation in the region of low excitation frequency; (b) great reduction of the basin of attraction of limited attractor in the center of the phase portrait and consequent inversion of its motions, representing the frontier between low and high excitation frequency; (c) the creation of a new periodic attractor through an inverse saddle-node bifurcation, in the region of high excitation frequency. Some geometrical properties of the phase portrait are identified by using some points of the transient response on the Poincare map, showing the existence of a sensitive geometrical structure with a

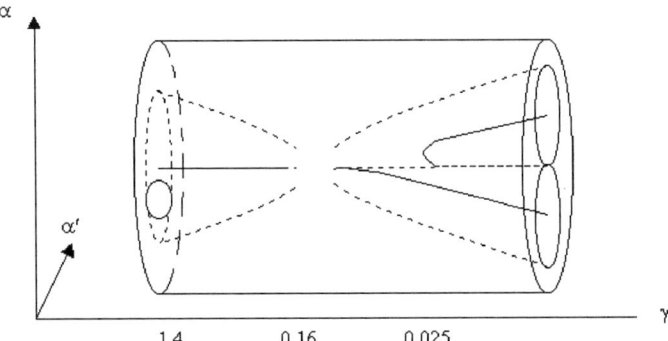

Figure 10  Summary of the bifurcational behavior of the phase portrait's evolution of the crank-pendulum mechanism.

high complexity level in the lower excitation frequencies. Therefore, only a deeper study of the geometry of these functions can reveal whether there is some standard among the solutions of this class of nonlinear equations.

## Acknowledgment

The authors would like to thank FAPESP (Fundação de Amparo à Pesquisa do Estado de São Paulo), for its financial support.

## References

1. Weibel, S., Kaper, T.J. and Baillieul, J. (1997) Global Dynamics of a Rapidly Forced Cart and Pendulum. *Nonlinear Dynamics*, **13**, 131–170.

2. Belato, D. (2002) Nonlinear Analysis of Non Ideal Holonomic Dynamic Systems. Ph.D. Thesis, Fac. Mech. Engineering, UNICAMP, Brazil, 186 p. (in Portuguese).

3. Belato, D., Weber, H.I., Balthazar, J.M. and Mook, D.T. (2001) Chaotic Vibrations of a Nonideal Electro-Mechanical System. *International Journal of Solids and Structures*, **38**, 1699–1706.

4. Belato, D., Balthazar, J.M. and Weber, H.I. (2003) A Note about the Appearance of Non Hyperbolic Solutions in a Mechanical Pendulum System. *Nonlinear Dynamics*, to appear.

5. Belato, D., Weber, H.I., Balthazar, J.M., Mook, D.T. and Rosário, J.M. (2000) Hamiltonian Dynamic of a Nonlinear Mechanical System. In *Nonlinear Dynamics, Chaos, Control and Their Applications to Engineering Sciences, vol. 4: Recent Developments in Nonlinear Phenomena*, J.M. Balthazar, P.B. Gonçalves, R.M.F.L.R.F. Brasil, I.L. Caldas, F.B. Rizatto (eds), pp. 359-368.

6. Weber, H.I., Balthazar, J.M. and Belato, D. (2003) Behaviour Analysis of a Nonlinear Mechanical System using Transient Response. In *Proceedings of 5$^{th}$ Int. Conference on Modern Practice in Stress and Vibration Analysis*, Trans Tech Publications Ltd.

7. Feudel, U. and Grebogi, C. (1997) Multistability and the Control of Complexity. *Chaos*, **7** (4), 597–604.

# A Review of Rigid-Body Collision Models in the Plane

Edson Cataldo [1] and Rubens Sampaio [2]

[1] *Universidade Federal Fluminense (UFF), Departamento de Matemática Aplicada*
*PGMEC-Programa de Pós-Graduação em Engenharia Mecânica*
*Rua Mário Santos Braga, S/No-24020, Centro, Niterói, RJ, Brasil*
*Tel: 55 21 2717-8269, E-mail: ecataldo@mec.uff.br*
[2] *Pontifícia Universidade Católica do Rio de Janeiro (PUC-Rio)*
*Departamento de Engenharia Mecânica*
*Rua Marquês de São Vicente, 225, 22453-900, Gávea, Rio de Janeiro, Brasil*
*Tel: 55 21 2529-9330, E-mail: rsampaio@mec.puc-rio.br*
*Internet: http://www.mec.puc-rio.br/prof/rsampaio/rsampaio.html*

In general the motion of a body takes place in a confined environment and collisions of the body with the containing wall are possible. In order to predict the dynamics of a body in these conditions one must know what happens in a collision. Therefore, the problem is: if one knows the pre-collision dynamics of the body and the properties of the body and the wall, one wants to predict the post-collision dynamics. This problem is quite old and appeared in the literature in 1668. Up to 1984 it seemed that Newton's model would be enough to solve the problem. But it was found that this was not the case and a renewed interest in the problem appeared. The aim of this paper is to treat the problem of plane collisions of rigid bodies, to make a review and to classify the different models found in the literature, and to present a new model, called the C-S model, that is a generalization of most of these models.

## 1 Introduction

From the simplest observation, we can say that the dynamics of a body, or of a system with more than one particle, can be modeled properly only if collisions are taken into account. In the works of Galileo and Descartes there are references to the collision between particles, but the first published works on this problem seem to be due to John Wallis and Christopher Wren, independently, in 1668. Some great scientists such as Newton, Huygens, Coriolis, Darboux, Routh, Apple, Carnot and Poisson have also treated the problem. At the beginning of the twentieth century the problem generated some discussion, as we can see in the works of Painlevé [1] and Klein [2]. But, up to 1984, all of these works used the theory developed by Newton or by Poisson and the difficulty was to include friction in the modeling, as was pointed out by Painlevé in his paper "Sur les lois de frottement de glissement."

In 1984, Kane [3] published a work, in a journal with limited circulation, where he pointed out an apparent paradox: the application of Newton's theory with Coulomb's friction, universally accepted, in a problem of collisions of a double pendulum, conducted to generation of energy. What could be wrong?

In 1986, Keller [4] presented a solution to Kane's paradox, but the solution was not easy to generalize. Keller's work was published in a journal with a large circulation and aroused widespread interest. Since then, interest has increased and there are some books totally dedicated to this topic, such as those of Glocker and Pfeiffer [5], Brach [6], Brogliato [7] and Monteiro-Marques [8].

Brach [6] presented a model with linear equations containing some nondimensional parameters that characterize the collision and he defined "ratio between impulses" instead of coefficient of friction. However, he did not give clear solutions to the problem when one considers *reverse sliding* during the collision. Stronge [9] suggested a coefficient of restitutions relating the energy during the compression phase to the energy during the expansion phase. Smith [10] presented a model with nonlinear equations. Wang and Mason [11] applied Routh's technique and compared the coefficients of restitutions given by Newton and by Poisson. Sabine Durand [12] studied the dynamics of systems with unilateral restrictions and included some systems related to collisions. Chatterjee [13] presented new laws based on simple algorithms. He did not use many parameters and obtained good results. Soianovici and Hermuzlu [14] have shown the limits of validity of some rigid bodies collision models. As their main interest was in robotics, they focused on collisions of slender bodies at low velocities. Cathérine Cholet [15] presented a new theory of rigid bodies collisions in the context of continuum mechanics following the ideas of Michel Frémond: a system formed by a set of rigid bodies is deformable because the relative positions between each pair of bodies vary. They discussed the theory and showed that it is coherent from the mathematical point of view and also experimentally validated.

## 2   Motion Equations

The collision is modeled as instantaneous. The generalized position of the system in the instant $t$ is defined by $\mathbf{q} = (q_1, q_2, \ldots, q_n)^T$. Let $C_1$ and $C_2$ be two bodies, and let $\mathbf{R}$ be the force of reaction exerted by $C_1$ on $C_2$. Then we write $\mathbf{R} = (\ R_N \ \ R_T \ )^T$.

The dynamics of the system is given by the Lagrangean equations:

$$\frac{d}{dt}\left(\frac{\partial T}{\partial \dot{q}_i}\right) - \frac{\partial T}{\partial q_i} = Q_i + r_i \tag{1}$$

with $Q_i$ the contribution of the external generalized forces, $r_i$ the generalized force

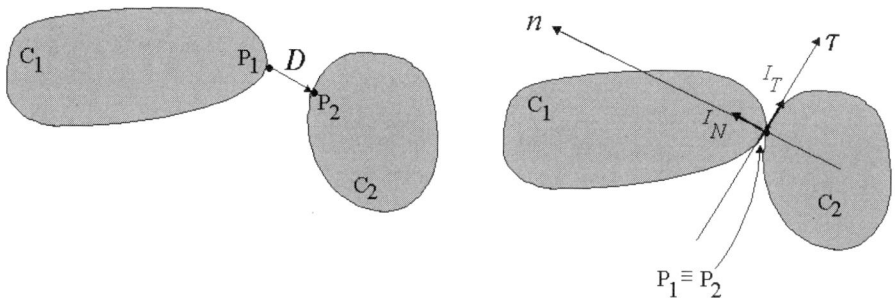

Figure 1  Collision between two bodies.

due to the reaction at the contact and $T$ the kinetic energy of the system. We should note that $r_i$ is only present when there is contact, otherwise it is null.

Considering only a planar situation, we have $n$ parameters of position and two reactions in the contact ($R_N$ and $R_T$) also unknown. Then, we need not only the $n$ equations obtained from Lagrange's equations but also two more equations, given by the collision laws that will be discussed later.

We consider collisions of only two bodies and at only one point (this is the case if one of the bodies is strictly convex).

Let $P_1$ and $P_2$ be the points of the bodies $C_1$ and $C_2$, respectively, that will be in contact in the collision. We denote by $\mathbf{D}$ the vector that represents the relative displacement between the two bodies and by $\dot{\mathbf{D}}$ the vector that represents the relative velocity between the bodies, as shown in Fig. 1.

In the point of contact we represent the impulses in the normal and tangential directions by $I_N$ and $I_T$. We use $\mathbf{u}_N$ and $\mathbf{u}_\tau$, the unitary vectors of the normal direction (given by $N$) and tangential direction (given by $\tau$), in a frame which we will call the collision frame, shown in Fig. 1.

Evaluating the relative velocity between the contact points we have:

$$\dot{\mathbf{D}} = \sum_{i=1}^{n} \frac{\partial P_2}{\partial q_i} \dot{q}_i + \frac{\partial P_2}{\partial t} - \sum_{i=1}^{n} \frac{\partial P_1}{\partial q_i} \dot{q}_i - \frac{\partial P_1}{\partial t} = \sum_{i=1}^{n} \left( \frac{\partial P_2}{\partial q_i} - \frac{\partial P_1}{\partial q_i} \right) \dot{q}_i + \frac{\partial P_2}{\partial t} - \frac{\partial P_1}{\partial t}.$$

We use the notations

$$W_T^i = \left( \frac{\partial P_2}{\partial q_i} - \frac{\partial P_1}{\partial q_i} \right).\mathbf{u}_\tau \quad , \quad W_N^i = \left( \frac{\partial P_2}{\partial q_i} - \frac{\partial P_1}{\partial q_i} \right).\mathbf{u}_n,$$

$$\tilde{w}_T = \left( \frac{\partial P_2}{\partial t} - \frac{\partial P_1}{\partial t} \right).\mathbf{u}_\tau \quad \text{and} \quad \tilde{w}_N = \left( \frac{\partial P_2}{\partial t} - \frac{\partial P_1}{\partial t} \right).\mathbf{u}_n.$$

We consider, then, $\mathbf{W}_T$, the column vector in which the components are $W_T^i$, and $\mathbf{W}_N$, the column vector in which the components are $W_N^i$.

We can write the normal ($\dot{D}_N$) and tangential ($\dot{D}_T$) components of $\dot{\mathbf{D}}$ as

$$\begin{cases} \dot{D}_N = \mathbf{W}_N^T \dot{\mathbf{q}} + \tilde{w}_N \\ \dot{D}_T = \mathbf{W}_T^T \dot{\mathbf{q}} + \tilde{w}_T. \end{cases}$$

Or we can write

$$\dot{\mathbf{D}} = [W]^T \dot{\mathbf{q}} + \tilde{\mathbf{w}}$$

with $\dot{\mathbf{D}} = \begin{pmatrix} \dot{D}_N \\ \dot{D}_T \end{pmatrix}$, $[W]^T = \begin{pmatrix} \mathbf{W}_N^T \\ \mathbf{W}_T^T \end{pmatrix}$ a matrix $(2, n)$ and $\tilde{w} = \begin{pmatrix} \tilde{w}_N \\ \tilde{w}_T \end{pmatrix}$.

The generalized force $\mathbf{r}$ can be written in terms of $[W]$ and $\mathbf{R}$ as

$$\mathbf{r} = \begin{pmatrix} \mathbf{W}_N & \mathbf{W}_T \end{pmatrix} \begin{pmatrix} R_N \\ R_T \end{pmatrix} \qquad \text{or} \qquad \mathbf{r} = [W]\mathbf{R}.$$

Integrating Eq. 1 in the interval $(t - \epsilon, t + \epsilon)$, with $t$ the instant of collision, we have

$$\left( \frac{\partial T}{\partial \dot{q}_i} \right) \Big|_{t-\epsilon}^{t+\epsilon} - \int_{t-\epsilon}^{t+\epsilon} \frac{\partial T}{\partial q_i} d\tau = \int_{t-\epsilon}^{t+\epsilon} Q_i d\tau + \int_{t-\epsilon}^{t+\epsilon} r_i d\tau.$$

We consider $Q_i$ continuous and limited and we make $\epsilon \to 0$. Then

$$\Delta \left( \frac{\partial T}{\partial \dot{q}_i} \right) = \lim_{\epsilon \to 0} \int_{t-\epsilon}^{t+\epsilon} r_i d\tau \quad \text{with} \quad \Delta \left( \frac{\partial T}{\partial \dot{q}_i} \right) = \left( \frac{\partial T}{\partial \dot{q}_i} \right) \Big|_E - \left( \frac{\partial T}{\partial \dot{q}_i} \right) \Big|_A.$$

We use the index $E$ to represent the right limit and $A$ to represent the left limit. We know that $\mathbf{r} = [W]\mathbf{R}$. Then

$$\Delta \left( \frac{\partial T}{\partial \dot{q}_i} \right) = \lim_{\epsilon \to 0} \int_{t-\epsilon}^{t+\epsilon} r_i d\tau = \lim_{\epsilon \to 0} \int_{t-\epsilon}^{t+\epsilon} \begin{pmatrix} W_N^i & W_T^i \end{pmatrix} \begin{pmatrix} R_N \\ R_T \end{pmatrix} d\tau.$$

We write the impulse $\mathbf{I}$ caused by the reaction $\mathbf{R}$ as

$$\mathbf{I} = \lim_{\epsilon \to 0} \int_{t-\epsilon}^{t+\epsilon} \mathbf{R} d\tau = \begin{pmatrix} I_N \\ I_T \end{pmatrix}.$$

Then

$$\Delta \left( \frac{\partial T}{\partial \dot{q}_i} \right) = \begin{pmatrix} W_N^i & W_T^i \end{pmatrix} \mathbf{I}.$$

We denote $\dfrac{\partial \mathbf{T}}{\partial \dot{\mathbf{q}}}$ as the vector in which the components are $\dfrac{\partial T}{\partial \dot{q}_i}$, so that we can write

$$\Delta \left( \frac{\partial \mathbf{T}}{\partial \dot{\mathbf{q}}} \right) = [W]\mathbf{I}.$$

But,

$$T = \frac{1}{2} \dot{\mathbf{q}} [M] \dot{\mathbf{q}}^T \Rightarrow [M](\dot{\mathbf{q}}_E - \dot{\mathbf{q}}_A) = [W]\mathbf{I} = \begin{pmatrix} W_N & W_T \end{pmatrix} \begin{pmatrix} I_N \\ I_T \end{pmatrix}. \qquad (2)$$

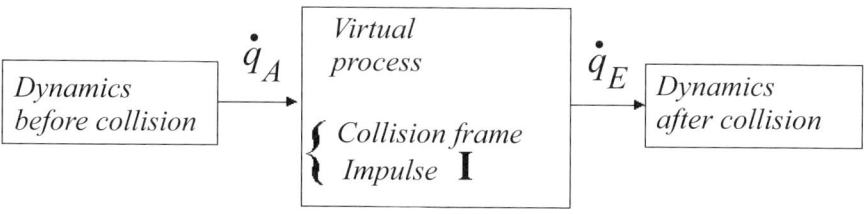

Figure 2 Virtual process scheme.

Our problem is to find $\dot{\mathbf{q}}_E$ and $\mathbf{I}$ given $[M]$, $[W]$ and $\dot{\mathbf{q}}_A$. Then, there are $n$ equations, and we want to find $n + 2$ unknowns. Therefore, we need two more equations. These two equations are given by the restitution laws discussed later.

If we consider an impulse of moment denoted by $\mathbf{I}_\theta$, and not only the normal and the tangential impulses, the equation will be given by

$$[M]\left(\dot{\mathbf{q}}_E - \dot{\mathbf{q}}_A\right) = \left( \begin{array}{ccc} \mathbf{W}_N & \mathbf{W}_T & \mathbf{W}_\theta \end{array} \right) \left( \begin{array}{c} I_N \\ I_T \\ I_\theta \end{array} \right). \tag{3}$$

In this case, we have $n$ equations to find $n + 3$ unknowns. We need three more equations. A collision theory is a prescription of these two, or three, conditions that are called laws of restitution.

We construct a collision model when we join the $n$ equations that describe the motion of the system with the equations given by the restitution laws.

Collision is a discontinuous process. However, to understand the several models that appear in the literature, and to see why some of them do not work, it is interesting to define a *virtual process* that connects $(q_A, \dot{q}_A)$ to $(q_E, \dot{q}_E)$. This *virtual process* has nothing to do with time. We show a scheme in Fig. 2 to illustrate this idea.

## 3   The Local Matrix Mass

Instead of writing the equations in terms of $\dot{\mathbf{q}}$ we can use $\dot{\mathbf{D}}$. The vector $\mathbf{D}$ was shown in Fig. 1, and it is important because it monitors when the collision occurs.

We can write

$$\dot{\mathbf{D}} = [W]^T \dot{\mathbf{q}} + \tilde{\mathbf{w}} = \left( \begin{array}{c} W_N^T \\ W_T^T \end{array} \right) \dot{\mathbf{q}} + \left( \begin{array}{c} \tilde{w}_N \\ \tilde{w}_T \end{array} \right). \tag{4}$$

Then,

$$\dot{\mathbf{D}}_{\mathbf{E}} - \dot{\mathbf{D}}_{\mathbf{A}} = [W]^T(\dot{\mathbf{q}}_{\mathbf{E}} - \dot{\mathbf{q}}_{\mathbf{A}}).$$

But,

$$[M](\dot{\mathbf{q}}_{\mathbf{E}} - \dot{\mathbf{q}}_{\mathbf{A}}) = [W]\mathbf{I} \Rightarrow (\dot{\mathbf{q}}_{\mathbf{E}} - \dot{\mathbf{q}}_{\mathbf{A}}) = [M]^{-1}[W]\mathbf{I}.$$

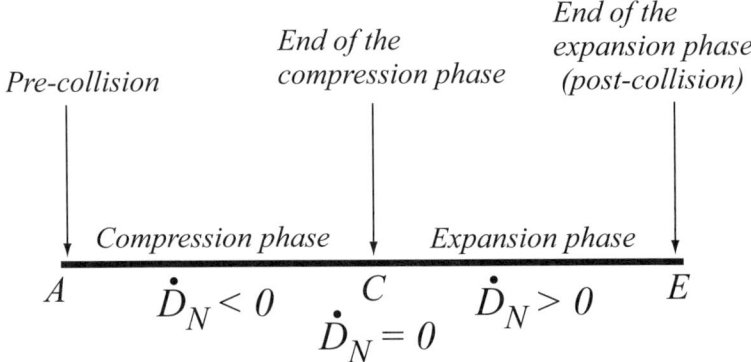

Figure 3  Compression phase and expansion phase.

Then,

$$\dot{\mathbf{D}}_\mathbf{E} - \dot{\mathbf{D}}_\mathbf{A} = [W]^T [M]^{-1} [W] \mathbf{I} = [\tilde{M}_L] \mathbf{I} .$$

So,

$$\mathbf{I} = [M_L](\dot{\mathbf{D}}_E - \dot{\mathbf{D}}_A) , \tag{5}$$

when $[\tilde{M}_L]$ is invertible and $[\tilde{M}]_L^{-1} = [M_L]$. We call $[M_L]$ the local matrix mass.

## 4  Compression Phase and Expansion Phase

In order to describe some of the collision models we will assume, formally, that the change from the pre-collision velocity to the post-collision velocity occurs in two phases: the compression phase and the expansion phase. The virtual process will be composed of these two phases, as shown schematically in Fig. 3.

## 5  Classification of Collision Models

The collision models can be classified in four groups. To complete a collision model we need the equations presented here plus the restitution laws, which will be discussed later.

We consider the index $A$ to represent the moment pre-collision, the index $C$ to represent the end of the compression phase and the index $E$ to represent the moment post-collision. To describe the equations that will be used we consider the following symbols:

$\dot{\mathbf{q}}$ – Vector of the velocity components of the generalized coordinates

$[M]$ – Mass matrix

$[W]$ – Vector that relates the coordinates at the contact point to the generalized coordinates

$\dot{\mathbf{D}}$ – Vector representing the relative velocity among the contact points. Its components in the normal direction, the tangential direction and the angular direction are respectively $\dot{D}_N$, $\dot{D}_T$ and $\dot{D}_\theta$

$\mathbf{I}$ – Vector representing the impulse at the contact point. Its components in the normal direction and the tangential direction and the moment of impulse are respectively $I_N$, $I_T$ and $I_\theta$

$\mu$ – Coefficient of friction used in restitution's law

$J$ – Moment of inertia

$I_{TS}$ – Parameter used to consider the reversible portions of the tangential impulse

$e_{np}$ – Poisson's normal coefficient of restitutions

$e_t$ – Tangential coefficient of restitutions

## 5.1 First group

This group does not consider the compression phase. It also does not consider the impulse of moment.

Then the equations are given by

$$\begin{pmatrix} \dot{D}_{NA} \\ \dot{D}_{TA} \end{pmatrix} = \begin{pmatrix} \mathbf{W}_N^T \\ \mathbf{W}_T^T \end{pmatrix} \dot{\mathbf{q}}_A + \begin{pmatrix} \tilde{w}_N \\ \tilde{w}_T \end{pmatrix}, \quad \begin{pmatrix} \dot{D}_{NE} \\ \dot{D}_{TE} \end{pmatrix} = \begin{pmatrix} \mathbf{W}_N^T \\ \mathbf{W}_T^T \end{pmatrix} \dot{\mathbf{q}}_E + \begin{pmatrix} \tilde{w}_N \\ \tilde{w}_T \end{pmatrix},$$

$$[M](\dot{\mathbf{q}}_E - \dot{\mathbf{q}}_A) - \begin{pmatrix} \mathbf{W}_N & \mathbf{W}_T \end{pmatrix} \begin{pmatrix} I_N \\ I_T \end{pmatrix} = \mathbf{0}.$$

## 5.2 Second group

This group does not consider the compression phase (as the first group does) but it considers the impulse of moment.

The equations are given by

$$\begin{pmatrix} \dot{D}_{NA} \\ \dot{D}_{TA} \\ \dot{D}_{\theta A} \end{pmatrix} = \begin{pmatrix} \mathbf{W}_N^T \\ \mathbf{W}_T^T \\ \mathbf{W}_\theta^T \end{pmatrix} \dot{\mathbf{q}}_A + \begin{pmatrix} \tilde{w}_N \\ \tilde{w}_T \\ \tilde{w}_\theta \end{pmatrix},$$

$$\begin{pmatrix} \dot{D}_{NE} \\ \dot{D}_{TE} \\ \dot{D}_{\theta E} \end{pmatrix} = \begin{pmatrix} \mathbf{W}_N^T \\ \mathbf{W}_T^T \\ \mathbf{W}_\theta^T \end{pmatrix} \dot{\mathbf{q}}_E + \begin{pmatrix} \tilde{w}_N \\ \tilde{w}_T \\ \tilde{w}_\theta \end{pmatrix}$$

and

$$[M](\dot{\mathbf{q}}_E - \dot{\mathbf{q}}_A) - \begin{pmatrix} \mathbf{W}_N & \mathbf{W}_T & \mathbf{W}_\theta \end{pmatrix} \begin{pmatrix} I_N \\ I_T \\ I_\theta \end{pmatrix} = \mathbf{0}.$$

## 5.3   Third group

This group considers the compression phase and the expansion phase. But it does not consider the impulse of moment.

The equation, in this case, are given by

$$\begin{pmatrix} \dot{D}_{NA} \\ \dot{D}_{TA} \end{pmatrix} = \begin{pmatrix} \mathbf{W}_N^T \\ \mathbf{W}_T^T \end{pmatrix} \dot{\mathbf{q}}_A + \begin{pmatrix} \tilde{w}_N \\ \tilde{w}_T \end{pmatrix},$$

$$\begin{pmatrix} \dot{D}_{NC} \\ \dot{D}_{TC} \end{pmatrix} = \begin{pmatrix} \mathbf{W}_N^T \\ \mathbf{W}_T^T \end{pmatrix} \dot{\mathbf{q}}_C + \begin{pmatrix} \tilde{w}_N \\ \tilde{w}_T \end{pmatrix},$$

$$\begin{pmatrix} \dot{D}_{NE} \\ \dot{D}_{TE} \end{pmatrix} = \begin{pmatrix} \mathbf{W}_N^T \\ \mathbf{W}_T^T \end{pmatrix} \dot{\mathbf{q}}_E + \begin{pmatrix} \tilde{w}_N \\ \tilde{w}_T \end{pmatrix}$$

and

$$[M](\dot{\mathbf{q}}_C - \dot{\mathbf{q}}_A) - \begin{pmatrix} \mathbf{W}_N & \mathbf{W}_T \end{pmatrix} \begin{pmatrix} I_{NC} \\ I_{TC} \end{pmatrix} = \mathbf{0},$$

$$[M](\dot{\mathbf{q}}_E - \dot{\mathbf{q}}_C) - \begin{pmatrix} \mathbf{W}_N & \mathbf{W}_T \end{pmatrix} \begin{pmatrix} I_{NE} \\ I_{TE} \end{pmatrix} = \mathbf{0}.$$

## 5.4   Fourth group

This group considers the compression phase and the expansion phase. It also considers the impulse of moment.

The equations are given by

$$\begin{pmatrix} \dot{D}_{NA} \\ \dot{D}_{TA} \\ \dot{D}_{\theta A} \end{pmatrix} = \begin{pmatrix} \mathbf{W}_N^T \\ \mathbf{W}_T^T \\ \mathbf{W}_\theta^T \end{pmatrix} \dot{\mathbf{q}}_A + \begin{pmatrix} \tilde{w}_N \\ \tilde{w}_T \\ \tilde{w}_\theta \end{pmatrix},$$

$$\begin{pmatrix} \dot{D}_{NC} \\ \dot{D}_{TC} \\ \dot{D}_{\theta C} \end{pmatrix} = \begin{pmatrix} \mathbf{W}_N^T \\ \mathbf{W}_T^T \\ \mathbf{W}_\theta^T \end{pmatrix} \dot{\mathbf{q}}_C + \begin{pmatrix} \tilde{w}_N \\ \tilde{w}_T \\ \tilde{w}_\theta \end{pmatrix},$$

$$\begin{pmatrix} \dot{D}_{NE} \\ \dot{D}_{TE} \\ \dot{D}_{\theta E} \end{pmatrix} = \begin{pmatrix} \mathbf{W}_N^T \\ \mathbf{W}_T^T \\ \mathbf{W}_\theta^T \end{pmatrix} \dot{\mathbf{q}}_E + \begin{pmatrix} \tilde{w}_N \\ \tilde{w}_T \\ \tilde{w}_\theta \end{pmatrix}$$

and

$$[M](\dot{\mathbf{q}}_C - \dot{\mathbf{q}}_A) - \begin{pmatrix} \mathbf{W}_N & \mathbf{W}_T & \mathbf{W}_\theta \end{pmatrix} \begin{pmatrix} I_{NC} \\ I_{TC} \\ I_{\theta C} \end{pmatrix} = \mathbf{0},$$

$$[M](\dot{\mathbf{q}}_E - \dot{\mathbf{q}}_C) - \left( \begin{array}{ccc} \mathbf{W}_N & \mathbf{W}_T & \mathbf{W}_\theta \end{array} \right) \left( \begin{array}{c} I_{NE} \\ I_{TE} \\ I_{\theta E} \end{array} \right) = \mathbf{0} \, .$$

# 6  Restitution Laws

To solve the problem, that is, to find $\dot{\mathbf{q}}_E$ and $\mathbf{I}$ given $\dot{\mathbf{q}}_A$, $[M]$ and $[W]$, we need the $n$ equations given by the jump conditions from Lagrange's equations plus two equations (or three equations if we consider the impulse of moment). These equations are given by the restitution laws, which we will divide into restitution laws in the normal direction and restitution laws in the tangential direction. These equations model the constitutive behavior of the materials.

## 6.1  Restitution laws in the normal direction

The most-used restitution laws in the normal direction are those given by Newton and Poisson. Each of these laws defines a coefficient of restitutions that will be used in the models.

### 6.1.1  Newton's coefficient of restitutions

Newton's coefficient of restitutions, given by $e_n$, is defined as the ratio between the normal relative velocity post-collision $(\dot{D}_{NE})$ and the normal relative velocity pre-collision $(\dot{D}_{NA})$. Then, $e_n = -\dot{D}_{NE}/\dot{D}_{NA}$.

As we can see, this coefficient of restitutions takes into consideration only the kinematics of the system in collision.

### 6.1.2  Poisson's coefficient of restitutions

Poisson's coefficient of restitutions, $e_{np}$, is given by the ratio between the normal impulse in the expansion phase $(I_{NE})$ and the normal impulse in the compression phase $(I_{NC})$.

$$e_{np} = \frac{I_{NE}}{I_{NC}} \, .$$

This coefficient of restitutions takes into consideration the dynamics of the system in the virtual process of collision.

## 6.2  Restitution laws in the tangential direction

In the tangential direction, the first restitution law to be considered is the perfect collision; that is, when the tangential impulse is null $(I_T = 0)$. This is the case where we do not consider friction in the collision.

When we consider friction we use restitution's law, but modified. In collision problems it is expressed in terms of impulses and not in terms of forces. The form used is

$$I_T < \mu I_N \quad \text{se} \quad \dot{D}_T = 0 \, ,$$

and

$$I_T = -s\mu I_N \;\; \text{sendo} \;\; s = \frac{\dot{D}_T}{\mid \dot{D}_T \mid} \; , \; \text{se} \; \dot{D}_T \neq 0 \,.$$

$\mu$ is the coefficient of friction.

There are other coefficients of friction in the normal and tangential directions. But we are only talking about those given here.

# 7   Some Collision Models

A collision model is formed when we use some equation from one of the fourth groups plus the restitution laws. The models treated here are Newton's model, Kane's model, Brach's model, Glocker–Pfeiffer's model and Wang–Mason's model.

In Table 1 we show a general view of these models.

Table 1   General view of the collision models.

| Model | Groups of equations | Law of normal rest. | Law of tangential rest. |
|---|---|---|---|
| Newton | first | Newton | $I_T = 0$ |
| Kane | first | Newton | Coulomb |
| Brach | second | Newton | *ratio* between impulses |
| Glocker–Pfeiffer | third | Poisson | Coulomb |
| Wang–Mason (using Newton) | second | Newton | Coulomb |
| Wang–Mason (using Poisson) | third | Poisson | Coulomb |
| Modelo C-S | fourth | Poisson | Coulomb |

# 8   A New Collision Model: the C-S Model

A new collision model is proposed: the C-S model. This model generalizes some of the principal models from the literature. It considers both the compression phase and the expansion phase and also considers the impulse of moment.

The equations used are given in the following

In the compression phase:

$$[M](\dot{\mathbf{q}}_{\mathbf{C}} - \dot{\mathbf{q}}_{\mathbf{A}}) = \left( \begin{array}{ccc} \mathbf{W}_N & \mathbf{W}_T & \mathbf{W}_\theta \end{array} \right) \begin{pmatrix} I_{NC} \\ I_{TC} \\ I_{\theta C} \end{pmatrix} . \tag{6}$$

$$\begin{pmatrix} \dot{D}_{NC} \\ \dot{D}_{TC} \\ \dot{D}_{\theta C} \end{pmatrix} = \begin{pmatrix} \mathbf{W}_N^T \\ \mathbf{W}_T^T \\ \mathbf{W}_\theta^T \end{pmatrix} (\dot{\mathbf{q}}_C - \dot{\mathbf{q}}_A) + \begin{pmatrix} \dot{D}_{NA} \\ \dot{D}_{TA} \\ \dot{D}_{\theta A} \end{pmatrix}. \tag{7}$$

For restitutions in the tangential direction we use restitution's law in the form

$$\begin{cases} \mid I_{TC} \mid < \mu I_{NC} \Rightarrow \dot{D}_{TC} = 0 \\ I_{TC} = \mu I_{NC} \Rightarrow \dot{D}_{TC} \leq 0 \\ I_{TC} = -\mu I_{NC} \Rightarrow \dot{D}_{TC} \geq 0 \end{cases}.$$

We use the coefficient of moment in the compression phase given by

$$e_{mC} I_{\theta C} = -(1 + e_{mC}) \overline{J} \dot{D}_{\theta C}, \qquad \overline{J} = \frac{J_1 J_2}{J_1 + J_2}.$$

In the expansion phase:

$$[M](\dot{\mathbf{q}}_{\mathbf{E}} - \dot{\mathbf{q}}_{\mathbf{C}}) = \begin{pmatrix} \mathbf{W}_N & \mathbf{W}_T & \mathbf{W}_\theta \end{pmatrix} \begin{pmatrix} I_{NE} \\ I_{TE} \\ I_{\theta E} \end{pmatrix}. \tag{8}$$

$$\begin{pmatrix} \dot{D}_{NE} \\ \dot{D}_{TE} \\ \dot{D}_{\theta E} \end{pmatrix} = \begin{pmatrix} \mathbf{W}_N^T \\ \mathbf{W}_T^T \\ \mathbf{W}_\theta^T \end{pmatrix} (\dot{\mathbf{q}}_E - \dot{\mathbf{q}}_C) + \begin{pmatrix} \dot{D}_{NC} \\ \dot{D}_{TC} \\ \dot{D}_{\theta C} \end{pmatrix}. \tag{9}$$

We use the coefficient of moment in the expansion phase given by

$$e_{mE} I_{\theta E} = -(1 + e_{mE}) \overline{J} \dot{D}_{\theta E}, \qquad \overline{J} = \frac{J_1 J_2}{J_1 + J_2}.$$

For the tangential restitutions we consider

$$I_{TS} = \frac{1}{2} \left[ \mu \nu I_{NE} \operatorname{sign}(I_{TC}) + e_n e_t I_{TC} \right], \qquad 0 \leq \nu \leq 1, \qquad 0 \leq e_t \leq 1.$$

If $I_{TC} \geq 0$ , $I_{TS} \geq 0 \Rightarrow -\mu I_{NE} + 2 I_{TS} \leq I_{TE} \leq \mu I_{NE}$

$$\begin{cases} -\mu I_{NE} + 2 I_{TS} < I_{TE} < \mu INE \Rightarrow \dot{D}_{TE} = 0, \\ I_{TE} = +\mu I_{NE} \Rightarrow \dot{D}_{TE} \leq 0, \\ I_{TE} = -\mu I_{NE} + 2 I_{TS} \Rightarrow \dot{D}_{TE} \geq 0. \end{cases}$$

If $I_{TC} \leq 0$ , $I_{TS} \leq 0 \Rightarrow -\mu I_{NE} \leq I_{TE} \leq \mu I_{NE} + 2 I_{TS}$

$$\begin{cases} -\mu I_{NE} < I_{TE} < \mu I_{NE} + 2 I_{TS} \Rightarrow \dot{D}_{TE} = 0, \\ I_{TE} = +\mu I_{NE} + 2 I_{TS} \Rightarrow \dot{D}_{TE} \leq 0, \\ I_{TE} = -\mu I_{NE} \Rightarrow \dot{D}_{TE} \geq 0. \end{cases}$$

For the normal restitutions we use Poisson's coefficient given by

$$e_{np} = \frac{I_{NE}}{I_{NC}}.$$

$I_{NC}$ is the normal impulse at the end of the compression phase, and $I_{NE}$ is the normal impulse at the end of the expansion phase.

### 8.1.1  Particular cases

As we have said, the C-S model generalizes some of the models from the literature. We will describe briefly three of these models and show what should be done, in the C-S model, to particularize the respective model.

### 8.1.2  Newton's model

This is the simplest model. It considers the coefficient of restitutions given by Newton:

$$e_n = -\frac{\dot{D}_{NE}}{\dot{D}_{NA}}.$$

It does not consider friction (there will be no tangential restitutions), and it does not consider the impulse of moment.

If in the C-S model we consider $e_{mC} = e_{mE} = -1$, $e_t = \nu = 0$ and $\mu = 0$ then we will obtain Newton's model.

It is important to note that when we do not consider friction, Newton's coefficient of restitutions and Poisson's coefficient of restitutions are equivalent.

### 8.1.3  Wang–Mason's model (considering Poisson's coefficient of restitutions)

Wang–Mason use Routh's [16] technique.

This model uses the coefficient of restitutions in the normal direction given by Poisson:

$$e_{np} = \frac{I_{NE}}{I_{NC}}.$$

It considers friction given by restitution's law (modified, as we have presented), and it does not consider the impulse of moment.

If in the C-S model we consider $e_{mC} = e_{mE} = -1$ and $e_t = \nu = 0$ then we will obtain Wang–Mason's model.

### 8.1.4  Glocker–Pfeiffer's model

This model uses the coefficient of restitutions in the normal direction given by Poisson. It considers friction and also the possible reversible portions of the tangential impulse. It does not consider the impulse of moment. If in the C-S model we consider $e_{mC} = e_{mE} = -1$ then we will obtain Glocker–Pfeiffer's model.

# 9 New Results using the C-S Model

Using the C-S model we can observe some behaviors that could not be described using other models. As an example, we consider the problem of a ball colliding with two barriers as shown in Fig. 4.

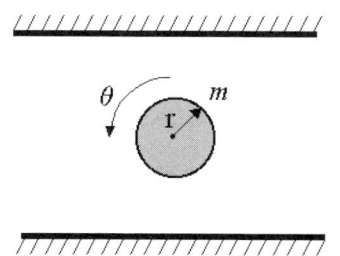

Figure 4 Collision of a ball with two barriers.

We consider the following values for the parameters and initial conditions: mass of the ball $= 1$ kg, $\theta_0 = 0$, $\dot{\theta} = 0$, $x_0 = 0$, $\dot{x}_0 = 1$, $y_0 = 0.9$, $\dot{y}_0 = -1$, $e_{np} = 1$, distance between the barriers $= 1.01$, $\mu = 1$ and $r = 0.1$.

If we consider $e_{mC} = -1$, $e_{mE} = -1$, $e_t = 0$ and $\nu = 0$ we obtain the same prediction obtained from Wang–Mason's model or Glocker–Pfeiffer's model. It is the behavior of a ball used, for example, in a table tennis game. We show the trajectory of the center of the mass in Fig. 5.

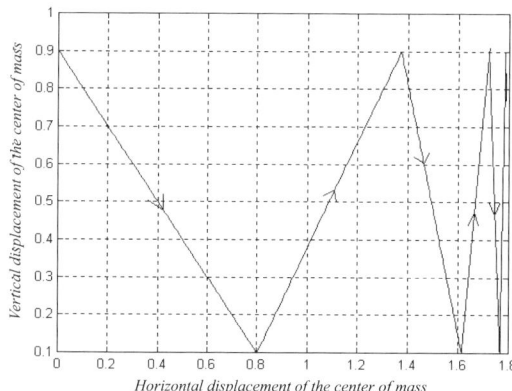

Figure 5 The C-S model using $e_{mC} = e_{mE} = -1$, $e_t = 0$ and $\nu = 0$.

If we consider $e_{mC} = -1$, $e_{mE} = -1$, $e_t = 1$ and $\nu = 1$ we obtain the same prediction obtained from Glocker–Pfeiffer's model in the case of superball-like behavior. It is the behavior of a ball made of steel and not hollow. We show the trajectory of the center of the mass in Fig. 6.

If we consider $e_{mC} = -1$, $e_{mE} = -0.5$, $e_t = 1$ and $\nu = 1$ we obtain a new behavior that could not be observed if we had used other models. We show the trajectory of the center of mass in Fig. 7.

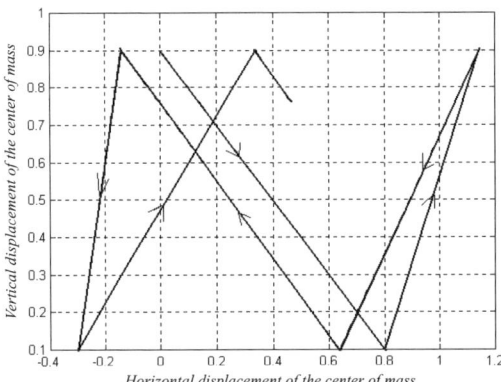

Figure 6  The C-S model using $e_{mC} = e_{mE} = -1$, $e_t = 1$ and $\nu = 1$.

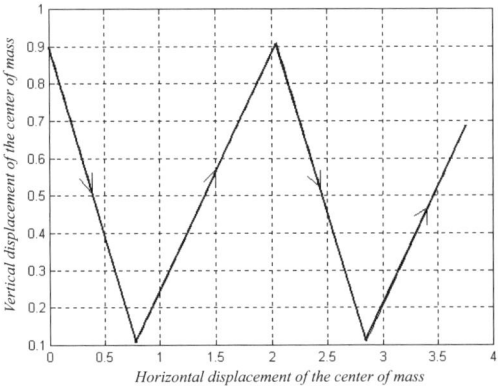

Figure 7  The C-S model using $e_{mC} = e_{mE} = -0.5$, $e_t = 1$ and $\nu = 1$.

# 10   Conclusions

We studied rigid body collisions, considering that these collisions are instantaneous. After making a systematic study of some rigid body collision models we could formulate a new model: the C-S model. Using this model we could show some comparisons between models, and we could present some behaviors that could not be obtained using other models. We showed simulations and animations using the C-S model in a way that would make us understand what was happening. Other discussions about comparisons between models are in [17]–[19].

# References

1. Painlevé, P. (1905) Sur les Lois de Frottement de Glissement. *C. R. Acad. Sci. Paris*, **121**, 112–115; **141**, 401–405.

2. Klein, F. (1910) Zu Painlevés Kritik des Coulombschenreibungsgestze. *Zeitsch Math. Phys.*, **58**, 186–191.

3. Kane, T.R. (1984) A Dynamic Puzzle. *Stanford Mechanics Alumni Club Newsletter*, 6.

4. Keller, J.B. (1986) Impact with Friction. *ASME Journal of Applied Mechanics*, **53**, 1–4.

5. Pfeiffer, F. and Glocker, C. (1996) *Multibody Dynamics with Unilateral Contacts*. John Wiley & Sons, New York.

6. Brach, R.M. (1991) *Mechanical Impact Dynamics – Rigid Bodies Collisions*. John Wiley & Sons, New York, p. 260.

7. Brogliato, B. (1996) Nonsmooth Impact Mechanics: Models, Dynamics and Control. In *Lectures Notes in Control and Information Sciences*, Springer, Berlin.

8. Monteiro-Marques, M.D.P. (1993) *Differential Inclusions in Nonsmooth Mechanical Problems. Shocks and Dry Friction*. Birkhauser, Boston.

9. Stronge, W.J. (1990) Rigid Body Collisions with Friction. *Proceedings of the Royal Society of London A*, **431**, 169–181.

10. Smith, C.E. and Liu, P.P. (1992) Coefficients of Restitutions. *Journal of Applied Mechanics*, **59**, 963.

11. Mason, M.T. and Wang, Y. (1992) Two-Dimensional Rigid-Body Collisions with Friction. *ASME Journal of Applied Mechanics*, **59**, 635–642.

12. Durand, S. (1996) Dynamique des Systèmes á Liaisons Unilatérales avec Frottement Sec. Ph.D. Thesis, L'École Nationale des Ponts et Chaussées, Paris, France (in French).

13. Chatterjee, A. (1997) Rigid Body Collisions: Some General Considerations, New Collision Laws, and Some Experimental Data. Ph.D. Thesis, Cornell University.

14. Stoianovici, D. and Hurmuzlu, Y. (1996) A Critical Study of the Applicability of Rigid-Body Collision Theory. *Journal of Applied Mechanics*, **63**, 307–316.

15. Cholet, C. (1997) Chocs de Solides Rigides. Ph.D. Thesis, Université Paris VI, Paris, France.

16. Routh, E.J. (1877) *An Elementary Treatise on the Dynamics of a System of Rigid Bodies*. (3rd edition), Mac Millan and C°, London.

17. Cataldo, E. (1999) Simulation and Modeling of Plan Rigid Bodies Collisions. Ph.D. Thesis, Pontifícia Universidade Católica do Rio de Janeiro (PUC-Rio), RJ, Brazil, p. 295 (in Portuguese).

18. Cataldo, E. and Sampaio, R. (1999) Comparing Some Models of Collisions between Rigid Bodies. *Proceedings of PACAM VI/DINAME*, **8**, 1301–1304.

19. Cataldo, E. and Sampaio, R. (2001) Comparación entre Modelos de Colisión de Cuerpos. *Revista Internacional de Métodos Numéricos y Diseño en Ingenieria*, **17** (2), 149–168, Espanha.

# PART II

# Optimal Round-Trip Earth–Mars Trajectories for Robotic Flight and Manned Flight

A. Miele, T. Wang and S. Mancuso

*Rice University, Houston, Texas, USA*

This paper deals with LEO–LMO–LEO trajectories, with LEO denoting a low Earth orbit and LMO denoting a low Mars orbit, within the frame of the restricted four-body problem, the four bodies being Sun, Earth, Mars, spacecraft. To assess the interplay between flight time, characteristic velocity, and mass ratio, we study optimal trajectories under a variety of boundary conditions: (T1) minimum energy trajectory with free total time and free stay time in LMO; (T2) compromise trajectory, namely, minimum energy trajectory with free total time and 30-day stay time in LMO; (T3) fast transfer trajectory with free total time and 30-day stay time in LMO; (T4) fast transfer trajectory with 440-day total time and 30-day stay time in LMO. These trajectories are ordered in the sense of decreasing values of the flight time, increasing values of the characteristic velocity, and increasing values of the mass ratio. In particular, trajectories T1 and T2 belong to Class C1 (the phase angle travels of Earth and spacecraft differ by 360 deg), while trajectories T3 and T4 belong to Class C2 (the phase angle travels of Earth and spacecraft are the same).

For robotic missions, trajectory T1 is the best. For manned missions, a substantial shortening of the flight time is needed, but this translates into stiff penalties in characteristic velocity and mass ratio. Comparing the extreme trajectories T1 and T4, we see that, while the total time has been reduced from 970 days to 440 days, the characteristic velocity has been increased from 11.30 km/s to 20.79 km/s, and the mass ratio has been increased from 20 to 304.

The above mass ratios refer to the interplanetary portion of an Earth–Mars–Earth voyage; they do not include the contributions due to the terminal portions of the voyage, which are flown in a planetary atmosphere. When the planetary flight contributions are included, one obtains overall mass ratio of order 1000 for a minimum energy trajectory and 10,000 for a fast transfer trajectory.

At this time, the best that can be done is to continue the exploration of Mars via robotic spacecraft. The exploration of Mars via manned spacecraft must be subordinated to advances yet to be achieved in the areas of spacecraft structural factors, engine specific impulses, and life support systems.

# 1    Introduction

Various studies on trajectory feasibility and optimization relevant to Mars missions have appeared in recent years (Refs. [1–10]), the major objective being to contain the characteristic velocity and the mission time.

This paper continues the studies of Refs. [6–10] by the same authors; it considers trajectories leading from a low Earth orbit (LEO) to a low Mars orbit (LMO) and back after a stay in LMO. Four types of optimal trajectories are studied: (T1) minimum energy trajectory with free mission time and free stay time in LMO; (T2) compromise trajectory, namely, minimum energy trajectory with free mission time and 30-day stay time in LMO; (T3) fast transfer trajectory with free mission time and 30-day stay time in LMO; (T4) fast transfer trajectory with 440-day mission time and 30-day stay time in LMO.

The assumed physical model is the restricted four-body model: the spacecraft is considered subject to the gravitational fields of Earth, Mars, and Sun along the entire trajectory; this is done in order to achieve increased accuracy with respect to the method of patched conics (Refs. [11, 12]). The optimal trajectory problem is formulated as a mathematical programming problem, which is solved via the sequential gradient-restoration algorithm (Refs. [13–16]).

The trajectories investigated can be grouped into two classes depending on whether the round-trip phase angle travels of Earth and spacecraft differ by 360 deg (Class C1) or are the same (Class C2). Trajectories T1 and T2 are slow transfer trajectories of Class C1; trajectories T3 and T4 are fast transfer trajectories of Class C2.

For the slow transfer trajectories of Class C1, the velocity impulses are applied only at LEO and LMO. For the fast transfer trajectories of Class C2, the average angular velocity of the spacecraft with respect to Sun must be increased; this requires the application of a supplementary midcourse velocity impulse in either the outgoing trip (trajectory T4) or return trip (trajectory T3). *Vis-à-vis* Class C1, Class C2 leads to shorter flight times, higher values of the characteristic velocity (sum of the velocity impulses), and even higher values of the mass ratio (ratio of departing mass to returning mass).

# 2    System Equations

Let LEO denote a low Earth orbit, and let LMO denote a low Mars orbit. The missions considered in this paper involve the spacecraft transfer from LEO to LMO for the outgoing trip and from LMO to LEO for the return trip. To study these missions, we employ the restricted four-body model (Sun, Earth, Mars, spacecraft) and the following assumptions:

(A1) the trajectories of Earth, Mars, and spacecraft belong to the same plane;

(A2) the orbits of Earth and Mars around the Sun are circular;

(A3) LEO and LMO are circles centered in their respective planets; circularization of the spacecraft motion is assumed prior to departure and after arrival;

(A4) for a round trip executed via a slow transfer trajectory of Class C1, four velocity impulses are assumed at the departure from LEO, arrival to LMO, departure from LMO, and arrival to LEO;

(A5) for a round trip executed via a fast transfer trajectory of Class C2, an additional midcourse impulse might be needed in either the outgoing trip or return trip.

We employ three coordinate systems nonrotating with respect to one another: a Sun-centered system, an Earth-centered system, and a Mars-centered system. Depending on need, Cartesian coordinates or polar coordinates are used. In Cartesian coordinates, the pair $(x,y)$ identifies the position vector, while the pair $(u,w)$ identifies the velocity vector. In polar coordinates, the pair $(r,\theta)$ identifies the position vector, while the pair $(V,\eta)$ identifies the velocity vector.

Subscripted notation is used: the inertial motions of Earth, Mars, and spacecraft with respect to the Sun S are characterized by the single subscripts E, M, P; the relative motions of the spacecraft with respect to the planets Earth and Mars are characterized by the double subscripts PE and PM. For details of the transformation between systems and conversion between coordinates, see Ref. [17].

In this paper, the inertial motion of the spacecraft is formulated in the Sun-centered system and Cartesian coordinates; the boundary conditions are formulated in the planet-centered systems and polar coordinates.

## 2.1   Motion with respect to Sun

In the Sun-centered system and using Cartesian coordinates, the inertial motion of the spacecraft is described by the following differential equations:

$$\dot{x}_P = u_P, \tag{1a}$$

$$\dot{y}_P = w_P, \tag{1b}$$

$$\dot{u}_P = -\frac{\mu_S}{r_P^3} x_P - \frac{\mu_E}{r_{PE}^3}(x_P - x_E) - \frac{\mu_M}{r_{PM}^3}(x_P - x_M), \tag{1c}$$

$$\dot{w}_P = -\frac{\mu_S}{r_P^3} y_P - \frac{\mu_E}{r_{PE}^3}(y_P - y_E) - \frac{\mu_M}{r_{PM}^3}(y_P - y_M), \tag{1d}$$

with

$$r_P = \sqrt{x_P^2 + y_P^2}, \tag{1e}$$

$$r_{PE} = \sqrt{(x_P - x_E)^2 + (y_P - y_E)^2}, \tag{1f}$$

$$r_{PM} = \sqrt{(x_P - x_M)^2 + (y_P - y_M)^2}\,, \tag{1g}$$

where $\mu_S$, $\mu_E$, and $\mu_M$ are the gravitational constants of the Sun, Earth, and Mars, and where $r_P$, $r_{PE}$, and $r_{PM}$ denote the radial distances of the spacecraft from the Sun, Earth, and Mars. In light of Assumption (A2), the Earth coordinates $x_E$, $y_E$ and Mars coordinates $x_M$, $y_M$ are known trigonometric functions of the time; see Ref. [6].

The functions $x_P(t)$, $y_P(t)$, $u_P(t)$, $w_P(t)$ obtained by integrating Eqs. (1) can be converted into the corresponding functions in polar coordinates via the relations

$$r_P = \sqrt{x_P^2 + y_P^2}\,, \tag{2a}$$

$$\theta_P = \tan^{-1}\left(\frac{y_P}{x_P}\right)\,, \tag{2b}$$

$$V_P = \sqrt{u_P^2 + w_P^2}\,, \tag{2c}$$

$$\eta_P = \tan^{-1}\left(\frac{w_P}{u_P}\right)\,. \tag{2d}$$

## 2.2   Motion with respect to planets

In the planet-centered systems and using Cartesian coordinates, the relative motion of the spacecraft can be computed via the relations

$$x_{PE} = x_P - x_E\,, \qquad\qquad x_{PM} = x_P - x_M\,, \tag{3a}$$

$$y_{PE} = y_P - y_E\,, \qquad\qquad y_{PM} = y_P - y_M\,, \tag{3b}$$

$$u_{PE} = u_P - u_E\,, \qquad\qquad u_{PM} = u_P - u_M\,, \tag{3c}$$

$$w_{PE} = w_P - w_E\,, \qquad\qquad w_{PM} = w_P - w_M\,. \tag{3d}$$

In light of assumption (A2), the Earth velocity components $u_E$, $w_E$ and Mars velocity components $u_M$, $w_M$ are known trigonometric functions of the time. In polar coordinates, the relations corresponding to (3) are

$$r_{PE} = \sqrt{x_{PE}^2 + y_{PE}^2}\,, \qquad\qquad r_{PM} = \sqrt{x_{PM}^2 + y_{PM}^2}\,, \tag{4a}$$

$$\theta_{PE} = \tan^{-1}\left(\frac{y_{PE}}{x_{PE}}\right)\,, \qquad\qquad \theta_{PM} = \tan^{-1}\left(\frac{y_{PM}}{x_{PM}}\right)\,, \tag{4b}$$

$$V_{PE} = \sqrt{u_{PE}^2 + w_{PE}^2}\,, \qquad\qquad V_{PM} = \sqrt{u_{PM}^2 + w_{PM}^2}\,, \tag{4c}$$

$$\eta_{PE} = \tan^{-1}\left(\frac{w_{PE}}{u_{PE}}\right)\,, \qquad\qquad \eta_{PM} = \tan^{-1}\left(\frac{w_{PM}}{u_{PM}}\right)\,. \tag{4d}$$

# 3   Boundary Conditions

## 3.1   Outgoing trip

In the Earth-centered system and using polar coordinates, the spacecraft conditions at the departure from LEO (time $t = t_0$) are given by

$$r_{\text{PE}}(t_0) = r_{\text{LEO}}, \tag{5a}$$

$$V_{\text{PE}}(t_0) = V_{\text{LEO}} + \Delta V_{\text{LEO}}(t_0), \qquad V_{\text{LEO}} = \sqrt{\frac{\mu_{\text{E}}}{r_{\text{LEO}}}}, \tag{5b}$$

$$\eta_{\text{PE}}(t_0) = \theta_{\text{PE}}(t_0) + \frac{\pi}{2}, \tag{5c}$$

with $\theta_{\text{PE}}(t_0)$ free. Relative to Earth, $V_{\text{LEO}}$ is the spacecraft velocity in the low Earth orbit prior to application of the tangential, accelerating velocity impulse; $\Delta V_{\text{LEO}}(t_0)$ is the accelerating velocity impulse at LEO; $V_{\text{PE}}(t_0)$ is the spacecraft velocity after application of the accelerating velocity impulse.

If an accelerating velocity impulse is needed at midcourse of the outgoing trip, in the Sun-centered system and using polar coordinates, the spacecraft midcourse condition (time $t = t_1$) is given by

$$V_{\text{P}}(t_{1+}) = V_{\text{P}}(t_{1-}) + \Delta V_{\text{MID}}(t_1), \tag{6}$$

with $r_{\text{P}}(t_1)$, $\theta_{\text{P}}(t_1)$, $\eta_{\text{P}}(t_1)$ continuous. In Eq. (6), $\Delta V_{\text{MID}}(t_1)$ is the accelerating velocity impulse at $t_1$; the subscript $-$ denotes a quantity evaluated before application of the velocity impulse, and the subscript $+$ denotes a quantity evaluated after application of the velocity impulse.

In the Mars-centered system and using polar coordinates, the spacecraft conditions at the arrival to LMO (time $t = t_2$) are given by

$$r_{\text{PM}}(t_2) = r_{\text{LMO}}, \tag{7a}$$

$$V_{\text{PM}}(t_2) = V_{\text{LEO}} + \Delta V_{\text{LEO}}(t_2), \qquad V_{\text{LMO}} = \sqrt{\frac{\mu_{\text{M}}}{r_{\text{LMO}}}}, \tag{7b}$$

$$\eta_{\text{PM}}(t_2) = \theta_{\text{PM}}(t_2) + \frac{\pi}{2}, \tag{7c}$$

with $\theta_{\text{PM}}(t_2)$ free. Relative to Mars, $V_{\text{LMO}}$ is the spacecraft velocity in the low Mars orbit after application of the tangential, decelerating velocity impulse; $\Delta V_{\text{LEO}}(t_2)$ is the decelerating velocity impulse at LMO; $V_{\text{PM}}(t_2)$ is the spacecraft velocity before application of the decelerating velocity impulse.

## 3.2   Return trip

In the Mars-centered system and using polar coordinates, the spacecraft conditions at the departure from LMO (time $t = t_3$) are given by

$$r_{\text{PM}}(t_3) = r_{\text{LMO}}, \tag{8a}$$

$$V_{\text{PM}}(t_3) = V_{\text{LMO}} + \Delta V_{\text{LMO}}(t_3), \qquad V_{\text{LMO}} = \sqrt{\frac{\mu_{\text{M}}}{r_{\text{LMO}}}}, \tag{8b}$$

$$\eta_{\mathrm{PM}}(t_3) = \theta_{\mathrm{PM}}(t_3) + \frac{\pi}{2}\,, \tag{8c}$$

with $\theta_{\mathrm{PM}}(t_3)$ free. Formally, Eqs. (8) can be obtained from Eq. (7) by simply replacing the time $t = t_2$ with the time $t = t_3$. However, there is a difference of interpretation: $V_{\mathrm{LMO}}$ is now the spacecraft velocity in the low Mars orbit before application of the tangential, accelerating velocity impulse; $\Delta V_{\mathrm{LMO}}(t_3)$ is the accelerating velocity impulse at LMO; $V_{\mathrm{PM}}(t_3)$ is the spacecraft velocity after application of the accelerating velocity impulse.

If a decelerating velocity impulse is needed at midcourse of the return trip, in the Sun-centered system and using polar coordinates, the spacecraft midcourse condition (time $t = t_4$) is given by

$$V_{\mathrm{P}}(t_{4+}) = V_{\mathrm{P}}(t_{4-}) - \Delta V_{\mathrm{MID}}(t_4)\,, \tag{9}$$

with $r_{\mathrm{P}}(t_4)$, $\theta_{\mathrm{P}}(t_4)$, $\eta_{\mathrm{P}}(t_4)$ continuous. In Eq. (9), $\Delta V_{\mathrm{MID}}(t_4)$ is the decelerating velocity impulse at $t_4$; the subscript $-$ denotes a quantity evaluated before application of the velocity impulse, and the subscript $+$ denotes a quantity evaluated after application of the velocity impulse.

In the Earth-centered system and using polar coordinates, the spacecraft conditions at the arrival to LEO (time $t = t_5$) are given by

$$r_{\mathrm{PE}}(t_5) = r_{\mathrm{LEO}}\,, \tag{10a}$$

$$V_{\mathrm{PE}}(t_5) = V_{\mathrm{LEO}} + \Delta V_{\mathrm{LEO}}(t_5) \qquad V_{\mathrm{LEO}} = \sqrt{\frac{\mu_{\mathrm{E}}}{r_{\mathrm{LEO}}}}\,, \tag{10b}$$

$$\eta_{\mathrm{PE}}(t_5) = \theta_{\mathrm{PE}}(t_5) + \frac{\pi}{2}\,, \tag{10c}$$

with $\theta_{\mathrm{PE}}(t_5)$ free. Formally, Eqs. (10) can be obtained from Eqs. (5) by simply replacing the time $t = t_0$ with the time $t = t_5$. However, there is a difference of interpretation: $V_{\mathrm{LEO}}$ is now the spacecraft velocity in the low Earth orbit after application of the tangential, decelerating velocity impulse; $\Delta V_{\mathrm{LEO}}(t_5)$ is the decelerating velocity impulse at LEO; $V_{\mathrm{PE}}(t_5)$ is the spacecraft velocity before application of the decelerating velocity impulse.

## 4   Supplementary Relations

The boundary conditions and midcourse conditions (5)–(10) must be completed by a number of supplementary relations.

### 4.1   Time intervals

Let $\tau_{\mathrm{OUT}}$, $\tau_{\mathrm{STAY}}$, $\tau_{\mathrm{RET}}$, $\tau$ denote the transfer time of the outgoing trip, stay time in LMO, transfer time of the return trip, and total time. By definition, the following time intervals can be introduced:

$$\tau_{\mathrm{OUT}} = t_2 - t_0\,, \quad \tau_{\mathrm{STAY}} = t_3 - t_2\,, \quad \tau_{\mathrm{RET}} = t_5 - t_3\,, \quad \tau = t_5 - t_0\,, \tag{11}$$

so that

$$\tau = \tau_{\mathrm{OUT}} + \tau_{\mathrm{STAY}} + \tau_{\mathrm{RET}}\,. \tag{12}$$

If there are midcourse impulses during the outgoing trip (time $t_1$) or return trip (time $t_4$), the following supplementary time intervals can be introduced:

$$\tau_{\mathrm{MID}}(t_1) = t_1 - t_0, \qquad \tau_{\mathrm{MID}}(t_4) = t_4 - t_3. \tag{13}$$

## 4.2 Mars/Earth inertial phase angle difference

Let

$$\Delta\theta(t) = \theta_{\mathrm{M}}(t) - \theta_{\mathrm{E}}(t) \tag{14}$$

denote the phase angle difference between Mars and Earth at any time. Use of (14) at the end times $t_2$ and $t_3$ of the stay in LMO in conjunction with Assumption (A2) yields the relation

$$\Delta\theta(t_3) = \Delta\theta(t_2) + (\omega_{\mathrm{M}} - \omega_{\mathrm{E}})\tau_{\mathrm{STAY}}, \tag{15}$$

where $\omega_{\mathrm{M}}$ and $\omega_{\mathrm{E}}$ are the angular velocities of Mars and Earth around the Sun. The relation (15) connects the stay time in LMO and the Mars/Earth phase angle differences at the arrival to LMO and departure from LMO.

# 5 Optimal Trajectory Problems

For a round-trip Mars mission, four trajectories are studied: (T1) minimum energy trajectory with free mission time and free stay time on LMO; (T2) compromise trajectory, namely, minimum energy trajectory with free mission time and 30-day stay time on LMO; (T3) fast transfer trajectory with free mission time and 30-day stay time on LMO; (T4) fast transfer trajectory with 440-day mission time and 30-day stay time on LMO.

The first two trajectories are slow transfer trajectories of Class C1, meaning that the Earth and spacecraft angular travels differ by 360 deg. These trajectories can be realized via four velocity impulses at LEO and LMO, whose nature is accelerating on departure and braking on arrival.

The last two trajectories are fast transfer trajectories of Class C2, meaning that the Earth and spacecraft angular travels are the same. These trajectories can be realized via five velocity impulses, thereby requiring one more velocity impulse than the trajectories of Class C1. The extra velocity impulse occurs at midcourse of the return trip for trajectory T3 and at midcourse of the outgoing trip for trajectory T4; its nature is accelerating for trajectory T4 and braking for trajectory T3.

## 5.1 Approach

The optimization problems of this paper can be solved in either optimal control format (Refs. [13–15]) or mathematical programming format (Ref. [16]). Because there are no controls in the differential system (1), while there are parameters in the boundary conditions and midcourse conditions (5)–(10), it is natural to prefer the mathematical programming format. In this view, the main function of the differential system (1) is the generation of the gradients of the final conditions with respect to the parameters appearing in the initial conditions and midcourse conditions.

## 5.2 Performance index

The most general performance index governing the optimization of trajectories T1–T4 is the total characteristic velocity,

$$\Delta V = \Delta V_{\mathrm{LEO}}(t_0) + \Delta V_{\mathrm{MID}}(t_1) + \Delta V_{\mathrm{LMO}}(t_2) + \Delta V_{\mathrm{LMO}}(t_3) + \Delta V_{\mathrm{MID}}(t_4) + \Delta V_{\mathrm{LEO}}(t_5),\tag{16}$$

which is the sum of the magnitudes of the velocity impulses.

## 5.3 Essential constraints

The optimization problems can be simplified to a great degree if the computation is organized in such a way that the differential system (1), departure conditions (5) and (8), and midcourse conditions (6) and (9) are satisfied at each step of the computational process. Under this scenario, the essential constraints include the six arrival conditions (7) and (10) plus the two supplementary conditions (12) and (15). Hence, the total number of essential constraints is $q = 8$.

## 5.4 Parameters

In the most general case, there are $n = 16$ parameters in the problem, more precisely:
basic velocity impulses

$$\Delta V_{\mathrm{LEO}}(t_0), \quad \Delta V_{\mathrm{LMO}}(t_2), \quad \Delta V_{\mathrm{LMO}}(t_3), \quad \Delta V_{\mathrm{LEO}}(t_5),\tag{17a}$$

corresponding time intervals

$$\tau_{\mathrm{OUT}} = t_2 - t_0, \quad \tau_{\mathrm{STAY}} = t_3 - t_2, \quad \tau_{\mathrm{RET}} = t_5 - t_3, \quad \tau = t_5 - t_0,\tag{17b}$$

Mars/Earth phase angle differences at departure from LEO and LMO

$$\Delta\theta(t_0) = \theta_{\mathrm{M}}(t_0) - \theta_{\mathrm{E}}(t_0), \quad \Delta\theta(t_3) = \theta_{\mathrm{M}}(t_3) - \theta_{\mathrm{E}}(t_3),\tag{17c}$$

spacecraft/planet relative phase angles at departure from LEO and LMO

$$\theta_{\mathrm{PE}}(t_0), \quad \theta_{\mathrm{PM}}(t_3),\tag{17d}$$

midcourse velocity impulses and corresponding time intervals

$$\Delta V_{\mathrm{MID}}(t_1), \qquad \tau_{\mathrm{MID}}(t_1) = t_1 - t_0,\tag{17e}$$

$$\Delta V_{\mathrm{MID}}(t_4), \qquad \tau_{\mathrm{MID}}(t_4) = t_4 - t_3.\tag{17f}$$

## 5.5 Degrees of freedom

In the most general case, there are $n = 16$ parameters, $q = 8$ essential constraints, and hence $n - q = 8$ degrees of freedom. These degrees of freedom must be saturated in such a way that the total characteristic velocity (16) is minimized.

## 5.6 Particular trajectories

The number of degrees of freedom can be reduced if additional conditions are imposed on the trajectory.

(T1) Minimum Energy Trajectory.
This is the trajectory of Class C1 computed for free mission time and free stay time in LMO, while excluding the midcourse velocity impulses in both the outgoing trip and return trip. If we set

$$\tau = \text{free}, \qquad \tau_{\text{STAY}} = \text{free}, \tag{18a}$$

$$\Delta V_{\text{MID}}(t_1) = 0, \qquad \tau_{\text{MID}}(t_1) = 0, \tag{18b}$$

$$\Delta V_{\text{MID}}(t_4) = 0, \qquad \tau_{\text{MID}}(t_4) = 0, \tag{18c}$$

we see that the number of free parameters reduces to $n = 12$; hence, the number of degrees of freedom reduces to $n - q = 4$, to be saturated so that the total characteristic velocity (16) is minimized.

(T2) Compromise Trajectory.
This is the trajectory of Class C1 computed for free mission time and fixed stay time in LMO, while excluding the midcourse velocity impulses in both the outgoing trip and return trip. If we set

$$\tau = \text{free}, \qquad \tau_{\text{STAY}} = 30 \text{ days}, \tag{19a}$$

$$\Delta V_{\text{MID}}(t_1) = 0, \qquad \tau_{\text{MID}}(t_1) = 0, \tag{19b}$$

$$\Delta V_{\text{MID}}(t_4) = 0, \qquad \tau_{\text{MID}}(t_4) = 0, \tag{19c}$$

we see that the number of free parameters reduces to $n = 11$; hence, the number of degrees of freedom reduces to $n - q = 3$, to be saturated so that the total characteristic velocity (16) is minimized.

(T3) Fast Transfer Trajectory.
This is the trajectory of Class C2 computed for free mission time and fixed stay time in LMO, excluding a midcourse velocity impulse in the outgoing trip, but including a midcourse velocity impulse in the return trip. If we set

$$\tau = \text{free}, \qquad \tau_{\text{STAY}} = 30 \text{ days}, \tag{20a}$$

$$\Delta V_{\text{MID}}(t_1) = 0, \qquad \tau_{\text{MID}}(t_1) = 0, \tag{20b}$$

$$\Delta V_{\text{MID}}(t_4) = \text{free}, \qquad \tau_{\text{MID}}(t_4) = \text{free}, \tag{20c}$$

we see that the number of free parameters reduces to $n = 13$; hence, the number of degrees of freedom reduces to $n - q = 5$, to be saturated so that the total characteristic velocity (16) is minimized.

(T4) Fast Transfer Trajectory.
This is the trajectory of Class C2 computed for fixed mission time and fixed stay time in LMO, including a midcourse velocity impulse in the outgoing trip, but excluding a midcourse velocity impulse in the return trip. If we set

$$\tau = 440 \text{ days}, \qquad \tau_{\text{STAY}} = 30 \text{ days}, \qquad (21a)$$

$$\Delta V_{\text{MID}}(t_1) = \text{free}, \qquad \tau_{\text{MID}}(t_1) = \text{free}, \qquad (21b)$$

$$\Delta V_{\text{MID}}(t_4) = 0, \qquad \tau_{\text{MID}}(t_4) = 0, \qquad (21c)$$

we see that the number of free parameters reduces to $n = 12$; hence, the number of degrees of freedom reduces to $n - q = 4$, to be saturated so that the total characteristic velocity (16) is minimized.

# 6  Mission Data

## 6.1  Planetary data

The gravitational constants for Sun, Earth, Mars are given by

$$\mu_{\text{S}} = 1.327\text{E}11 \text{ km}^3/\text{s}^2, \qquad \mu_{\text{E}} = 3.986\text{E}05 \text{ km}^3/\text{s}^2, \qquad \mu_{\text{M}} = 4.283\text{E}04 \text{ km}^3/\text{s}^2. \qquad (22)$$

Earth and Mars travel around the Sun along orbits with average radii

$$r_{\text{E}} = 1.496\text{E}08 \text{ km}, \qquad r_{\text{M}} = 2.279\text{E}08 \text{ km}. \qquad (23a)$$

The associated average translational velocities and angular velocities are given by

$$V_{\text{E}} = 29.78 \text{ km/s}, \qquad V_{\text{M}} = 24.13 \text{ km/s}, \qquad (23b)$$

$$\omega_{\text{E}} = 0.9855 \text{ deg/day}, \qquad \omega_{\text{M}} = 0.5241 \text{ deg/day}. \qquad (23c)$$

In particular, the angular velocity difference between Earth and Mars is

$$\Delta\omega = \omega_{\text{E}} - \omega_{\text{M}} = 0.4614 \text{ deg/day}. \qquad (24)$$

## 6.2  Orbital data

For the outgoing trip, the spacecraft is to be transferred from a low Earth orbit to a low Mars orbit; for the return trip, the spacecraft is to be transferred from a low Mars orbit to a low Earth orbit. The radii of the terminal orbits are

$$r_{\text{LEO}} = 6841 \text{ km}, \qquad r_{\text{LMO}} = 3597 \text{ km}, \qquad (25a)$$

corresponding to the altitudes

$$h_{\text{LEO}} = 463 \text{ km}, \qquad h_{\text{LEO}} = 200 \text{ km}, \qquad (25b)$$

since the Earth and Mars surface radii are

$$R_{\text{E}} = 6378 \text{ km}, \qquad R_{\text{M}} = 3397 \text{ km}. \qquad (25c)$$

The circular velocities (subscript c) at LEO and LMO are

$$(V_c)_{\text{LEO}} = 7.633 \text{ km/s}, \qquad (V_c)_{\text{LMO}} = 3.451 \text{ km/s}, \qquad (25d)$$

and the corresponding escape velocities (subscript *) are

$$(V_*)_{\text{LEO}} = 10.795 \text{ km/s}, \qquad (V_*)_{\text{LMO}} = 4.880 \text{ km/s}. \qquad (25e)$$

# 7 Numerical Results

For trajectories T1–T4, the numerical results obtained via the sequential gradient-restoration algorithm for mathematical programming problems are shown in Figs. 1–4 and Tables 1–4.

Figures 1–4 show the geometry of the optimal trajectories for the outgoing trip (top) and return trip (bottom).

Table 1 shows the major parameters for a round-trip LEO–LMO–LEO, namely, total time, stay time in LMO, characteristic velocity (sum of the velocity impulses), and mass ratio (ratio of departing mass to returning mass). While the characteristic velocity can be computed without specifying the spacecraft configuration, the computation of the mass ratio requires the specification of the spacecraft structural factor ($\varepsilon = 0.1$) and engine specific impulse ($I_{\text{SP}} = 450$ s).

Table 2 shows the outgoing trip time, stay time in LMO, return trip time, and total time for a round trip LEO–LMO–LEO. Table 3 compares the phase angle travels of spacecraft, Earth, Mars (loops around the Sun). Table 4 shows the distribution of characteristic velocities for the outgoing trip, return trip, and round trip.

(T1) This is the minimum energy trajectory with free total time and free stay time in LMO. There is no midcourse impulse in either the outgoing trip or return trip. The total flight time of 970 days includes 258 days for the outgoing trip, 454 days for the stay in LMO, and 258 days for the return trip. The total characteristic velocity is 11.30 km/s and the corresponding mass ratio is 20.1.

Trajectory T1 belongs to Class C1. For a round trip, *vis-à-vis* Earth, the spacecraft performs one loop less around the Sun; hence, the phase angle travel of the spacecraft is 360 deg less than that of Earth. Geometrically, trajectory T1 is nearly a Hohmann transfer trajectory, which is the elliptical trajectory bitangent to the Earth and Mars orbits around the Sun. Trajectory T1 would be exactly a Hohmann transfer trajectory, should one neglect the gravitational constants of Earth and Mars *vis-à-vis* the gravitational constant of the Sun.

Trajectory T1 is flown in the region of space between the orbits of Earth and Mars. Because of angular momentum conservation, the average angular velocity of the spacecraft is smaller than that of Earth, but larger than that of Mars.

(T2) This is the minimum energy trajectory with free total time and fixed stay time in LMO of 30 days. There is no midcourse velocity impulse in either the outgoing trip or return trip. The total flight time of 841 days includes 412 days for the outgoing trip, 30 days for the stay in LMO, and 399 days for the return trip. The total characteristic velocity is 15.61 km/s and the corresponding mass ratio is 68.8, about 3.4 times that of trajectory T1.

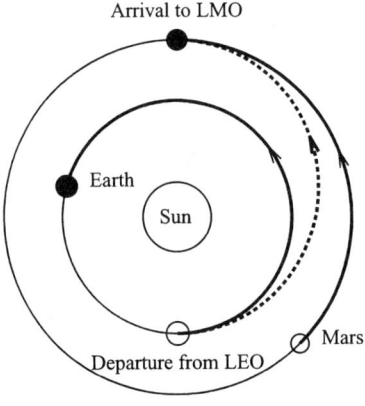

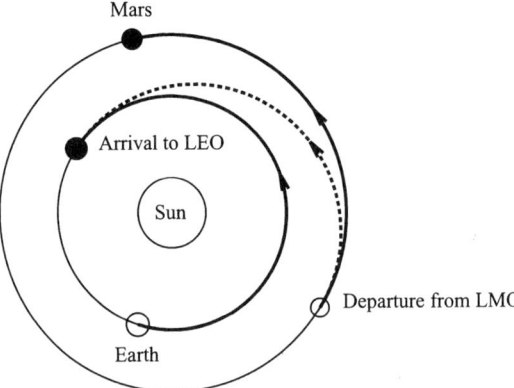

Figure 1 Minimum energy trajectory in interplanetary space, Sun coordinates, total time = 970.0 days, stay time in LMO = 454.3 days, total velocity impulse = 11.30 km/s, no midcourse impulse.

Trajectory T2 also belongs to Class C1. For the round trip, the spacecraft performs one loop less around the Sun; hence, the phase angle travel of the spacecraft is 360 deg less than that of Earth. Geometrically, trajectory T2 deviates considerably from a Hohmann transfer trajectory. Indeed, it overshoots the Mars orbit in both the outgoing trip and return trip, with consequent reduction of the average angular velocity with respect to that of trajectory T1. However, the total flight time is less than that of trajectory T1 due to the considerable shortening of the stay time in LMO.

(T3) This is a fast transfer trajectory with free total time and fixed stay time in LMO of 30 days. A braking midcourse impulse takes place in the return trip at a point preceding the perihelion of the spacecraft trajectory by about 45 deg. The total time of 546 days includes 208 days for the outgoing trip, 30 days for the stay in LMO, and 308 days for the return trip. The total characteristic velocity is 18.52 km/s and the corresponding mass ratio is 150, about 7.5 times that of trajectory T1.

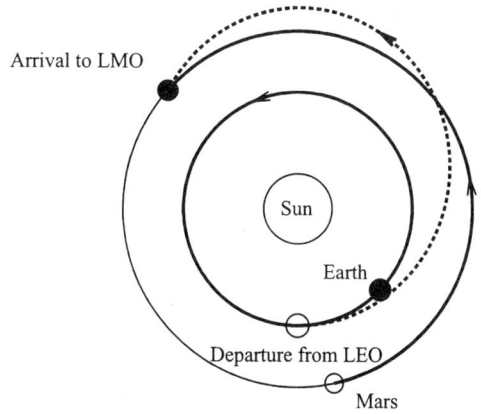

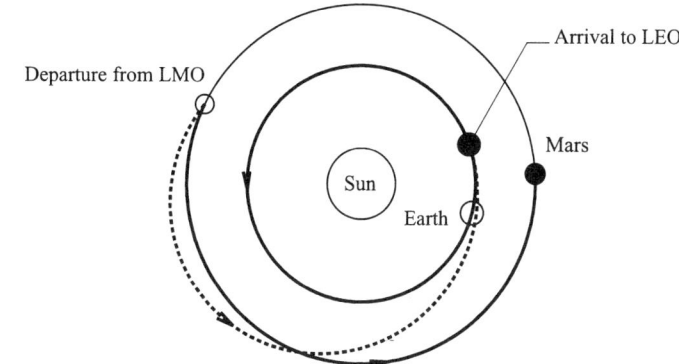

Figure 2   Compromise trajectory in interplanetary space, Sun coordinates, total time = 841.7 days, stay time in LMO = 30.0 days, total velocity impulse = 15.61 km/s, no midcourse impulse.

Trajectory T3 belongs to Class C2. For the round trip, the spacecraft performs the same angular travel as Earth, and hence the same number of loops around the Sun. For the outgoing trip, the average angular velocity of the spacecraft is slower than that of Earth; the opposite occurs in the return trip, where the spacecraft trajectory undershoots the Earth orbit for a considerable portion of the return trip. As a result, for the round trip, the average angular velocities of the spacecraft and Earth are the same.

(T4) This is a fast transfer trajectory with fixed total time of 440 days and fixed stay time in LMO of 30 days. An accelerating velocity impulse takes place in the outgoing trip at a point near the perihelion of the spacecraft trajectory. The total flight time of 440 days includes 261 days for the outgoing trip, 30 days for the stay in LMO, and 149 days for the return trip. The total characteristic velocity is 20.79 km/s and the corresponding mass ratio is 304, about 15 times that of trajectory T1.

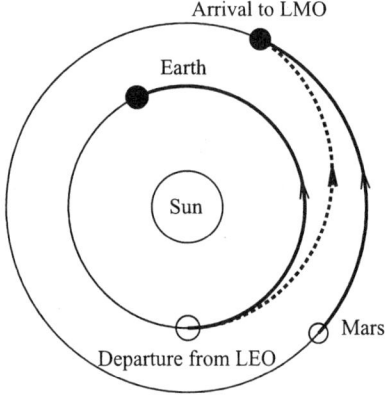

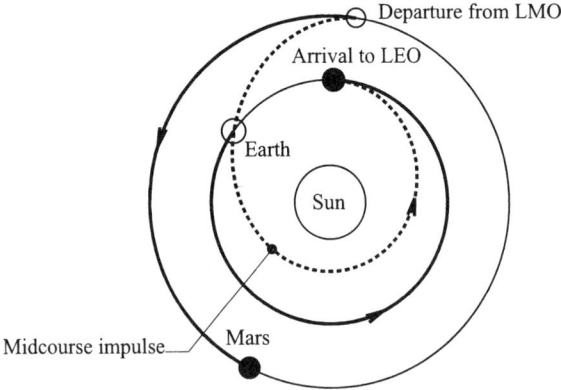

Figure 3 Fast trajectory in interplanetary space, Sun coordinates, total time = 545.9 days, stay time in LMO = 30.0 days, total velocity impulse = 18.52 km/s, return trip midcourse impulse.

Trajectory T4 belongs also to Class C2. For the round trip, the spacecraft performs the same angular travel as Earth, and hence the same number of loops around the Sun. For the return trip, the average angular velocity of the spacecraft is slower than that of Earth; the opposite occurs in the outgoing trip where the spacecraft trajectory undershoots the Earth orbit for a considerable portion of the outgoing trip. As a result, for the round trip, the average angular velocities of the spacecraft and Earth are the same.

# 8    Discussion and Conclusions

In this paper, four optimal round-trip Earth–Mars trajectories were studied under a variety of boundary conditions with the following objectives in mind: to assess the interplay between flight time, characteristic velocity, and mass ratio. There are

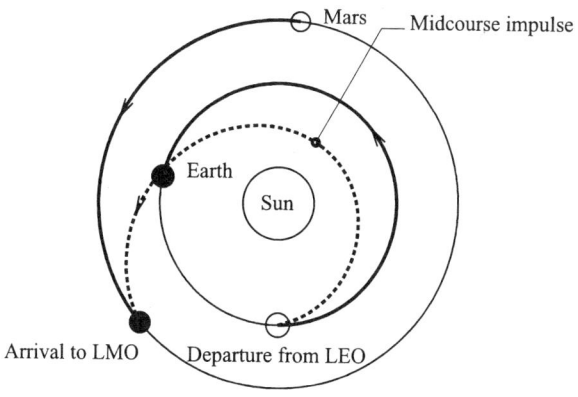

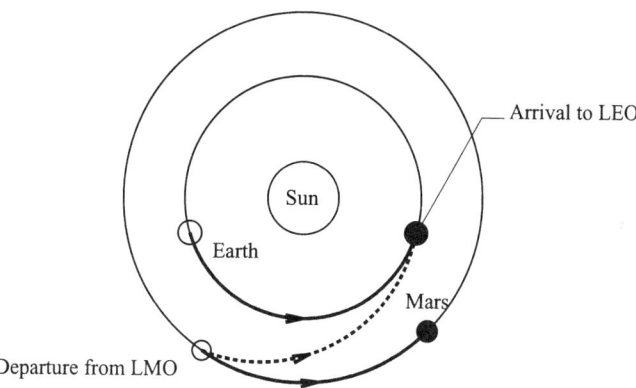

Figure 4  Fast trajectory in interplanetary space, Sun coordinates, total time = 440.0 days, stay time in LMO = 30.0 days, total velocity impulse = 20.79 km/s, outgoing trip midcourse impulse.

significant differences between these trajectories. T1 and T2 are slow transfer trajectories such that the spacecraft phase angle travel is a cycle behind that of Earth and no midcourse impulse is applied. T3 and T4 are fast transfer trajectories such that the spacecraft phase angle travel is the same as that of Earth and a midcourse impulse is applied. As a result, the fast transfer trajectories can reduce considerably the mission time at the expense of a considerable increase in characteristic velocity and an even larger increase in mass ratio.

For robotic missions, T1 is the best trajectory. For manned missions, a substantial shortening of the flight time is needed, but this translates into stiff penalties in the characteristic velocity and mass ratio. Comparing the slowest trajectory T1 and the fastest trajectory T4, we see that the flight time has been reduced by 55 % from 970 days to 440 days, while the mass ratio has been increased by a multiplicative factor of 15 from 20 to 304.

Table 1  Major parameters for round trip LEO–LMO–LEO.

| Quantity | T1 | T2 | T3 | T4 |
|---|---|---|---|---|
| $\tau$ [days] | 970.0 | 841.7 | 545.9 | 440.0 |
| $\tau_{STAY}$ [days] | 454.3 | 30.0 | 30.0 | 30.0 |
| $\Delta V$ [km/s] | 11.30 | 15.61 | 18.52 | 20.79 |
| $m(t_0)/m(t_5)$ | 20.1 | 68.8 | 150.0 | 304.1 |

Table 2  Travel/stay times [days] for round-trip LEO–LMO–LEO.

| Quantity | T1 | T2 | T3 | T4 |
|---|---|---|---|---|
| $\tau_{OUT}$ | 257.9 | 412.3 | 207.7 | 261.3 |
| $\tau_{STAY}$ | 454.3 | 30.0 | 30.0 | 30.0 |
| $\tau_{RET}$ | 257.8 | 399.3 | 308.2 | 148.7 |
| $\tau$ | 970.0 | 841.7 | 545.9 | 440.0 |

Table 3  Phase angle travel of spacecraft, Earth, and Mars (loops around the Sun) for round trip LEO–LMO–LEO.

| Quantity | T1 | T2 | T3 | T4 |
|---|---|---|---|---|
| $[\theta_P(t_5) - \theta_P(t_0)]/2\pi$ | 1.656 | 1.304 | 1.494 | 1.204 |
| $[\theta_E(t_5) - \theta_E(t_0)]/2\pi$ | 2.656 | 2.304 | 1.494 | 1.204 |
| $[\theta_M(t_5) - \theta_M(t_0)]/2\pi$ | 1.412 | 1.225 | 0.795 | 0.641 |

The above mass ratios refer to the interplanetary portion of the voyage. They do not include the effects of the planetary portions of the voyage (ascent from/descent to the Earth surface and descent to/ascent from the Mars surface). When the planetary components of the mass ratio are considered, one obtains overall mass ratios of order 1000 for Earth–Mars–Earth transfer via a minimum energy trajectory and of order 10,000 for Earth–Mars–Earth transfer via a fast transfer trajectory.

At this time, the best that can be done is to continue the exploration of Mars via robotic spacecraft. We are simply not ready for the exploration of Mars via manned spacecraft. This requires advances yet to be achieved in the areas of spacecraft structural factors, engine specific impulses, and life support systems. Also, for the interplanetary portion of the voyage, consideration should be given to either the use of electrical engines instead of chemical engines or the combined use of chemical engines and electrical engines.

Table 4  Characteristic velocities [km/s] for a round-trip LEO–LMO–LEO.

|              | $\Delta V$          | T1    | T2    | T3    | T4    |
|--------------|---------------------|-------|-------|-------|-------|
| Outgoing trip | $\Delta V_{\rm LEO}$ | 3.55  | 3.79  | 3.62  | 5.07  |
|              | $\Delta V_{\rm MID}$ | 0.00  | 0.00  | 0.00  | 2.68  |
|              | $\Delta V_{\rm LMO}$ | 2.10  | 4.15  | 2.72  | 5.06  |
| Return trip  | $\Delta V_{\rm LMO}$ | 2.10  | 3.92  | 4.62  | 3.80  |
|              | $\Delta V_{\rm MID}$ | 0.00  | 0.00  | 3.48  | 0.00  |
|              | $\Delta V_{\rm LEO}$ | 3.55  | 3.76  | 4.08  | 4.18  |
| Round trip   | $\Delta V_{\rm OUT}$ | 5.65  | 7.93  | 6.34  | 12.81 |
|              | $\Delta V_{\rm RET}$ | 5.65  | 7.68  | 12.17 | 7.97  |
|              | $\Delta V_{\rm INT}$ | 11.30 | 15.61 | 18.52 | 20.79 |

# References

1. Niehoff, J.C. (1988) Pathways to Mars: New Trajectory Opportunities. In *NASA Mars Conference*, D.B. Reiber (ed.), Univelt, San Diego, California, pp. 381–401.

2. Hoffman, S.J., McAdams, J.V. and Niehoff, J.C. (1989) Round-Trip Trajectory Options for Human Exploration of Mars. In *Advances in the Astronautical Sciences*, J. Teles (ed.), Univelt, San Diego, California, vol. 69, pp. 663–679.

3. Walberg, G. (1993) How Shall We Go to Mars? A Review of Mission Scenarios. *Journal of Spacecraft and Rockets*, **30** (2), 129–139.

4. Braun, R.D., Powell, R.W., Engelund, W.C., Gnoffo, P.A., Weilmuenster, K.J. and Mitcheltree, R.A. (1995) Mars Pathfinder Six-Degree-of-Freedom Entry Analysis. *Journal of Spacecraft and Rockets*, **32** (6), 993–1000.

5. Crain, T., Bishop, R.H., Fowler, W. and Rock, K. (2000) Interplanetary Flyby Mission Optimization using a Hybrid Global–Local Search Method. *Journal of Spacecraft and Rockets*, **37** (4), 468–474.

6. Miele, A. and Wang, T. (1999) Optimal Transfers from an Earth Orbit to a Mars Orbit. *Acta Astronautica*, **45** (3), 119–133.

7. Miele, A. and Wang, T. (1999) Optimal Trajectories and Mirror Properties for Round-Trip Mars Missions. *Acta Astronautica*, **45** (11), 655–668.

8. Miele, A. and Wang, T. (1999) Optimal Trajectories and Asymptotic Parallelism Property for Round-Trip Mars Mission. In *Proceedings of the 2nd International Conference on Nonlinear Problems in Aviation and Aerospace*, Daytona Beach, Florida, 1998, S. Sivasundaram (ed.), European Conference Publications, Cambridge, England, vol. 2, pp. 507–539.

9. Miele, A., Wang, T. and Mancuso, S. (2000) Optimal Free-Return Trajectories for Moon Missions and Mars Missions. *Journal of the Astronautical Sciences*, **48** (2–3), 183–206.

10. Miele, A., Wang, T. and Mancuso, S. (2001) Assessment of Launch Vehicle Advances to Enable Human Mars Excursions. *Acta Astronautica*, **49** (11), 563–580.

11. Wercinski, P.F. (1996) Mars Sample Return: A Direct and Minimum-Risk Design. *Journal of Spacecraft and Rockets*, **33** (3), 381–385.

12. Casalino, L., Colasurdo, G. and Pastrone, D. (1998) Optimization Procedure for Preliminary Design of Opposition-Class Mars Missions. *Journal of Guidance, Control, and Dynamics*, **21** (1), 134–140.

13. Miele, A., Pritchard, R.E. and Damoulakis, I.N. (1970) Sequential Gradient-Restoration Algorithm for Optimal Control Problems. *Journal of Optimization Theory and Applications*, **5** (4), 235–282.

14. Miele, A., Wang, T. and Basapur, V.K. (1986) Primal and Dual Formulations of Sequential Gradient-Restoration Algorithms for Trajectory Optimization Problems. *Acta Astronautica*, **13** (8), 491–505.

15. Rishikof, B.H., McCormick, B.R., Pritchard, R.E. and Sponaugle, S.J. (1992) SEGRAM: A Practical and Versatile Tool for Spacecraft Trajectory Optimization. *Acta Astronautica*, **26** (8–10), 599–609.

16. Miele, A., Huang, H.Y. and Heideman, J.C. (1969) Sequential Gradient-Restoration Algorithm for the Minimization of Constrained Functions: Ordinary and Conjugate Gradient Versions. *Journal of Optimization Theory and Applications*, **4** (4), 213–243.

17. Miele, A. (1962) *Flight Mechanics, Vol. 1: Theory of Flight Paths*. Addison-Wesley Publishing Company, Reading, Massachusetts.

# Aircraft Take-Off in Windshear: A Viability Approach

N. Seube,[1] R. Moitie[1] and G. Leitmann[2]

[1] *Ecole Nationale Supérieure des Ingénieurs des Etudes et Techniques d'Armement*
*29806 Brest Cedex, France*
[2] *College of Engineering, University of California*
*Berkeley CA 94720, USA*

This paper is devoted to the analysis of aircraft dynamics during take-off in the presence of windshear. We formulate the take-off problem as a differential game against Nature. Here, the first player is the relative angle of attack of the aircraft (considered as the control variable), and the second player is the disturbance caused by a windshear. We impose state constraints on the state variables of the game, which represents aircraft safety constraints (minimum altitude, given altitude rate). By using viability theory, we address the question of the existence of an open loop control assuring a viable trajectory (i.e., satisfying the state constraints) no matter what the disturbance is, i.e., for all admissible disturbances caused by the windshear. Through numerical simulations of the viability kernel algorithm, we demonstrate the capabilities of this approach for determining safe flight domains of an aircraft during take-off within windshear.

## 1   Introduction

In view of the potentially severe consequences of atmospheric conditions such as windshear during take-off and landing of aircraft, there has been and continues to be considerable interest in designing aircraft guidance schemes (possibly for use with an autopilot) to enhance the chances for survival when encountering windshear.

Some aircraft control schemes, among many others, may be found in the following references and references therein. In [11] and many other related papers, Miele *et al.* consider an optimal control approach whereas Bryson *et al.* in [3] employ an inverse scheme to obtain control laws assuring survival for a range of windshear intensities. In both of these studies, knowledge of the structure of the windshear, that is, a model, is assumed. In [9] and [10], Leitmann *et al.* employ robust control theory to derive guidance schemes for crash avoidance. No windshear model is assumed in [9] and [10]; however, in [9] some of the windshear properties are assumed to have known bounds, whereas in [10] these properties are assumed to be bounded but the bounds need not be known. The aircraft dynamics employed in what follows here is the same as in [3], [9], [10] and [11].

In some of the references mentioned above, thrust and relative angle of attack or its time derivative are taken as control variables. In all, the thrust during take-off is assumed to be set at its maximum value. Both the angle of attack and its derivative are subject to prescribed bounds, the latter to account for unmodelled dynamics. In [9] and [10], the bounds on the angle of attack and its rate are not taken into account in the controller design; however, they are employed in all numerical simulation studies. Here we shall use angle of attack rate as control, and bound both it and the angle of attack itself.

While the earlier work quoted above deals with obtaining control strategies for survival enhancing, here we do not derive such control schemes, but rather address the question of the existence of such controls. In particular, we impose constraints on the state variables of the system to define an acceptable flight domain. We then employ tools from viability theory [1] to obtain numerically the region of initial states, or an approximation thereof, from which viable trajectories emanate. In other words, we obtain the initial states for which there exists at least one control, subject to the given bounds, which generates a trajectory which remains within the prescribed flight domain in the presence of all possible windshear conditions considered. We consider windshear conditions in which only bounding information on windshear properties is assumed.

We express the dynamics of the aircraft, travelling in a vertical plane, by a state equation of an uncertain system

$$y' = f(y, u, d),$$

where $y$ is the state, $u$ is the control, and $d$ is a disturbance. The state is subject to a given constraint

$$y \in K,$$

defining the flight domain, while the control and disturbance are subject to

$$u \in U \quad \text{and} \quad d \in D,$$

expressing the control constraint and the disturbance constraint alluded to above. In particular, we shall take control and disturbance to be measurable and, of course, bounded functions of time $t$.

The paper is organized into sections as follows: Section 2 deals with the dynamics of the aircraft, including the state and control constraints. The three situations to be treated are outlined and, in particular, depending on the situation considered, the disturbances together with the assumed information concerning them are introduced. In Sections 3 and 4, we present the details of the three situations mentioned above together with numerical simulation results. The Appendix presents relevant results from viability theory as well as a discussion of a scheme for obtaining approximations of the "winning" region of initial states, that is, for which there exist controls resulting in viable trajectories. For an earlier discussion of "winning" regions, albeit in a somewhat different context, see [8].

## 2   Aircraft Dynamics and Problem Outline

We consider the aircraft dynamics together with the assumptions employed earlier, e.g., in [9] and [11]:

$$
\begin{cases}
x' &= V\cos(\gamma) + W_x & x \in [\underline{x}, +\infty) \\[2ex]
h' &= V\sin(\gamma) + W_h & h \in [\underline{h}, +\infty) \\[2ex]
V' &= \dfrac{1}{m}(T(V)\cos(\alpha+\delta) - D(\alpha,V) - mg\sin(\gamma) - \\
& \quad - (W_x'\cos(\gamma) + W_h'\sin(\gamma)) & V \in [\underline{V}, \overline{V}] \\[2ex]
\gamma' &= \dfrac{1}{mV}(T(V)\sin(\alpha+\delta) + L(\alpha,V) - mg\cos(\gamma) + \\
& \quad + \dfrac{1}{V}(W_x'\sin(\gamma) - W_h'\cos(\gamma)) & \gamma \in [\underline{\gamma}, \overline{\gamma}] \\[2ex]
\alpha' &= u & \alpha \in [\underline{\alpha}, \overline{\alpha}]
\end{cases}
\tag{1}
$$

The following notation is used:

*State variables:*

| | | |
|---|---|---|
| $x$ | : | Horizontal coordinate of aircraft center of mass (ft) |
| $h$ | : | Vertical coordinate of aircraft center of mass (ft) |
| $V$ | : | Aircraft speed relative to wind-based reference frame (ft/sec) |
| $\gamma$ | : | Relative path inclination (rad) |
| $\alpha$ | : | Relative angle of attack (rad) |

*Windshear disturbances*

| | | |
|---|---|---|
| $W_x$ | : | Horizontal wind component (ft/sec) |
| $W_h$ | : | Vertical wind component (ft/sec) |
| $W_x'$ | : | Time derivative of the horizontal wind (ft/sec$^2$) |
| $W_h'$ | : | Time derivative of the vertical wind (ft/sec$^2$) |

*Drag, lift, and thrust forces*

| | | |
|---|---|---|
| $D(\alpha,V)$ | : | Drag force (lb) |
| $L(\alpha,V)$ | : | Lift force (lb) |
| $T(V)$ | : | Thrust (lb) |

Constant parameters

$m$   :   Aircraft mass (lb sec$^2$/ft)
$g$   :   Gravitational force per unit mass (lb ft/sec$^2$)
$\delta$   :   Thrust inclination (rad)

Here it should be noted that, as in [11], the relative angle of attack is introduced as a state variable and its derivative $\alpha'$ is taken as the control variable $u$. This is done in order to account for the constraint imposed on $\alpha'$. The control $u$ will be therefore supposed to range within

$$U := [-C, +C], \qquad (2)$$

where $C$ is the maximum time derivative of $\alpha$.

We shall say that a viable trajectory is one which is generated by an admissible control (measurable and satisfying the control constraint) and remains in the flight domain for all time, no matter what the admissible disturbances (considered wind-shears) are. The admissible flight domain is defined by the state constraints in (1), and will be denoted by

$$F_D := [\underline{x}, +\infty) \times [\underline{h}, +\infty) \times [\underline{V}, \overline{V}] \times [\underline{\gamma}, \overline{\gamma}] \times [\underline{\alpha}, \overline{\alpha}] \,.$$

The components of wind velocity and its time derivative $W_x$, $W_h$, $W_x'$, and $W_h'$ will be considered as measurable functions of time, playing the role of disturbances. They will be deemed admissible if they satisfy the assumed bounds.

The aerodynamic forces and the thrust are taken as

$$T(V) = A_0 + A_1 V + A_2 V^2 \,, \qquad (3)$$

$$D(V, \alpha) = \frac{1}{2} C_D(\alpha) \rho S V^2 \,, \qquad C_D(\alpha) = B_0 + B_1 \alpha + B_2 \alpha^2 \,, \qquad (4)$$

$$L(V, \alpha) = \frac{1}{2} C_L(\alpha) \rho S V^2 \,, \qquad C_L(\alpha) = \begin{cases} C_0 + C_1 \alpha & \text{if } \alpha \leq \alpha_{**} \,, \\ C_0 + C_1 \alpha + C_2 (\alpha - \alpha_{**})^2 & \text{if } \alpha_{**} \leq \alpha \,, \end{cases}$$
$$(5)$$

where $(A_i, B_i, C_i)_{i=0,1,2}$ and $\alpha_{**}$ depend on the aircraft under consideration, $S$ is a reference surface, and $\rho$ is the air density at the surface (for take-off).

In this paper, we address the problem of determining the subsets of the flight domain $F_D$ which are the sets of initial states for which there exists at least one viable trajectory (and hence at least one control resulting in survival, i.e., no crash). We do not address the determination of such control functions, but only their existence.

For the dynamical system described above, we shall consider the following three problems:

*Problems with altitude rate constraint*

1) For a 3-dimensional case with state variables $V$, $\gamma$, and $\alpha$, we consider a bound on the vertical acceleration $h''$ which leads to a state variable constraint (Section 3.1),

2) For a 4-dimensional case with state variables $V$, $\gamma$, $\alpha$, and $h'$, we impose a constraint assuring positivity of $h'$ (Section 3.3).

*Problem with altitude constraint*

3) For a 4-dimensional case with state variables $h$, $V$, $\gamma$, and $\alpha$, the altitude constraint is imposed directly (Section 4).

In Problems 1, 2, and 3 we do not assume a windshear structure, that is a model of the windshear, but only impose bounds on various windshear parameters.

In all subsequent numerical simulations, we use the data given in [11], corresponding to a Boeing 727 aircraft, with three JT8D-17 turbofan engines:

$$A_0 = 44564.0 \text{ lb}, \quad A_1 = -23.98 \text{ lb ft}^{-1}, \quad A_2 = 0.01442 \text{ lb ft}^{-2}\text{s}^2,$$

$$B_0 = 0.07351, \quad B_1 = -0.08617, \quad B_2 = 1.996,$$

$$C_0 = 0.1667, \quad C_1 = 6.231, \quad C_2 = -21.65,$$

$$C = 3°/\text{s} \, \delta = 2°, \quad \rho = 0.002203 \text{ lb ft}^{-4}\text{s}^2, \quad S = 1560 \text{ ft}^2, \quad mg = 180000 \text{ lb}.$$

The flight domain $F_D$ we will use is defined by

$$\underline{x} = 0, \quad \underline{h} = 50 \text{ ft},$$
$$\underline{V} = 184 \text{ ft/s}, \quad \overline{V} = 350 \text{ ft/s},$$
$$\underline{\gamma} = -10°, \quad \overline{\gamma} = +10°,$$
$$\underline{\alpha} = 0°, \quad \overline{\alpha} = +16°,$$

and the control bound in (2) is

$$C = 3°/\text{s}.$$

# 3   Problems with Altitude Rate Constraint

## 3.1   Viability analysis of a 3-D model

For this problem, we consider that the state variables are $V$, $\gamma$, and $\alpha$, and we shall study the regulation of an altitude rate constraint. The vertical acceleration $h''$ is supposed to range in a given interval to assure keeping the altitude rate "close" to some nominal value.

As pointed out in [9], it follows from equation (1) that $h''(t)$ does not depend on the windshear disturbance; namely,

$$h'' = \frac{T(V)}{m}(\cos(\alpha + \gamma)\sin(\gamma) + \sin(\alpha + \gamma)\cos(\gamma)) - \frac{D(\alpha, V)}{m}\sin(\gamma) +$$

$$+ \frac{L(\alpha, V)}{m}\cos(\gamma) - g := \phi(V, \gamma, \alpha). \tag{6}$$

We now consider the state equations and constraints

$$
\begin{cases}
V' = \dfrac{1}{m}(T(V)\cos(\alpha+\delta) - D(\alpha,V) - mg\sin(\gamma) - \\
\qquad - (W_x'\cos(\gamma) + W_h'\sin(\gamma)) & V \in [\underline{V},\overline{V}] \\[2ex]
\gamma' = \dfrac{1}{mV}(T(V)\sin(\alpha+\delta) + L(\alpha,V) - mg\cos(\gamma) + \\
\qquad + \dfrac{1}{V}(W_x'\sin(\gamma) - W_h'\cos(\gamma)) & \gamma \in [\underline{\gamma},\overline{\gamma}] \\[2ex]
\alpha' = u & \alpha \in [\underline{\alpha},\overline{\alpha}]
\end{cases}
\tag{7}
$$

for which we assume the time derivatives of the windshear components to be bounded with known bounds; that is,

$$
|W_x'(t)| \le \overline{W_x'}, \quad |W_h'(t)| \le \overline{W_h'}, \quad \text{for all} \quad t \ge 0.
\tag{8}
$$

The vertical windshear component is assumed to have a known lower bound; that is,

$$
\underline{W_h} \le W_h(t), \quad \text{for all} \quad t \ge 0.
$$

Here we wish to keep the altitude rate constant, that is, $h'(t) = h'(0) > 0$, namely, we would like a control $u$ such that $h''(t) = 0$ for all $t \ge 0$.

Since (7) is an uncertain system, one cannot find a constrained control $u = \alpha'$ which assures $h'' = \phi(V,\gamma,\alpha) = 0$ in the presence of any admissible disturbance $(W_x',W_h')$. Thus, we consider the relaxed acceleration constraint:

$$
\phi(V,\gamma,\alpha) \in [-\varepsilon,+\varepsilon],
\tag{9}
$$

for a specified $\varepsilon > 0$. Since this no longer assures a non-negative altitude rate, we add another constraint to assure this; namely,

$$
V\sin(\gamma) + \underline{W_h} \ge 0,
\tag{10}
$$

where $\underline{W_h}$ denotes the minimum value of the vertical wind velocity.

Therefore, we suppose that the windshear disturbance time derivatives satisfy (8), and its vertical component is bounded from below. The constraints we consider impose a non-negative climb rate $h'$ whilst bounding the value of $h''$.

In order to investigate the existence of admissible control u which generates flight trajectories with a climb rate which is "close" to a prescribed reference climb rate $h_R'$ (namely, the climb rate at the time control is initiated), we constrain the state $(V,\gamma,\alpha)$ to lie in the following subset of the state space:

$$
D_1 := \Pi_{V,\gamma,\alpha}(F_D) \cap \{(V,\gamma,\alpha) \mid V\sin(\gamma) + \underline{W_h} \ge 0 \text{ and } \phi(V,\gamma,\alpha) \le \varepsilon\},
\tag{11}
$$

where $\Pi_{V,\gamma,\alpha}(F_D)$ denotes the projection of the desired flight domain on the $(V,\gamma,\alpha)$-space.

**Problem 1** — The problem now is to approximate the largest subset of $D_1$ containing states $(V_0,\gamma_0,\alpha_0)$ from which there emanate trajectories generated by admissible controls such that the state $(V(t),\gamma(t),\alpha(t))$ remains within the given flight domain defined by $D_1$ in the presence of all admissible disturbances. If this subset of $D_1$ is not empty, then there exists an admissible control $u$ which assures both "closeness" of $h'$ to $h'_R$ and non-negativity of $h'$ regardless of windshear derivative components satisfying (8). In the Appendix it is shown that the subset of $D_1$ which we are seeking is the so-called leadership kernel of $D_1$, denoted by $\text{Lead}(D_1)$.

## 3.2 Problem 1: numerical simulation results

Here, we take the parameters defining the climb rate constraints to be

$$\underline{W_h} = -35 \text{ ft/s}, \qquad \varepsilon = 0.5 \text{ ft/s}^2.$$

As shown in the Appendix, we employ numerical simulation tools for discriminating kernels to obtain an approximation of $\text{Lead}(D_1)$, the set of initial states from which there emanates at least on viable trajectory for all admissible disturbances (windshear subject to (8)). Thus we are able to check that the bound on control $u$ is "compatible" with the admissible windshear in the sense that $V(t)$, $\gamma(t)$, and $\alpha(t)$ remain within their prescribed bounds for all $t \geq 0$.

Results are to be interpreted in terms of minimal take-off velocity, and minimal relative path inclination required for keeping the velocity and path inclination within acceptable bounds.

Figures 1 and 2 present a two-dimensional representation of the leadership kernel of system (7) under the following disturbances bounds:

- Figure 1: $\overline{W'_x} = 2.61$ ft/s, $\overline{W'_h} = 0.78$ ft/s,

- Figure 2: $\overline{W'_x} = 3.45$ ft/s, $\overline{W'_h} = 1.05$ ft/s.

Figures 1 and 2 contain two plots. The first plot represents the projection of $\text{Lead}(D_1)$ onto the $(V,\alpha)$-space, i.e., the defined by $\Pi_{V,\alpha}(\text{Lead}(D_1))$.

The second plot shows the projection of $\text{Lead}(D_1)$ onto the $(V,\alpha)$-space, i.e., the defined by $\Pi_{V,\gamma}(\text{Lead}(D_1))$.

In Figure 2 we observe that at maximum relative aircraft speed ($\overline{V} = 350$ ft/sec), the leardership kernel is reduced to $\alpha = 6.2°$ and $\gamma \in [9.5°, 10°]$. For the situation of lower windshear rate bounds, one can see from Figure 1 that $\alpha = 6.2°$ and $\gamma \in [7.8°, 10°]$.

It appears that the resulting leadership kernel is very small. This is due to the fact that the state constraints we imposed on the system are very restrictive in that we require the climb rate to be "near" a given value ($h'' \approx 0$) and $h' \geq 0$. However, we know that there exist viable trajectories, in terms of minimal aircraft altitude, that do not satisfy the constraint (10) (see [9] for simulation results of such trajectories). In the next section, we shall relax these constraints in order to assure a minimum altitude. The price for this relaxation is the addition of a state variable by adding the altitude $h$ to the state vector.

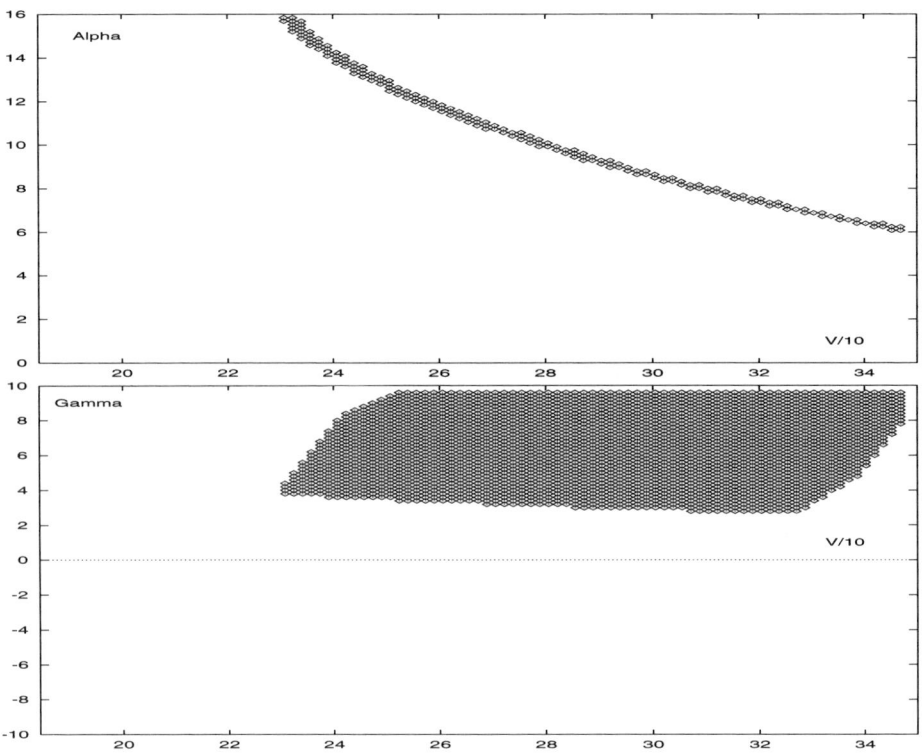

Figure 1   Projection onto the $(V, \alpha)$ and $(V, \gamma)$-spaces of the leadership kernel of $D_1$ for Problem 1 with wind acceleration bounds 2.61 and 0.78 ft/s$^2$.

## 3.3   Viability analysis of a 4-D model

In this section we drop constraint (9) and relax constraint (10) by introducing the climb rate $h'$ as a state variable subject to a constraint. Here then the state variables are $V$, $\gamma$, $h'$, and $\alpha$. The state equations and constraints are now:

$$
\begin{cases}
V' = \dfrac{1}{m}(T(V)\cos(\alpha + \delta) - D(\alpha, V) - mg\sin(\gamma) - \\
\qquad - (W'_x\cos(\gamma) + W'_h\sin(\gamma))) & V \in [\underline{V}, \overline{V}] \\[2ex]
\gamma' = \dfrac{1}{mV}(T(V)\sin(\alpha + \delta) + L(\alpha, V) - mg\cos(\gamma) + \\
\qquad + \dfrac{1}{V}(W'_x\sin(\gamma) - W'_h\cos(\gamma)) & \gamma \in [\underline{\gamma}, \overline{\gamma}] \quad (12) \\[2ex]
\alpha' = u & \alpha \in [\underline{\alpha}, \overline{\alpha}] \\[2ex]
(h')' = \dfrac{T(V)}{m}(\cos(\alpha + \delta)\sin(\gamma) + \sin(\alpha + \delta)\cos(\gamma)) - \\
\qquad - \dfrac{D(\alpha, V)}{m}\sin(\gamma) + \dfrac{L(\alpha, V)}{m}\cos(\gamma) - g\,.
\end{cases}
$$

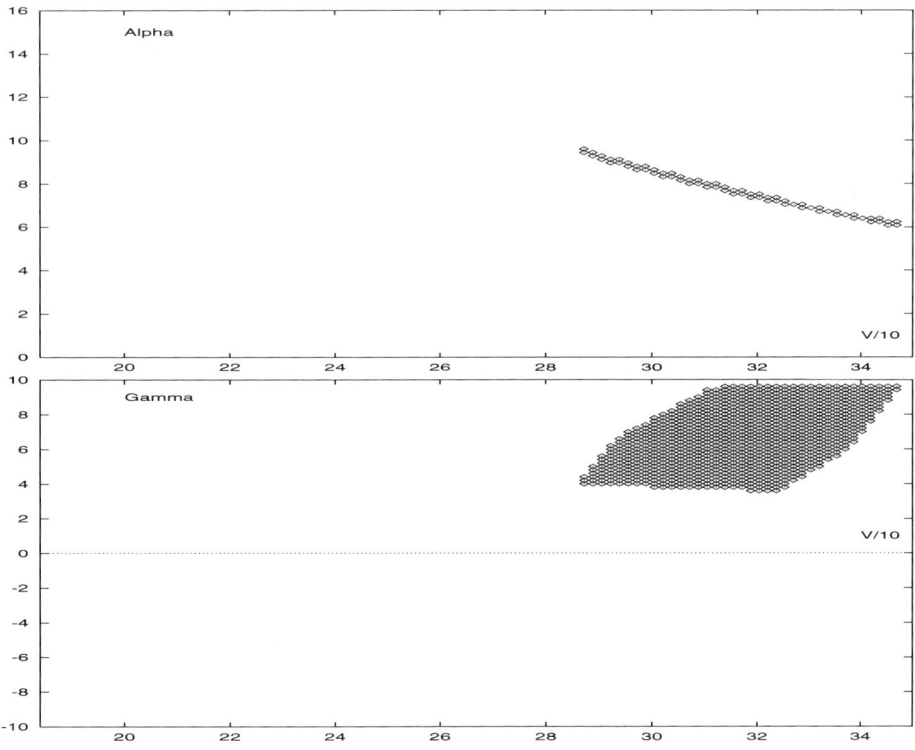

Figure 2  Projection onto the $(V, \alpha)$ and $(V, \gamma)$-spaces of the leadership kernel of $D_1$ for Problem 1 with wind acceleration bounds 5.63 and 1.57 ft/s$^2$.

As in Section 3.1 the wind acceleration components are subject to given bounds (8). Now, we introduce a climb rate constraint

$$h' \in [h'_R - \eta, h'_R + \eta]$$

for given constants $h'_R > 0$ and $\eta < h'_R$, thereby removing the need for constraint (10). Thus, the state is constrained to the set

$$D_2 := \Pi_{V,\gamma,\alpha}(F_D) \times [h'_R - \eta, h'_R + \eta]. \tag{13}$$

**Problem 2** — The problem is now to approximate the largest subset of $D_2$, containing states $(V_0,\gamma_0,\alpha_0,h'_0)$ from which there emanate trajectories of system (12) generated by admissible controls such that the state $V(t)$, $\gamma(t)$, $\alpha(t)$, $h'(t)$ remains within $D_2$ in the presence of all admissible windshear disturbances. If this subset is not empty, then there exists an admissible control $u(\cdot)$ which assures that the climb rate $h'$ remains "near" a prescribed one and, in particular, remains positive. As in Problem 1, we seek the leadership kernel, Lead$(D_2)$.

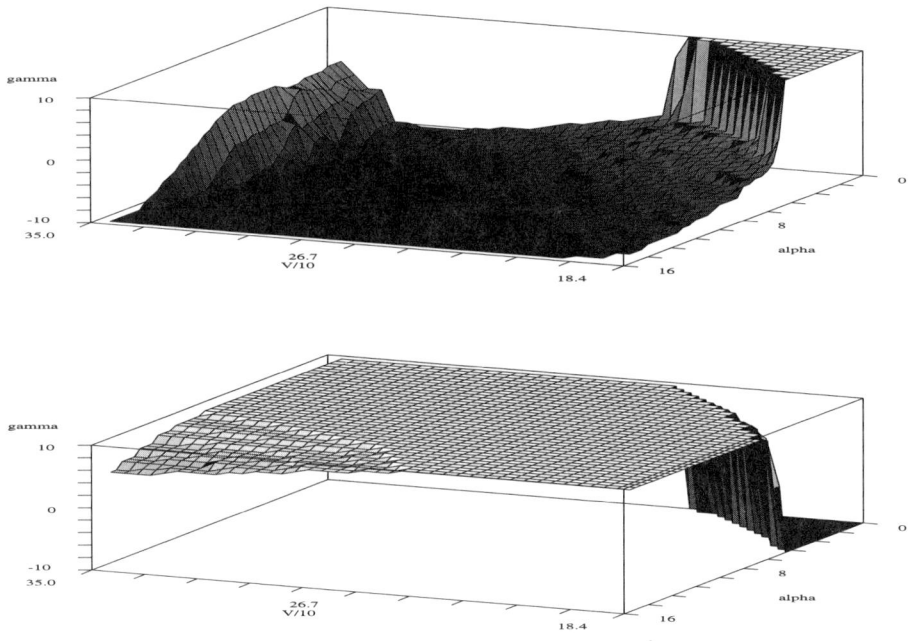

Figure 3 Projection onto the $(V, \gamma, \alpha)$-space of the viability kernel of $D_2$ (leadership kernel in the absence of wind disturbance) for Problem 2. Minimum value of $\gamma$ in the first figure, and maximum value of $\gamma$ in the second figure.

## 3.4   Problem 2: numerical simulation results

For numerical simulations we take $h'_0 = 35$ ft/s and $\eta = 0.5$ ft/s.

First of all, we consider the situation of no windshear disturbance, that is, $W'_x(t) = W'_h(t) = 0$. In that case, the leadership kernel $\mathrm{Lead}(D_2)$ is simply the viability kernel of $D_2$ (the definition of the viability kernel is postponed in the Appendix). Figure 3 shows a projection along the $h'$ axis of the viability kernel of $D_2$, namely, $\Pi_{V,\gamma,\alpha}(\mathrm{Viab}(D_2))$.

Using Figure 3 we can check, for example, whether the initial relative angle of attack $\alpha$ and speed $V$ are too low of too high to allow a viable trajectory.

Next we allow for wind disturbances subject to (8) with $\overline{W'_x} = 2.61$ ft/s and $\overline{W'_x} = 0.78$ ft/s in Figure 4, and with $\overline{W'_x} = 5.63$ ft/s and $\overline{W'_x} = 1.57$ ft/s in Figure 5. These figures show the projections onto the $(V,\gamma,\alpha)$-space of the leadership kernel of $D_2$ for the system (12) and all wind disturbances satisfying (8), for $\gamma$ only the minimum and maximum values are shown. As expected, we see that the leadership kernel decreases in size with increasing wind disturbance. Diminution of the leadership kernel with increasing wind disturbance is primarily due to a decrease in allowable values of $\alpha$. Furthermore, the largest $V$-section of the leadership kernel seems to occur at $V = 285$ ft/s; at that speed the angle of attack is relatively low (about $6°$). Thus, one may consider this relative aircraft speed to be the safest in terms of survival.

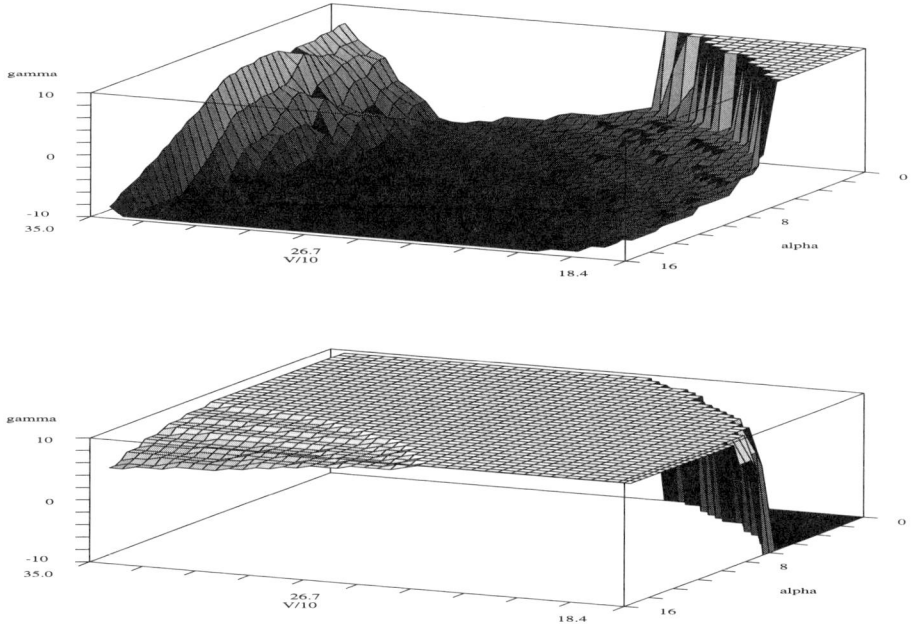

**Figure 4** Projection onto the $(V, \gamma, \alpha)$-space of the leadership kernel of $D_2$ for Problem 2 with wind acceleration bounds 2.61 and 0.78 ft/s$^2$. Minimum value of $\gamma$ in the first figure, and maximum value of $\gamma$ in the second figure.

## 4 Problem with Altitude Constraint

In this section we replace the climb rate constraints (9) and (10) by constraining the altitude directly, that is, by imposing a positive lower bound on $h$. We shall consider a system in which the windshear components and their derivative are subject to known bounds.

In order to impose a lower bound on the altitude, we must consider $h$ as a component of the state vector. Thus, the state variables of the resultant 4-D system are $h$, $V$, $\gamma$, and $\alpha$. The state equations and constraints are

$$
\left\{
\begin{aligned}
h' &= V\sin(\gamma) + W_h && h \in [\underline{h}, +\infty) \\[2mm]
V' &= \frac{1}{m}(T(V)\cos(\alpha+\delta) - D(\alpha, V) - mg\sin(\gamma) - \\
    &\quad - (W'_x\cos(\gamma) + W'_h\sin(\gamma)) && V \in [\underline{V}, \overline{V}] \\[2mm]
\gamma' &= \frac{1}{mV}(T(V)\sin(\alpha+\delta) + L(\alpha, V) - mg\cos(\gamma) + \\
    &\quad + \frac{1}{V}(W'_x\sin(\gamma) - W'_h\cos(\gamma)) && \gamma \in [\underline{\gamma}, \overline{\gamma}] \\[2mm]
\alpha' &= u && \alpha \in [\underline{\alpha}, \overline{\alpha}]
\end{aligned}
\right.
\tag{14}
$$

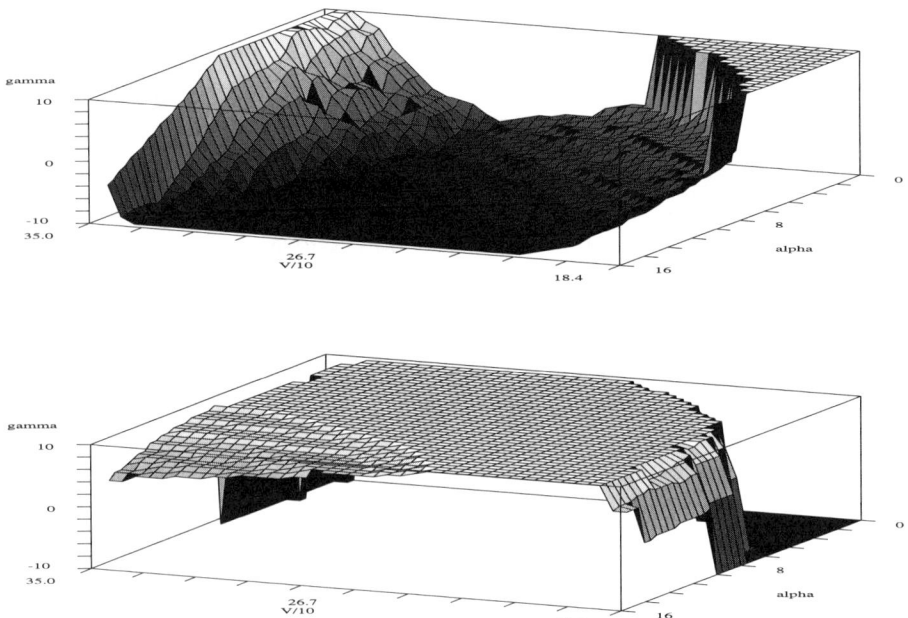

Figure 5  Projection onto the $(V, \gamma, \alpha)$-space of the leadership kernel of $D_2$ for Problem 2 with wind acceleration bounds 5.63 and 1.57 ft/s$^2$. Minimum value of $\gamma$ in the first figure, and maximum value of $\gamma$ in the second figure.

where, of course, $\underline{h}$ is the minimum allowable altitude. Here, we assume that the vertical wind speed component is bounded

$$|W_h'(t)| \leq \overline{W_h'} \tag{15}$$

and the windshear derivative is subject to constraint (8). Thus, windshear disturbance is said to be admissible if $W_h$, $W_h'$, and $W_x'$ are measurable functions of time, and subject to constraints (8) and (15). For the problem considered here, the state constraint set is

$$D_3 := \Pi_{h,V,\gamma,\alpha}(F_D). \tag{16}$$

**Problem 3** — The problem is now to determine the largest subset of $D_3$ containing the states $(h_0, V_0, \gamma_0, \alpha_0)$ from which emanates trajectories generated by admissible controles such that the state $(h(t), V(t), \gamma(t), \alpha(t))$ remains within the flight domain defined by $D_3$ in the presence of all admissible windshear disturbances. As before, we seek the leadership kernel, here Lead $(D_3)$.

## 4.1 Problem 3: numerical simulation results

Figure 6 portrays the projection onto the $(V, \gamma, \alpha)$-space of Lead($D_3$) where, as before, only the minimum and maximum values of $\gamma$ for all $V$ and $\alpha$ are plotted. Bounds on the windshear velocities are:

$$\overline{W_h} = 15, \qquad \overline{W_h'} = 2.61, \qquad \overline{W_x'} = 0.78.$$

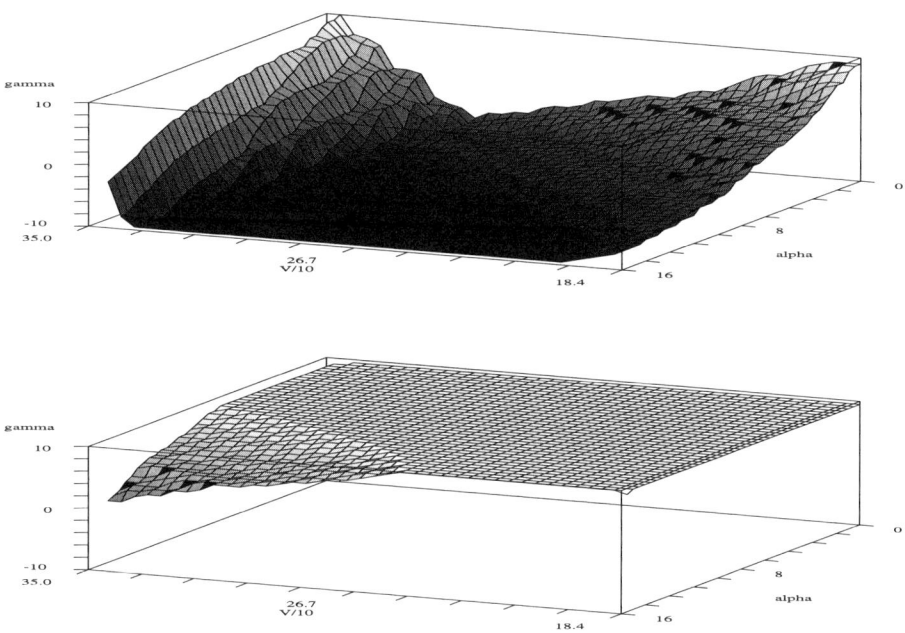

Figure 6 Projection of the leadership kernel for Problem 3 minimal value of $\gamma$ (above), maximal value of $\gamma$ (below).

Again, initial relative speed $V_0 = 285$ ft/s is of interest since the corresponding section of Lead$(D_3)$ appears to be maximal with respect to the range of $\alpha$ and $\gamma$.

# 5  Conclusion

We presented a new methodology for obtaining safe flight domains for an aircraft during take-off in presence of windshear. Our approach checks the consistency among windshear intensity, state constraints, and bounds on the time derivative of the relative angle of attack.

In particular, using viability theory we utilize a numerical procedure to calculate the set of initial flight conditions for which there exist viable trajectories generated by an angle of attack with a constrained time derivative, no matter what the considered windshear intensity are, and without information about the structure of the windshear.

In this sense, our methodology can be used for predicting the safety of aircraft take-off guidance schemes during windshear conditions, in checking that the closed loop system state lies within the "safe" flight domain.

# 6  Appendix: Results from Viability Theory

## 6.1  Control systems

We consider the following control system, for $y \in \Re^n$ and $u \in \Re^m$,

$$y'(t) = f(y(t), u(t)), \quad y(0) = y_0, \quad u(t) \in U, \quad \forall t \geq 0. \tag{17}$$

We make the following assumptions: $f(\cdot, \cdot)$ is affine in $u$ and is supposed to be Lipschitz. $U$ is supposed to be compact and convex.[1] We will denote by $F$ the following set-valued map:

$$F(y) = \bigcup_{u \in U} f(y, u).$$

Let us consider a closed set $K$ of $\Re^n$. We shall say that a solution $y(\cdot)$ to (17) is *viable in K* if $y(t) \in K$ for all time $t \geq 0$, and that $K$ is a viability domain, if from each $y(0)$ in $K$ there originates a viable trajectory of (17).

### 6.1.1  The viability kernel

Let $K$ be a closed subset of $\Re^n$. The viability kernel of $K$ under the dynamics (17) (denoted by $\mathrm{Viab}_F(K)$) is the set of initial conditions $y_0 \in \Re^n$ from which there exist a control $u(\cdot)$ with $u(t) \in U$ generating a trajectory satisfying the constraint $y(t) \in K$, $\forall t \geq 0$.

One of the major results from viability theory is that the viability kernel of $K$ under the dynamics (17) is the largest closed viability domain contained in $K$. Every viable solution in $K$ must remain in $\mathrm{Viab}_F(K)$ for all $t \geq 0$.

In particular, the viability kernel of a set $K$ under (17) is empty if and only if every solution starting in $K$ leaves $K$ in finite time. Every solution originating from $K \setminus \mathrm{Viab}_F(K)$ also leaves $K$ in finite time.

### 6.1.2  Approximation of the viability kernel

In order to approximate the viability kernel, we define a discrete dynamics associated to (17). Under the above assumptions, $F$ is a Lipschitz map, and we denote by $l$, its Lipschitz constant. We assume that $F$ is bounded by $M$, namely, $\forall y \in \Re^n$, $\forall z \in F(y) \|z\| \leq M$.

Let us consider the explicit discretization of $F$:

$$G_\rho(y) := y + \rho F(y) + M l \rho^2 B \tag{18}$$

for some $\rho \geq 0$ ($B$ denotes the unit ball of $\Re^n$). The subset $\mathrm{Viab}_{G_\rho}^N(K)$ of initial states $y_0 \in K$ such that at least one solution of (18) starting at $y_0$ is viable in $K$ for all $n \in [0, N]$ is called the *N-viability kernel of K* and

$$\overrightarrow{\mathrm{Viab}}_{G_\rho}(K) := \bigcap_{N > 0} \mathrm{Viab}_{G_\rho}^N(K)$$

is called the *discrete viability kernel of K under* $G_\rho$.

---

[1] General results about viability theory can be found in [2] under less restrictive assumptions.

From [12], we know that under our assumptions on $F$, the Painlevé–Kuratowski set-limit of $\overrightarrow{\text{Viab}}_{G_\rho}(K)$ when $\rho \to 0$ is the "continuous" viability kernel $\text{Viab}_F(K)$. Therefore, we can approximate continuous viability kernels by discrete ones by using the following algorithm:

$$\begin{cases} K^0 & = & K \\ K^{n+1} & = & \{y \in K^n, \ G_\rho(y) \cap K^n \neq \emptyset\} \end{cases} \tag{19}$$

and the sequence of subset $K^n$ satisfies:

$$\bigcap_{n=0}^{\infty} K^n = \overrightarrow{\text{Viab}}_{G_\rho}(K).$$

In order to implement the above algorithm, we associate to $G_\rho$ a set-valued map defined on a finite set. We consider $Y_h$ a grid of the state space. For a given $h$ we associate to any $y \in \Re^n$, its projection $y_h$ on the grid $Y_h$. For any set $K$, we define its projection onto the grid by:

$$K_h = (K + hB) \cap Y_h.$$

Let us now consider the following set valued-map:

$$G_\rho^h(y_h) := \left(y_h + \rho F(y_h) + (2h + lh\rho + Ml\rho^2)B\right) \cap Y_h.$$

It is already known (see [14]) that the sequence of discrete viability kernels of the finite discretization of $F$ converges, in the sense of Painlevé–Kuratowski, to the viability kernel of $F$:

$$\lim_{\substack{\rho \to 0^+ \\ h/\rho \to 0^+}} \overrightarrow{\text{Viab}}_{G_\rho^h}(K_h) = \text{Viab}_F(K).$$

In practice, we have to choose the time step $\rho$ such that $h/\rho \to 0^+$, which expresses a compatibility condition with the space step discretization h. Approximations of the viability kernel of any closed set $K$ can therefore be computed by the viability kernel algorithm.

## 6.2 Uncertain control systems

We consider the state equation of an uncertain system

$$y'(t) = f(y(t), u(t), d(t)), \quad \forall t \geq 0, \quad u(t) \in U, d(t) \in D, \tag{20}$$

where $u(\cdot)$ is the control and $d(\cdot)$ a measurable disturbance. We suppose that the dynamics $f$ are affine with respect to $u$ and $d$, that $y \to f(y, u, d)$ is $l$-Lipschitz, and that $f(y, u, d)$ is bounded by a constant $M$:

$$\sup_{y \in \Re^n} \sup_{u \in U} \sup_{d \in D} \|f(y, u, d)\| \leq M.$$

The sets $D$ and $U$ are supposed to be compact and convex. We also suppose that system (20) meets the Isaacs' condition:

$$\forall p \in \Re^n, \quad \forall y \in \Re^n, \quad \inf_{u \in U} \sup_{d \in D} \langle f(y, u, d), p \rangle = \sup_{d \in D} \inf_{u \in U} \langle f(y, u, d), p \rangle,$$

### 6.2.1  Discriminating and leadership kernels

Let us define

$$F(y, u) = \bigcup_{d \in D} f(y, u, d).$$

According to [5], we shall say that a closed set $K$ is a *discriminating domain* for (20) if for any $y_0 \in K$ and for any $u \in U$, there exists a solution of $y'(t) \in F(y(t), u)$ originating from $y_0$, which remains in $K$.

We consider the largest closed discriminating domain of $K$, contained in $K$: this subset is called the *discriminating kernel* [2] of $K$ under $f$ and will be denoted by $\text{Disc}_f(K)$. This discriminating kernel can be characterized in terms of trajectories:

> The discriminating kernel of $K$ for the system (20) is the closed set of initial conditions $y(0)$ such that for any time $t$, there exists a function $S_{u|[0,t]} : d(\cdot) \to u(\cdot)$ (which depends only on the values of $d$ over the time interval $[0, t]$), satisfying the following property: for any $d(\cdot)$, every solution of $y'(t) = f(y(t), S_u(d)(t), d(t))$ is viable in $K$.

The discriminating kernel of a set $K$ can also be expressed by an infinite intersection of viability kernel for an auxiliary system (see [4]):

$$\text{Disc}_f(K) = \bigcap_{i=0}^{\infty} (K_i),$$

where the sequence $K_i$ is defined by:

$$K_0 = K, \quad K_{i+1} = \bigcap_{d \in D} \text{Viab}_d(K_i).$$

Here, $\text{Viab}_d(K_i)$ stands for the viability kernel of $K_i$ for the dynamics $y'(t) = f(y(t), u(t), d)$ with a constant $d \in D$.

In order to characterize the existence of a control such that for all admissible disturbances $d(\cdot)$, system (20) admits viable trajectories in a set closed set $K$, we use the concept of leadership kernels:

> The leadership kernel of a closed set $K \in \Re^n$ (that is denoted by $\text{Lead}_f(K)$) is the largest closed set of initial conditions $y(0) \in K$ such that for any $\varepsilon > 0$ and any $T > 0$, there exists a control $u(\cdot)$ such that at any time $t \geq T$ and for any function $S_{d|[0,t]} : u \to d$, the solution of $y'(t) = f(y(t), u(t), S_d(u(t)))$ is quasi-viable in $K$ in the sense that $y(t) \in K \oplus \varepsilon B, \forall t \in [0, T]$.

As stated in [4], when $f$ enjoys the Isaacs' property, the discriminating and leadership kernels for $f$ coincide. As a consequence, we can use the numerical schemes devoted to discriminating kernel approximation for computing leadership kernel approximations.

---

[2]Existence of the discriminating kernel is proved in [4].

### 6.2.2 Discriminating kernel approximation

Consider the discrete set-valued map associated to $F(y, u)$:

$$G_\rho(y, u) := y + \rho F(y, u) + Ml\rho^2 B \qquad (21)$$

for some $\rho \geq 0$. A closed set $D$ is a discrete discriminating domain for the dynamics (21) if

$$\forall y \in D, \quad \forall u \in U, G_\rho(y, u) \cap D \neq \emptyset.$$

Under the above assumption on $f$, we know that, for any $K \in \Re^n$, there exists a largest closed discrete discriminating domain contained in $K$. This set is called the discrete discriminating kernel of $K$ under $G_\rho$, and will be denoted by $\overrightarrow{\mathrm{Disc}}_{G_\rho}(K)$.

Let us consider the following sequence of closed subsets of $K$:

$$\begin{cases} K^0 & = \ K\,, \\ K^{n+1} & = \ \{y \in K^n\,, \ \forall u \in U, \ G_\rho(y, u) \cap K^n \neq \emptyset\}\,. \end{cases} \qquad (22)$$

Then,

$$\bigcap_{n=0}^{\infty} K^n = \overrightarrow{\mathrm{Disc}}_{G_\rho}(K)\,.$$

Again, we consider $Y_h$ a grid of the state space, and $U_h$ a grid of the space $U$ of admissible controls.

Let us consider the following set valued-map:

$$G_\rho^h(y_h, u_h) := \left(y_h + \rho F(y_h, u_h) + (2h + lh\rho + Ml\rho^2)B\right) \cap Y_h\,.$$

The sequence of discrete discriminating kernels of $K_h$ converge to the discriminating kernel of $F$ (see [5] for details):

$$\lim_{\substack{\rho \to 0^+ \\ h/\rho \to 0^+}} \overrightarrow{\mathrm{Disc}}_{G_\rho^h}(K_h) = \mathrm{Disc}_F(K)\,.$$

This result leads to an algorithm that allows one to approximate the discriminating kernel of a set $K$ over a finite number of steps. When the Isaacs' condition is fulfilled (it is the case for the dynamics studied in Sections 3.1, 3.3, 4) this algorithm provide a tool for approximating leadership kernels.

# References

1. Aubin, J.-P. (1991) *Viability Theory*. Birkhauser, Boston.
2. Aubin, J.-P. (1998) *Dynamic Economy Theory*. Springer Verlag, Heidelberg.
3. Bryson, A.E. and Zhao, Y. (1987) Feedback Control for Penetrating a Downburst. *AIAA Paper*, No. 87-2343.
4. Cardaliaguet, P. (1993) Domaines Discriminants en jeux Différentiels. Thèse, Université de Paris-Dauphine.

5. Cardaliaguet, P., Quincampoix, M. and Saint-Pierre, P. (1994) Some Algorithms for Differential Games with Two Players and One Target. *Mathematical Modelling and Numerical Analysis*, **28** (4), 441–461.

6. Chen, Y.H. and Pandey, S. (1989) Robust Control Strategy for Take-off Performance in a Windshear. *Optimal Control Applications & Methods*, **10**, 65–79.

7. Doyen, L. and Seube, N. (1998) Control of Uncertain Systems under Bounded Chattering. *Dynamics and Control*, **8**, 163–176.

8. Getz, W.M. and Leitmann, G. (1979) Qualitative Differential Games with Two Targets. *Journal of Mathematical Analysis and Applications*, **68** (2).

9. Leitmann, G. and Pandey, S. (1991) Aircraft Control for Flight in an Uncertain Environment: Takeoff in Windshear. *Journal of Optimization Theory and Applications*, **70** (1).

10. Leitmann, G., Pandey, S. and Ryan, E. (1993) Adaptive Control of Aircraft in Windshear. *Int. Journ. of Robust and Nonlinear Control*, **3**, 133–153.

11. Miele, A., Wang, T. and Melvin, W.W. (1986) Guidance Strategies for Near-Optimum Take-off Performance in a Windshear. *Journal of Optimization Theory and Applications*, **50** (1).

12. Quincampoix, M. and Saint-Pierre, P. (1995) An Algorithm for Viability Kernels in Holderian Case: Approximation by Discrete Dynamical Systems. *Journal of Mathematical Systems, Estimation and Control*, **5** (1).

13. Quincampoix, M. and Seube, N. (1998) Stabilization of Uncertain Systems Through Piecewise Constant Controls. *Journal of Mathematical Analysis and Applications*, **218**, 240–255.

14. Saint-Pierre, P. (1994) Approximation of the Viability Kernel. *Applied Mathematics & Optimization*, **29**, 187–209.

# Stability of Torsional and Vertical Motion of Suspension Bridges Subject to Stochastic Wind Forces

## N.U. Ahmed

*School of Information Technology and Engineering*
*Department of Mathematics, University of Ottawa, Ottawa, Ontario*
*E-mail: ahmed@site.uottawa.ca*

The objective of this paper is to discuss numerical results based on the recent theoretical work [1,2] on the stability analysis of suspension bridges subject to random aerodynamic forces. This opens up the prospect of accurate analysis and design of suspension bridges which may be subject to frequent stochastic wind or seismic forces.

## 1    Introduction

In this paper we consider the question of stability of suspension bridges subject to random wind forces. We use the model proposed in Ahmed and Harbi [4] which constitutes an extension of the deterministic models of Roseau [9] and Jacover and McKenna [8] to stochastic models. We present interesting simulation results with physical interpretation.

## 2    Suspension Bridge Models

### 2.1    Suspension bridge model A

A simplified model of a suspension bridge is given by a coupled system of partial differential equations of the form:

$$\begin{cases} m_b z_{tt} + \alpha D^4 z - F_0(y - z) = m_b g + f_1, & x \in \Sigma \equiv (0, L), \quad t \geq 0, \\ m_c y_{tt} - \beta D^2 y + F_0(y - z) = m_c g + f_2, & x \in \Sigma \equiv (0, L), \quad t \geq 0, \end{cases} \tag{1}$$

where the first equation describes the vibration of the road bed in the vertical plane, and the second equation describes that of the main cable from which the road bed is suspended by tie cables (stays). Here $m_b$, $m_c$ are the masses per unit length; $\alpha$, $\beta$ are the flexural rigidity of the road bed and coefficient of elasticity of the cable respectively, and $D^k$ denotes the spatial derivative of order $k$. The function $F_0$ represents the restraining force experienced both by the road bed and the suspension cable as transmitted through the tie lines (stays), thereby producing the coupling between the two. The functions $f_1$ and $f_2$ represent external as well as nonconservative forces generally time dependent. Considering displacement around the static equilibrium $\{z_s, y_s\}$, we obtain the following system of equations:

$$\begin{cases} m_b \tilde{z}_{tt} + \alpha D^4 \tilde{z} - F(\tilde{y} - \tilde{z}) = m_b g + f_1 \,, & x \in \Sigma \equiv (0, L) \,, \quad t \geq 0 \,, \\ m_c \tilde{y}_{tt} - \beta D^2 \tilde{y} + F(\tilde{y} - \tilde{z}) = m_c g + f_2 \,, & x \in \Sigma \equiv (0, L) \,, \quad t \geq 0 \,, \end{cases} \tag{2}$$

where $\tilde{z} \equiv z - z_s$, $\tilde{y} \equiv y - y_s$ and the function $F$ is given by

$$F(\zeta) \equiv F_0(\zeta + z_s - y_s) - F_0(z_s - y_s) \,.$$

Note that $F(0) = 0$. Throughout the rest of the paper we assume that the displacements, again denoted by $z$, $y$ instead of $\tilde{z}$, $\tilde{y}$, are measured relative to the static positions. We use equation (2) as the general model. Assuming that the structure (beam) is clamped at both ends the boundary conditions are given by

$$\begin{cases} z(t,0) = z(t,L) = 0, \quad Dz(t,0) = Dz(t,L) = 0, \\ y(t,0) = y(t,L) = 0 \,. \end{cases} \tag{3}$$

In case the beam is hinged at both ends the boundary conditions are

$$\begin{cases} z(t,0) = z(t,L) = 0, \quad D^2 z(t,0) = D^2 z(t,L) = 0, \\ y(t,0) = y(t,L) = 0 \,. \end{cases} \tag{4}$$

Other combinations, such as hinged on one side and clamped on the other, are also used. The initial conditions are given by

$$\begin{cases} z(0,x) = z_1(x), \quad z_t(0,x) = z_2(x), \quad x \in (0, L) \,, \\ y(0,x) = y_1(x), \quad y_t(0,x) = y_2(x), \quad x \in (0, L) \end{cases} \tag{5}$$

where $z_1$, $z_2$, $y_1$, $y_2$ are suitable real valued functions defined on $\Sigma = (0, L)$.

## 2.2  Suspension bridge model B

It is known that wind forces acting on the bridge depend on the velocity of the wind and also the orientation of the deck (i.e., torsional angles and their rates) and hence the state of the system. This, however, requires inclusion of the dynamics of both transverse and torsional motion of the road bed. One can use the approximate model involving beam instead of plate as proposed in [8, 9] provided the deck is narrow or equivalently the deck is sufficiently rigid across. Let $L$ denote the span of the bridge and $2l$ the width of the decks. Then we have the following model:

$$\begin{cases} m z_{tt} + \beta_1 D^4 z - \gamma_1 D^2 z + K F_1(z, \theta) = f_1\,, & x \in \Sigma \equiv (0, L)\,, \quad t \geq 0\,, \\ m \theta_{tt} + \beta_2 D^2 \theta - \gamma_2 D^2 \theta + K l F_2(z, \theta) = f_2\,, & x \in \Sigma \equiv (0, L)\,, \quad t \geq 0\,. \end{cases} \tag{6}$$

where $F_1$ and $F_2$ are given by

$$\begin{cases} F_1(z, \theta) = F_0(z + l \sin \theta) + F_0(z - l \sin \theta)\,, \\ F_2(z, \theta) = \{F_0(z + l \sin \theta) + F_0(z - l \sin \theta)\} \cos \theta\,, \end{cases} \tag{7}$$

with $F_0$ being $F_0(\xi) = \begin{cases} \xi & \text{for } \xi < 0\,, \\ 0 & \text{otherwise}\,. \end{cases}$ The positions of the suspension cables are given by

$$\begin{cases} y_1 = F_0(z + l \sin \theta)\,, \\ y_2 = F_0(z - l \sin \theta)\,. \end{cases} \tag{8}$$

The boundary conditions of $\{z, \theta\}$ are similar to those of $z$ of the model equation (1) as given by the expressions (3) or (4). This model describes both the vertical and the torsional motions of the structure. This is a very simplified model of the complex dynamics of a suspension bridge. In any case for all engineering applications this is a very good approximation. The functions $\{f_1, f_2\}$ represent aerodynamic and viscous forces, including forces exerted by structural elastic resistances (structural damping).

## 3 Abstract Deterministic Model

In this section we consider the system models (A) and (B) with either of the boundary conditions (3) or (4) and reformulate them as ordinary differential equations on appropriate Hilbert spaces. For rigorous mathematical analysis see [1, 2].

**Model (A):** First we consider the model (A) given by equation (2). The abstract formulation has many advantages, as we shall see later in the sequel. First of all we write equation (2) in a slightly more general form as follows:

$$\begin{cases} z_{tt} + a D^4 z = F_1(t, x; z, Dz, D^2 z, z_t; y, Dy, y_t)\,, & x \in \Sigma\,, \quad t \geq 0\,, \\ y_{tt} - b D^2 y = F_2(t, x; z, Dz, D^2 z, z_t; y, Dy, y_t)\,, & x \in \Sigma\,, \quad t \geq 0\,, \end{cases} \tag{9}$$

where

$$a \equiv \frac{\alpha}{m_b} > 0\,, \qquad b \equiv \frac{\beta}{m_c} > 0\,,$$

$$F_1(t, x; z, Dz, D^2 z, z_t; y, Dy, y_t) \equiv \frac{1}{m_b}(F(y - z) + f_1(t, x; z, Dz, D^2 z, z_t; y, Dy, y_t))\,,$$

$$F_2(t, x; z, Dz, D^2 z, z_t; y, Dy, y_t) \equiv \frac{1}{m_c}(-F(y - z) + f_2(t, x; z, Dz, D^2 z, z_t; y, Dy, y_t))\,.$$

We have written equation (1) in a fairly general form to include viscous and aerodynamic damping and other forms of nonconservative forces. In fact later in the paper we have also included structural damping. For the purpose of analysis it is often convenient to formulate the system as an ordinary differential equation on an appropriate Banach space.

Let $\Sigma \subset R^n$ be a bounded open connected set with smooth boundary $\partial\Sigma$, and let $L_2(\Sigma)$ denote the space of equivalence classes of Lebesgue measurable and square integrable functions with the standard norm topology. For $m \in N$, let $H^m(\Sigma) = H^m$ denote the standard Sobolev space with the usual norm topology and $H_0^m(\Sigma) = H_0^m \subset H^m$ denote the completion in the topology of $H^m$ of $C^\infty$ functions on $\Sigma$ with compact supports. From classical results on Sobolev spaces it is well known that the elements of $H_0^m$ are those of $H^m$ which along with their conormal derivatives up to order $m - 1$ vanish on the boundary $\partial\Sigma$. We introduce the Hilbert spaces $H$ and $V$ as follows:

$$
\begin{aligned}
H &\equiv L_2(\Sigma) \times L_2(\Sigma)\,, \\
V &\equiv H_0^2 \times H_0^1\,,
\end{aligned}
\tag{10}
$$

with the first space equipped with the standard scalar product and the second space equipped with the scalar product and norms as follows:

$$
\begin{aligned}
\langle \phi, \psi \rangle_V &\equiv (D^2\phi_1, D^2\psi_1)_{L_2(\Sigma)} + (D\phi_2, D\psi_2)_{L_2(\Sigma)}\,, \\
\|\phi\|_V &\equiv \left( \|D^2\phi_1\|_{L_2(\Sigma)}^2 + |D\phi_2|_{L_2(\Sigma)}^2 \right)^{1/2}\,.
\end{aligned}
\tag{11}
$$

By virtue of Poincaré inequality the following norms

$$
\begin{aligned}
\|\phi\|_{H^m} &\equiv \left( \sum_{|\alpha| \leq m} \|D^\alpha v\|_{L_2(\Sigma)}^2 \right)^{1/2}\,, \\
\|\phi\|_{H_0^m} &\equiv \left( \sum_{|\alpha| = m} \|D^\alpha v\|_{L_2(\Sigma)}^2 \right)^{1/2}
\end{aligned}
\tag{12}
$$

are equivalent, and hence the space $V$ endowed with the scalar product and norm as defined by (11) is also a Hilbert space. We wish to present a semigroup formulation of the system which is equivalent to the formulation given in [1, 2]. We introduce the Hilbert space

$$
E \equiv V \times H
$$

with the scalar product and the associated norms given by

$$
\langle \phi, \psi \rangle_E \equiv \alpha(D^2\phi_1, D^2\psi_1) + \beta(D\phi_2, D\psi_2) + m_b(\phi_3, \psi_3) + m_c(\phi_4, \psi_4)\,,
\tag{13}
$$

and

$$
\|\phi\|_E^2 \equiv \alpha\|D^2\phi_1\|_{L_2(\Sigma)}^2 + \beta\|D\phi_2\|_{L_2(\Sigma)}^2 + m_b\|\phi_3\|_{L_2(\Sigma)}^2 + m_c\|\phi_4\|_{L_2(\Sigma)}^2\,,
\tag{14}
$$

respectively.

Note that $E$ is actually the physical energy space and (14) denotes the sum of elastic potential energies and the kinetic energies. We consider $E$ for the state space and

$$\phi \equiv \{\phi_1, \phi_2, \phi_3, \phi_4\}' \equiv \{z, y, z_t, y_t\}' \tag{15}$$

for the state. Define the operator $\mathcal{A}$ as the realization of the formal differential operator

$$\mathcal{A}(D)\phi \equiv \{\phi_3, \phi_4, -aD^4\phi_1, bD^2\phi_2\}' \tag{16}$$

with either one of the boundary conditions, say (3). Note that the domain of the operator $\mathcal{A}$ is given by

$$D(\mathcal{A}) \equiv (H^4 \cap H_0^2) \times (H^2 \cap H_0^1) \times H_0^2 \times H_0^1 \tag{17}$$

with

$$\mathcal{A}\phi = \mathcal{A}(D)\phi, \qquad \text{for} \ \ \phi \in \mathcal{D}(A). \tag{18}$$

Clearly both the domain and the range of the operator $\mathcal{A}$ are in $E$.

Define the operator

$$\mathcal{F}(t, \phi) \equiv \{0, 0, F_1(t, \phi), F_2(t, \phi)\}, \tag{19}$$

where

$$F_i(t, \phi) \equiv F_i(t; \phi_1, D\phi_1, D^2\phi_1, \phi_3; D\phi_2, \phi_4), \quad i = 1, 2.$$

Occasionally we have to distinguish the coupling term provided by the stay cables which support the roadbed from the main suspension cables. For this we write $\mathcal{F} = \mathcal{F}_c + \mathcal{F}_e$, where

$$\mathcal{F}_c(t, \phi) \equiv \left\{ 0, 0, \frac{1}{m_b} F(\phi_2 - \phi_1), -\frac{1}{m_c} F(\phi_2 - \phi_1) \right\}', \tag{20}$$

$$\mathcal{F}_e(t, \phi) \equiv \left\{ 0, 0, \frac{1}{m_b} f_1, \frac{1}{m_c} f_2 \right\}'. \tag{21}$$

The first component represents the (elastic) coupling operator, and the second component represents all other nonconservative and external forces. Thus we can formulate the system (9) as an abstract first order differential equation on the Hilbert space $E$, given by

$$\begin{cases} \dfrac{d\phi}{dt} = \mathcal{A}\phi + \mathcal{F}(t, \phi) = \mathcal{A}\phi + \mathcal{F}_c(t, \phi) + \mathcal{F}_e(t, \phi), \\[2mm] \phi(0) = \phi_0, \end{cases} \tag{22}$$

where $\phi(0) = \{\phi_1(0), \phi_2(0), \phi_3(0), \phi_4(0)\}' \equiv \phi_0$ denotes the initial state given by (5).

**Model (B):** Here we point out only the significant modifications necessary to construct a similar abstract model for the system (B). For this model we choose $H$ same as in (10) and

$$V \equiv H_0^2 \times H_0^2.$$

The state denoted by $\phi$ is given by the vector

$$\phi \equiv \{\phi_1, \phi_2, \phi_3, \phi_4\}' \equiv \{z, \theta, z_t, \theta_t\}' , \tag{23}$$

and the differential operator $\mathcal{A}(D)$ is given by

$$\mathcal{A}(D)\phi \equiv \{\phi_3, \phi_4, -a_1 D^4\phi_1 + a_2 D^2\phi_1, -b_1 D^2\phi_2 + b_2 D^2\phi_2\}' \tag{24}$$

where $\{a_1, a_2, b_1, b_2\}$ are positive numbers given by

$$a_1 = \frac{\beta_1}{m}, \quad a_2 = \frac{\gamma_1}{m}, \quad b_1 = \frac{\beta_2}{\mathcal{I}}, \quad b_2 = \frac{\gamma_2}{\mathcal{I}} .$$

Again we choose the energy space $E \equiv V \times H$ as the state space and introduce the scalar product

$$\langle\phi, \psi\rangle_E \equiv \beta_1(D^2\phi_1, D^2\psi_1) + \gamma_1(D\phi_1, D\psi_1) + \beta_2(D^2\phi_2, D^2\psi_2) + \gamma_2(D\phi_2, D\psi_2) +$$

$$+ m(\phi_3, \psi_3) + \mathcal{I}(\phi_4, \psi_4) , \tag{25}$$

and denote the associated norm by $\|\cdot\|_E$ which is given by the square root of

$$\|\phi\|_E^2 \equiv \beta_1\|D^2\phi_1\|_{L_2(\Sigma)}^2 + \gamma_1\|D\phi_1\|_{L_2(\Sigma)}^2 + \beta_2\|D^2\phi_2\|_{L_2(\Sigma)}^2 + \gamma_2\|D\phi_2\|_{L_2(\Sigma)}^2 +$$

$$+ m\|\phi_3\|_{L_2(\Sigma)}^2 + \mathcal{I}\|\phi_4\|_{L_2(\Sigma)}^2 . \tag{26}$$

In this case the abstract differential operator $\mathcal{A}$ is given by the realization of $\mathcal{A}(D)$ with the chosen boundary conditions. The domain of the operator is given by

$$D(\mathcal{A}) \equiv (H^4 \cap H_0^2) \times (H^4 \cap H_0^2) \times H_0^2 \times H_0^2 \tag{27}$$

with

$$\mathcal{A}\phi = \mathcal{A}(D)\phi, \qquad \text{for} \quad \phi \in \mathcal{D}(A) . \tag{28}$$

The nonlinear operator $\mathcal{F}$ is again given by the sum of a coupling operator and the operator associated with all the external and nonconservative forces is as follows:

$$\mathcal{F} = \mathcal{F}_c + \mathcal{F}_e ,$$

where

$$\mathcal{F}_c(t, \phi) \equiv \left\{0, 0, -\frac{K}{m}F_1(\phi), -\frac{Kl}{\mathcal{I}}F_2(\phi)\right\}' , \tag{29}$$

$$\mathcal{F}_e(t, \phi) \equiv \left\{0, 0, \frac{1}{m}f_1, \frac{1}{\mathcal{I}}f_2\right\}' , \tag{30}$$

and $F_1$ and $F_2$ are given by the expressions (7). The functions $\{f_i, i = 1, 2\}$ absorb all external and nonconservative forces which include viscous and aerodynamic forces and structural elastic resistances. For example $f_i$ may be given by

$$f_i \equiv f_i(t, x; z, Dz, D^2z, z_t; \theta, D\theta, D^2\theta, \theta_t), \quad t \geq 0, \quad x \in \Sigma .$$

In any case the torsional dynamics coupled with vertical motion given by equation (6) can be represented again by the same abstract form (22) by redefining the operators $\mathcal{A}$, $\mathcal{F}$ and the space $E \equiv V \times H$ accordingly as given above. Thus the abstract model for this system is also given by

$$
\begin{cases}
\dfrac{d\phi}{dt} = \mathcal{A}\phi + \mathcal{F}(t, \phi) = \mathcal{A}\phi + \mathcal{F}_c(t, \phi) + \mathcal{F}_e(t, \phi), \\
\phi(0) = \phi_0,
\end{cases}
\tag{31}
$$

where the operators $\mathcal{A}$ and $\mathcal{F}$ are given by the expressions (27)–(30). Using this result and the variation of constants formula [3] we can rewrite both the evolution equations (22) and (31) as an integral equation on $E$,

$$
\phi(t) = S(t)\phi_0 + \int_0^t S(t - s)\mathcal{F}(s, \phi(s))\, ds, \quad t \geq 0,
\tag{32}
$$

where $S(t)$, $t \geq 0$ is the semigroup generated by the operator $\mathcal{A}$. For details see [1,2]. It was shown in [2], under suitable assumptions on $\mathcal{F}$, that for each initial state $\phi_0 \in E$, this equation has a unique solution in $C(I, E)$ see [2].

# 4 Abstract Stochastic Model

In addition to normal vehicular load, a bridge is also occasionally subject to random loads such as seismic and wind forces. Such random forces can be modeled by adding two noise terms $N_1$ and $N_2$ on the right-hand side of the basic equations (2) and (6) as follows. These random forces are distributed both spatially and temporally, leading to the modified versions

$$
\begin{cases}
z_{tt} + aD^4 z = F_1(t, x; z, Dz, D^2 z, z_t; y, Dy, y_t) + \dfrac{1}{m_b}\tilde{\sigma}_1 N_1, & x \in \Sigma, \quad t \geq 0, \\
y_{tt} - bD^2 y = F_2(t, x; z, Dz, D^2 z, z_t; y, Dy, y_t) + \dfrac{1}{m_c}\tilde{\sigma}_2 N_2, & x \in \Sigma, \quad t \geq 0,
\end{cases}
\tag{33}
$$

for equation (2) and

$$
\begin{cases}
z_{tt} + a_1 D^4 z - a_2 D^2 z = -\dfrac{K}{m} F_1(z, \theta) + \dfrac{1}{m} f_1 + \dfrac{1}{m}\tilde{\sigma}_1 N_1, & x \in \Sigma, \quad t \geq 0, \\
\theta_{tt} + b_1 D^4 \theta - b_2 D^2 \theta = -\dfrac{Kl}{\mathcal{I}} F_2(z, \theta) + \dfrac{1}{\mathcal{I}} f_2 + \dfrac{1}{\mathcal{I}}\tilde{\sigma}_2 N_2, & x \in \Sigma, \quad t \geq 0,
\end{cases}
\tag{34}
$$

for equation (6). The variables $\tilde{\sigma}_1$ and $\tilde{\sigma}_2$ are certain multiplicative operators whose physical significance will be clear later. One natural way of modeling the noise processes is to use the modal expansion as follows:

$$
\begin{cases}
N_1(t, x) \equiv \displaystyle\sum_{i \geq 1} \sqrt{\gamma_i}\, u_i(x) \dfrac{d}{dt} w_i(t), \\
N_2(t, x) \equiv \displaystyle\sum_{i \geq 1} \sqrt{\beta_i}\, v_i(x) \dfrac{d}{dt} \tilde{w}_i(t),
\end{cases}
\tag{35}
$$

where $\{u_i, v_i\} \in H \equiv L_2(\Sigma) \times L_2(\Sigma)$ denote the modes of either of the linear systems. Using the state as defined by (15) or (23), the abstract stochastic counterpart of system (22) or (31) is given by

$$\begin{cases} \dfrac{d\phi}{dt} = \mathcal{A}\phi + \mathcal{F}(t,\phi) + \sigma \dfrac{dW}{dt}\,, \\[2mm] \phi(0) = \phi_0\,, \end{cases} \qquad (36)$$

where $\dfrac{d}{dt}W$ denotes the generalized derivative of $W$ and

$$\sigma \equiv \begin{pmatrix} 0 & 0 \\ 0 & 0 \\ \sigma_1 & 0 \\ 0 & \sigma_2 \end{pmatrix}\,,$$

with $\sigma_1 = \tilde{\sigma}_1/m_b$, $\sigma_2 = \tilde{\sigma}_2/m_c$ and $\sigma_1 = \tilde{\sigma}_1/m$, $\sigma_2 = \tilde{\sigma}_2/\mathcal{I}$ for systems (22) and (31) respectively. One can bypass the generalized derivatives by using the so-called Ito calculus and write equation (36) as a stochastic integral equation

$$\phi(t) = S(t)\phi_0 + \int_0^t S(t-s)\mathcal{F}(s,\phi(s))\,ds + \int_0^t S(t-s)\sigma(s)\,dW(s)\,, \qquad (37)$$

where $S(t)$, $t \geq 0$, is the semigroup as defined in Section 3. The first integral is the standard Lebesgue integral, and the last integral is defined in the Wiener–Ito sense [3]. Symbolically equation (36) is also written as

$$\begin{cases} d\phi(t) = \mathcal{A}\phi(t)dt + \mathcal{F}(t,\phi(t))dt + \sigma(t)dW(t)\,, \\[2mm] \phi(0) = \phi_0\,. \end{cases} \qquad (38)$$

Considering $\sigma$ as a linear operator (or an operator valued function) taking values from $\mathcal{L}(H, E)$, this model is sufficient for additive random perturbation. However, for a suspension bridge multiplicative perturbation is more natural since the amount of wind energy injected into the structure at any time depends not only on the wind velocity but also on the inclination of the deck and hence the state. Thus the operator $\sigma$ is also dependent on the state $\phi$, and consequently a more complete model is given by the following stochastic evolution equation,

$$\begin{cases} d\phi = \mathcal{A}\phi\,dt + \mathcal{F}(t,\phi)dt + \sigma(t,\phi)dW(t)\,, \\[2mm] \phi(0) = \phi_0\,. \end{cases} \qquad (39)$$

From now on we consider these abstract stochastic models without referring to the specific model (A) or (B) unless it is essential.

For further analysis of the system we must now formally introduce the probability space $(\Omega, \mathcal{B}, \mathcal{B}_t \uparrow \subset B, P)$. Let $\{W(t), t \geq 0\}$ be an $H$ valued Wiener process adapted (or measurable with respect) to the sigma algebra $\mathcal{B}_t$ satisfying

$P\{W(0) = 0\} = 1$ and having mean zero and (incremental) covariance $Q \in \mathcal{L}(H)$ so that

$$\mathcal{E}(W(t), h)^2 = t(Qh, h), \qquad \text{for } h \in H, \ t \geq 0,$$

where $\mathcal{E}(Z)$ denotes the mathematical expectation of the random variable $Z$.

For details on stochastic integrals see [3, 2].

Now we are prepared to consider the system (39). There are several concepts of solutions: strong, mild, weak and martingale, see [3]. We are mainly interested in strong and mild solutions.

Let $I \equiv [0, T]$, $T < \infty$, and $M_\infty(I, E)$ denote the vector space of all $\mathcal{B}_t$, $t \geq 0$, measurable $E$ valued random processes $\{\phi(t), t \in I\}$ satisfying

$$ess - \sup_{t \in I} \mathcal{E}\left(\|\phi(t)\|_E^2\right) < \infty.$$

Furnished with the norm topology,

$$\|\phi\|_{M_\infty} \equiv \left(ess - \sup_{t \in I} \mathcal{E}\left(\|\phi(t)\|_E^2\right)\right)^2,$$

where $M_\infty$ is a Banach space. Again under suitable assumptions on the nonlinear operators $\mathcal{F}$ and $\sigma$ one can prove the existence of solutions of these equations in the Banach space $M_\infty(I, E)$, see [2, Theorem 4.2].

# 5 Stability in the Presence of Stochastic Load

In a recent paper [1] we considered stability properties of suspension bridges in the Lyapunov sense which was later extended in several directions to the stochastic case [2].

## 5.1 Additive perturbation

First we consider the case of additive stochastic perturbation as represented by the model (38).

### 5.1.1 Linear model (undamped)

Here the operator $\mathcal{F} = \mathcal{F}_c$. This reduces the problem to the situation where both $f_1$ and $f_2$ are zero and $F(\xi) = K\xi$, $\xi \in R$ (see equation (2)). For the Lyapunov function we choose the total energy functional given by

$$\mathcal{V} \equiv \frac{1}{2}\|\phi\|_E^2 + \frac{K}{2}\|\phi_2 - \phi_1\|_{L_2(\Sigma)}^2, \tag{40}$$

where, by definition of the norm topology of the space $E$ (see equation (14)), the first term represents the sum of kinetic and (elastic) potential energies of the road bed and the suspension (main) cable. The second term represents the elastic energy stored in the stay cables. The assumption of linear coupling holds only for model (A).

Using Ito differential rules one can justify (see [2, Theorem 5.1]) that the expected value of the energy functional $\mathcal{V}$ along the solution of (38) is given by

$$\mathcal{E}\mathcal{V}(\phi(t)) = \mathcal{E}\mathcal{V}(\phi_0) + \frac{1}{2} \int_0^t \mathcal{E}\{\mathrm{tr}(\sigma(s)Q\sigma(s)^*)\} \, ds, \qquad \text{for } t \ge 0, \qquad (41)$$

and $\mathcal{E}\{\mathcal{V}(\phi(t))\} < \infty$, for all finite $t \ge 0$.

### 5.1.2  Nonlinear model (undamped)

We consider both the models (A) and (B). For model (A), $\mathcal{F} = \mathcal{F}_c$ (see equation (20)) with the typical nonlinearity

$$F(\xi) = \begin{cases} K\xi & \text{for } \xi < 0, \\ 0 & \text{otherwise} \end{cases} \qquad (42)$$

where $K$ stands for the stiffness coefficient of the (stay) vertical cables. In fact physically one can admit any real valued nondecreasing function $F$ with its graph lying in the first and third quadrant of the plane. Define

$$\begin{cases} G(\zeta) \equiv \displaystyle\int_0^\zeta F(\xi) \, d\xi, \\[2mm] V_v(\eta) \equiv \displaystyle\int_\Sigma G(\eta(x)) \, dx. \end{cases} \qquad (43)$$

For the Lyapunov function $\mathcal{V}$ we choose again the total energy functional given by

$$\mathcal{V}(\phi) \equiv \frac{1}{2}\|\phi\|_E^2 + V_v(\phi_2 - \phi_1), \qquad (44)$$

where $V_v$ denotes the component of elastic energy contributed by the vertical cables. For model (B), $\mathcal{F} = \mathcal{F}_c$, and it is given by the expression (29) where $F_1$ and $F_2$ are given by equation (7). For the Lyapunov function we choose the energy functional given by

$$\mathcal{V}(\phi) \equiv \frac{1}{2}\|\phi\|_E^2 + V_v(\phi), \qquad (45)$$

where the square of the $E$-norm in this case is given by (26), and the elastic energy of the vertical cables denoted by $V_v$ is given by

$$V_v(\phi) \equiv K \int_0^L \{G_0(\phi_1 + l\sin\phi_2) + G_0(\phi_1 - l\sin\phi_2)\} \, dx, \qquad (46)$$

where

$$G_0(\eta) \equiv -\int_0^\eta F_0(-\xi) \, d\xi.$$

For the nonlinear coupling the following result holds for both the models (A) and (B) (see [2, Theorem 5.2]):

$$\mathcal{E}\mathcal{V}(\phi(t)) = \mathcal{E}\mathcal{V}(\phi_0) + \frac{1}{2}\int_0^t \mathcal{E}\{\operatorname{tr}(\sigma(s)Q\sigma(s)^*)\}\,ds\,, \qquad \text{for } t \geq 0\,, \qquad (47)$$

and $\mathcal{E}\{\mathcal{V}(\phi(t))\} < \infty$, for all finite $t \geq 0$.

**Remark:** Note that even though the two results (41) and (47), for linear and nonlinear coupling respectively, look alike, their $\mathcal{V}$ functions are different, and hence the numerical results are expected to differ.

**Remark:** Another interesting point related to wind actions is that if the operator valued function $\sigma$ vanishes after a period of time say, $[0, t]$, for example a wind gust persists over this period and then ceases completely, both the systems are left behind with the expected energy

$$\mathcal{E}\mathcal{V}(\phi(t)) = \mathcal{E}\mathcal{V}(\phi_0) + \frac{1}{2}\int_0^\tau \mathcal{E}\{\operatorname{tr}(\sigma(s)Q\sigma(s)^*)\}\,ds\,, \qquad \text{for } t \geq \tau\,, \qquad (48)$$

Hence we observe that even if the bridge were calm at time zero implying $\phi_0 = 0$ and hence $\mathcal{E}\mathcal{V}(\phi_0) = 0$, the wind gust delivers into the bridge structure an amount of energy equal to the second component of the expressions (41) and (47). Thus in the absence of sufficient damping this can eventually damage the bridge due to sustained oscillation and the associated mechanical fatigue.

### 5.1.3 *Viscous and Structural Damping*

In reality every bridge structure has some viscous damping of varying degrees depending on the geometry of the structure and the geographical location (atmospheric temperatures and pressures, etc.). Further, the materials used for construction of the decks may possess structural damping. In this case the function $f_1$ appearing in the operator $\mathcal{F}_e$ given by (21) or (30) has the form $f_1 = f_1(\phi_3, D^2\phi_3)$. In the linear case
$$f_1(\phi_3, D^2\phi_3) = -a\phi_3 + bD^2\phi_3\,,$$
where $a > 0$ is the viscous damping coefficient, and $b > 0$ is the coefficient of structural damping.

In any case for both the models (A) and (B) we have (see [2, Theorem 5.3])

$$\mathcal{E}\mathcal{V}(\phi(t)) = \mathcal{E}\mathcal{V}(\phi_0) + \mathcal{E}\int_0^t \{\langle \phi, \mathcal{F}_e(\phi)\rangle + \frac{1}{2}\operatorname{tr}(\sigma(s)Q\sigma(s)^*)\}\,ds\,. \qquad (49)$$

Further if

$$\mathcal{E}\mathcal{V}(\phi_0) + \frac{1}{2}\int_0^\infty \mathcal{E}\{\operatorname{tr}(\sigma(s)Q\sigma(s)^*)\}\,ds \equiv \mathcal{E}_w < \infty\,,$$

then the total kinetic energy of the system decays asymptotically to zero.

**Remark:** The implication of the preceding result is as follows. In the presence of viscous or structural damping, initial energy plus the wind energy delivered to the bridge structure is eventually damped out provided the total energy delivered is finite and has not exceeded a threshold value at which structural failure is likely to occur.

## 5.2   Multiplicative perturbation

It is known that wind forces acting on the bridge depend on the velocity of wind and also the relative orientation of the deck (i.e., torsional angles and their rates) and hence the state of the system. Thus the model (B) is appropriate. For stochastic analysis we consider the wind velocity $V_w$ to be given by the sum of two components $V_w = v + \tilde{v}$, where $v$ denotes the mean wind velocity and $\tilde{v}$ the fluctuation around the mean. This is a stochastic process given by $d\tilde{v} = \kappa dw$, where $w$ is a standard Brownian motion, and $\kappa^2$ denotes the fluctuation energy density. On the basis of small noise (small fluctuation) assumption it was shown in [4] that the aerodynamic forces are given by

$$\begin{cases} f_{1a} \equiv \bar{f}_{1a} + 4l\rho v\kappa |\sin\theta| \sin\theta \dfrac{dw}{ds} \,, \\[2mm] f_{2a} \equiv \bar{f}_{2a} + 2l^2\rho v\kappa |\sin\theta| \sin\theta \dfrac{dw}{ds} \,, \end{cases} \tag{50}$$

where

$$\bar{f}_{1a} \equiv 2\pi l\rho v^2 \left( \theta - \frac{1}{|v|} z_t \right) \,,$$
$$\bar{f}_{2a} \equiv \pi l^2\rho v^2 \left( \theta - \frac{1}{|v|} z_t - \frac{l}{|v|}\theta_t \right) \,, \tag{51}$$

are the approximations of the mean aerodynamic forces [4,9]. Defining

$$c_1 \equiv \frac{2\pi l\rho v^2}{m} \,, \qquad c_2 \equiv \frac{\pi l^2\rho v^2}{\mathcal{I}} \,,$$

and disregarding viscous or structural damping it follows from (30) that $\mathcal{F}_e$ is given by

$$\mathcal{F}_e(t,\phi) = \left\{ 0, 0, c_1 \left( \phi_2 - \frac{1}{|v|}\phi_3 \right), c_2 \left( \phi_2 - \frac{1}{|v|}\phi_3 - \frac{l}{|v|}\phi_4 \right) \right\}' . \tag{52}$$

In this case the two components of the dispersion operator $\sigma$ are given by

$$\sigma_1 \equiv \frac{4l\rho v\kappa}{m} |\sin\theta| \sin\theta = \frac{4l\rho v\kappa}{m} |\sin\phi_2| \sin\phi_2 \,,$$
$$\sigma_2 \equiv \frac{2l^2\rho v\kappa}{\mathcal{I}} |\sin\theta| \sin\theta = \frac{2l^2\rho v\kappa}{\mathcal{I}} |\sin\phi_2| \sin\phi_2 \,. \tag{53}$$

Again it follows from [2, Theorem 5.4] that the expected value of the energy functional $\mathcal{V}$ defined by equation (45) is given by

$$\mathcal{E}\mathcal{V}(\phi(t)) = \mathcal{E}\mathcal{V}(\phi_0) + \mathcal{E} \int_0^t \left\{ \langle \phi, \mathcal{F}_e(\phi) \rangle + \frac{1}{2}\mathrm{tr}(\sigma(s,\phi)Q\sigma(s,\phi)^*) \right\} ds \,. \tag{54}$$

Further for each given (mean) wind velocity $v$, the domain of stability

$$\mathcal{S}^v \equiv \left\{ e \in E \ : \ \langle e, \mathcal{F}_e(e) \rangle + \frac{1}{2} \mathrm{tr}(\sigma(s) Q \sigma(s)^*) \leq 0 \right\} \tag{55}$$

is nonempty.

This holds for both models. For the model (A), $\sigma$ is generally independent of the state while for the model (B) it is dependent. Indeed for the model (B)

$$\frac{1}{2} \mathrm{tr}(\sigma(s) Q \sigma(s)^*) = \left( \frac{16}{m^2} + \frac{4l^2}{\mathcal{I}^2} \right) (l \rho v \kappa)^2 \int_\Sigma (\sin e_2)^4 dx \equiv c_3 \int_\Sigma (\sin e_2)^4 dx \,,$$

where

$$c_3 \equiv \left( \frac{16}{m^2} + \frac{4l^2}{\mathcal{I}^2} \right) (l \rho v \kappa)^2 \,.$$

Thus, for any mean wind velocity $v$, the domain of stability is given by

$$\mathcal{S}^v \equiv \left\{ e \in E \ : \ \langle e, \mathcal{F}_e(e) \rangle + \frac{1}{2} \mathrm{tr}(\sigma(s) Q \sigma(s)^*) = (c_1 e_3 + c_2 e_4, e_2)_{L_2} - \right.$$

$$\left. - \frac{c_2}{|v|} (e_4, e_3)_{L_2} - \frac{c_1}{|v|} \|e_3\|_{L_2}^2 - \frac{c_2 l}{|v|} \|e_4\|_{L_2}^2 + c_3 \int_\Sigma (\sin e_2)^4 dx \leq 0 \right\} . \tag{56}$$

Note that the parameters $\{c_1, c_2, c_3\}$ are all positive and quadratic in $v$, and hence the first and the last components of the expression in (56) are quadratic and the remaining terms linear in $v$. Clearly for $v = 0$, $\mathcal{S}_0 = E$. Note that the third and fourth components have stabilizing effects (aerodynamic damping), and they are linear in $|v|$ while the fifth component has a destabilizing influence. The combined effect of the first, second and fifth components may supersede the stabilizing influence of the third and fourth components and hence may induce instability at sufficiently large wind velocities. Thus there exists a critical wind velocity $v_c$ such that for all $v \in \{v \ : \ |v| < |v_c|\}$ the set $\mathcal{S}^v \neq \emptyset$. These results are illustrated by numerical examples given in the following section.

**Remark:** The constants $\{c_1, c_2, c_3\}$ are functions of the basic bridge parameters $\{m, \mathcal{I}, l\}$ and the wind velocity $v$, air density $\rho$ and the parameter $\kappa$ determining the intensity of wind fluctuation energy. Clearly the size of the domain of stability as discussed above is determined by these parameters. It follows from the expression for the domain that, for a given mean wind velocity, the more massive the structure is (large $m$, $\mathcal{I}$), the larger is the domain of stability. Similarly the narrower the structure is (small $l$) the larger is the domain of stability. Further, as mentioned earlier, appropriate choice of the so-called smart materials for decks and girders may enlarge the stability boundary. In any case given the survey data such as the mean wind velocity, air density and the gusty factor $\kappa$ of the construction site, one can choose the design materials so as to achieve a domain of stability which may be considered safe from an engineering point of view. This result may be used as a guideline to avoid over- and under-design as contemplated by Doole and Hogan in their paper [6].

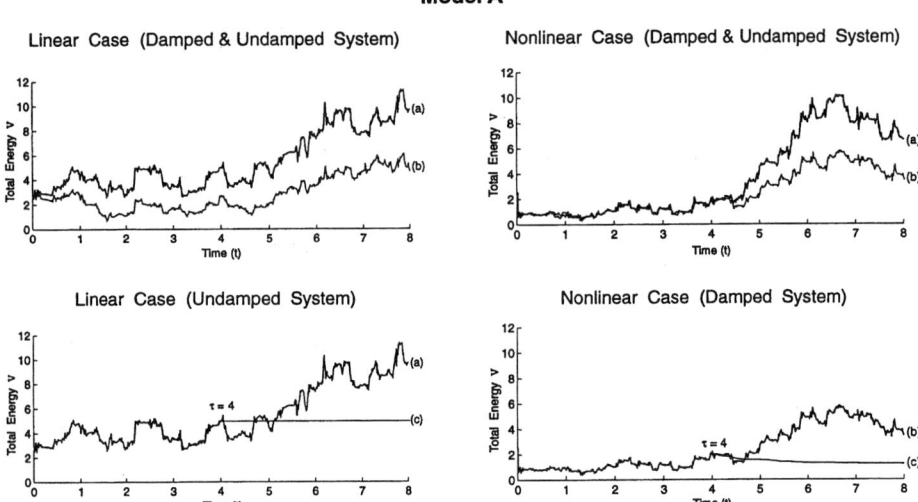

Figures 1 (left) and 2 (right).

# 6    Numerical Results and Discussion

In this section we present some simulation results and discussion thereof. For numerical solution we use a FORTRAN program using a finite difference scheme on a SUN SPARC station 5.    For numerical stability the ratio $\Delta t/\Delta x$ was chosen as $2 \times 10^{-3}$.

## 6.1    Linear case (model A)

Consider system (38) with $\mathcal{F} = \mathcal{F}_c$ linear, that is, the operator $\mathcal{F}$ consists of only the coupling operator, and it is linear. In the absence of damping and external forces (random or deterministic, $\sigma \equiv 0$), this system is conservative and $\mathcal{E}\mathcal{V}(\phi(t)) = \mathcal{E}\mathcal{V}(\phi_0)$ for all $t \geq 0$. If the wind forces persist, that is $\sigma \not\equiv 0$, the expected energy delivered to the bridge structure continues to grow as evidenced by the expression (41). Figure 1a shows the energy flow $\{\mathcal{V}(\phi(t)), t \geq 0\}$ as a positive random process. Since this is not the mean (expected value), the fluctuation observed is expected. Figure 1b shows energy flow corresponding to the damped case. If the wind blows for a period of time, say $[0, \tau]$, implying that the operator $\sigma \not\equiv 0$ for $t \in [0, \tau]$, and $\sigma \equiv 0$ for $t > \tau$, the total expected energy delivered to the system is given by (48). In other words the system is left behind with some excess energy beyond the initial energy and in the absence of damping it will continue to oscillate at that enhanced energy level which may cause failure due to (material) fatigue. This is illustrated by the total energy plot as shown in Figure 1c, which consists of a part of Figure 1a up to the cut-off time $\tau = 4$ and the flat line beyond it.

**Model B**

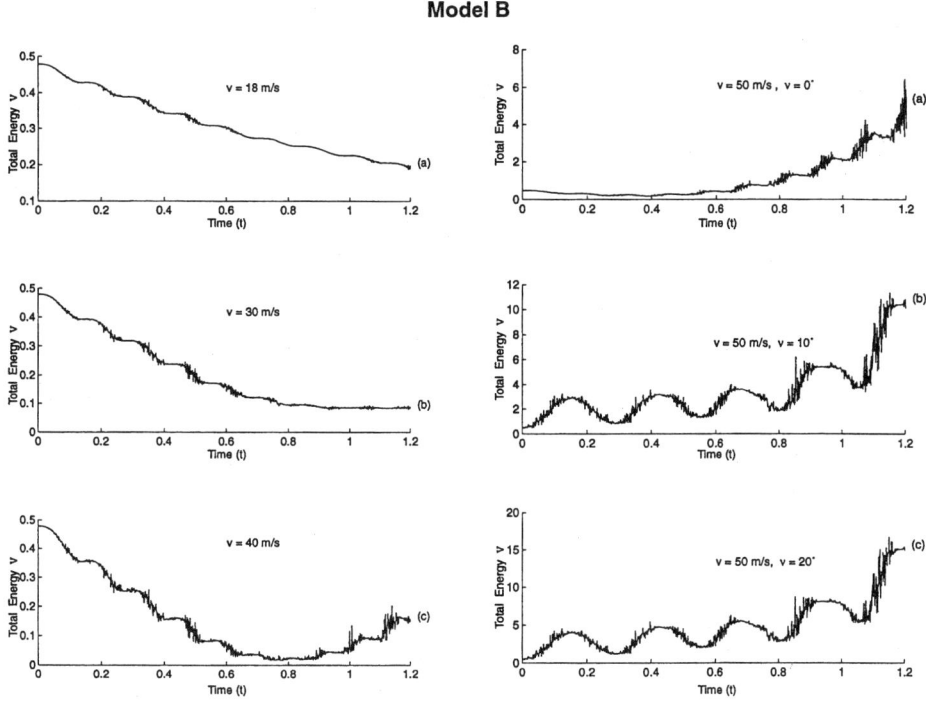

Figures 3 (left) and 4 (right).

## 6.2 Nonlinear case (model A)

The nonlinear system is also subject to the same initial data as in the linear case. We have seen that the nonlinear system is also conservative in the absence of any damping and external forces. In the presence of wind forces, as in the linear case, the energy tends to build up until cutoff. This is illustrated by Figures 2a, 2b (damped) and 2c (damped). The intensity of the residual energy depends not only on the intensity of wind forces during the period but also on the state of coupling between the cable and the deck at the start of the wind. Note the decay of energy as shown in Fig. 2c.

## 6.3 Aerodynamic forces (model B)

It is clear from the model equations (34) and (51) that some viscous damping is always provided by the presence of wind motion. This is evident from expression (51). The terms with negative signs represent negative aerodynamic rate feedback which promotes damping, while the remaining terms represent destabilizing forces increasing quadratically with the mean wind velocity v. For the simulation experiment we neglected additional (viscous) damping. In the presence of random wind forces with mean velocities ($v = 18, 30, 40$), it is clear from Figures 3a, 3b and 3c that as long as the mean wind velocity does not exceed a critical value the system

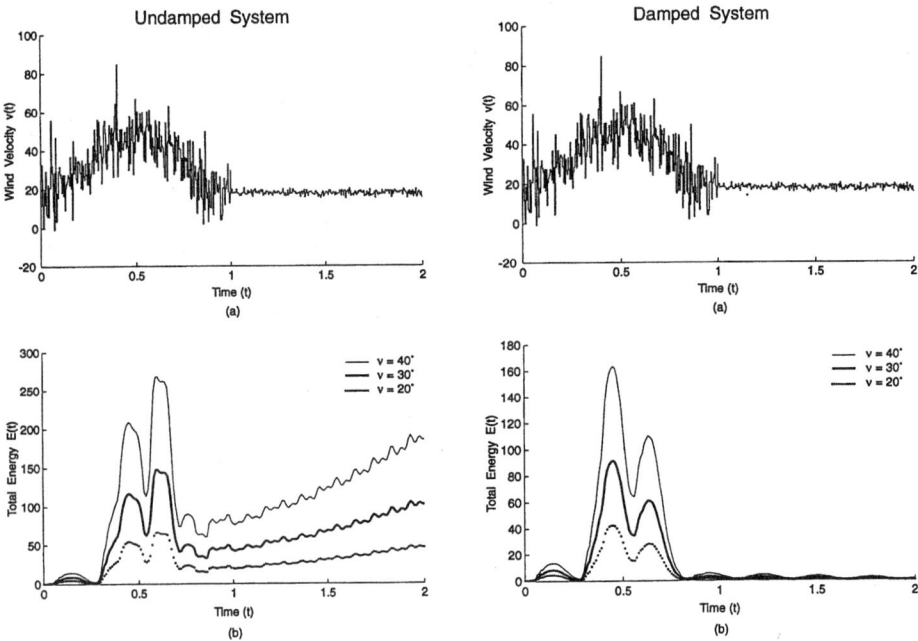

Figures 5 (left) and 6 (right).

remains stable. Beyond this critical value the system becomes unstable. One may recall that the critical value depends on the system parameters as discussed in the remark following Section 5. These results correspond to angle of attack $\nu = 0$.

The results shown in Fig. 4 correspond to $v = 50$ and $\nu = 0, 10°, 20°$. In this case $\theta$ appearing in equations (50) is replaced by $\theta - \nu$. It is clear from these results that, with increasing angle of attack (relative to the horizontal plane), the structure becomes increasingly unstable, which is expected.

## 6.4   Multiplicative perturbation

Consider the wind forces as modeled in Section 5.2, which represent a mean wind velocity with a random fluctuation term generated by scaled-down Brownian motion. The mean is generated by integrating the function $a(t) \equiv a\phi(t)\sin(2\pi t)$, where $\phi(t) = 1$ for $t \in [0,1]$ and zero after. The fluctuation energy is $\kappa^2$ representing the variance. The physical implication is that the wind starts at time 0 and ends at time 1, returning to the ambient condition as shown in the top graphs of Figures 5, 6. The response of the bridge structure without and with damping is shown in Figures 5–10. For the undamped case, it is clear from the figures that the structure becomes increasingly unstable as the angle of attack increases. However, by providing some material or structural damping (through smart materials), the situation is drastically improved. The system remains stable during the severe wind conditions and returns to the equilibrium state after the wind has passed. This is shown in the graphs of Figs. 6, 8 and 10.

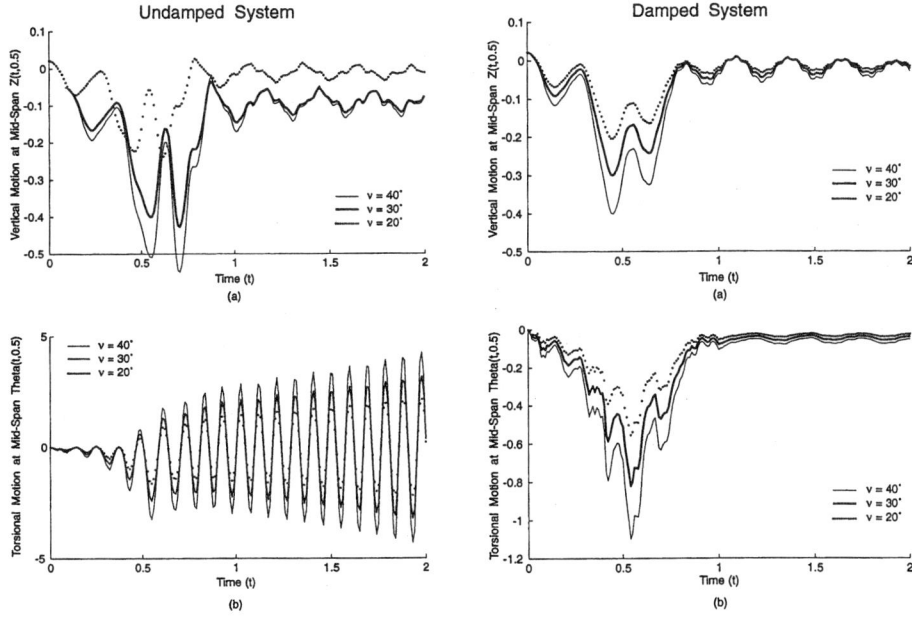

Figures 7 (left) and 8 (right).

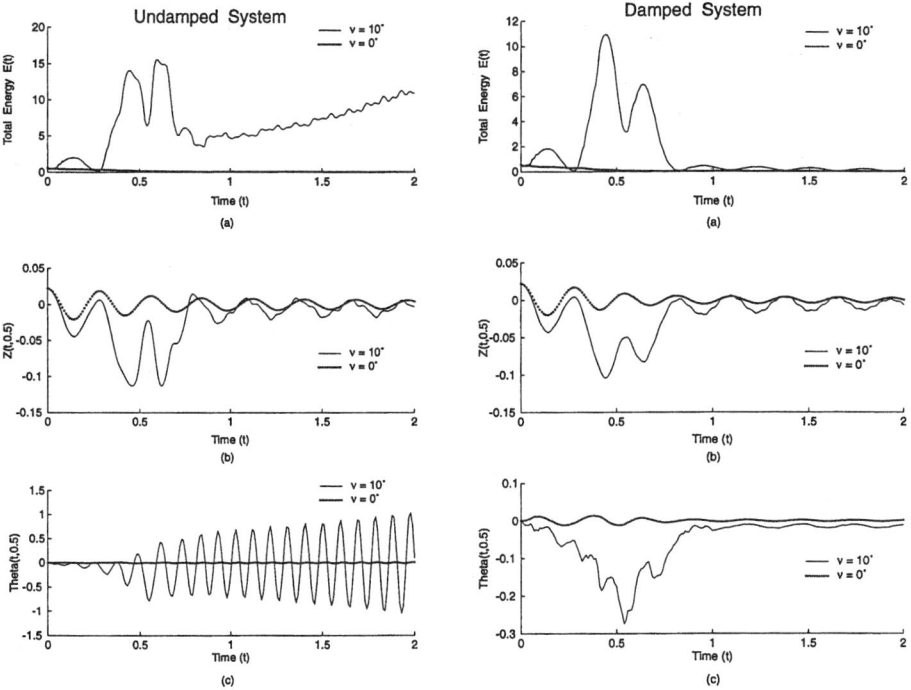

Figures 9 (left) and 10 (right).

# 7  Concluding Remarks

In this paper we have presented simulation results for two models of suspension bridges subject to stochastic wind load. The numerical results coincide with the conclusions based on purely theoretical studies [1, 2].

# References

1. Ahmed, N.U. and Harbi, H. (1998) Mathematical Analysis of Dynamic Models of Suspension Bridge. *SIAM Journal of Applied Mathematics*, **58** (3), 853–874.

2. Ahmed, N.U. (2000) A General Mathematical Framework for Stochastic Analysis of Suspension Bridges. *Nonlinear Analysis*, Series B, **1**, 451–483.

3. Ahmed, N.U. (1991) *Semigroup Theory with Applications to Systems and Control*. Pitman Research Notes in Mathematics Series No. 246. Longman Scientific and Technical, UK, John Wiley & Sons, Inc., New York.

4. Ahmed, N.U. and Harbi, H. (1998) Torsional and Longitudinal Vibration of Suspension Bridges Subject to Aerodynamic Forces. *Journal of Mathematical Problems in Engineering*, **3** (4), 1–29.

5. Ahmed, N.U. and Harbi, H. (1998) Stability of Suspension Bridge II: Aerodynamic vs Structural Damping. In *Proc. of SPIE, Smart Structures and Materials 1998, Smart Systems for Bridges, Structures, and Highways*, San Diego, California, pp. 276–289.

6. Doole, S.H. and Hogan, S.J. (1996) A Piecewise Linear Suspension Bridge Model: Nonlinear Dynamics and Orbit Continuation. *Dynamics and Stability of Systems*, **11** (1), 19–47.

7. Glover, J., Lazer, A.C. and McKenna, P.J. (1989) Existence and Stability of Large Scale Nonlinear Oscillations in Suspension Bridges. *Journal of Applied Mathematics and Physics (ZAMP)*, **40**, 172–200.

8. Jacover, D. and McKenna, P.J. (1994) Nonlinear Torsional Flexings in a Periodically Forced Suspended Beam. *Journal of Computational and Applied Mathematics*, **52**, 241–265.

9. Roseau, M. (1984) *Vibration in Mechanical Systems*. Springer-Verlag, Berlin, Heidelberg, New York, London, Paris, Tokyo.

# Time Delayed Control
# of Structural Systems

Firdaus E. Udwadia,[1] Hubertus F. von Bremen,[2]
Ravi Kumar[3] and Mohamed Hosseini[2]

[1] *Department of Aerospace and Mechanical Engineering*
*Civil Engineering, Mathematics, and Decision Systems*
*University of Southern California*
*Los Angeles, CA 90089-1453, USA*
[2] *Department of Aerospace and Mechanical Engineering*
*University of Southern California*
*Los Angeles, CA 90089-1453, USA*
[3] *Structural Analysis and Research Corporation*
*5000 McKnight Road, Pittsburgh, PA 15237, USA*

Time delays are ubiquitous in control systems. They usually enter because of the sensors and actuators used in them. Traditionally, time delays have been thought to have a deleterious effect on both the stability and the performance of controlled systems, and much research has been done in attempting to eliminate them, compensate for them, or nullify their presence. In this paper we take a different view. We investigate whether purposefully injected time delays can be used to *improve* both the system's stability and performance. Our analytical, numerical, and experimental investigation shows that this can indeed be done. Analytical results of the effects of time delays on collocated and non-collocated control of classically damped and non-classically damped systems are given. Experimental and numerical results confirm the theoretical expectations. Issues of system uncertainties and robustness of time delayed control are addressed. The results are of practical value in improving the performance and stability of controllers because these characteristics (performance and stability) improve *dramatically* with the intentional injection of small time delays in the control system. The introduction of such time delays constitutes a 'minimal change' to a controller already installed in a structural system for active control. Hence, from a practical standpoint, time delays can be implemented in a nearly costless and highly reliable manner to improve control performance and stability, an aspect that cannot be ignored when dealing with the economics and safety of large structural systems subjected to strong earthquake ground shaking.

# 1   Introduction

The active control of large-scale structural systems usually requires the generation of large control forces which often need to be provided at high frequencies. Actuator and sensor dynamics do not permit the instantaneous generation of such forces, and hence the effective control gets delayed in time. Thus the presence of time delays in the control are inevitable when controlling building structures subjected to dynamic loads, such as those caused by strong earthquake ground shaking. In order to accommodate for the time delays, the mathematical formulations of the problem of controlling building structures are usually more complicated than the formulations without time delays. The fact that the models are more complicated when time delays are included and that in some cases the presence of time delays destabilize the control has fueled the predominant view that time delays are an undesirable element in the active control of structures. With this view in mind, methods to cancel out, reduce, or change the effect of time delays have been developed. This paper proposes a different viewpoint: its central theme is that instead of considering time delays as always being injurious, one could aim to exploit their presence, especially since they are ubiquitous. We show that the proper intentional introduction of time delays can: (1) stabilize even a non-collocated control system (which may be unstable in the absence of time delays), and (2) improve control performance.

Extensive work has been done on the control of structural systems where no consideration to time delays is given. The fact that the results do not include time delays does not undermine their importance when controlling large building structures, since several of the concepts can be used as a starting point when dealing with time-delayed problems. Feedback control of structural systems yields different stability characteristics depending on whether collocated or non-collocated control is used. Direct velocity feedback (no time delay) control of a discrete dynamical system with collocation of actuators and sensors is known to be stable for all values of the control gain (Auburn 1980, Balas 1979a). Balas (1979b) has investigated the potential of direct output feedback control for systems where sensors and actuators need not be collocated. When actuators' dynamics are considered, Goh and Caughey (1985) and Fanson and Caughey (1990) have shown that position feedback is preferable to velocity feedback under collocated control. Cannon and Rosenthal (1984) deal with experimental studies of collocated and non-collocated control of flexible structures. Based on these studies, it has been concluded that it is very difficult to achieve robust non-collocated control of flexible structures.

In practical feedback control systems, small time delays in the control action are inevitable because of the involved dynamics of the actuators and sensors. As stated before, these time delays become particularly significant when the control effort demands large control forces and/or high frequencies. It is therefore crucial to understand the effect of time delays on the control of structural systems. Several papers in the literature treat the presence of time delays as a negative factor. Yang *et al.* (1990) show that time delays worsen performance for their proposed controllers. Agrawal *et al.* (1993) indicated methods of compensation for time delay in the active control of structural systems. On the other hand, some previous studies suggest that time delays can be used to good advantage. Kwon *et al.* (1989) show that the intentional use of time delays may improve the performance of the control system. Udwadia and Kumar (1994) show that dislocated velocity control, which leads to instability in the absence of time delays, can even be used to stabilize an MDOF system (for small gains) by an appropriate choice of time delays. In the present paper (Section 3) we show that the intentional injection of time delays can increase the maximum gain for stability of a non-classically damped system (when compared with the system with no time delays). Experimental results on a two-degree-of-freedom torsional system (Section 4) confirm our analytical findings and show that it is possible to choose time delays which improve the controller's performance when compared to the controlled system with no time delays.

This paper is organized as follows. Section 2 deals with the time-delayed control of classically damped systems. Results for collocated as well as for non-collocated control of undamped and underdamped systems are given. The theoretical expectations are confirmed numerically when applied to a building structure model which is subjected to an earthquake. Numerical results on the sensitivity of the control methodology to perturbations (a) of the parameters of a building structure, and (b) of the time delays used are also presented. Section 3 deals with more general, non-classically damped, linear systems. Section 4 presents experimental data on a two-degree-of-freedom non-classically damped torsional system. Numerical results which corroborate the theoretical findings obtained in the previous sections are also presented for comparison. Here we also show that it is possible for a non-system pole (a pole whose root locus does not start at an open loop pole of the structural system) to dictate the maximum gain for stability of the time-delayed, controlled system. Section deals 5 with robustness issues. It deals with the control of uncertain systems with uncertain time varying delays in the control input. Our conclusions are presented in Section 6.

# 2  Classically Damped Structural Systems

This section deals with the time-delayed control of classically damped structures. Collocated as well as non-collocated control of undamped and underdamped structures are considered. Most of the results apply to controllers of the PID type (proportional, integral, derivative). Numerical computations are used to confirm some of the theoretical expectations for a building structure subjected to an earthquake. Most of the results apply only to the case when system poles are considered (that is, for poles whose root locus originates at the open loop poles of the structural system). The sensitivity of the stability of the building structure when the parameters of the structure and the time delay are varied is explored numerically. It is shown numerically that even under considerable perturbations (7% to 50%) of the parameter values the system remains stable. Details of some of the analytical results presented in this section can be found in Udwadia and Kumar (1994) and part of the numerical results can be found in Udwadia and Kumar (1996).

## 2.1  System model and general formulation

Consider a linear classically damped structural system whose response $x(t)$ is described by the matrix differential equation

$$Mx''(t) + Cx'(t) + Kx(t) = g(t); \qquad x(0) = 0, \ x'(0) = 0, \qquad (2.1)$$

where $M$ is the $n$ by $n$ positive definite symmetric mass matrix, $C$ is the $n$ by $n$ symmetric damping matrix, and $K$ is the $n$ by $n$ positive definitive symmetric stiffness matrix. The forcing function is given by the $n$-vector $g(t)$.

Since the system is classically damped, it can be transformed to the diagonal system

$$z''(t) + \Xi z'(t) + \Lambda z(t) = T^T M^{-1/2} g(t); \qquad z(0) = 0, \ z'(0) = 0, \qquad (2.2)$$

where $\Xi = \text{diag}(2\xi_1, 2\xi_2, \ldots, 2\xi_n)$, $\Lambda = \text{diag}(\lambda_1^2, \lambda_2^2, \ldots, \lambda_n^2)$, and the matrix $T = [t_{ij}]$ is the orthogonal matrix of eigenvectors of $M^{-1/2} K M^{-1/2}$. Taking the Laplace transform of (2.2) and solving, we get

$$\tilde{x}(s) = M^{-1/2} T \Xi T M^{-1/2} \tilde{g}(s), \qquad (2.3)$$

where the wiggles indicate the transformed functions, and $\Xi$ is given by

$$\Xi = \text{diag}((s^2 + 2\xi_1 s + \lambda_1^2)^{-1}, (s^2 + 2\xi_2 s + \lambda_2^2)^{-1}, \ldots, (s^2 + 2\xi_n s + \lambda_n^2)^{-1}).$$

The open loop poles are given by the zeros of the equations

$$s^2 + 2\xi_q s + \lambda_q^2 = (s - \gamma_{+q})(s - \gamma_{-q}) = 0, \quad q = 1, 2, \ldots, n. \qquad (2.4)$$

The poles have been denoted by $\gamma_{\pm q}$, the sign indicating the sign in front of the radical of the quadratic equations given in (2.4). In this paper we assume the system to be generic and all poles to have multiplicity one (no repeated poles).

The feedback control uses a linear combination of $p$ responses $x_{s_k}(t)$, $k = 1, 2, \ldots, p$, which are fed to a controller. In general, the responses are time-delayed by $T_{s_k}(t)$. The actuator will apply a force to the system, affecting the $j$-th equation of (2.1). When $j \in \{s_k : k = 1, 2, \ldots, p\}$ the sensors and the actuator are collocated, and if $j \notin \{s_k : k = 1, 2, \ldots, p\}$ the sensors and the actuator are non-collocated (dislocated). The control methodology applied to a shear frame building structure is shown in Figure 2.1.

Denoting the non-negative control gain by $\mu$ and the controller transfer function by $\mu \tau_c(s)$, the closed loop system poles are given by

$$\tilde{A}(s)\tilde{x}(s) = [Ms^2 + Cs + K]\tilde{x}(s) = \tilde{g}(s) - \mu \tau_c(s) \sum_{k=1}^{p} a_{s_k} \tilde{x}_{s_k}(s) \exp[-sT_{s_k}]e_j,$$

$$(2.5)$$

where $e_j$ is the unit vector with unity in the $j$-th element and zeros elsewhere. The numbers $a_{s_k}$ are the coefficients of the linear combination of the responses fed to the controller. Moving the last term on the right of (2.5) to the left gives

$$\tilde{A}_1(s)\tilde{x}(s) = \tilde{g}(s), \qquad (2.6)$$

where $\tilde{A}_1(s)$ is obtained by adding $\mu \tau_c(s) a_{s_k} \exp[-sT_{s_k}]$ to the $(j, s_k)$-th element of $\tilde{A}(s)$, for $k = 1, 2, \ldots, p$.

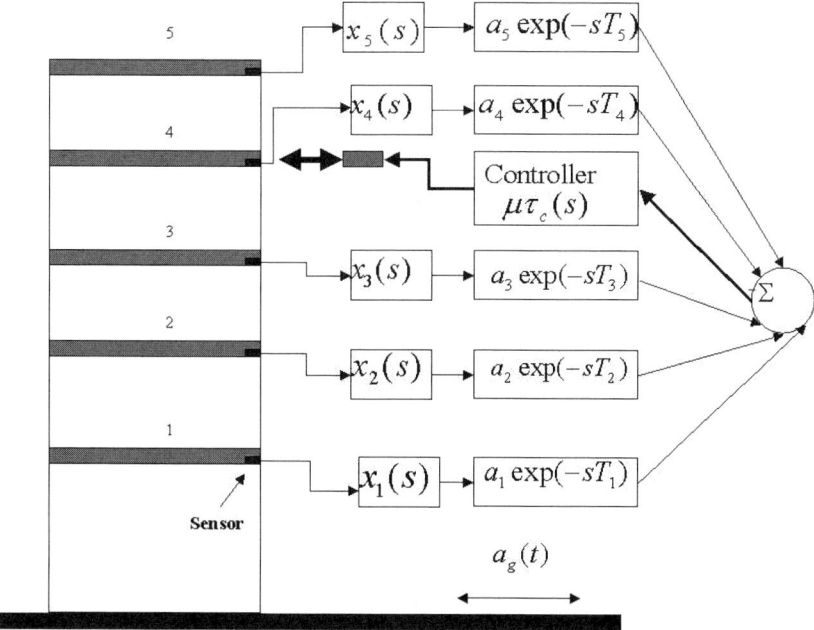

Figure 2.1 Shear frame building structure and control methodology.

The closed loop poles are then given by the relation

$$\det[\tilde{A}_1(s)] = \det[\tilde{A}(s)]\left\{1 + \mu\tau_c(s)\sum_{k=1}^{p}a_{s_k}\exp[-sT_{s_k}]\tilde{x}_{s_k,j}^{(\delta)}(s)\right\}, \qquad (2.7)$$

where $\tilde{x}_{s_k,j}^{(\delta)}(s)$ is the Laplace transform of the open loop response to an impulsive force applied at node $j$ at time $t = 0$. The open loop response to the impulsive force is given by $\tilde{x}_{s_k,j}^{(\delta)}(s) = \sum_{i=1}^{n}\dfrac{t_{s_k,j}^{(M)}t_{i,j}^{(M)}}{s^2 + 2\xi_i s + \lambda_i^2}$, where $t_{s_k,r}^{(M)} = \sum_{u=1}^{n}m_{s_k,u}^{-1/2}t_{u,r}$, with $m_{i,j}^{-1/2}$ being the $(i, j)$-th element of $M^{-1/2}$ and $T^{(M)} = M^{-1/2}T = [t_{i,j}^{(M)}]$.

The following set of conditions will be referred to as condition set C1. Given that the open loop poles of the system are $\gamma_{\pm q}$, we have

$$(1) \quad \tau_c(\gamma_{\pm q}) \neq 0 \qquad\qquad \text{for} \quad q = 1, 2, \ldots, n\,,$$

$$(2) \quad \sum_{k=1}^{p}a_{s_k}\exp[-\gamma_{\pm q}T_{s_k}]t_{s_k,q}^{(M)} \neq 0 \qquad \text{for} \quad q = 1, 2, \ldots, n\,,$$

$$(3) \quad t_{j,q}^{(M)} \neq 0 \qquad\qquad\qquad \text{for} \quad q = 1, 2, \ldots, n\,.$$

The first condition means that the open loop poles of the system are not also zeros of the controller transfer function. The second condition is a generalized observability condition which requires that all mode shapes are observable from the summed, time-delayed sensor measurements. The third condition is a controllability condition which requires that the controller cannot be located at any node of any mode of the system. Observe that if any of the three conditions are not satisfied for one open loop pole $\gamma_q$, then by (2.7) the open loop pole $\gamma_q$ is also a closed loop pole. However, if C1 is satisfied, we have the following result.

**Result 2.1:** When the open loop system has distinct poles and condition set C1 is satisfied, the open loop and the closed loop systems have no poles in common.

If condition set C1 is satisfied, then by Result 2.1 and (2.7), the closed loop poles of the system are given by the values of $s$ that satisfy the equation

$$1 + \mu \tau_c(s) \sum_{k=1}^{p} \sum_{i=1}^{n} a_{s_k} \exp[-sT_{s_k}] \left[ \frac{t_{s_k,j}^{(M)} t_{i,j}^{(M)}}{s^2 + 2\xi_i s + \lambda_i^2} \right] = 0. \tag{2.8}$$

In general, equation (2.8) may have an infinite number of zeros due to the time delay term. As the parameter $\mu$ is varied, we obtain the root locus of the closed loop poles. The poles that are on a root locus that starts at an open loop pole of the structural system will be called system poles. The poles that do not originate at an open loop pole of the system will be called non-system poles. To simplify matters, some of the following results deal with the system poles only. The simplification of dealing with the system poles only, allows us to obtain bounds on the gain and the time delay to guarantee stability. These results should be viewed with caution since in general there are an infinite number of poles, and as is shown in Section 4, a non-system pole may determine what is the maximum gain for stability for some systems. It will be made clear when our results apply to all the poles considered, and when only to the system poles.

The following result and all of its consequences apply to the case when only system poles are considered. Multiplying (2.8) by $s^2 + 2\xi_r s + \lambda_r^2$, then differentiating with respect to $\mu$ and letting $s \to \gamma_{\pm r} = -\xi_r \pm i(\lambda_r^2 - \xi_r^2)^{1/2}$ and $\nu \to 0$, we obtain

$$\left. \frac{ds}{d\mu} \right|_{\substack{\mu \to 0 \\ s \to \gamma_{\pm r}}} = -\frac{\tau_c(\gamma_{\pm r})}{2 \pm i(\lambda_r^2 - \xi_r^2)^{1/2}} \left[ \sum_{k=1}^{p} a_{s_k} \exp[-\gamma_{\pm r} T_{s_k}] t_{s_k,r}^{(M)} \right] t_{j,r}^{(M)}. \tag{2.9}$$

**Result 2.2:** A sufficient condition for the closed loop system to remain stable for infinitesimal gains is that

$$\text{Re} \left\{ \left. \frac{ds}{d\mu} \right|_{\substack{\mu \to 0 \\ s \to \gamma_{\pm r}}} \right\} < 0, \qquad r = 1, 2, \ldots, n. \tag{2.10}$$

Again, Result 2.2 applies only when system poles are considered, and in general the result may not be true when non-system poles are also considered.

We now particularize the controller to be of the proportional, integral and derivative (PID) form. The transfer function of the controller is then given by

$$\tau_c(s) = K_0 + K_1 s + \frac{K_2}{s}, \qquad \text{with } K_0, K_1, K_2 \geq 0.$$

The term $K_0$ corresponds to proportional control, the term $K_1 s$ corresponds to derivative control and the term $\dfrac{K_2}{s}$ corresponds to integral control. Next we specialize results for the undamped case.

## 2.2 Undamped systems

The following set of results apply to undamped systems. When the damping matrix $C = 0$, the open loop poles lie on the imaginary axis, and by (2.10), the next result follows.

**Result 2.3:** For undamped systems $(C = 0)$, condition (2.10) is a necessary and sufficient condition for stability for small gains.

For an undamped system, the open loop poles are of the form $\gamma_{\pm r} = \pm i\gamma_r$. Using (2.9 and 2.10) we can derive the following requirement for stability for small gains (this result applies to system poles only).

$$\operatorname{Re}\left\{\left(\frac{K_0}{\pm i\lambda_r} + K_1 - \frac{K_2}{\lambda_r^2}\right)\left(\sum +k = 1^p a_{s_k}\exp[\mp\lambda_r T_{s_k}]t_{s_k,r}^{(M)}t_{j,r}^{(M)}\right)\right\} > 0, \quad (2.11)$$

for $r = 1, 2, \ldots, n$.

Using the relation (2.11), we can derive the following result.

**Result 2.4:** When using one sensor collocated with the actuator for an undamped system, the PID feedback control is stable (for small gains) if and only if

$$a_j\left\{-\frac{K_0}{\lambda_r}\sin(\lambda_r T_j) + \left(K_1 - \frac{K_2}{\lambda_r^2}\right)\cos(\lambda_r T_j)\right\} > 0 \qquad \text{for } r = 1, 2, \ldots, n. \tag{2.12}$$

**Result 2.5:** For undamped systems with one sensor collocated with the actuator, we have the following conditions for stability for small gains.

(a) Velocity feedback $(K_0 = K_2 = 0)$ is stable as long as the time delay is such that $T_j < \dfrac{\pi}{2\lambda_{\max}}$, where $\lambda_{\max}$ is the highest undamped natural frequency of the system.

(b) Integral control $(K_0 = K_1 = 0)$ is stable as long as the delay is such that $T_j < \dfrac{\pi}{2\lambda_{\max}}$.

(c) Proportional control $(K_1 = K_2 = 0)$ is stable as long as the time delay is such that $0 < T_j < \dfrac{\pi}{2\lambda_{\max}}$.

(d) When $K_0 = 0$ and the time delay is such that $T_j < \dfrac{\pi}{2\lambda_{\max}}$, the undamped system will be stabilized when $K_1 > \dfrac{K_2}{\lambda_{\min}^2}$ and $a_j > 0$, or $K_1 < \dfrac{K_2}{\lambda_{\max}^2}$ and $a_j < 0$.

**Result 2.6:** When the system is undamped, and

(1) condition set C1 is satisfied,

(2) one sensor is used and is collocated with the actuator, and,

(3) no time delay is used,

then the PID control, if stable for $\mu \to 0^+$, is stable for all $\mu > 0$ provided

$$\det\left[\tilde{A}\left(-\frac{K_2}{K_1}\right)\right] + \mu a_j \det\left[\tilde{A}_2\left(-\frac{K_2}{K_1}\right)\right] \neq 0, \qquad (2.13)$$

for any positive $\mu$, where $\tilde{A}_2$ is obtained by deleting the $j-$th row and the $j-$th column of the matrix $\tilde{A}$.

When velocity (or integral) feedback control is used, condition (2.13) is always satisfied, hence we get the well-known result that stability is guaranteed for $\mu > 0$. The upper bound for the stability of the system described in result 2.6 can be obtained to be

$$\mu < \frac{-\det\left[\tilde{A}\left(-\frac{K_2}{K_1}\right)\right]}{a_j K_0 \det\left[\tilde{A}_2\left(-\frac{K_2}{K_1}\right)\right]},$$

provided the right-hand side in the inequality is positive. If not, the system is stable for all $\mu > 0$, provided it is stable for small gains. Result 2.6 and the above bound on the gain for stability apply to the case when all the poles are considered. The next result, however, only applies when system poles are considered.

**Result 2.7:** When the system is undamped, and

(1) condition set C1 is satisfied,

(2) one sensor is used and is collocated with the actuator, and,

(3) time delay $T_j < \dfrac{\pi}{2\lambda_{\max}}$,

then velocity feedback control will be stable as long as

$$\mu < \left(a_j K_1 \eta_0 \sum_{i=1}^{n} \frac{\left[t_{j,i}^{(M)}\right]^2}{\eta_0^2 - \lambda_i^2}\right)^{-1}, \qquad \text{where} \ \ \eta_0 = \frac{\pi}{2T_j}.$$

The previous results have focused on collocated control. In the following results we will consider the control of non-collocated (dislocated) systems. Results dealing with no time delay are presented first and are followed by results for time-delayed systems.

**Result 2.8:** When using a PID controller, where

(1) the sensors and the actuator are not collocated,

(2) the time delays, $T_{s_k}$, $k = 1, 2, \ldots, p$, are all zero,

(3) the matrix $M$ is diagonal, and

(4) $K_1 > \dfrac{K_2}{\lambda_{\min}^2}$ or $K_1 < \dfrac{K_2}{\lambda_{\max}^2}$,

it is impossible to stabilize the undamped system for small gains.

The next result gives necessary conditions for the stability of a system using a PID controller.

**Result 2.9:** When using a PID controller, where

(1) the sensors and the actuator are not collocated,

(2) the time delays, $T_{s_k}$, $k = 1, 2, \ldots, p$, are all zero,

(3) the matrix $M$ is non-diagonal,

a necessary condition for the undamped system to be stabilized for small gains is

(a) $\displaystyle\sum_{k=1}^{p} a_{s_k} m_{s_k,j}^{-1} > 0$, when $K_1 > \dfrac{K_2}{\lambda_{\min}^2}$,

and

(b) $\displaystyle\sum_{k=1}^{p} a_{s_k} m_{s_k,j}^{-1} < 0$, when $K_1 < \dfrac{K_2}{\lambda_{\min}^2}$.

Often building structures are modeled by tridiagonal stiffness matrices. The following result applies to such structures.

**Result 2.10:** If $M$ and $K$ are positive definite, $M$ is diagonal and $K$ is tridiagonal, having negative subdiagonal elements, it is possible to find a location $j$ (for the actuator) and a location $s_1$ (for the sensor), $j \neq s_1$, so that the sequence $\left\{ t_{s_1,i}^{(M)} t_{j,1}^{(M)} \right\}_{i=1}^{n}$ will only have one sign change.

**Result 2.11:** When using an ID controller, for a system as defined in Result 2.10, where

(1) condition set C1 is satisfied,

(2) one sensor is used and it is not collocated with the actuator,

(3) the sign change in the sequence $\left\{ t_{s_1,i}^{(M)} t_{j,1}^{(M)} \right\}_{i=1}^{n}$ occurs when $i = m$,

(4) time delay $T_{s_1} = \dfrac{\pi}{2\lambda_{m-1}} - \varepsilon$, where $\epsilon$ is a small positive quantity and,

$T_{s_1} \lambda_m > \dfrac{\pi}{2}$,

(5)  $\dfrac{\lambda_{\max}}{\lambda_{\min}} \leq 3$,     and

(6)  $K_1 > \dfrac{K_2}{\lambda_{\min}^2}$  or  $K_1 < \dfrac{K_2}{\lambda_{\max}^2}$,

it is possible to stabilize an undamped (open loop) system for small gains.

**Result 2.12:** For the undamped system described in Result 2.11, velocity feedback control will be stable as long as $\mu < G$, where $G$ is the minimum of all positive $B_l$, for $l = 0, 1, 2, \ldots$, where

$$b_l = -\left( a_{s_1} K_1 \eta_l \sin(\eta_l T_{s_1}) \sum_{i=1}^{n} \frac{t_{s_1,i}^{(M)} t_{j,1}^{(M)}}{\lambda_i^2 - \eta_l^2} \right)^{-1} \qquad \text{and} \qquad \eta_l = \frac{(2l+1)\pi}{2T_{s_1}}.$$

## 2.3   Underdamped systems

This section deals with underdamped systems. Specializing equation (2.9) to underdamped systems, and utilizing Result 2.2 we get the following result which is again applicable to the case when only system poles are considered.

**Result 2.13:** When using PID control for underdamped systems, $\xi_i < \lambda_i$, $i = 1, 2, \ldots, n$ a sufficient condition for the closed loop system to be stable for small gains is

$$-\frac{1}{(\lambda_r^2 - \xi_r^2)^{1/2}} \left[ K_0 - \left( K_1 + \frac{K_2}{\lambda_r^2} \right) \xi_r \right] \times$$

$$\times \left( \sum_{k=1}^{n} a_{s_k} \exp[\xi_r T_{s_k}] \sin\left( (\lambda_r^2 - \xi_r^2)^{1/2} T_{s_k} \right) t_{s_k,r}^{(M)} t_{j,r}^{(M)} \right) +$$

$$+ \left[ K_1 - \frac{K_2}{\lambda_r^2} \right] \left( \sum_{k=1}^{n} a_{s_k} \exp[\xi_r T_{s_k}] \cos\left( (\lambda_r^2 - \xi_r^2)^{1/2} T_{s_k} \right) t_{s_k,r}^{(M)} t_{j,r}^{(M)} \right) > 0$$

for $r = 1, 2, \ldots, n$.

**Result 2.14:** When the sensor and actuator are collocated and only one sensor is used, for PID control, if $K_1 - \dfrac{K_2}{\lambda_r^2} \neq 0$, for all $r$, a sufficient condition for small gains stability is

$$a_j \left[ K_1 - \frac{K_2}{\lambda_r^2} \right] \cos\left( (\lambda_r^2 - \xi_r^2)^{1/2} T_{s_k} + \phi \right) > 0, \qquad \text{for } r = 1, 2, \ldots, n,$$

where

$$\phi = \tan^{-1} \left[ \frac{\left( K_1 + \dfrac{K_2}{\lambda_r^2} \right) \xi_r}{\left( K_1 - \dfrac{K_2}{\lambda_r^2} \right) (\lambda_r^2 - \xi_r^2)^{1/2}} \right].$$

**Result 2.15:** When using one sensor, collocation of the sensor with an actuator of the given feedback control type will cause the closed loop system poles to move to the left in the $s$-plane, as long as the given condition on the time delay is satisfied.

(a) For velocity feedback, the time delay needs to satisfy $T_j < \min\limits_{\forall r} \left[ \dfrac{\frac{\pi}{2} + \phi}{(\lambda_r^2 - \xi_r^2)^{1/2}} \right]$,

where $\phi = \tan^{-1} \left[ \dfrac{\xi_r}{(\lambda_r^2 - \xi_r^2)^{1/2}} \right]$.

(b) For integral feedback, the time delay needs to satisfy $T_j < \min\limits_{\forall r} \left[ \dfrac{\frac{\pi}{2} - \phi}{(\lambda_r^2 - \xi_r^2)^{1/2}} \right]$,

where $\phi$ is as in part (a).

(c) For proportional feedback, the time delay needs to satisfy $0 < T_j < \min\limits_{\forall r} \left[ \dfrac{\pi}{(\lambda_r^2 - \xi_r^2)^{1/2}} \right]$.

(d) For a PID controller, the time delay needs to satisfy $T_j < \min\limits_{\forall r} \left[ \dfrac{\frac{\pi}{2} - \phi}{(\lambda_r^2 - \xi_r^2)^{1/2}} \right]$,

where $\phi$ is as defined in Result 2.14, when $K_1 > \dfrac{K_2}{\lambda_{\min}^2}$ and $a_j > 0$, or $K_1 < \dfrac{K_2}{\lambda_{\max}^2}$ and $a_j < 0$.

## 2.4 Numerical results

The numerical results are obtained for an undamped shear frame building structure (five-degree-of-freedom system) shown in Fig. 2.1. The mass and the stiffness of each storey are 1 and 1,600 (taken in SI units), respectively. The root loci presented only show the system poles. The controller's gain $\mu$ has been varied from 0 to 100 units. Velocity feedback control is used for all the results.

The first example deals with the collocated control of the structure with and without time delay. The controller and the sensor are collocated at the fourth storey. Figure 2.2a shows the root loci of the closed loop system poles for velocity feedback control with no time delay ($T_4 = 0$ sec). As expected from Result 2.6, the system is stable since the system's closed loop poles have negative real parts. Figure 2.2b shows the root loci for the closed loop system poles with velocity feedback control when a time delay of $T_4 = 0.025$ sec is used. Introducing a time delay of $T_4 = 0.025$ sec makes the system unstable even for very small gains, as predicted by Result 2.4.

In the second example, the non-collocated control of the structure is studied with and without time delay. The actuator is placed at mass 4 (fourth storey) and is fed the velocity signal from location 5 (fifth storey). Figure 2.3a shows the root loci for the closed loop system poles for velocity feedback control with no time delay ($T_5 = 0$ sec) and $j = 4$, $s_1 = 5$, $a_5 = 1$. As guaranteed by Result 2.8, even for

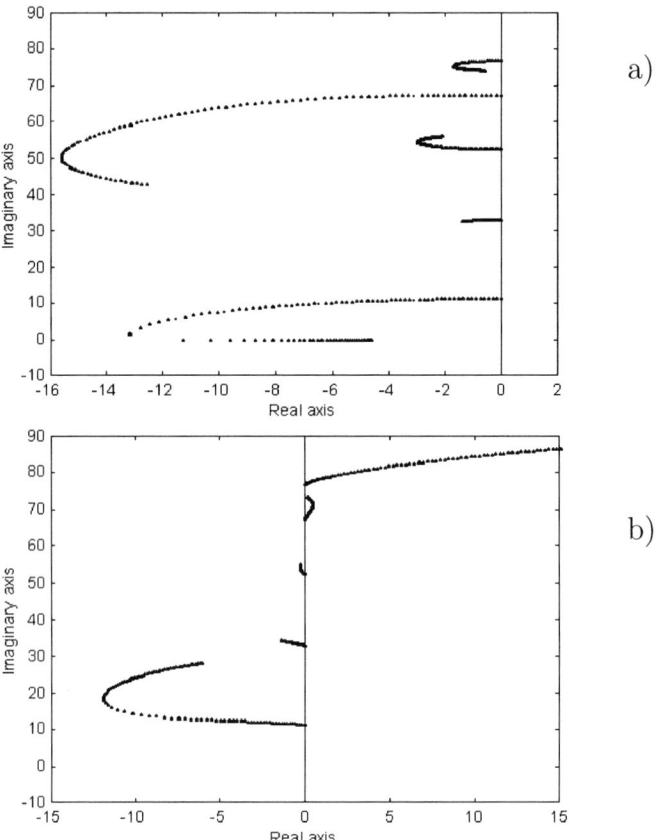

Figure 2.2  Root loci of closed loop system poles for collocated velocity feedback control with $j = 4$, $s_1 = 4$, $a_4 = 1$, and time delay (a) $T_4 = 0$ sec, (b) $T_4 = 0.025$ sec.

vanishingly small gains, the third, fourth and fifth closed loop system poles are in the right-half $s$-plane, hence causing instability. However, the introduction of an appropriate time delay, such as $T_5 = 0.04$ sec, makes the closed loop system poles remain in the left-half plane until a certain value of the controller gain. This is illustrated in Figure 2.3b, where the closed loop system poles are shown for $j = 4$, $s_1 = 5$, $a_5 = 1$, and a time delay of $T_5 = 0.04$ sec. The upper bound on the gain for stability obtained by tracing the root loci is the same as the one predicted by Result 2.12. This example shows that by appropriately injecting time delay into a system, it is possible to stabilize a system which is unstable for zero time delay.

Figure 2.4 shows the displacement time history of mass 5 relative to the base, when the structure is subjected to the ground motion of the S00E component of the Imperial Valley Earthquake of 1940. The structural responses are numerically computed using a fourth-order Runge-Kutta scheme. The response time histories are shown only for the first 10 sec. The response is shown for no control ($\mu = 0$), and for $\mu = 10$ units, using non-collocated velocity control with $j = 1$, $s_1 = 5$, $a_5 = 1$,

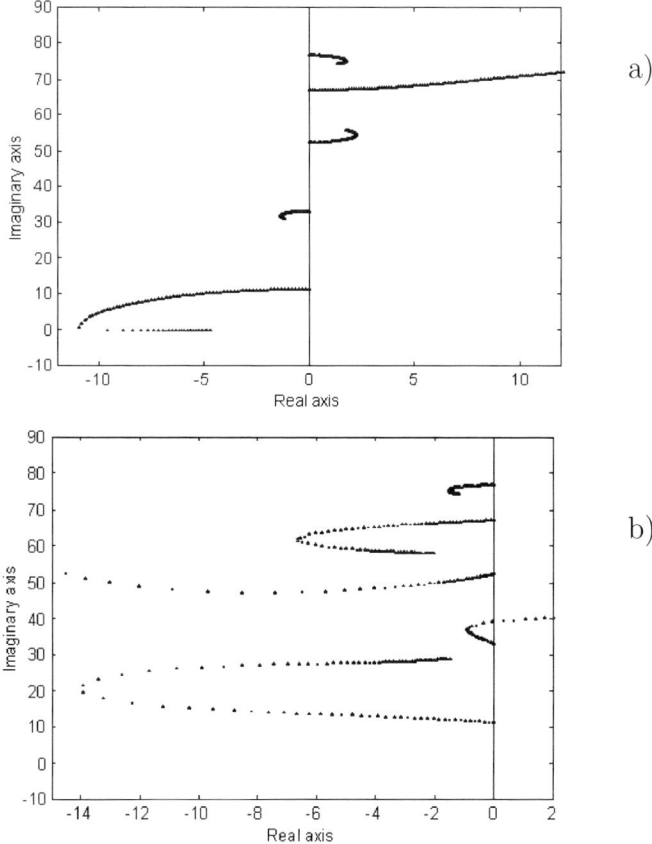

Figure 2.3 Root loci of closed loop system poles for non-collocated velocity feedback control with $j = 4$, $s_1 = 5$, $a_4 = 1$, and time delay (a) $T_5 = 0$ sec, (b) $T_5 = 0.04$ sec.

and $T_5 = 0.04$ sec. The results on Figure 2.4 show that by using intentional time-delayed velocity feedback, the displacement of the 5-th mass is significantly smaller than the displacement of the mass when no control is used. Figure 2.5 shows the time histories of the incoming force per storey (i.e., negative of storey mass times ground acceleration), and the control force required when the controller's gain is $\mu = 10$ units.

To explore the robustness of this control methodology, numerical results of the sensitivity of the closed loop system poles to changes in the mass, stiffness and time delay parameters (for the nominal system presented in the last example) are presented in the next section.

## 2.5   Sensitivity of the stability to perturbed parameters

There is always uncertainty about the exact values of the system's parameters. The uncertainty might come from not being able to measure or estimate the system

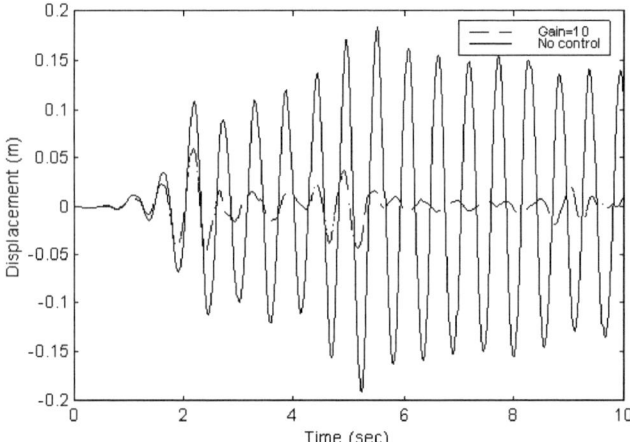

Figure 2.4 Relative displacement response of mass 5 ($j = 1$, $s_1 = 5$, $a_5 = 1$, and $T_5 = 0.04$ sec) for non-collocated velocity control.

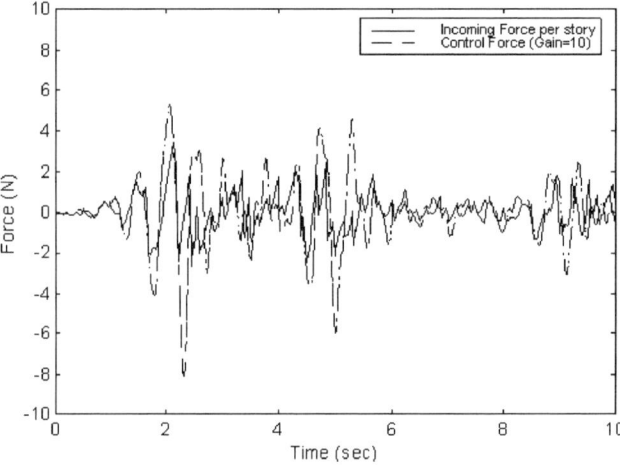

Figure 2.5 Incoming force per storey and control force time histories ($j = 1$, $s_1 = 5$, $a_5 = 1$, and $T_5 = 0.04$ sec) for non-collocated velocity control.

parameters accurately. It may also come from variations of the system parameters caused by fatigue, structural degradation, etc. Changes in the parameter values will lead to changes in the closed loop poles, and thus changes in the performance of the system. This stresses the importance of knowing how sensitive the time-delayed control is to parameter variations. Furthermore, we have shown that the purposeful injection of time delays in non-collocated systems brings about stability. As the injection of such time delays in actual systems can at best be chosen only approximately (because of the uncertainties in actuator dynamics, etc.), it is important to

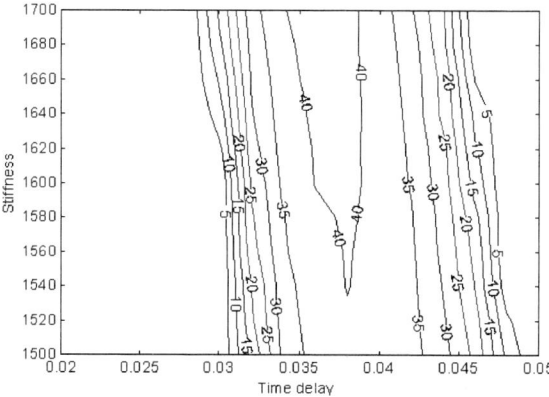

Figure 2.6 Sensitivity of the maximum gain for stability to changes in stiffness and time delay for $j = 1$, $s_1 = 5$, $a_5 = 1$, and mass $= 1$.

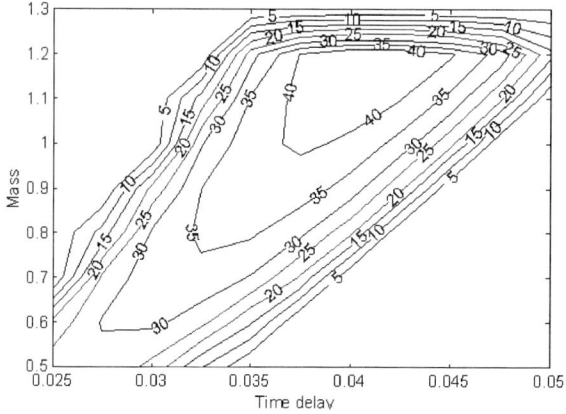

Figure 2.7 Sensitivity of the maximum gain for stability to changes in mass and time delay for $j = 1$, $s_1 = 5$, $a_5 = 1$, and stiffness $= 1,600$.

determine the sensitivity of such a control technique to uncertainties in the time delays.

These sensitivities are studied for the undamped shear frame building structure described in Section 2.4 (see Fig. 2.1). The results are for the non-collocated velocity feedback control, where the actuator is placed at mass 4, and the sensor takes delayed velocity readings from mass 5 ($j = 4$, $s_1 = 5$, $a_5 = 1$). By varying the stiffness (or mass) and the time delay parameters of the shear frame structure, the changes of the maximum gain for stability of the system are obtained. The maximum gain for stability is numerically computed, and it is the largest gain that assures stability when closed loop system poles are considered. The results are presented through contour maps with constant gain curves.

Level curves of maximum gain for stability for the above-described system are presented in Figures 2.6 and 2.7. In Fig. 2.6 the stiffness and the time delay are varied, keeping the mass constant at 1 (SI units). Sensitivity to variations of the mass and the time delay are presented on Fig. 2.7, when the stiffness is kept constant at 1,600 (SI units).

Both graphs show a continuous dependence of the maximum gain for stability on the chosen parameters. The contour plots can help us in the design process by allowing us to select the system parameters that yield a desired performance. The figures show that stable dislocated control brought about by the purposeful injection of time delays could be made effective even in the presence of considerable uncertainties in the parameters that model the structural system. Further results on the stability of controlled systems with uncertainties in the system parameters and time delay are presented in Section 5. The next section deals with the control of more general linear systems which include both non-classically and classically damped structures.

# 3   Non-Classically Damped Systems

In the last section we presented results dealing with classically damped systems, where several of the results apply when only system poles are considered. In this section we present a formulation that deals with more general systems, and the results apply to both, non-classically damped as well as to classically damped systems. Furthermore, the results apply when all the poles are considered; they are not just limited to considering system poles. The importance of having a formulation that deals with "all" the poles of the control system with time delays will be illustrated in section 4 where it is shown that a pole not originating from an open loop pole dictates the stability of the system.

We show that when all the open loop poles have negative real parts and the controller transfer function is an analytic function, then given any time delays there exists a range in the gain (which could depend on the time delays) for which the closed loop feedback system is stable. The results are then specialized to systems with a single sensor. We show that under some conditions there exists a range in the gain for which the closed loop system is stable for all time delays. The section ends with the application of some of the results to a single degree of freedom oscillator.

## 3.1   General formulation

The same general formulation given for classically damped systems (Sec. 2) will be used in this section, except that the systems considered here are more general. They include both classically damped and non-classically damped systems. Consider the following matrix equation corresponding to a structural system with the response $x(t)$ given as

$$Mx'' + Cx' + Kx = g(t), \qquad x(0) = 0 \text{ and } x'(0) = 0, \qquad (3.1)$$

where $M$ is an $n$ by $n$ mass matrix, $C$ is the $n$ by $n$ damping matrix and $K$ is the $n$ by $n$ stiffness matrix. The $n$-vector $g(t)$ is the distributed applied force.

The Laplace transform of the above system is

$$\tilde{A}(s)\tilde{x}(s) = (Ms^2 + Cs + K)\tilde{x}(s) = \tilde{g}(s)\,. \tag{3.2}$$

As in Section 2, we use $p$ responses in the feedback control. A linear combination of the sensed responses is fed to the controller which generates the control force. In general, the system may have several actuators; in this paper we will deal with one actuator. Suppose we have a control effort affecting the $j-$th equation of (3.2) given by $\mu(f(s), \tilde{x}(s))e_j$, where $(f(s), \tilde{x}(s)) = \sum_{i=1}^{n} f_i(s), \tilde{x}_i(s)$, $e_j$ is a vector with 1 in the $j-$th location and zero for all other entries, and $\mu$ is the control gain. The function $f_i(s)$ contains the controller transfer function which uses the signal from $\tilde{x}_i(s)$, and in the presence of a time delay $T_i$, it includes the term $\exp[-sT_i]$ as a factor. Equation (3.2) with the control effort becomes

$$\tilde{A}(s)\tilde{x}(s) = \tilde{g}(s) - \mu(f(s), \tilde{x}(s))e_j\,. \tag{3.3}$$

Moving the term $\mu(f(s), \tilde{x}(s))e_j$ to the left-hand side of (3.3), we get the equation

$$\tilde{A}_1(s)\tilde{x}(s) = \tilde{g}(s)\,. \tag{3.4}$$

The open loop poles of (3.4) are given by equation (3.5a), and the closed loop poles are given by equation (3.5b) as follows:

$$\det\left[\tilde{A}(s)\right] = 0\,, \tag{3.5a}$$

$$\det\left[\tilde{A}_1(s)\right] = 0\,. \tag{3.5b}$$

The equation for the closed loop poles can be written as

$$\det\left[\tilde{A}_1(s)\right] = \det\left[\tilde{A}(s)\right] + \mu\sum_{i=1}^{n} f_i(s)\tilde{A}_{ij}(s)\,, \tag{3.6}$$

where $\tilde{A}_{ij}(s)$ is the cofactor of $\tilde{a}_{ij}(s)$, in other words

$$\tilde{A}_{ij}(s) = (-1)^{(i+j)}\{Minor\,(\tilde{a}_{ij}(s))\}\,,$$

and the matrix $\tilde{A}$ is the one defined in equation (3.2).

## 3.2   General analytical results

The stability of the system described in equation (3.3) is dictated by the sign of the real part of the closed loop poles. Using the argument principle, we can determine bounds on the gain so that the system described in (3.3) is stable in the presence of time delays, provided the open loop poles of the system have negative real parts.

The next result is similar to Result 2.1. It gives conditions so that the open and closed loop systems have no poles in common.

**Result 3.1:** Suppose the open loop poles are given by $\lambda$ with $k = 1, 2, \ldots, 2n$ and the condition

$$q(\lambda_k) = \sum_{i=1}^{n} f_i(\lambda_k) \tilde{A}_{ij}(\lambda_k) \neq 0 \qquad \text{for} \quad k = 1, 2, \ldots, 2n$$

is satisfied. Then the closed loop system and the open loop system have no poles in common.

**Proof:** For any $k$, clearly

$$\det\left[\tilde{A}_1(\lambda_k)\right] = \det\left[\tilde{A}_1(\lambda_k)\right] + \mu \sum_{i=1}^{n} f_i(\lambda_k) \tilde{A}_{ij}(\lambda_k) = \mu \sum_{i=1}^{n} f_i(\lambda_k) \tilde{A}_{ij}(\lambda_k) \neq 0.$$

So no solution of (3.5a) is shared by (3.6), establishing the claim.                     $\square$

For the stability of the closed loop system, we need the real part of the closed loop poles to be negative. The next result gives bounds on the gain so that the closed loop system remains stable.

**Result 3.2:** For any given set of time delays, suppose the open loop poles $\lambda_1, \lambda_2, \ldots, \lambda_{2n}$ all have negative real parts, and the following two conditions are satisfied:

(a) $q(s) = \displaystyle\sum_{i=1}^{n} f_i(s) \tilde{A}_{ij}(s) \neq 0$, and

(b) $q(s)$ is analytic,

for all $s$ in the right-half complex plane and along the imaginary axis. Then there exists an interval $I_{\mu^*} = [-\mu^*, \mu^*]$, with $\mu^* > 0$, such that for any gain $\mu \in I_{\mu^*}$, the closed loop poles are in the left-half complex plane (i.e., the system is stable).

**Proof:** $p(s) = \det\left[\tilde{A}(s)\right]$.
We will use the argument principle (see pages 152–154 in Ahlfors, 1979). Equation (3.6) can be visualized as

$$\det\left[\tilde{A}_1(s)\right] = h(s) = p(s) + \mu q(s),$$

where $p(s) = \det\left[\tilde{A}(s)\right]$, and $q(s)$ is as above. Consider the contour given by the half-circle of radius $R$ in the right-half plane with boundary $\Gamma_R$ and enclosing the region $\Omega_R$, see Figure 3.1.

For any fixed $R$ we have

$$\frac{h'(s)}{h(s)} = \frac{p'(s) + \mu q'(s)}{p(s) + \mu q(s)} = \frac{p'(s) + \mu q'(s)}{p(s)\left(1 + \dfrac{\mu q(s)}{p(s)}\right)}.$$

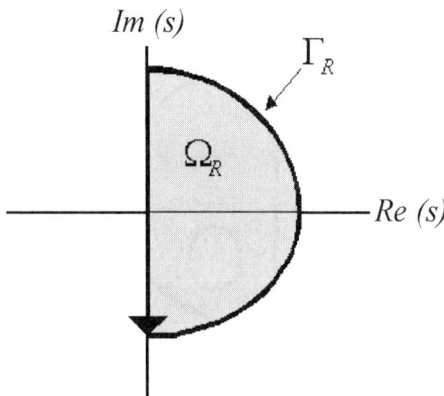

Figure 3.1 Contour of integration.

Now $\dfrac{1}{1+\dfrac{\mu q(s)}{p(s)}} = \displaystyle\sum_{k=0}^{\infty}\left(-\dfrac{\mu q(s)}{p(s)}\right)^{k}$ for all $s \in \Omega_R$, provided $\left|\dfrac{\mu q(s)}{p(s)}\right| < 1$, or

$|\mu| < \left|\dfrac{p(s)}{q(s)}\right|$. By conditions (a) and (b), and the fact that $p(s)$ is a polynomial,

we have that $\dfrac{p(s)}{q(s)}$, is analytic on $\Omega_R$; also, $p(s) \neq 0$ in $\Omega_R$. Thus by the minimum

modulus principle, $\left|\dfrac{p(s)}{q(s)}\right|$ has a nonzero minimum on the boundary $\Gamma_R$ of $\Omega_R$. Let

$\mu^*$ be such that $\mu^* < \min\limits_{s\in\Omega_R} \left|\dfrac{p(s)}{q(s)}\right| = \min\limits_{s\in\Gamma_R} \left|\dfrac{p(s)}{q(s)}\right|$. Therefore, for any $\mu \in [-\mu^*, \mu^*]$,

the infinite series converges. Let $\mu \in [-\mu^*, \mu^*]$, we then have

$$\int_{\Gamma_R} \frac{h'(s)}{h(s)} ds = \int_{\Gamma_R} \frac{p'(s) + \mu q'(s)}{p(s)} \left(\sum_{k=0}^{\infty}\left(-\frac{\mu q(s)}{p(s)}\right)^{k}\right) ds =$$

$$= \sum_{k=0}^{\infty}\left(\int_{\Gamma_R} \frac{p'(s) + \mu q'(s)}{p(s)} \left(-\frac{\mu q(s)}{p(s)}\right)^{k} ds\right) = 0.$$

Note that for each $k$, the integral is an integral of an analytic function over a closed curve, and thus equal to zero. $R$ is arbitrary, so by the argument principle (see pages 152-154 in Ahlfors, 1979) there exists a range $[-\mu^*, \mu^*]$ in $\mu$ for which all the closed loop poles are in the left-half plane. $\qquad\square$

Result 3.2 can be strengthened so that condition (a) is not needed. That is, the function $q(s)$ is allowed to have zeros in the right-half plane.

**Result 3.3:** For a given set of time delays, suppose the open loop poles $\lambda_1, \lambda_2,$ $\ldots, \lambda_{2n}$ all have negative real parts, the function $q(s)$ is analytic for all $s$ in the right-half complex plane and along the imaginary axis, and $q(s)$ has zeros at $s_1, s_2, \ldots, s_m$

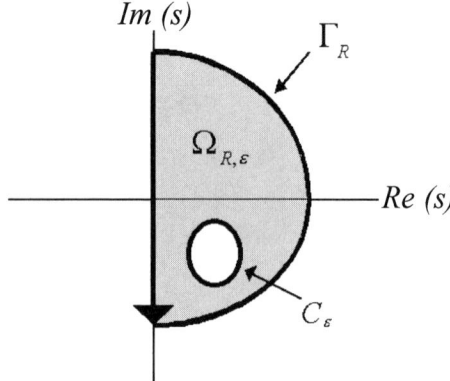

Figure 3.2  Contour of integration.

in the right-half complex plane (including perhaps zeros on the imaginary axis). Then there exists an interval $I_{\mu^*} = [-\mu^*, \mu^*]$, with $\mu^* > 0$, such that for any gain $m \in I_{\mu^*}$ the closed loop poles are in the left-half complex plane (i.e., the system is stable).

**Proof:** The proof is similar to the one given for Result 3.2. First suppose that $q(s)$ has only one zero $s_1$ in the right-half complex plane.

Let $\Gamma_{R,\varepsilon}$ be the contour given by $\Gamma_{R,\varepsilon} = \Gamma_R \cup C_\varepsilon$, where $\Gamma_R$ is the contour given by the half circle of radius $R$ in the right-half plane, and $C_\varepsilon$ is the circle of radius $\varepsilon$ centered about $s_1$. Let $\Omega_{R,\varepsilon}$ be the region bounded by the circle $C_\varepsilon$ and the half circle $\Gamma_R$, see Figure 3.2.

As in the proof of Result 3.2, $\det[\tilde{A}_1(s)] = h(s) = p(s) + \mu q(s)$. We need to show that $\dfrac{1}{1 + \dfrac{\mu q(s)}{p(s)}} = \displaystyle\sum_{k=0}^{\infty} \left( -\dfrac{\mu q(s)}{p(s)} \right)^k$ converges for $\mu \in [-\mu^*, \mu^*]$, where $\mu^*$ needs to be determined. In the region $\Omega_{R,\varepsilon}$, the function $\dfrac{p(s)}{q(s)}$ is analytic, and thus $\left| \dfrac{p(s)}{q(s)} \right|$ has a minimum on the boundary $\Gamma_{R,\varepsilon}$ of $\Omega_{R,\varepsilon}$ (recall that $p$ has no zeros in the left-half plane, nor on the imaginary axis). The question is, what will occur when $\varepsilon \to 0$? Since the minimum occurs on the boundary $\Gamma_{R,\varepsilon}$, it must either occur on $\Gamma_R$ or on $C_\varepsilon$. It will be shown that it must occur on $\Gamma_R$.

The function $q(s)$ can be expressed as $q(s) = (s - s_1)g(s)$. The function $\dfrac{p(s)}{q(s)}$ is analytic inside the closed disk bounded by $C_\varepsilon$, so it has a nonzero minimum $M_\varepsilon = \min_{s \in C_\varepsilon} \left| \dfrac{p(s)}{q(s)} \right|$. Using the last observations we get

$$M_\varepsilon = \min_{s \in C_\varepsilon} \left| \frac{p(s)}{q(s)} \right| = \min_{s \in C_\varepsilon} \left| \frac{p(s)}{(s - s_1)g(s)} \right| \geq \min_{s \in C_\varepsilon} \left| \frac{p(s)}{g(s)} \right| \min_{s \in C_\varepsilon} \frac{p(s)}{|s - s_1|} = M_\varepsilon \frac{1}{\varepsilon} .$$

Thus, if $\varepsilon \to 0$, then $\min_{s \in C_\varepsilon} \left| \dfrac{p(s)}{q(s)} \right| \to \infty$. Therefore, the minimum of $\left| \dfrac{p(s)}{q(s)} \right|$ over the region $\Omega_{R,\varepsilon}$ must occur on the boundary $\Gamma_R$. Using the fact the minimum occurs on $\Gamma_R$, we see that the series $\dfrac{1}{1 + \dfrac{\mu q(s)}{p(s)}} = \sum\limits_{k=0}^{\infty} \left( -\dfrac{\mu q(s)}{p(s)} \right)^k$ converges. We can apply exactly the same argument given in the proof of Result 3.2, establishing the result.

When $q(s)$ has more than one zero (even repeated zeros), a similar argument as the one presented can be used to show that the minimum will occur on the boundary $\Gamma_R$, and hence the series $\dfrac{1}{1 + \dfrac{\mu q(s)}{p(s)}} = \sum\limits_{k=0}^{\infty} \left( -\dfrac{\mu q(s)}{p(s)} \right)^k$ converges, and again the result follows. □

Note that Results 3.2 and 3.3 apply for systems with time delay, and the function $q(s)$ would contain all the expressions containing time delay.

## 3.3 Results for one actuator and one sensor

In this section we present results that apply to the system described in equation (3.3), when one sensor and one actuator are used.

Since we have one sensor at the $i$-th location and the actuator at the $j$-th location, then the closed loop poles are given by

$$\det\left[ \tilde{A}_1(s) \right] = \det\left[ \tilde{A}(s) \right] + \mu f_i(s)\tilde{A}_{ij}(s) = p(s) + \mu q(s). \qquad (3.7)$$

Here $q(s) = f_i(s)\tilde{A}_{ij}(s)$, and $p(s) = \det\left[ \tilde{A}(s) \right]$. The function $f_i(s)$ is the product of the controller transfer function $\tau_c(s)$ and the term $\exp[-sT_i]$ due to the time delay. Let $g(s) = \tau_c(s)\tilde{A}_{ij}(s)$, then $q(s) = g(s)\exp[-sT_i]$. Since $\tilde{A}_{ij}(s)$ is a polynomial, it is analytic, and if we assume that the controller transfer function $\tau_c(s)$ is analytic in the right-half plane, then $g(s)$ is analytic in the right-half plane. In most situations it is reasonable to expect that $p(s)$ goes to infinity faster than $q(s)$ does as $|s| \to \infty$, so here we shall assume that $\left| \dfrac{p(s)}{q(s)} \right| \to \infty$ as $|s| \to \infty$.

Assuming that the open loop poles have negative real parts, then Result 3.3 applies. Following the proof of the result, for large enough $R$, the bound on the gain for stability is given by

$$\mu^* < \min_{s \in \Omega_R} \left| \frac{p(s)}{q(s)} \right| = \min_{s \in \Gamma_R} \left| \frac{p(s)}{q(s)} \right|.$$

The boundary $\Gamma_R$ is composed of a semicircle $C_\varepsilon$ of radius $R$ and a segment of the imaginary axis $I_R$ of length $2R$. The minimum must occur on the segment $I_R$, provided $R$ is large enough and $\left| \dfrac{p(s)}{q(s)} \right| \to \infty$ as $|s| \to \infty$. Since $\left| \dfrac{p(s)}{q(s)} \right| \to \infty$ as $|s| \to \infty$, then there exists an $R^*$ so that the minimum occurs for this $R^*$, and for any $R^* < R$ we have $\min\limits_{s \in \Omega_{R^*}} \left| \dfrac{p(s)}{q(s)} \right| = \min\limits_{s \in \Omega_R} \left| \dfrac{p(s)}{q(s)} \right|$. For $R^* < R$, $C_{R^*}$ is in $\Omega_R$, by the minimum modulus principle, the minimum cannot occur in the interior of $\Omega_R$, thus it occurs on $I_{R*}$. The above observations are used in the proof of the following result.

**Result 3.4:** When using one sensor and one actuator, there exists an interval $[-\mu^*, \mu^*]$ in the gain $\mu$, with $\mu^* > 0$, which is *independent of time delay*, such that the closed loop poles have negative real parts, provided all the open loop poles have negative real parts, and $\left|\dfrac{p(s)}{q(s)}\right| \to \infty$ as $|s| \to \infty$.

**Proof:** By the previous argument, for large enough $R$, the minimum $\min\limits_{s \in \Gamma_R}\left|\dfrac{p(s)}{q(s)}\right|$ will occur on $I_R$. Say it occurs at $s = iw^{**}$ ($i = \sqrt{-1}$) then for large enough $R$,

$$\mu^* < \mu^{**} = \min_{s \in \Omega_R}\left|\frac{p(s)}{q(s)}\right| = \min_{w \in I_R}\left|\frac{p(iw)}{g(iw)\exp[-iwT_i]}\right| = \min_{w \in I_R}\left|\frac{p(iw)}{g(iw)}\right| = \left|\frac{p(iw^{**})}{g(iw^{**})}\right|.$$
(3.8)

This establishes the result. $\qquad\qquad\qquad\qquad\qquad\qquad\qquad\qquad\qquad\qquad\qquad$ $\square$

From the result, the bound $\mu^{**}$ can be obtained by considering the system with no time delay and finding the minimum of $\left|\dfrac{p(iw)}{q(iw}\right|$ for $w \in I_R$. For some systems it may be possible that $\mu^{**}$ is actually the maximum gain for stability for the system with no time delay. This will be the case in *Example 1* shown after Result 3.6 as well as for the control systems explored in Section 4. That is, for zero time delay one finds the maximum gain for stability using positive and negative feedback, and $\mu^{**}$ is the minimum of the two maximum gains. In general, however, it is not necessarily true that the minimum of the maximum gains for stability $\mu^{**}$ is achieved for the system with zero time delay. It is possible for some systems that the maximum gain for stability of the system with zero time delay is larger than the gain that is independent of time delay. This case will be illustrated in *Example 2*.

Since the bound $\mu^{**}$ occurs on the imaginary axis and it may not necessarily be the bound for the maximum gain for stability for the system with no time delay, and it is a bound on the gain for which the system is stable for all time delays, one might ask if $\mu^{**}$ will actually be the maximum gain for stability at some time delay. The following result answers this question.

**Result 3.5:** For systems with one sensor and one actuator, as described in Result 3.4, suppose $\mu^{**}$ occurs at $s = iw^{**}$, then there exists a time delay $T$ such that
$$p(iw^{**}) + \mu^{**}g(iw^{**})\exp(-iw^{**}T) = 0.$$

That is, the bound of the maximum gain for stability independent of time delay given in Result 3.4 is achieved at some time delay $T$. The value of $T$ will be given in the proof, and it is derived from the system properties without time delay.

**Proof:** Let the real and imaginary parts of $p(iw^{**})$ and $q(iw^{**})$ from Result 3.4 be given by
$$p(iw^{**}) = p_R + ip_I, \quad \text{and} \quad g(iw^{**}) = g_R + ig_I.$$
(3.9)

Consider
$$p_R + ip_I + \mu^{**}(g_R + ig_I)(x - iy) = 0.$$

Then

$$x - iy = -\frac{p_R g_R + p_I g_I}{\mu^{**}(g_R^2 + g_I^2)} - \frac{p_I g_R - p_R g_I}{\mu^{**}(g_R^2 + g_I^2)} . \qquad (3.10)$$

Note that $\mu^{**} = \sqrt{\dfrac{p_R^2 + p_I^2}{g_R^2 + g_I^2}}$. and $x^2 + y^2 = 1$. Therefore we can take $\cos(w^{**}T) = x$ and $\sin(w^{**}T) = y$, establishing the result. $\qquad \square$

Result 3.5 indicates that there is a time delay at which the system will achieve the bound on the maximum gain for stability given in Result 3.4. The result also indicates if the bound of the maximum gain for stability that is independent of time delay is actually achieved or not for zero time delay. If $x = 1$, then for the system with zero time delay the maximum gain for stability will be reached with gain $\mu = \mu^{**}$. Similarly, if $x = -1$, then for the system with zero time delay the maximum gain for stability will be reached with gain $\mu = -\mu^{**}$. If $|x| \neq 1$, then the maximum gain for stability of the system with zero time delay is larger than the maximum gain for stability that is independent of time delay. The question of the uniqueness of the occurrence of poles at a gain of $\mu^{**}$ (for positive and negative feedback) is dealt with in the next result.

**Result 3.6:** For the case of a single actuator and a single sensor, suppose that a closed loop pole given by (3.7) is at $s = iw^*$, $\mu = \mu^*$ and with a time delay of $T = T^*$. Then there exist closed loop poles at $s = iw^*$, $\mu = -\mu^*$, with time delays $T = T^* + \dfrac{(2n + 1)\pi}{w^*}$, for $n = 0, 1, 2, \ldots$, and at $s = iw^*$, $\mu = \mu^*$, with time delays $T = T^* + \dfrac{2n\pi}{w^*}$, for $n = 1, 2, \ldots,$.

**Proof:** For a pole on the imaginary axis, we have

$$p(iw) + \mu g(iw) \exp[-iwT] = p(iw) + \mu g(iw) \left( \cos(wT) - i \sin(wT) \right) . \qquad (3.11)$$

When the pole occurs at $w = w^*$, $\mu = \mu^*$ and $T = T^*$, then (3.11) is also satisfied with $w = w^*$, $\mu = \mu^*$, provided the time delay $T$ satisfies the equations

$$\cos(w^*T^*) = \cos(w^*T) \quad \text{and} \quad \sin(w^*T^*) = \sin(w^*T) .$$

These equations are satisfied when $T = T^* + \dfrac{2n\pi}{w^*}$, for $n = 1, 2, \ldots,$. Similarly, when the pole occurs at $w = w^*$, $\mu = \mu^*$ and $T = T^*$, then (3.11) is also satisfied with $w = w^*$, $\mu = -\mu^*$, provided the time delay $T$ satisfies the equations

$$\cos(w^*T^*) = -\cos(w^*T) \quad \text{and} \quad \sin(w^*T^*) = -\sin(w^*T) .$$

These equations are satisfied when $T = T^* + \dfrac{(2n + 1)\pi}{w^*}$, for $n = 0, 1, 2, \ldots,$. $\qquad \square$

One consequence of this result is that the maximum gain for stability which is independent of time delay (from Result 3.4) will be achieved for both positive and negative feedback, when an appropriate time delay is chosen. A time delay $T^*$ at which the minimum maximum gain for stability is achieved can be obtained using Result 3.6.

In the next example, the bound on the maximum gain for stability that is independent of time delay (given in Results 3.4) coincides with the maximum gain for stability of the system with zero time delay.

*Example 1:* As a simple illustration of Results 3.4, 3.5 and 3.6, consider a single degree of freedom oscillator with response $x(t)$, and with mass $m$, damping $c > 0$ and stiffness $k > 0$. The oscillator is controlled by using a time-delayed negative velocity feedback $-\mu x'(t - T)$, with time delay $T$ and using a control gain $\mu$. The motion of the oscillator is described by the scalar differential equation

$$mx''(t) + cx'(t) + kx(t) = -\mu x'(t - T), \qquad x(0) = 0, \quad x'(0) = 0. \qquad (3.12)$$

The Laplace transform of (3.12) is

$$(ms^2 + cs + k + \mu s \exp[-sT]) \, \tilde{x}(s) = 0. \qquad (3.13)$$

The closed loop poles of the system described by (3.12) are the zeros of the equation

$$ms^2 + cs + k + \mu s \exp[-sT] = 0. \qquad (3.14)$$

From Result 3.4, we have that the maximum gain for stability which is independent of time delay occurs on the imaginary axis at some $s = iw$. Evaluation of (3.14) at $s = iw$ gives

$$-mw^2 + icw + k + i\mu w \left( \cos(wT) + i\sin(wT) \right) = 0. \qquad (3.15)$$

The imaginary part of (3.15) is

$$cw + \mu w \cos(wT) = 0 \qquad \text{or} \qquad \cos(wT) = \frac{-c}{\mu}. \qquad (3.16)$$

Note that $w = 0$ is not a solution, since it would violate (3.15). Equation (3.16) has a solution only if $\left| \dfrac{c}{\mu} \right| \le 1$, that is $|\mu| \ge c$. Thus for $|\mu| < c$, equation (3.15) has all poles in the left-half complex plane (i.e., all solutions have negative real parts).

For zero time delay ($T = 0$) the closed loop poles are given by the zeros of the equation $ms^2 + (c + \mu)s + k = 0$. From the Routh stability criterion, for negative feedback ($\mu > 0$) the system will be stable for all gains. On the other hand, for positive feedback ($\mu < 0$), the system has a cross-over pole at $\mu = -c$ and $w = \sqrt{\dfrac{k}{m}}$.

Using Result 3.5 one can confirm this observation. From Result 3.5 we have

$$\mu^{**} = \min_{w \in I_R} \left| \frac{p(iw)}{g(iw)} \right| = \min_{w \in I_R} \left| \frac{-mw^2 + k + icw}{iw} \right| = c,$$

and the minimum occurs at $w^{**} = \sqrt{\dfrac{k}{m}}$.

Additionally from Result 3.5, for negative feedback we have that $x = \cos\left( \sqrt{\dfrac{k}{m}} T \right) = -1$ and $y = \sin\left( \sqrt{\dfrac{k}{m}} T \right) = 0$. The smallest time delay for which these equations are satisfied is $T = \pi\sqrt{\dfrac{m}{k}}$. Thus for negative feedback and zero time delay the bound on the maximum gain for stability independent of time delay is not achieved. For positive feedback we have $x = \cos\left(\sqrt{\dfrac{k}{m}}T\right) = 1$ and $y = \sin\left(\sqrt{\dfrac{k}{m}}T\right) = 0$, and thus the bound on the maximum gain for stability is reached at zero time delay.

Using result 3.6 we have that for negative feedback the system will reach the minimum bound for the gain for stability at $w = \sqrt{\dfrac{k}{m}}$ for the time delays $T = (2n+1)\pi\sqrt{\dfrac{m}{k}}$ (for $n = 0, 1, 2, \dots$) with a gain of $\mu = c$. For positive feedback the system will reach the minimum bound for the gain at $w = \sqrt{\dfrac{k}{m}}$ for the time delays $T = 2n\pi\sqrt{\dfrac{m}{k}}$ (for $n = 0, 1, 2, \dots$) with a gain of $\mu = -c$.

The next example shows the case where the maximum gain for stability of the system with zero time delay is larger than the gain for stability that is independent of time delay, given in Result 3.4.

*Example 2:* Consider again a single degree of freedom oscillator with response $x(t)$, and with mass $m$, damping $c > 0$ and stiffness $k > 0$. This time the oscillator is controlled by using the negative feedback time-delayed proportional control $-\mu x(t - T)$, with time delay $T$ and a control gain $\mu$. The motion of the oscillator is described by the scalar differential equation

$$mx''(t) + cx'(t) + kx(t) = -\mu x(t - T)\,, \qquad x(0) = 0\,, \quad x'(0) = 0\,. \qquad (3.17)$$

After taking the Laplace transform of (3.17), the poles of the system are the zeros of the equation

$$ms^2 + cs + k + \mu\exp[-sT] = 0\,. \qquad (3.18)$$

The bound on the maximum gain for stability that is independent of time delay can be obtained using (3.8). When $2mk - c^2 > 0$, we have

$$\mu^{**} = \min_{w \in I_R}\left|\frac{p(iw)}{g(iw)}\right| = \min_{w \in I_R}\left|-mw^2 + icw + k\right| = \sqrt{\frac{c^2(4mk - c^2)}{4m^2}}\,. \qquad (3.19)$$

Where the minimum in (3.19) occurs at $w^{**} = \sqrt{\dfrac{2mk - c^2}{2m^2}}$.

For zero time delay the closed loop poles are the zeros of the equation $ms^2 + cs + (k + \mu) = 0$. Based on the Routh stability criterion, the negative feedback system will be stable for all gains. In the case of positive feedback, the system will have a unique cross-over at $w = 0$ and a gain of $\mu = -k$. When $2mk - c^2 > 0$ we have that $\mu^{**} < k$. This indicates that the magnitude of the maximum gain for stability of the system with zero time delay for positive feedback is larger than the bound of the gain that is independent of time delay.

The values for $x$ and $y$ in (3.10) are

$$x = -\frac{c}{\sqrt{4km - c^2}} \quad \text{and} \quad y = \sqrt{\frac{4km - 2c^2}{4km - c^2}}. \tag{3.20}$$

From Result 3.5 one can compute the time delay at which the bound on the gain for stability that is independent of time delay is reached by the system. The smallest time delay at which the system reaches the bound on the gain for stability that is independent of time delay is the smallest value of $T$ that satisfies the equations

$$T = \frac{\cos^{-1}(x)}{w^{**}} \quad \text{and} \quad T = \frac{\sin^{-1}(y)}{w^{**}}. \tag{3.21}$$

Using Result 3.6, one can now obtain all the possible instances in which the system reaches the bound on the gain that is independent of time delay for the system with positive and negative feedback.

Though the above simple examples illustrate Results 3.4, 3.5 and 3.6 for classically damped systems, the results presented in this section apply to non-classically damped system as well. We showed that when the open loop system has all poles with negative real parts, and the controller transfer function is an analytic function, then there is a range in the gain for which the closed loop system is stable (even in the presence of time delays).

For the case of a single sensor with a single actuator, we showed that there is a range in the gain for which the system is stable for all time delays. This is provided the open loop system has all poles with negative real parts, and the controller transfer function satisfies some conditions.

The results imply that for the systems considered, all poles originate in the left complex plane, and any pole in the right-half complex plane can be traced to the left-half complex plane by varying the gain.

The fact that for some systems the lower bound on the maximum gain for stability which is independent of time delay is achieved when no time delay is present (*Example* 1), suggests that the use of a time delay may actually increase the maximum gain for stability when compared to the system with zero time delay for such systems. Therefore with an appropriate choice of time delay one could increase the range in the gain for which the system is stable, thereby making the use of time delays desirable. From a control design point of view, this could improve the control performance dramatically for such systems.

# 4    Experiments with a Non-Classically Damped, 2–DOF Torsional System

The effect of time delay on the control of a two-degrees-of freedom (2–DOF) torsional bar is explored experimentally and numerically. The maximum gain for stability is determined experimentally over a range of time delays for collocated integral and non-collocated derivative control, and this is compared to analytical/numerical predictions based on a model of the system. For the case of collocated integral control, it is found both experimentally and numerically that a pole of the structural system, which *does not* originate from an open loop pole, dictates the maximum gain for stability for some time delays. In this section we first provide a description of the apparatus, including a mathematical model. Following this, the experimental procedure is summarized. Finally, the experimental and numerical results are presented. Additional experimental and numerical results on collocated and non-collocated proportional, derivative and integral control of the torsional bar can be found in von Bremen *et al.* (2001).

In the absence of any time delays, the system has 2–DOF (with two sets of system poles). However, in the presence of time delays in the control loop, this seemingly simple system has a complex behavior for it is no longer finite dimensional. It has an infinite number of poles, and as the control gain increases from zero these poles 'stream in' from $-\infty$ in the left half complex plane towards the imaginary axis. As we will show, their root loci can 'collide' with those of the system poles, thereby leading to interesting behavior and bifurcations.

## 4.1   Experimental setup and model

The setup (Fig. 4.1) consists of two discs that undergo torsional vibrations. The inertial properties of the disks can be altered by fastening additional weights to them. The control system has four primary components: (1) the real-time controller that generates the input trajectory and computes the control algorithm, (2) the software for defining the controller, (3) the actuator at the lower disc, and (4) the optical sensors. The real-time controller is a digital signal processor-based single-board computer. The servo loop closure involves the computation of the user-supplied control algorithm, and these computations occur at a rate of once every sampling period (0.00442 sec).

The actuator that actuates the lower disk utilizes a brushless DC motor with electrical commutation. Electrical commutation is accomplished by a sinusoidal switching scheme, which has the advantage of reducing the magnitude of torque ripple. A sensor is secured to the motor shaft and reads its position. There are four incremental rotary shaft optical encoders on the system. Three are used to sense the position of the rotating disks. They have a resolution of 4,000 pulses per revolution. The fourth encoder, with a resolution of 1,000 pulses per revolution, is connected to the motor.

Accompanying the experimental results is a numerical study of the system. A model of the experimental apparatus appears in Figure 4.2. The equations of motion of the two masses in Figure 4.1 are as follows:

$$\begin{cases} J_1\theta_1'' + c_1\theta_1' + k_1(\theta_1 - \theta_2) = T_c(t)\,, \\ J_2\theta_2'' + c_2\theta_2' + k_1(\theta_2 - \theta_1) + k_2\theta_2 = 0\,. \end{cases} \tag{4.1}$$

Here, $J_i$, $i = 1, 2$, are the mass moments of inertia of the disks; $c_i$, $i = 1, 2$, are the respective viscous damping coefficients; $k_i$, $i = 1, 2$, are the stiffness coefficients; $\theta_i$, $i = 1, 2$, are the angular displacements of the disks; and $T_c(t)$ is the actuator torque.

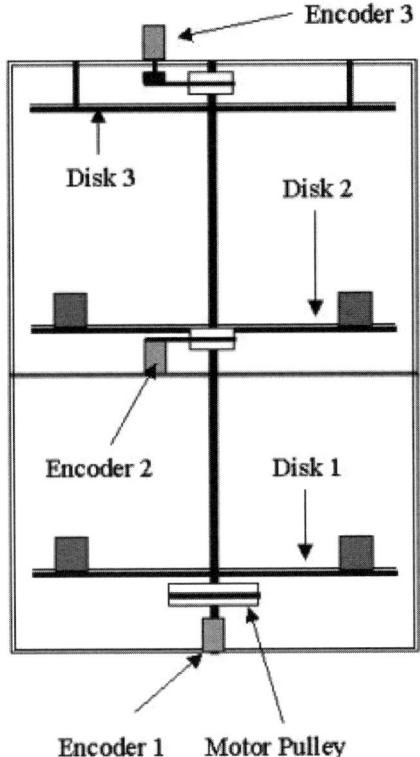

Figure 4.1  Experimental apparatus.

The actuator control torque for non-collocated derivative and for collocated integral control is respectively of the form

$$T_c(t) = -\mu\, \theta_2'(t-T) \qquad \text{and} \qquad T_c(t) = -\mu \int\limits_0^t \theta_1(\tau - T)\, d\tau. \qquad (4.2)$$

In both cases, $\mu$ is the control gain, and $T$ is the time delay in the control.

The system parameters are estimated by clamping each disk, in turn, and measuring the vibratory responses. These results are found in Table 4.1. Here, $\omega_i$, $i = 1, 2$, are the natural frequencies of the disks; $\zeta_i$, $i = 1, 2$, are the damping ratios of disks 1 and 2.

## 4.2   Control procedure

Derivative control requires the estimation of the response derivative in real time. The following numerical approximation was used to compute the derivative $\theta_2'(t-T)$:

$$\theta_2'(t-T) = \frac{1}{2h}\{\theta_2(t - (2h+T)) - 4\theta_2(t - (h+T)) + 3\theta_2(t-T)\} + O(h^2).$$

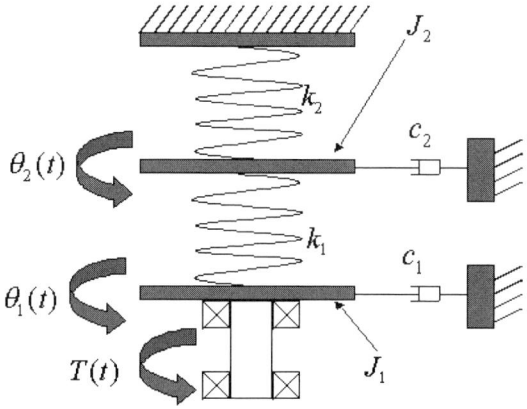

Figure 4.2  Model of experimental apparatus.

Table 4.1  Experimentally determined system parameters.

| System Parameter | Experimental Value |
| --- | --- |
| $J_1$ | 0.00252 kg m$^2$ |
| $J_2$ | 0.00194 kg m$^2$ |
| $k_1$ | 2.830 N m/rad |
| $k_2$ | 2.697 N m/rad |
| $c_1$ | 0.00659 N m sec/rad |
| $c_2$ | 0.00229 N m sec/rad |
| $\omega_1$ | 33.5 rad/s |
| $\omega_2$ | 37.3 rad/s |
| $\zeta_1$ | 0.0390 |
| $\zeta_2$ | 0.0158 |

Here $h = T_s$ (with $T_s = 0.00442$ sec), and $T$ is the time delay. Using the above expression for the derivative, the control effort with control gain $\mu$ for non-collocated derivative control using a time delay of $T$ becomes

$$T_c(t) = -\frac{\mu}{2T_s} \left\{ \theta_2(t) - (2T_s + T)) - 4\theta_2(t - (T_s + T)) + 3\theta_2(t - T) \right\}. \qquad (4.3)$$

Integral control requires the estimation of the integral. The trapezoidal rule is used to approximate the integral as

$$\int_0^{nh} \theta_1(\tau - T)\, d\tau = \frac{h}{2} \left\{ \theta_1(-T) + \sum_{j=1}^{n-1} \theta_1(jh - T) + \theta_1(nh - T) \right\} + O(h^2).$$

Again, $h = T_s$ and $T$ is the time delay. Recall that the system is initially at rest, so for $t < 0$, $\theta_1(t) \equiv 0$. Using the expression for the integral and the last observation, the control effort for collocated integral control with a time delay of $T$ is

$$T_c(t) = -\frac{\mu T_s}{2} \left\{ \sum_{j=1}^{n-1} \theta_1(jT_s - T) + \theta_1(nT_s - T) \right\}. \qquad (4.4)$$

In order to implement a time delay in the control, the control algorithm stores disk positions for previous sampling periods. Then, when defining the control law as in equations 4.3 and 4.4, this past data is utilized. This means, however, that one is limited to time delays that are multiples of the sampling period.

The system described above is equipped with a safety feature which aborts control if the flexible shaft is over deflected or if the speed of the motor is too high. This was taken into account when the stability of the system was to be determined. A stable system was one where the amplitudes of motion would decrease with time. An unstable system was one where the amplitudes of motion increased with time, often leading to the safety limits being exceeded.

## 4.3  Experimental results

The maximum gain for stability of the system over a range of time delays is experimentally determined when using collocated integral control and non-collocated derivative control. These experimental results are compared to numerical predictions of the maximum gain for stability of the system and, in general, there is close agreement for a wide range of time delays.

We also present numerical results that help us understand the effect that time delays have on the stability of the system. The numerical results presented deal with bifurcations, and the fact that a *non-system pole* (a pole whose root locus does not originate at an open loop pole of the structural system) can dictate the maximum gain for stability for the closed loop system. A system pole is simply a pole whose root locus originates at an open loop pole of the structural system. Non-system poles include poles that originate at the poles of the controller, and those caused by the presence of the time delay.

When increasing the gain from zero, the pole that first crosses the imaginary axis will be called the dominant pole. This pole in essence limits the range of the gain for which the system is stable. Intuitively, one might expect that the dominant pole to be a system pole, because system poles are directly related to the physical structure and the non-system poles are induced by the controller and the time delay.

The open loop poles of the system when using the experimental parameter values from Table 4.1 are at $s_{1,2} = -0.7482 \pm 59.4029i$ and at $s_{3,4} = -1.1495 \pm 21.0010i$. Note that the poles come in conjugate pairs, and this will be the case for all the poles including the non-system poles. On root loci plots, the location of the open loop poles will be denoted with the symbol "*" (making it easy to identify the root locus corresponding to a system pole). The system poles will usually be traced using a thick line, while the non-system poles with a thin line. In this subsection, all the time delays are given in units of seconds and frequencies in rads/sec.

Tracking the system poles as the gain changes is done simply by following the root loci of the poles starting at an open loop pole. However, the task of tracking non-system poles is a difficult one, since we may have infinitely many of them, and there is no systematic way of selecting an initial location for all the poles[1] and tracing them by varying the gain (as for system poles). The non-system poles that originate at the poles of the controller or end at the zeros of the controller can, however, be traced in the usual way.

### 4.3.1 Collocated integral control

Figure 4.3a represents a plot of the experimental results for time delay versus the maximum gain for stability under collocated integral control. The solid line on the plot is numerically determined using the system model described above. According to the numerical results, then, a coordinate pair (time delay, gain) which finds itself above the line is unstable, and one below the line is stable. The experimental maximum gain for stability is also depicted on this graph as data points.

The numerical results show a trend of increasing maximum gain for increasing time delay until about 0.12 sec where the gain begins to decrease. The curve of the maximum gain for stability suggests that one can properly choose a time delay that can give the system a *larger* maximum gain for stability than the one the system has for no time delay.

A root locus of a pole that starts (when the gain is close to zero) in the left-half complex plane may cross the imaginary axis several times (as the gain is increased). The value of the pole (on a root locus) in the complex plane when it crosses from the left-half (complex) plane to the right-half (complex) plane for the first time, as the gain is increased gradually from zero, will be called the cross-over frequency. Figures 4.3b and 4.3c show the maximum gain for stability and the cross-over frequency versus time delay for collocated integral control, when only system poles are considered. The points denoted by "o" correspond to system poles originating at $s_1$, and the points denoted by "*" correspond to the system pole originating at $s_3$ (due to symmetry, the conjugates of these two system poles are also present in the system). Similar plots are given on Figures 4.3d and 4.3e, which show the maximum gain for stability and the cross-over frequencies versus time delay for collocated integral control, when all (system and non-system) poles are considered. Even though the system does not satisfy all the required conditions of Result 3.4, the maximum gain for stability which is independent of time delay, given by Result 3.4 is 3.266, which is exactly the numerically expected value, and it is very close to the experimentally observed value. We note that at zero time delay, the smallest maximum gain for stability is achieved (see Figures 4.3a, 4.3b and 4.3d).

Figures 4.3d and 4.3e are smooth continuous curves, while Figures 4.3b and 4.3c seem discontinuous and present abrupt changes. The difference between the system-poles data plots and the all-poles data plots suggests that there is a range in the time delay where a non-system pole is the dominant pole. This expectation is confirmed by plots in Figures 4.3f and 4.3g, which show the root locus of the system poles (thick lines) and two non-system poles (thin lines) for time delays of 0.11 sec and 0.13 sec, respectively. The root loci are taken for values of the gain $\mu$ between 0 and 50, except for the portion of the root loci that lies on the horizontal axis,

---

[1]By means of Jensen's formula (see Rudin 1987, page 307) it is possible to determine the magnitude of the non-system poles at given gains and time delays for special systems.

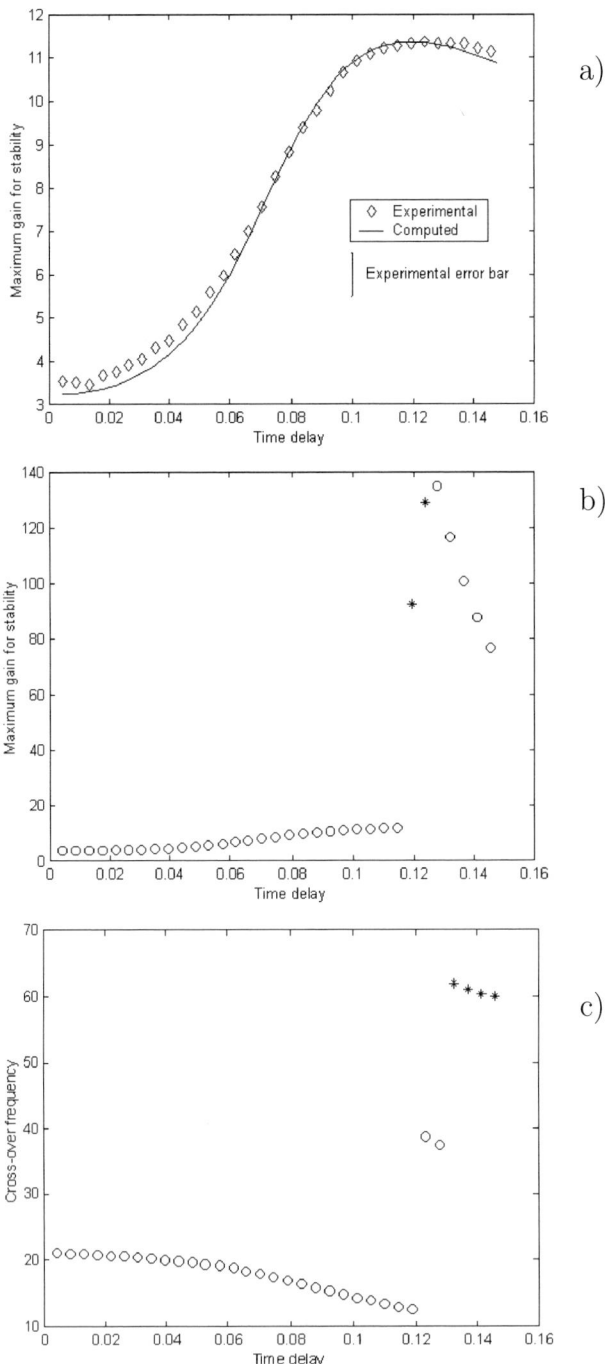

Figure 4.3  (a) Experimental maximum gain for stability versus time delay for integral collocated control. (b) Maximum gain for stability versus time delay for collocated integral control using system poles only. (c) Cross-over frequency versus time delay for collocated integral control system poles only data.

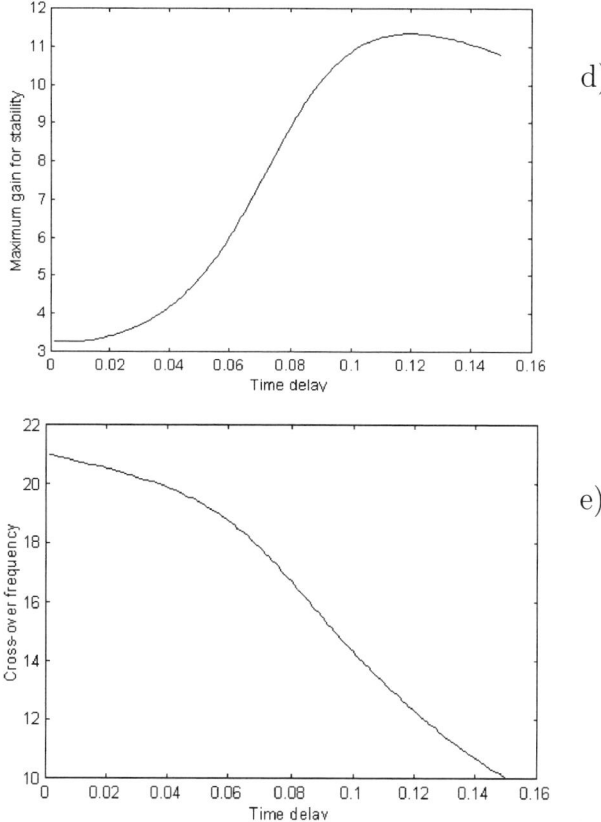

Figure 4.3 *(continued)* (d) Maximum gain for stability versus time delay for collocated integral control, including all poles. (e) Cross-over frequency versus time delay for collocated integral control, including all poles.

where the initial gain is larger than 0 in order to avoid problems of scaling. For the time delay of 0.11 sec, two system poles cross the imaginary axis first, that is, the dominant poles are two conjugate system poles (see Fig 4.3f), while for a time delay of 0.13 sec, two conjugate non-system poles have become dominant (see Fig. 4.3g).

The fact that there is an exchange of dominant pole status between a system pole and a non-system pole suggests the existence of a bifurcation. This bifurcation is presented on Figure 4.3h which shows the root loci for a system pole (thick line) and a non-system pole (thin line) at different time delays. The system pole root locus starts at a gain of zero, while the portion of the non-system pole displayed starts from the horizontal axis with a gain larger than zero. The maximum gain is 50 for both poles. A bifurcation occurs at a time delay of about 0.112 sec. The poles selected are such that the system pole is the dominant pole for time delays less than 0.112 sec, and the non-system pole is the dominant pole for time delays greater than 0.113 sec. Initially (for small time delays), the system pole is

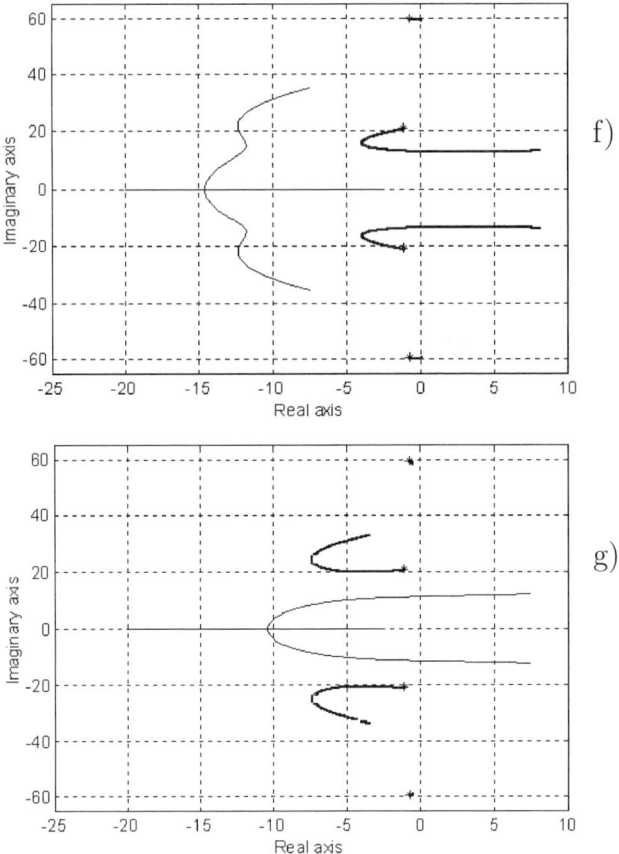

Figure 4.3 *(continued)* (f) Root locus of the system poles and two non-system poles for time a delay of 0.11 sec, using collocated integral control. (g) Root locus of the system poles and two non-system poles for time a delay of 0.13 sec, using collocated integral control.

the dominant pole, as the time delay is increased the root loci of the two poles move closer until they touch (approximately at a time delay of 0.11265 sec). This is the point where the bifurcation occurs. At the bifurcation, one "arm" of the root loci is exchanged among the two poles. The system pole gives the "arm" that dictates the maximum gain for stability to the non-system pole, and the non-system poles gives the slow-moving "arm" to the system pole. After the exchange of "arms" at the bifurcation, the root loci of the two poles move apart as the time delay is increased.

To confirm that a non-system pole is actually the dominant pole, two experiments are conducted on the 2–DOF torsional bar. The system is fed a sine wave with a given frequency, and the steady state amplitude of the response of disk 1 is recorded for different gains, while the control effort is active. The experiments are conducted using three different frequencies. These chosen frequencies are the

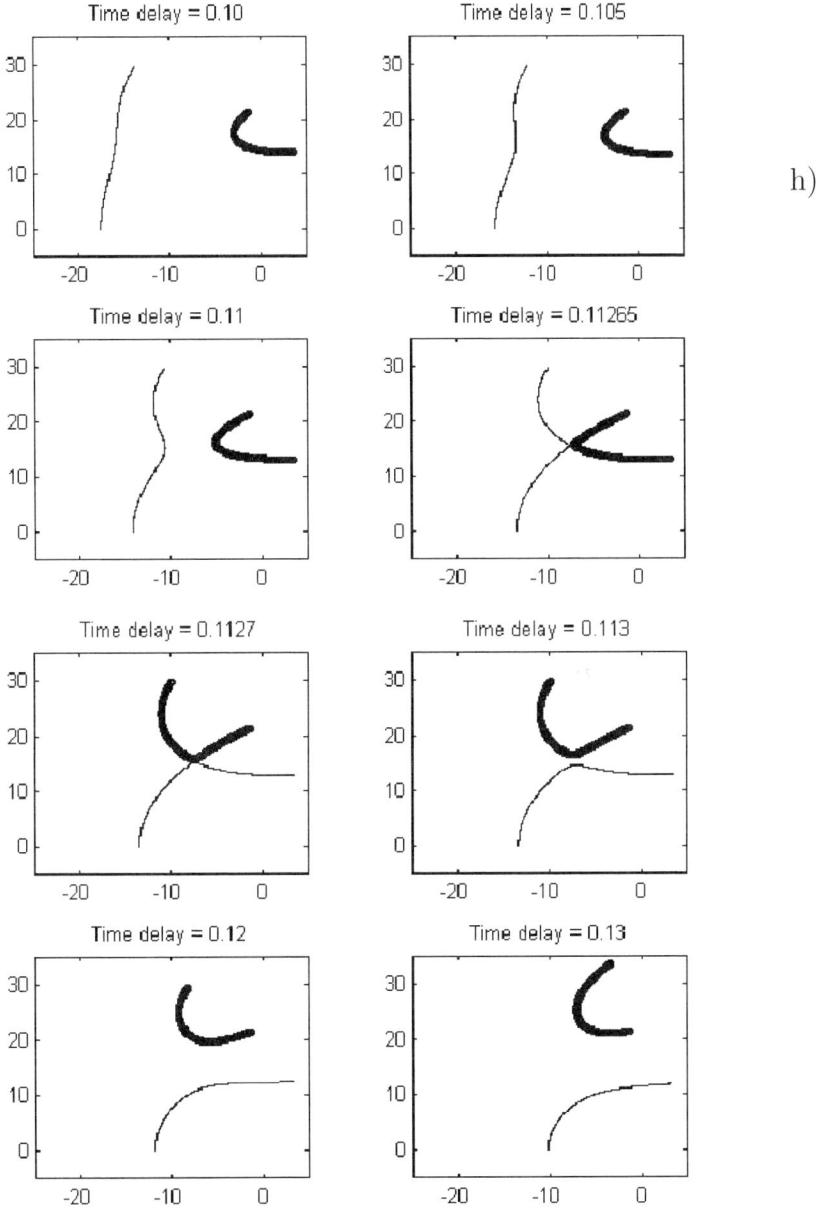

h)

Figure 4.3 *(continued)* (h) Root loci for different time delays showing bifurcation for collocated integral control.

cross-over frequencies for two of the system poles (originating from $s_1$ and $s_2$) and a non-system pole at a given time delay. Recall that the cross-over frequency is defined as the purely imaginary value of a pole (moving along a root locus and starting in the left-half complex plane) when it first crosses the imaginary axis, as

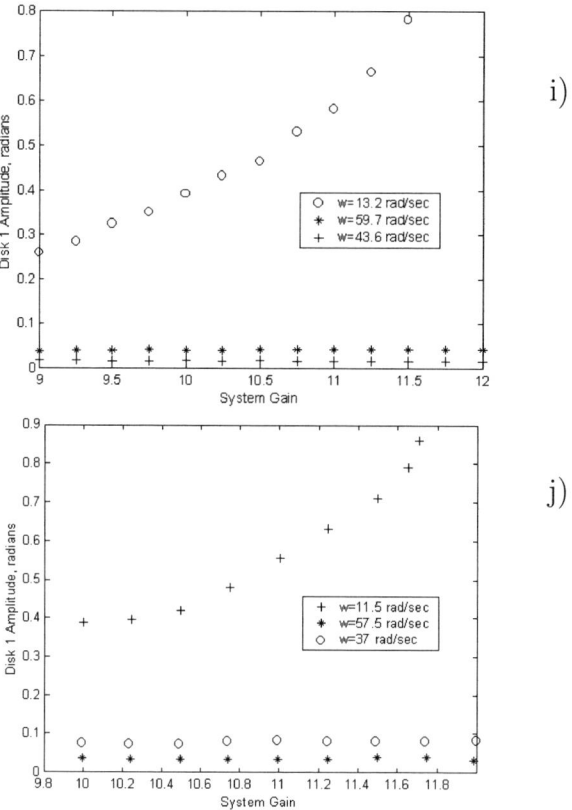

Figure 4.3 *(continued)* (i) Amplitude of the steady state response of disk 1 versus gain for different frequencies of the sine wave using collocated integral control and a time delay of 0.11 sec. (j) Amplitude of the steady state response of disk 1 versus gain for different frequencies of the sine wave using collocated integral control and a time delay of 0.13 sec.

the gain is gradually increased from zero. The time delays are taken to be 0.11 sec in the first experiment and 0.13 sec in the second. The objective is to experimentally confirm the location of points on the root locus.

The results for a time delay of 0.11 sec are shown in Figure 4.3i. The frequencies of 13.2 and 59.7 rad/sec correspond to the cross-over frequencies of the system poles, while the frequency of 43.6 rad/sec is the cross-over frequency for the non-system pole. As expected, when the system is excited near the frequency of the dominant pole, and at a gain close to the maximum gain for stability, the amplitude of the response increases drastically. However, when the system is excited at a frequency which is "far" from the cross-over frequency, the amplitude of the response does not increase greatly, even in the vicinity of the maximum gain for stability (as seen for the frequencies of 59.7 and 43.6 rad/sec). This experiment confirms that for a time delay of 0.11 sec, the dominant pole is a system pole which crossed-over at a frequency of about 13.2 rad/sec.

The results for a time delay of 0.13 sec are shown in Figure 4.3j. The frequencies of 57.5 and 37 rad/sec are the cross-over frequencies for the system poles, and the frequency of 11.5 rad/sec is the cross-over frequency of a non-system pole. The plot shows that the amplitude of the oscillations of disk 1 increases when the system is excited at the expected cross-over frequency of the non-system pole near the maximum gain for stability. On the other hand, when the system is excited at the cross-over frequencies of the system poles (which are not dominant), the amplitude of the response remains small and almost unchanged near the maximum gain for stability corresponding to the dominant pole. This behavior confirms the expectation that at a time delay of 0.13 sec, a non-system pole is the dominant pole.

The experimental results show that we that a non-system pole can be the dominant pole for collocated integral control, and that our analysis can predict this behavior. Also, the maximum gain for stability *with* time delays is much higher than in the absence of a time delay. The behavior of the system can be theoretically explained in terms of bifurcation diagrams.

### 4.3.2 Non-collocated derivative control

Figure 4.4a represents a plot of time delay versus the maximum gain for stability for non-collocated derivative control. As before, the solid line depicts a numerically generated estimate of the maximum gain for stability as a function of time delay. The data points represent experimentally determined values for the maximum gains for stability at various time delays.

The numerical results show a trend of increasing maximum gain for increasing time delay until about 0.04 sec where the gain decreases. Thus a proper choice of time delay can give the system a much *larger* maximum gain for stability than if there were no time delay.

Figure 4.4b shows the numerically determined maximum gain for stability versus time delay for the torsional bar using non-collocated derivative control. For the time delays shown, the dominant pole is always a system pole. Again, points denoted by "o" correspond to system poles originating at $s_1$, and the points denoted by "*" correspond to the system pole originating at $s_3$. In the range of time delays shown, there are three instances in which the dominant pole changes from one system pole to another. The changes occur at time delays of about 0.04, 0.08 and 0.125 sec. Figure 4.4c shows the expected cross-over frequency (frequency of the dominant pole at the maximum gain for stability) versus time delay. The fact that the dominant pole changes from one system pole to another suggests the presence of bifurcations at the locations where the changes occur.

The maximum gain for stability for non-collocated derivative control which is independent of time delay, given by Result 3.4 is 0.007988. When the time delay is zero, the above time-delay-independent maximum gain for stability is achieved (see Figures 4.4a and 4.4b). There is close agreement between the computational results from this section, and the results from the method of Section 3, with regard to the maximum gain for stability independent of time delay.

A bifurcation occurs at a time delay of about 0.0396 sec. This bifurcation is shown on Figure 4.4d, which shows the root loci of two of the system poles for

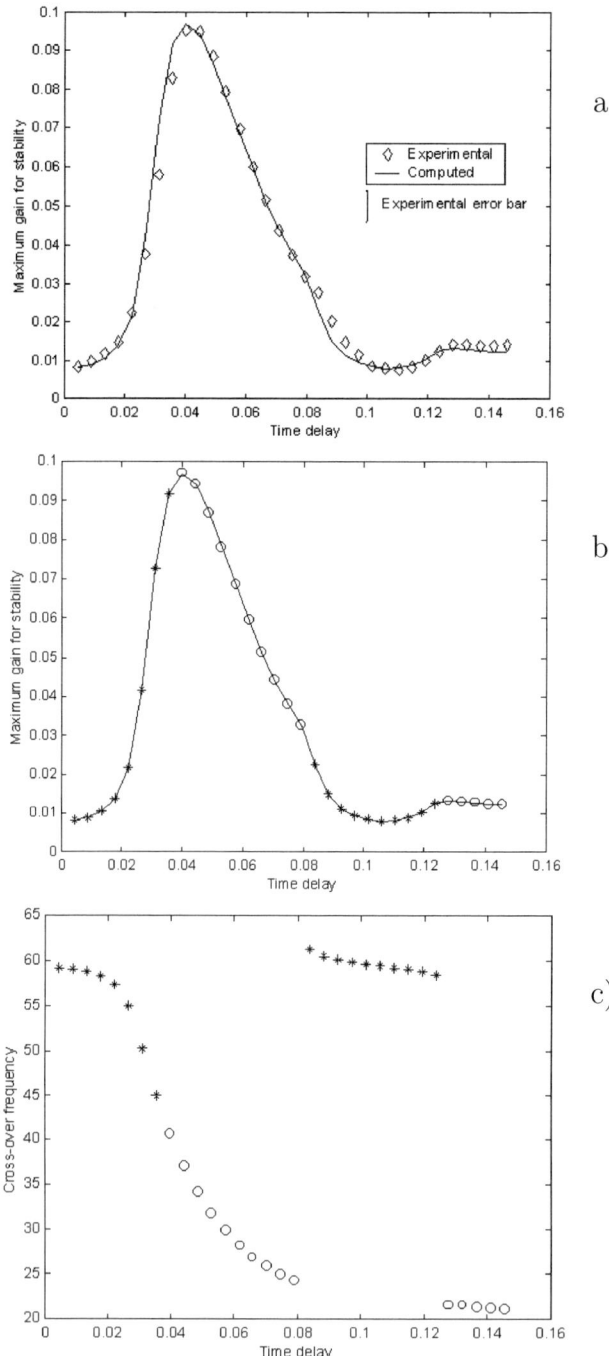

Figure 4.4  (a) Experimental maximum gain for stability versus time delay for non-collocated derivative control. (b) Expected maximum gain for stability versus time delay for non-collocated derivative control. (c) Expected cross-over frequency versus time delay for non-collocated derivative control.

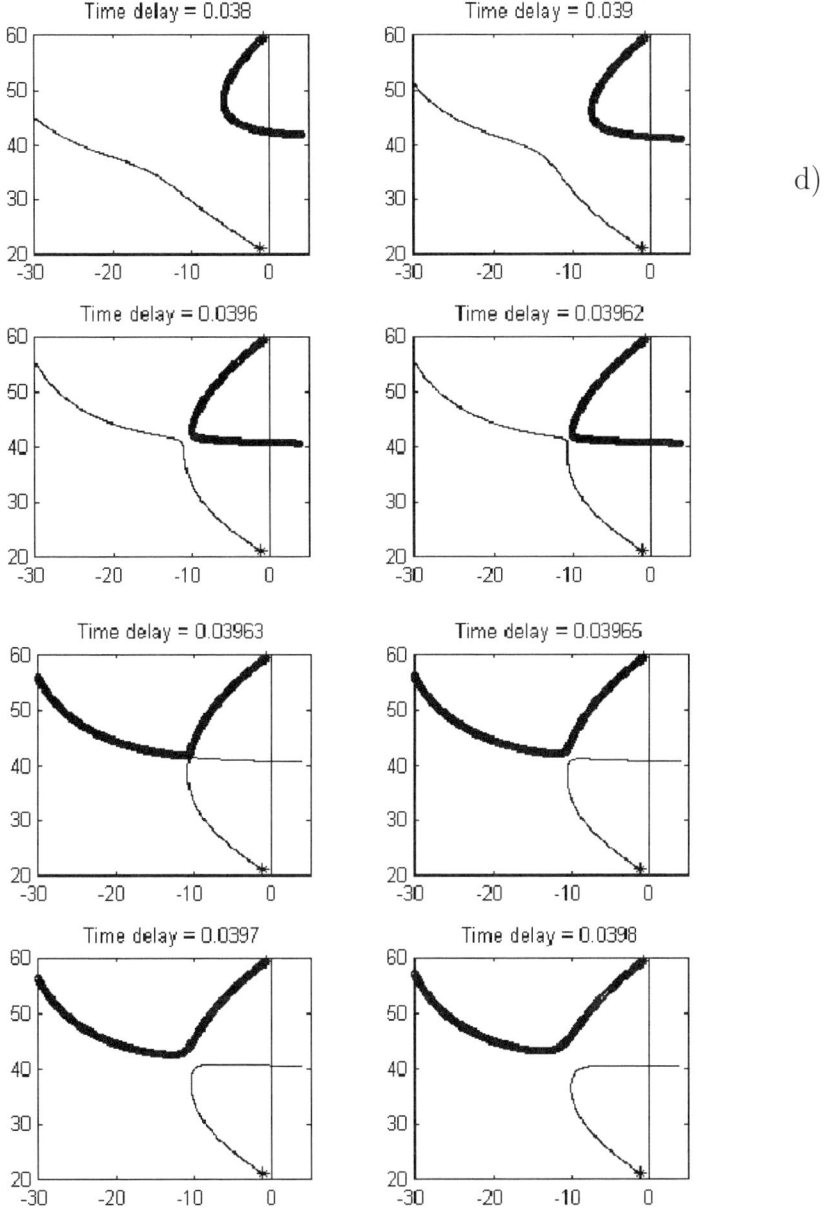

Figure 4.4 *(continued)* (d) Root loci for different time delays showing bifurcation for non-collocated integral control.

different time delays. At the bifurcation, one "arm" of the root loci is exchanged among the two system poles. The thick line represents the root locus for the pole originating at the open loop pole $s_1$, while the thin line is the root locus for the pole originating at $s_3$. The root loci correspond to gains between 0 and 0.12. The general behavior of the root loci of the poles at the bifurcation is similar to the one for collocated integral control (see Fig. 4.3h). The main difference between the two cases is that for collocated integral control, a system pole and a non-system pole are involved, while for the present case, only system poles are involved. At a time delay of about 0.08 sec we also have a bifurcation. However this bifurcation is less dramatic; there is no intersection of the root loci of the poles. The change of dominant pole occurs simply because there is a change in the rate at which the poles move across the complex plane.

The experimental and numerical results presented in this section show that for collocated integral control, there are time delays for which the dominant pole is a non-system pole. On the other hand, in the experiments with non-collocated derivative control, the dominant pole is a system pole (for the range in time delay considered). Good agreement between the experimental and the numerically expected maximum gain for stability is observed. For both collocated integral control and non-collocated derivative control (see Figures 4.3a and 4.4a), the introduction of a suitable time delay will yield a higher maximum gain for stability than the one for the system with no time delay. The next section deals with the control of time invariant systems with system uncertainties that include uncertainties in the time delay that is used. We model this uncertainty as a time-varying time delay.

# 5   Control of Uncertain Systems

The models presented in Sections 2 and 3 do not include system uncertainties in the formulation. However, the numerical results presented in Section 2.5 suggest that structural systems may remain stable after the parameters of the system are slightly perturbed (or uncertain). This section deals with the robust state feedback control of time invariant dynamic systems with time varying control delays and system uncertainties. Corless and Leitmann (1981), Gutman and Leitmann (1976) and Chen and Chen (1992) have worked on control systems which deal with uncertainties embedded in the system's structure or with externally introduced uncertainties (like measurement noise). In this section we introduce a new set of uncertainties in the system, in terms of the time delay in the control. Here the time delay in the control is allowed to be a bounded function of time.

Consider the system whose response $x(t)$ is governed by the matrix differential equation

$$x'(t) = Ax(t) + \Delta Ax(t) + Bu(t) + \Delta Bu(t) + \delta Bu(t - h(t)),  \qquad (5.1)$$

where the $m$-vector $u(t)$ is the control function, the matrices $\Delta A$, $\Delta B$, $\delta B$ have uncertain entries, and $h(t)$ is the time-varying time delay. The matrices $A$ and $\Delta A$ are $n$ by $n$, $B$, $\Delta B$ and $\delta B$ are $n$ by $m$, and $x(t)$ is an $n$-vector.

The elements of the uncertain matrices could either have deterministic or stochastic forms. Thus the stochastic uncertainties which are usually present in the stiffness and damping matrices can be included in the matrices $\Delta A$.

Assume that

$$\Delta A = BD \,, \quad \text{with} \quad \|D\| = k_1 \,, \quad k_1 \in [0, \infty) \,,$$

$$\Delta B = BE \,, \quad \text{with} \quad \|E\| = k_2 \,, \quad k_2 \in [0, 1) \,, \tag{5.2}$$

$$\delta B = BF \,, \quad \text{with} \quad \|E\| = \tilde{k}_2 \,, \quad \tilde{k}_2 \in [0, \infty) \,.$$

The above assumptions on the norms of the matrices can be interpreted as, the uncertainties in the system are bounded by known constants $k_1$, $k_2$ and $\tilde{k}_2$. The restriction that $k_2 \in [0, 1)$ is interpreted as follows: the uncertainty in the control cannot be so severe as to reverse the direction of the control action, for then one is not able to tell if the control is in the desired direction.

If the state feedback control $u(t)$ is chosen as

$$u(t) = -\frac{1}{2}\sigma B^T P x(t) \,, \tag{5.3}$$

where $P$ is the solution of the Lyapunov equation,

$$A^T P + P A = -(Q + H) \,, \qquad P, Q, H > 0 \,, \tag{5.4}$$

then the response of system (5.1) is stable, provided that the following conditions are satisfied.

$$h(t) < h_0 \,, \qquad h'(t) \le a < 1 \qquad \text{and} \qquad \rho > 4\eta^2 \,, \tag{5.5a}$$

where

$$\eta^2 = \frac{\tilde{k}_2^2 \|B^T P\|^2}{4(1-a)\lambda_{\min}(H)} \qquad \text{and} \qquad \rho = \frac{(1-k_2)^2 \lambda_{\min}(Q)}{k_1^2} \,.$$

The control gain is chosen as

$$\sigma \le \frac{1 - k_2}{\eta^2 + \theta} \,, \qquad \theta > 0 \,, \, , \tag{5.5b}$$

where $\theta$ is a positive constant which has to satisfy the condition,

$$\frac{\rho - 2\eta^2}{2}\left\{1 - \sqrt{1 - \Omega^2}\right\} < \theta < \frac{\rho - 2\eta^2}{2}\left\{1 + \sqrt{1 - \Omega^2}\right\} \,,$$

here $\Omega = \dfrac{2\eta^2}{\rho - 2\eta^2}$.

The result presented shows that systems with uncertainties in the parameters and in the time-delayed controller can be stabilized under special conditions. For a proof of the result and numerical simulations which validate it, see Udwadia *et al.* (1997) and Udwadia and Hosseini (1996).

# 6   Conclusions

In this paper we have studied time delayed control of structural systems from several viewpoints. We show that the *purposeful injection* of time delays can indeed improve both the stability and performance characteristics of controlled systems. Rather than developing complex schemes to eliminate, compensate for or nullify time delays (as has been the traditional view), we show analytically, numerically and experimentally that time delays can be purposefully and intentionally used to good advantage. This becomes more important because time delays are indeed ubiquitous in control systems. Furthermore, we show that such time delayed control can be made robust with respect to uncertainties in the system's parameters and to uncertainties in the implementation of the time delays themselves. We present next a detailed set of conclusions.

Section 2 deals with the control of classically damped structural systems. Results for collocated and non-collocated control of damped and undamped multi-degree-of-freedom systems were given. For undamped and underdamped systems, several analytical conditions on the time delay that guarantees that the closed loop system poles move to the left in the complex plane are given. Since PID control is one of the most commonly used control laws in practice, we use this controller to obtain conditions when time delays can stabilize undamped systems. We illustrate our results using numerical computations. We show that non-collocated derivative control that is unstable in the absence of time delays can be made stable through the simple introduction of suitable, intentional time delays. The robustness of the time-delayed control system to variations in structural parameter values (mass and stiffness), and the time delay is numerically explored for non-collocated derivative control. The results show that even under considerable perturbations (7% to 50%) of the parameter values the system remains stable.

General non-classically damped systems were treated in Section 3. The results presented deal with the case of one actuator, and the transfer function of the controller is assumed to be an analytic function. The main results can be summarized as follows: (1) When all the poles of the open loop system have negative real parts and the controller transfer function is an analytic function, then for any given time delays there exists a range in the gain (which is, in general, dependent on the given time delays) for which the closed loop system is stable. (2) For a single actuator and a sensor under the same conditions as before, there is an interval of gain which is independent of time delay for which the system is stable. This gain can be computed without knowledge of the time delay. (3) We show that through a proper choice of time delay, one may obtain a larger maximum gain for stability than the one for the system with no time delay. Thus the injection of a time delay can significantly improve the performance of the control of such a system.

Experimental and numerical results on the control of a 2–DOF torsional bar are presented in Section 4. The main results are summarized as follows: (1) Experimental results showing the maximum gain for stability versus time delay, for both collocated integral control and non-collocated derivative control, exhibit close agreement with the numerical/analytical expectations. Furthermore, the results show that the presence of a time delay will actually increase the maximum gain for stability, when compared with the maximum gain for stability when the system has no time delay. (2) For non-collocated derivative control, we numerically find that the dominant pole (the pole that dictates the maximum gain for stability) is always a system pole (for the range of time delays analyzed). For different time delays the dominant pole may change from one system pole to another. At these time delays we have bifurcations. (3) For collocated integral control, we numerically find that the dominant pole can be a non-system pole. This result is corroborated by actual experiments. As with the non-collocated case, the dominant pole changes as the time delay is varied, and now bifurcations involving system and non-system poles occur. We illustrate such a bifurcation experimentally and computationally.

Finally, Section 5 deals with the robust state feedback control of time invariant dynamic systems with time-varying time delays and system uncertainties. A state feedback control is suggested that guarantees the stability of the system, provided, principally, the so-called matching conditions are met.

Even though we have taken different approaches when dealing with the time delayed control of structural systems, throughout this paper there is the recurring theme that under an appropriate choice of time delay, the maximum gain for stability could be greater than the maximum gain for stability obtained when no time delay is used. Since time delays are ubiquitous in the control of large-scale structural systems, this strongly suggests the idea that instead of trying to eliminate/nullify time delays, we may actually want to introduce them appropriately in order to provide stable non-collocated control. We also show that this increase in the maximum gain for stability in the presence of delays can improve the control performance for collocated and non-collocated control.

Our results, both experimental and analytical, indicate that the research presented in this paper will be of practical value in improving the performance and stability of controllers. Through the proper purposive injection of small time delays, it is possible to dramatically improve the performance and stability of the control system. Furthermore, the injection of small time delays can be easily implemented in a reliable and nearly costless way on an already installed controller. Thus the performance and stability of active control systems that are already installed in building structures can be enhanced in a nearly costless way by including appropriate and intentional time delays in the feedback loop. This makes the method very attractive from an economic and safety retro-fit standpoint when dealing with the active control of structural systems subjected to strong earthquake ground shaking.

# References

1. Agrawal, A.K., Fujino, Y. and Bhartia, B.K. (1993) Instability Due to Time Delay and its Compensation in Active Control of Structures. *Earthquake Engineering and Structural Dynamics*, **22**, 211–224.

2. Ahlfors, L.V. (1979) *Complex Analysis*. McGraw-Hill, Inc., New York.

3. Auburn, J.N. (1980) Theory of the Control of Structures by Low-Authority Controllers. *Journal of Guidance and Control*, **3**, 444–451.

4. Balas, M.J. (1979a) Direct Velocity Feedback Control of Large Space Structures. *Journal of Guidance and Control*, **2**, 252–253.

5. Balas, M.J. (1979b) Direct Output Feedback Control of Large Space Structures. *The Journal of Astronomical Sciences*, **XXVII** (2), 157–180.

6. Cannon, R.H. and Rosenthal, D.E. (1984) Experiments in Control of Flexible Structures with Noncolocated Sensors and Actuators. *Journal of Guidance and Control*, **7**, 546–553.

7. Chen, Y.H. and Chen, J.S. (1992) Adaptive Robust Control of Uncertain Systems. *Control and Dynamic Systems*, **50**, 175–222.

8. Corless, M.J. and Leitmann, G. (1981) Continuous State Feedback Guaranteeing Uniform Ultimate Boundedness for Uncertain Dynamic Systems. *IEEE Transactions on Automatic Control*, **AC-26** (5), 1139–1144.

9. Fanson, J.L. and Caughey, T.K. (1990) Positive Position Feedback Control for Large Space Structures. *AIAA Journal*, **28** (4), 717–724.

10. Goh, C.J. and Caughey, T.K. (1985) On the Stability Problem Caused by Finite Actuator Dynamics in the Collocated Control of Large Space Structures. *International Journal of Control*, **41** (3), 787–802.

11. Gutman, S. and Leitmann, G. (1976) Stabilizing Control for Linear Systems with Bounded Parameter and Input Uncertainty. In *Proceedings of the 7th IFIP Conference on Optimization Techniques*. Springer Verlag, Berlin.

12. Kwon, W.H., Lee, G.W. and Kim, S.W. (1989) Delayed State Feedback Controller for the Stabilization of Ordinary Systems. In *Proceedings of the American Control Conference*, Pittsburgh, Pennsylvania, Vol. 1, 292–297.

13. Rudin, W. (1987) *Real and Complex Analysis*. McGraw-Hill Inc., New York.

14. Udwadia, F.E. and Hosseini, M.A.M. (1996) Robust Stabilization of Systems with Time Delays. In *ASCE Proceedings of the seventh specialty conference, Probabilistic Mechanics and Structural Reliability*, Worchester, Massachusetts, August 7–9, pp. 438–441.

15. Udwadia, F.E., Hosseini, M.A.M. and Chen, Y.H. (1997) Robust Control of Uncertain Systems with Time Varying Delays in Control Input. *Proceedings of the 1997 American Control Conference*, Albuquerque, New Mexico, June 4–6.

16. Udwadia, F.E. and Kumar, R. (1994) Time Delayed Control of Classically Damped Structural Systems. *International Journal of Control*, **60** (5), 687–713.

17. Udwadia, F.E. and Kumar, R. (1996) Time Delayed Control of Classically Damped Structures. *Proceedings of the 11th World Conference of Earthquake Engineering*. Acapulco, Mexico, June 23–28.

18. von Bremen, H.F. and Udwadia, F.E. (2000) Can Time Delays be Useful in the Control of Structural Systems? In *Proceedings of the 41$^{st}$ AIAA/ASME/ASCE/AHS/ASC Structures, Structural Dynamics, and Materials Conference*, Atlanta, Georgia, April 3–6.

19. von Bremen, H.F., Udwadia, F.E. and Silverman, M.C. (2001) Effect of Time Delay on the Control of a Torsional Bar. In *Proceedings of the 42$^{nd}$ AIAA/ASME/ASCE/AHS/ASC Structures, Structural Dynamics, and Materials Conference*, Seattle, Washington, April 16–19.

20. Yang, J.N., Akbarpour, A. and Askar, G. (1990) Effect of Time Delay on Control of Seismic-Excited Buildings. *Journal of Structural Engineering*, **116** (10), 2801–2814.

# Robust Real- and Discrete-Time Control of a Steer-by-Wire System in Cars

Eduard Reithmeier

*Institut für Meß- und Regelungstechnik*
*Universität Hannover, 30167 Hannover, Germany*
*Fax: +49-511-762-3234, Tel: +49-511-762-3331*
*E-mail: reithmeier@imr.uni-hannover.de*

There is a series of objectives to remove the mechanical coupling between the steering wheel and the wheel suspension system in cars and establish a so-called steer-by-wire system. One main reason is related to driving comfort, that is, torque feedback characteristics may be individually adjusted to the driver. Other reasons are focused on providing a higher security level during driving. For instance, the car's stabilizing control system could be enhanced by gaining more influence, particularly in critical driving situations. Other advantages are related to manufacturing aspects, like assembly improvements or an entire new layout of the engine compartment.

The idea of a steer-by-wire system is to bring a torque–actuator interface – one for the steering wheel and one for the turning wheel system – into place where the steer-shaft is cut in half. Both actuator systems will be managed separately. However, an information exchange system is established between the two systems. This work is concerned with the steering wheel part. We use steer-angle, steer-angle-velocity and the actual steer-shaft torque as measured variables in order to control the torque motor connected to the shaft according to a given or desirable torque vs. steer-angle characteristic respectively. The system contains a series of uncertainties which need to be taken into account. Hence, the controller design is primarily based on robust control techniques. The main problems are caused around a zero steer-angle due to changes in torque direction. The steering-behavior around the zero steer-angle becomes sloppy and undefined. This led us to an appropriate additional mechanical design to implement a discrete time control scheme for the desired torque characteristic. The author will report on the current state of our experimental and theoretical investigations.

# 1   Introduction

X-by-wire systems (such as brake-by-wire, shift-by-wire or even steer-by-wire) are becoming more and more important to car manufacturers. The main reasons are the advantages and particular properties (see below) which car makers have to take into consideration for prospective intelligent transportation systems in the future. Also, continuously falling costs for general electronic equipment as well as for computational hardware and software products are pushing this general tendency.

Figure 1 shows the wheel suspension and steering system of an average car. In a steer-by-wire system the steering shaft is cut into two independent parts. After separation, each interface will be equipped with an electric or electro hydraulic actuator. This results in a *Turning Wheel Actuator System* generating the torque $m_T$ and a *Steering Wheel Actuator System* (torque $m_S$). Both systems communicate via the so-called CAN-Bus. Due to the loss of the mechanical coupling there is also a loss of basically two items of information related to two physical constraints, namely on the one hand

- the kinematic condition that there is exactly one turning wheel angle $\psi$ for each steering wheel angle $\varphi$ and, on the other hand,
- the kinetic condition that the two interface torques are the same in magnitude.

The actuator torque $m_T$ necessary to turn the wheels depends on the dynamics of the whole suspension system and on the contact torque $m_C$. The torque $m_C$ again depends on parameters and variables like the turning angle $\psi$, speed $v$ of the car or the rolling resistance $\mu_R$ (cf. Figure 1). The driver employs an appropriate torque $m_D$ which finally constitutes the dynamical equilibrium of the whole system drawn in Figure 1. The main objective in designing the control of the steering wheel simulator is to establish a desired feedback characteristic to the driver. This feedback depends of course on the torque coming from the turning wheel actuator system. However, since the two systems are independent there might be more and different information processed therein.

A steer-by-wire system has some significant advantages in terms of security, comfort and manufacturing, and it is a necessary device for future transportation aspects. A traffic guided system, for instance, needs autonomous steering in order to apply appropriate control without being disturbed by the driver. It may be combined with automatic stability control (ASC) in order to support or to assist the driver in certain critical situations. And it can be easily used as a car locking device. Being able to adjust arbitrary torque feedback characteristics is considered to be a significant comfort aspect. The manufacturer is interested in more cost-effective assembly and exchange of spare parts. And, a separate steering system offers the possibility of redesigning the engine compartment in a completely new way.

# 2   Modeling of the Plant

In order to design a control scheme we need, of course, a model of the considered plant which, in our case, is the steering wheel simulator system. Our experimental setup consists basically of the hardware parts shown in Figure 2.

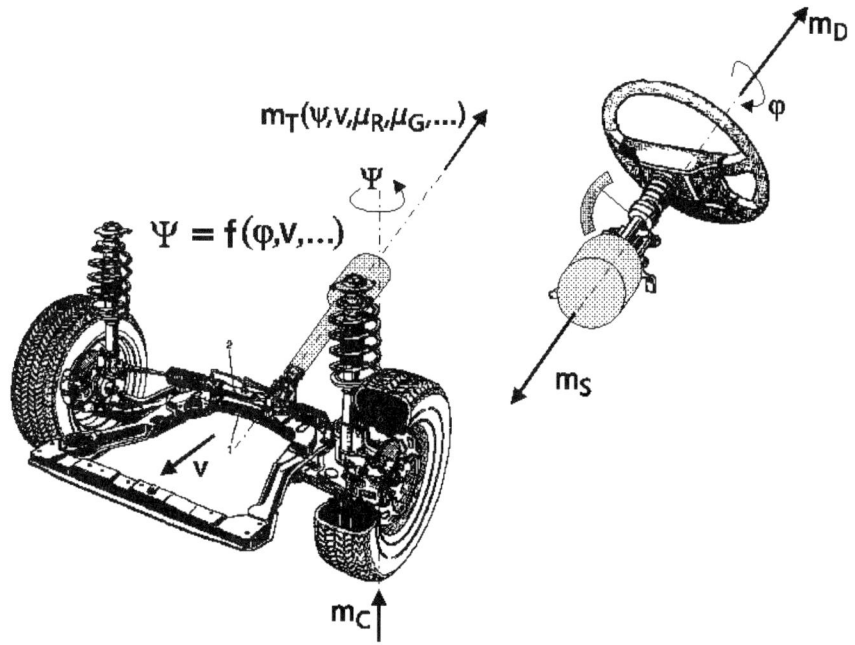

$$m_T(\Psi, v, \mu_R, \mu_G, \dots)$$

$$\Psi = f(\varphi, v, \dots)$$

$$m_S$$

$$m_C$$

Figure 1 Wheel suspension and steering system of an average car.

First of all, there is a current-controlled electronic power supply with plant input voltage $u_A$ for controlling a so-called torque motor. This motor, based on a three-phase-current system, supplies the power via the shaft torque and not the rotational velocity of the shaft. At the most we have only around $\pm 2$ turns of the steering wheel but it is necessary to have torques up to 60 [Nm] available. The rotor of the torque motor is directly coupled to the wheel shaft, and a torque sensor is mounted in between. An additional feature is a spring in parallel connection to the rotor shaft in order to support the torque feedback to the driver. $J$ is the moment of inertia of all turning parts, that is, rotor, shaft, torque sensor and steering wheel. They behave as one rigid body. The dynamics therefore are described by the steering angle $\varphi$ and its time derivative $\dot{\varphi}$. These two state variables will be a direct output of the equation of motion which is represented by a flow chart according to Figure 2. The torque employed by the driver is considered as an external but measured disturbance. Bearing friction and similar unknown effects are considered to be uncertainties. Power supply and torque motor act according to a first order element followed by some time delay $T_t$ (cf. Figure 3). The time constant $T_T$ of the first order element depends on the temperature of the motor and has therefore to be considered as an uncertain parameter.

The last step in modeling the plant is concerned with the human–machine interface, that is, with the driver. Instead of treating the driver as a structural uncertainty, it turns out to be more appropriate to transform him/her into a simple model involving parameter uncertainties. For that purpose Figure 4 shows a qualitative closed loop model of the driver.

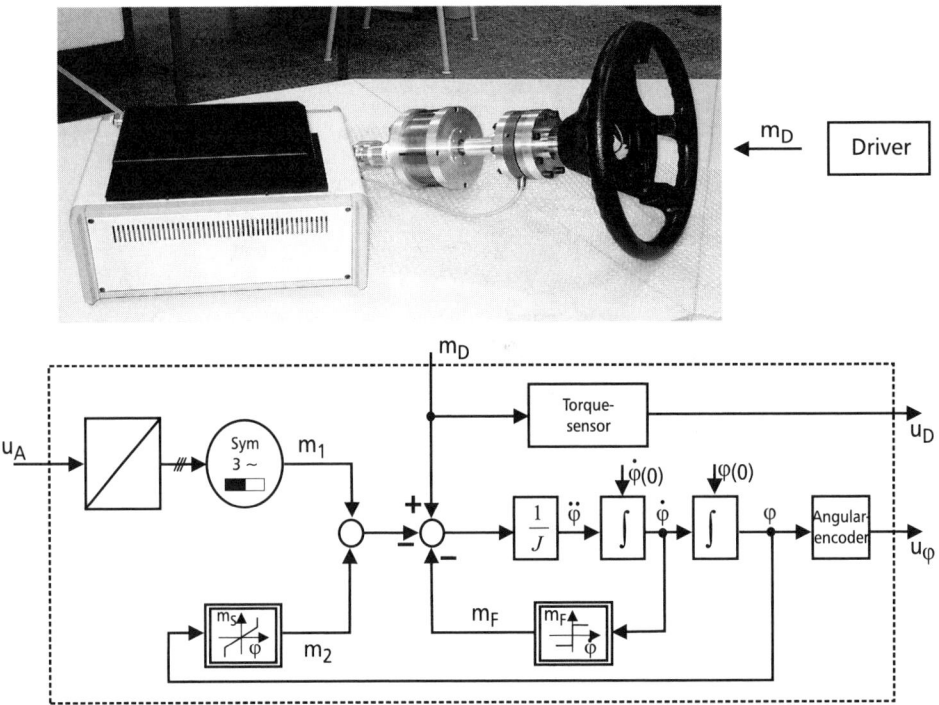

Figure 2  Model of the steering wheel simulator system.

Starting from some current road situation and some current location the driver will hopefully sense everything and will perform an analysis. The result will be some desired position of the car with respect to the road. After this, the driver will continue by giving a certain command to the actuators in order to steer the car into the desired position. In other words, the steering angle plays the role of the controlled variable, and the steering torque acts as input variable. Now, if the driver employs, say, a fictitious unit torque command to the steering wheel (cf. Figure 5), the system will respond similarly to a second order system (due to the damping $d_0$ and elasticity $\nu_0$ of the body parts). The strength of the muscles plays the role of the unknown amplification factor $K_D$. The magnitude and variation of the uncertain parameters involved can be estimated via experimental studies.

Employing the driver model in the plant leads to the flow chart given in Figure 6. It also shows the transfer function between the power supply input voltage $u_A$ and the torque gauge output voltage $u_D$. Of course, this transfer function contains all model uncertainties. The objective is to follow a desired reference torque $u_{D,Des}(\varphi)$ by properly commanding $u_A$.

In practice the control algorithm will be carried out by some micro controller, which requires digitized measurement values and which supplies digital input values. Within one sampling period there needs to be carried out a series of redundancy checks, AD and DA conversions as well as control algorithms with variable structure. This process takes a couple of milliseconds. That again makes it necessary to

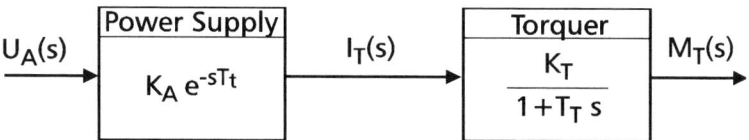

Figure 3  Transfer function of the torque motor.

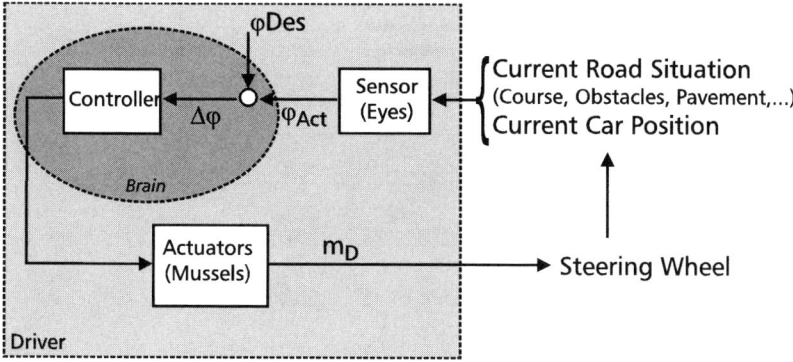

Figure 4  Simplified model of the driver.

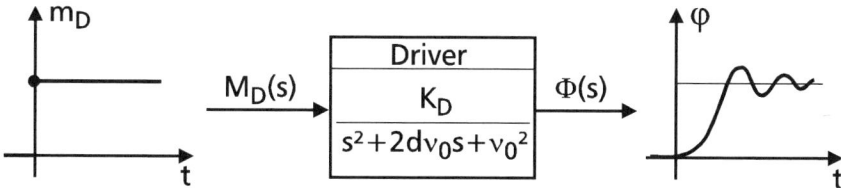

Figure 5  Transfer function of the simplified driver model.

consider a discrete time plant. If we describe the DA converter by the standard $H_0$ modulator and the plant $G_P$ by

$$G_P(s) = \left[ \frac{a}{s - s_1} + \frac{b}{s - s_2} + \frac{c}{s + T_T^{-1}} \right] \cdot \exp(-sT_t) \tag{1}$$

with some properly chosen constants $a$, $b$ and $c$, then the corresponding rational Z-transfer function $G$ is determined by

$$G(z) = \left[ \frac{a}{z - \exp(s_1 T)} + \frac{b}{s - \exp(s_2 T)} + \frac{c}{z - \exp(-T_T^{-1})} \right] \cdot z. \tag{2}$$

In this case we used the fact that the time delay $T_t$ of the power supply is much smaller than the sampling time $T$.

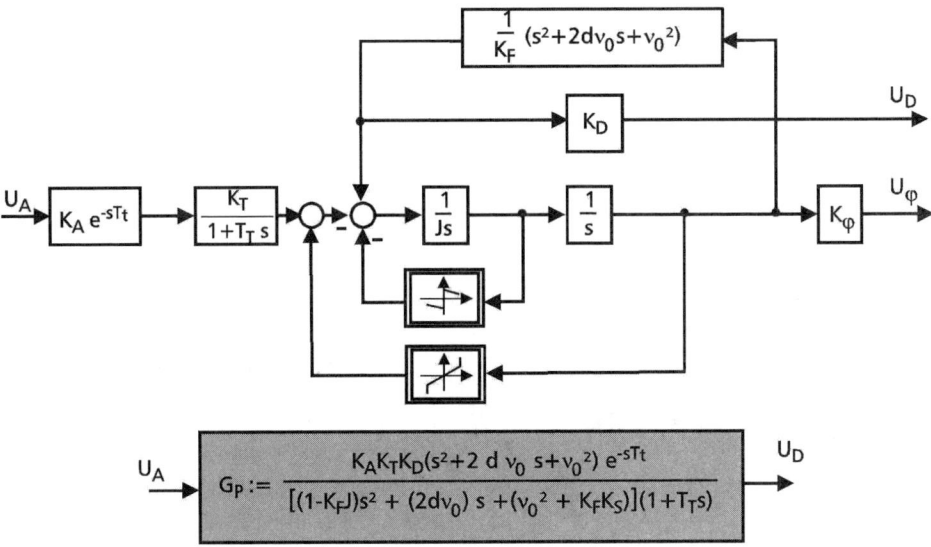

Figure 6  Transfer function of the simplified driver model.

# 3   Controller Design

Finally, the plant may be properly implemented into the closed loop system for the steering wheel simulator system according to Figure 7. The internal loop feeds back the torque directly. Micro controller board 2 contains the control algorithm and different administrative tasks such as communication with the power supply. The desired torque comes via the CAN bus from micro controller board 1 which is related to the turning wheel actuator system. In case there is a given arbitrary reference torque, we need also the actual angular position and its derivative which, in that case, will be also fed back via the CAN bus.

In this presentation we will concentrate on the design of a feedback controller which assures certain torque characteristics. The basic problem is that due to the structure of the given dynamical system it is not possible to follow any changes in the desired torque fast enough. A unit step response, for instance, looks qualitatively like the one in Figure 8.

We call $t_{off}$ the practical settling time once the unit step response enters a certain tolerance interval and stays therein. The closed loop control objective is to reduce the settling time and lead the system to a dead beat behavior according to Figure 9.

Hence, our arrangement for reaching this goal is to design a reference transfer function $G_R$ which behaves like a second order discrete time system with an additional damping term of the order integer $q$:

$$G_R(z) = (1 - z^{-1}) \cdot \mathcal{Z} \left\{ \frac{1}{s} \cdot \frac{K_0 \omega_0^2}{s^2 + 2D\omega_0 s + \omega_0^2} \cdot \left( \frac{\delta}{s + \delta} \right)^q \right\}. \tag{3}$$

The dead beat behavior is forced by the parameter $D > 1$. And the damping coefficient $\delta$ is chosen in such a way that the poles of the second order system stay

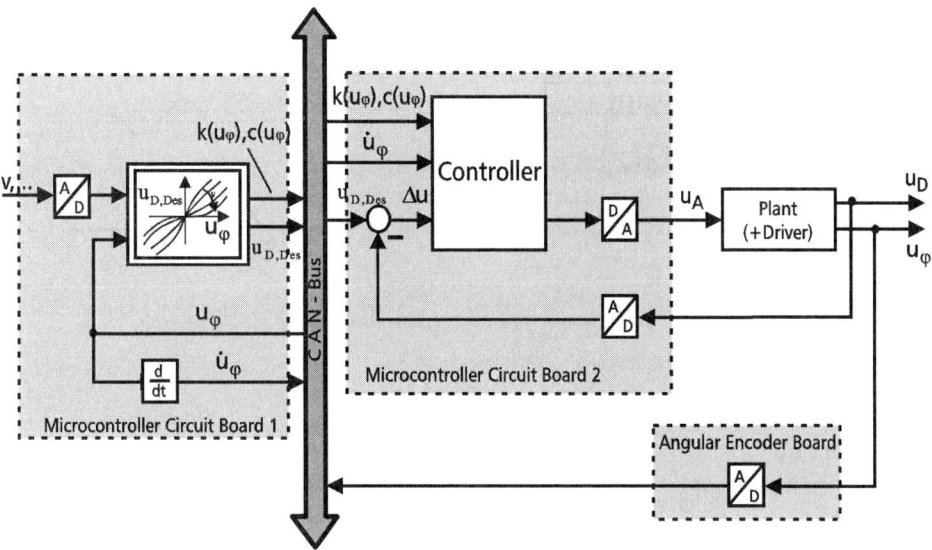

Figure 7  Discrete time closed loop system.

dominant. In order to get a feasible reference transfer function $G_R$ we need to choose the pole surplus $q$ in this case greater than or equal to 1. Once $G_R(z)$ is designed, the controller $G_C(z)$ based on plant $G(z)$ may be directly determined from the standard closed loop according to Figure 10. For $q = 1$, for instance, this yields

$$G_C = \frac{G_R}{G[1 - G_R]} \tag{4}$$

or, respectively

$$G_C(z) = \frac{z^{-1}[\beta_1 + \beta_2 z^{-1} + \beta_3 z^{-2}]}{\alpha_0 + \alpha_1 z^{-1} + \alpha_2 z^{-2} + \alpha_3 z^{-3} + \alpha_4 z^{-4}} \tag{5}$$

with coefficients $\alpha_i$ and $\beta_i$. These coefficients depend, in general, on a whole ensemble of parameters of the system. That is, uncertain parameters of the model and measured parameters as well as design parameters of the reference transfer function.

Of course, if the uncertain parameters vary in time or vary due to different drivers, the dominant poles will change their location and therefore the performance of the controller. To counter this situation we define a reference set of uncertain parameters and use, in addition, robustification procedures in order to improve the designed control scheme (see Section 4). Using the controller based on the reference set yields to a torque tracking behavior according to Figure 11 with respect to a feedback characteristic composed of two straight lines.

Figure 12a shows the actually measured torque feedback behavior of a real car. In that case, the driver turns the steering wheel at a speed of 50 km/h around 90 degrees and takes his/her hands off immediately after. At 90 degrees, the magnitude of the required torque is approximately 4 [Nm]. Figure 12b shows the same situation but produced by the steering wheel simulator.

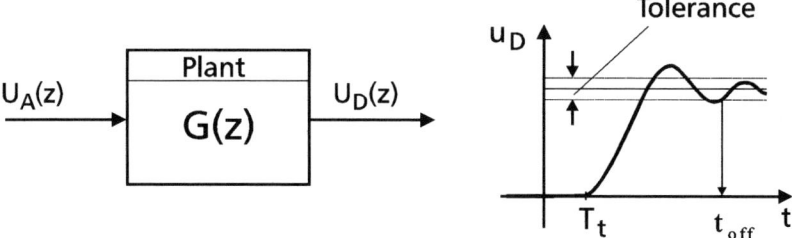

Figure 8  Unit step response of the plant (qualitatively).

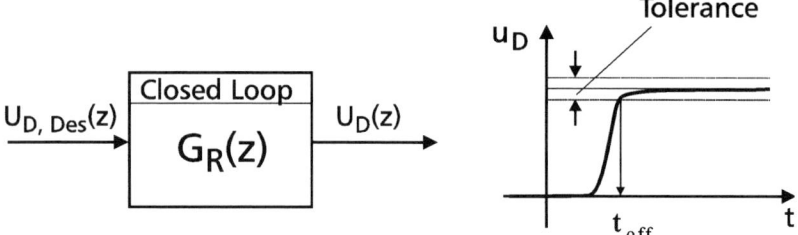

Figure 9  Unit step response of the closed loop (qualitatively).

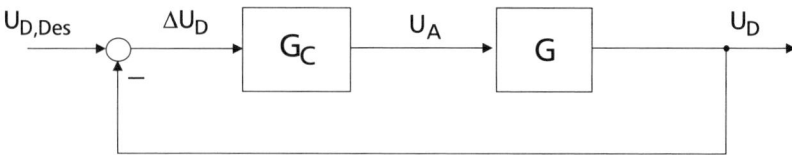

Figure 10  Standard closed loop.

## 4  Robustification

Despite the satisfying results stated in Figure 12, it turns out, in general, that it is better not to rely only on a nominal set of the uncertain parameters. To encounter this fact, it is necessary to improve the robustness of the closed loop with respect to these uncertainties. Since the controller was derived from the continuous time plant and a continuous time arrangement of the reference transfer function, the discrete time controller will also relate to some corresponding continuous time transfer element. That again makes it possible to analyze the system dynamics with respect to the corresponding continuous time elements of the discrete time open loop system.

A variation of the uncertain plant parameters results in a modified frequency response of the plant (see also Figure 13). The controller stays unchanged since it is based on the nominal set of plant parameters. The phase response of the

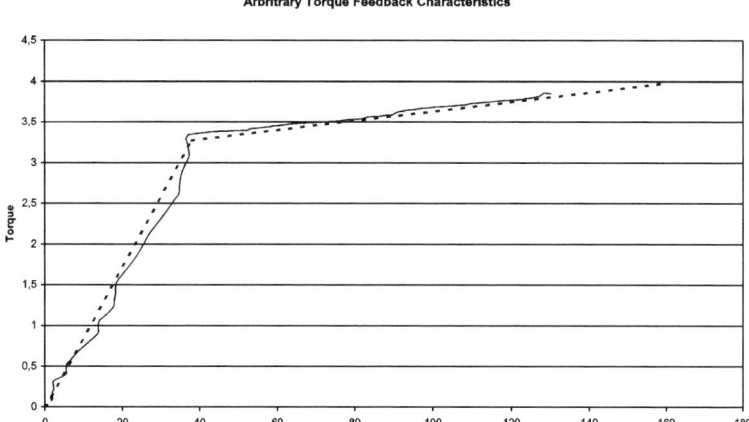

Figure 11  Actual torque (solid line), reference torque (dashed line).

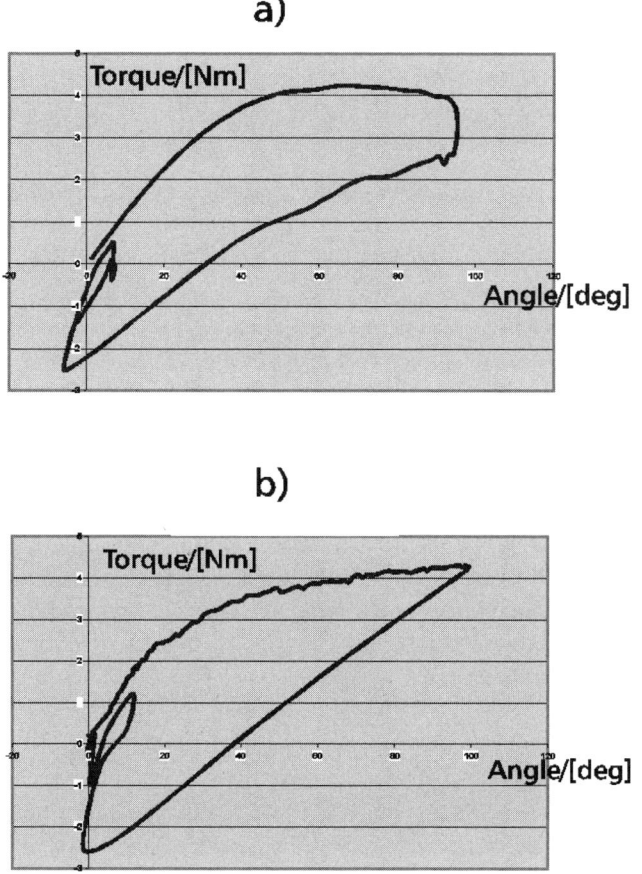

Figure 12  (a) Actual feedback characteristic. (b) Simulated feedback characteristic.

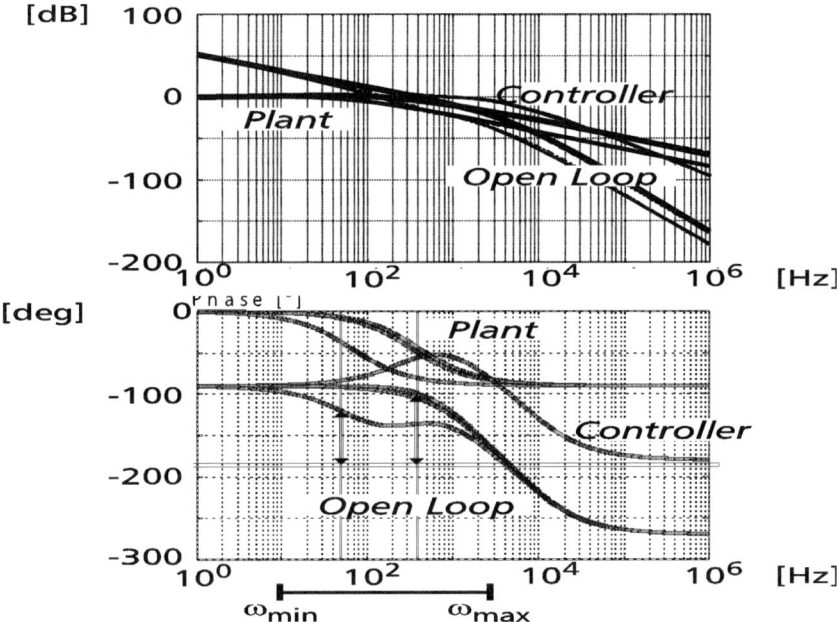

Figure 13  Frequency response of the plant, the nominal controller and the open loop.

plant is influenced only within a certain frequency range from $\omega_{min}$ to $\omega_{max}$ if the parameters are varied. Hence, it makes sense to lift the phase only inside that range in order to obtain a more robust phase margin. Using a transfer function

$$G_n(s) := k_0 \cdot \left[ \frac{1 + \dfrac{s}{\omega_L}}{1 + \dfrac{s}{\omega_H}} \right]^n \tag{6}$$

composed of $0 < n < 1$ lead elements with cutoff frequencies $\omega_L < 0.1\,\omega_{\min}$ and $\omega_H < 10\,\omega_{\max}$ as well as some appropriate amplification $k_0$ in serial connection to the controller will lift the phase of the open loop by $n \cdot \pi/2$ (see Figure 14a). Since $n$ is usually smaller than 1 we are forced to apply an approximation procedure by employing simple P, I or D elements. Using an approximation which consists of $N$ of these basic elements with cutoff frequencies $\omega_1, \ldots, \omega_N$ and $\omega_1', \ldots, \omega_N'$ yields a transfer function $G_N$ given by

$$G_N(s) := k_0 \cdot \prod_{j=1}^{N} \left[ \frac{1 + \dfrac{s}{\omega_j'}}{1 + \dfrac{s}{\omega_j}} \right] . \tag{7}$$

The corresponding frequency response is shown in Figure 15. This, again, results in a robust $PID_N T_{q+1}$ controller

$$G_C^*(s) := G_C(s) \cdot G_N(s) . \tag{8}$$

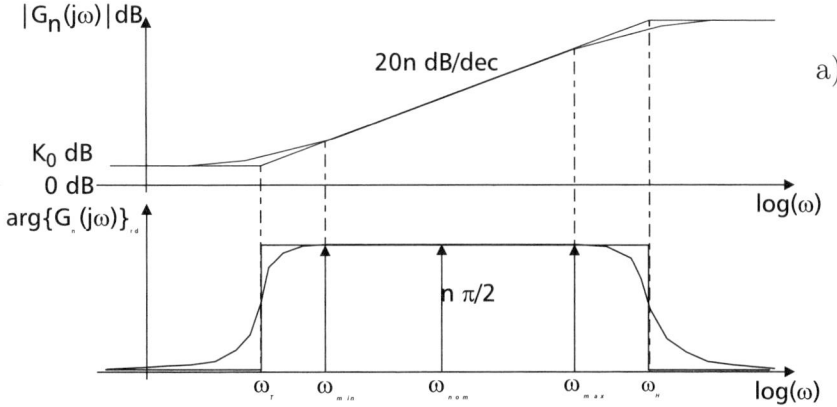

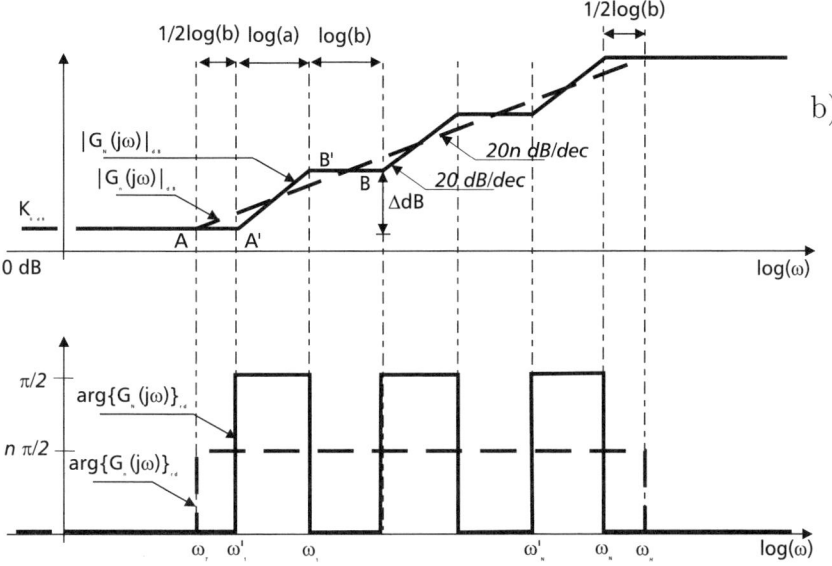

Figure 14 (a) Lift of the open loop phase margin. (b) Approximation $G_N(i\,\omega)$ of $G_n(i\,\omega)$.

where $q$, in our case, is chosen to be 1. The integer number $N$ relates to the approximation mentioned above (see also [1] for more details). The modified controller leads to the frequency response shown in Figure 15. The controller obviously lifts the phase inside the concerned frequency range and therefore leads to the desired result. Of course, the robust controller does not account for a better performance. On the contrary, the performance usually will deteriorate. This, however, is a general phenomenon which is encountered in the design of any robust concept. The right-hand side of Figure 16 shows the unit step response for different parameter settings. The robust controller leads to an extended settling time. On the other hand, overshooting will be significantly suppressed.

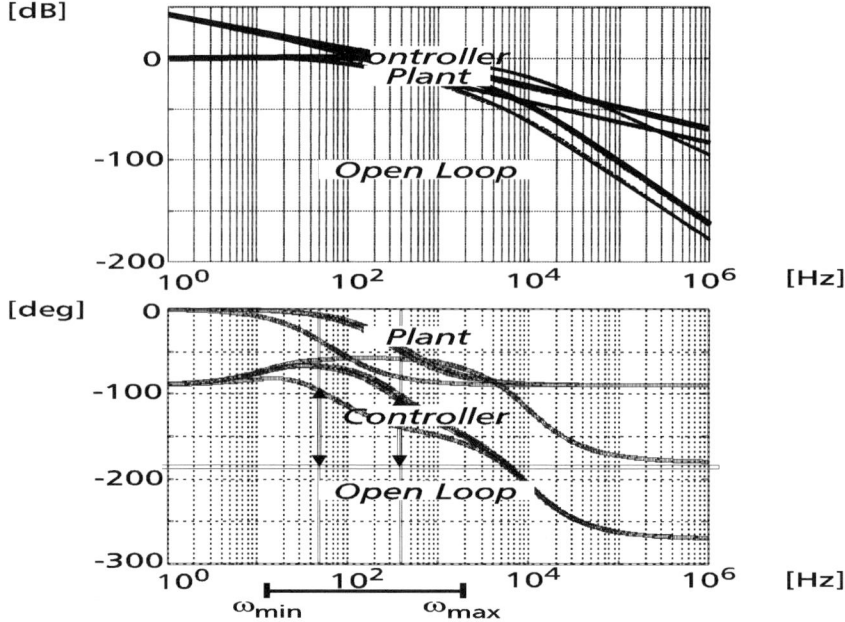

Figure 15  Frequency response of the plant, the robust controller and the open loop.

## Nichols-Diagram                                    Unit-Step Response

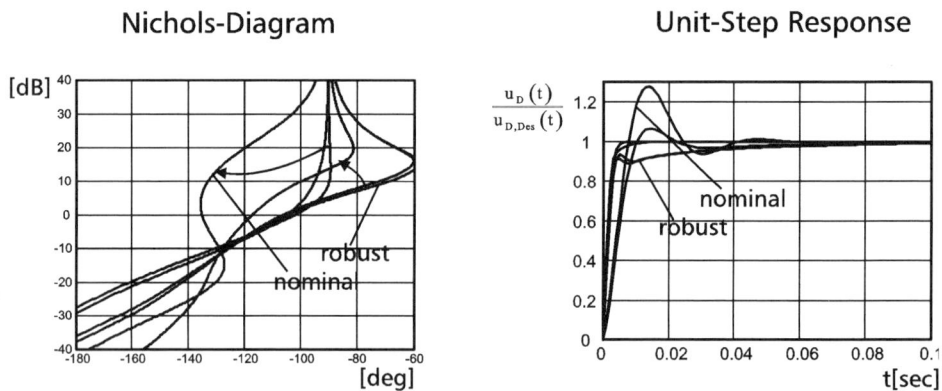

Figure 16  Closed loop behavior (nominal and robust control).

The left-hand side of Figure 16 shows the Nichols Diagram of the frequency response. Again, the results are shown for different parameter settings. The robust scheme leads obviously to a significant right shift of the characteristic curve and therefore to a more robust closed loop behavior.

In order to implement the control algorithm on some micro controller board, a time domain representation of the discrete time controller $G_C$ is needed. Fortunately the nominal controller $G_C$ as well as the robust controller $G_C^*$ are given by

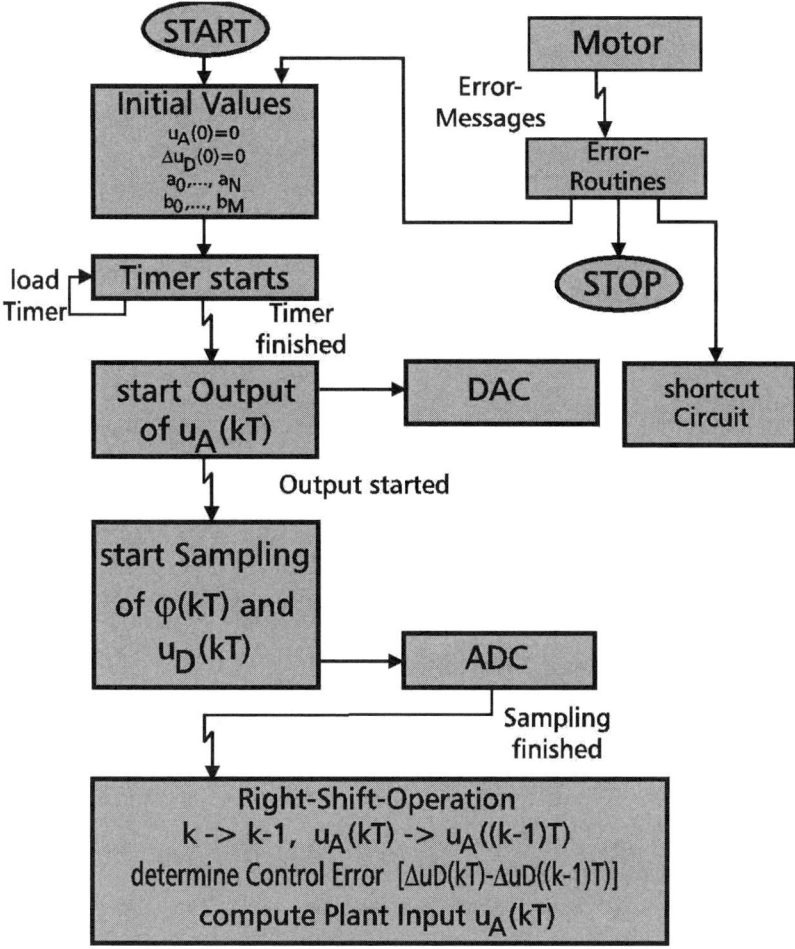

Figure 17  Implemented real time structure.

polynomial fractions in $z^{-1}$.

$$G_C(z) = \frac{U_A(z^{-1})}{\Delta U_D(z^{-1})} = \frac{\sum\limits_{m=0}^{M} b_m z^{-m}}{\sum\limits_{n=0}^{M} a_n z^{-n}}, \qquad N \geq M. \tag{9}$$

This representation allows an easy application of the inverse Z-transformation. The result is an algorithmic mapping which gives the $k$-th control input $u_A$ via the knowledge of the last $M$ measured torque values and the last $N$ computed control input values:

$$u_A(kT) = \frac{1}{a_0} \left[ \sum\limits_{m=0}^{M} b_m \Delta u_D((k-m)T) - \sum\limits_{n=0}^{N} a_n u_A((k-n)T) \right]. \tag{10}$$

See [2] for an optimization and real-time implementation of this algorithm. Compared to real-time programming, conventional programming does not allow for a fast response to external events, like a sensor or actuator failure. But this is exactly what is needed in security-related systems such as a steer-by-wire simulator. In addition, an interrupt-guided program eases the synchronization with respect to the sampling time $T$.

As shown in Figure 17, the implemented real-time structure is split into a series of independent parts. These parts are triggered via interrupts by external events. After setting the initial values a cycle timer will be loaded in time intervals $T$, where $T$ denotes the sampling time. After each cycle an output routine is started by the first interrupt. It shifts the actual control input to the DAC register. The corresponding voltage will be kept at the torque motor until the next interrupt starts. Immediately after, a second interrupt starts the acquisition of the measurement values. The digitized values will be transferred to the micro controller where the control algorithm resides. The new control input value stays in the register until the timer starts the next cycle. In case an external failure is going to happen, the appropriate failure routine will be able to interfere at certain interrupt locations, depending on their interrupt priority. Finally, the triggered failure routine will carry out appropriate emergency procedures.

# References

1. Thielking, O. (1999) Entwurf und Realisierung einer diskreten Drehmomentregelung für das Lenkrad eines Steer-by-Wire Fahrzeuges. Diplomarbeit am Institut für Meß- und Regelungstechnik, Universität Hannover.

2. Budde, T. (2000) Komplettierung der Software und der mechanischen Komponenten eines mikrokontrollergeregelten Steer-by-Wire Systems. Diplomarbeit am Institut für Meß- und Regelungstechnik, Universität Hannover.

3. Krautstrunk, A., Uhler, R., Zimmer, M. and Mutschler, P. (2000) Elektrisch lenken: Handkraftaktor für Steer-by-Wire. Thema FORSCHUNG 1/2000, Seite 104–113.

# Optimal Placement of Piezoelectric Sensor/Actuators for Smart Structures Vibration Control

Vicente Lopes, Jr.,[1] Valder Steffen, Jr.[2]
and Daniel J. Inman[3]

[1] Department of Mechanical Engineering – UNESP-Ilha Solteira
15385-000 Ilha Solteira, SP, Brazil
Tel: +(55) 18-763-8139, E-mail: vicente@dem.feis.unesp.br
[2] School of Mechanical Engineering Federal University of Uberlândia
38400-902 Uberlândia, MG, Brazil
Tel: +(55) 34-3239-4148, E-mail: vsteffen@mecanica.ufu.br
[3] Center for Intelligent Material Systems and Structures
Virginia Polytechnic Institute and State University
Blacksburg, VA 24061-0261, USA
Tel: (540) 231-4709, E-mail: dinman@vt.edu

Smart material technology has become an area of increasing interest for the development of lighter and stronger structures that are able to incorporate actuator and sensor capabilities for collocated control. In the design of actively controlled structures, the determination of the actuator locations and the controller gains is a very important issue. For that purpose, smart material modeling, modal analysis methods, and control and optimization techniques are the most important ingredients to be taken into account. The optimization problem to be solved in this context presents two interdependent aspects. The first is related to the discrete optimal actuator location selection problem, which is solved in this paper using genetic algorithms. The second is represented by a continuous variable optimization problem, through which the control gains are determined using classical techniques. A cantilever Euler–Bernoulli beam is used to illustrate the presented methodology.

# 1   Introduction

Vibration reduction is an important engineering goal in a variety of engineering applications. Today, static and dynamic analysis are not enough for design purposes and numerical simulation programs have to be coupled with optimization codes to perform automated optimal design. This procedure leads to competitive design configurations and guarantees that technological and economical constraints are respected. When mechatronic systems are taken into account, the problem becomes more complex because sensors, actuators and controls have to be considered together with the structure in all design steps.

A single piezoelectric element called a self-sensing actuator combines actuator and sensor capabilities for collocated control (Dosh *et al.* 1992). Models of the interaction between induced strain actuators and structures to which they are bonded have been presented by Crawley and De Luis (1987) and by Crawley and Anderson (1990).

Steffen and Inman (1999) used classical optimization techniques to determine the parameters related to the geometry and position of the piezoelectric element bonded to continuous Euler–Bernoulli beams, in such a way that the system poles are placed as far left as possible into the left half of the complex plane. Schiehlen and Schonerstedt (1998) used the finite element method to obtain a mathematical model of slender beams with piezoelectric actuators and sensors for vibration suppression.

Rao and Pan (1991) studied the discrete optimal actuator location problem in actively controlled structures. The zero-one optimization problem was solved using genetic algorithms. Kirby *et al.* (1994) has also used genetic algorithms to solve the optimal actuator size and location for multivariable control. Gabbert *et al.* (1997) presented a technique based on classical methods to determine actuator placement in smart structures by discrete-continuous optimization.

The design process of an adaptive structural system encompasses three principal phases: the structural design, the controller design and the placement of the actuators and sensors. The structure design is related to the static and dynamic behavior of the system. The finite element method together with modal analysis techniques is used in this first phase. The controller design involves the choice of the control law as well as the estimation of the control parameters. Different control laws can be used; however, in this research work we have chosen a proportional derivative control. The third phase leads to the placement of sensors and actuators and affects the signals available for the controller and the resulting influence on the structure through the actuators. Consequently, for optimal design purposes, the structure, the controller and the placement of actuators and sensors have to be considered simultaneously and constraint functions are taken into account to avoid undesired or unaffordable configurations (Lopes *et al.*, 2000a).

This paper presents the optimal design of smart structures using bonded piezo-electrics. The actuator placement is characterized by a discrete–continuous optimization problem that is solved using genetic algorithms. The continuous parameters of the control law are determined using classical optimization techniques. In the following, a brief review of the piezoelectricity phenomenon is presented, and the basic equations for the finite element model are shown. A proportional-derivative control is used, and the optimization strategy is discussed for the discrete and continuous cases. Finally, a numerical application using a cantilever beam with bonded PZTs is presented for illustration purposes.

# 2 Piezoelectricity Review

For a piezoceramic, the 3 direction (z-axis) is usually associated with the direction of poling, and the material is approximately isotropic in the other two directions. Materials that become electrically polarized when they are deformed present the direct piezoelectric effect, producing an electrical charge at the surface of the material. The converse piezoelectric effect results in a strain in the material when placed within an electric field. The direct and converse effects result in electromechanical coupling. While piezoelectric elements exhibit nonlinear hysteresis at high excitation levels, the response required in the current linear structural applications is approximately linear. In this work we will use the linear constitutive relations for piezoelectric materials as given, for instance, by Clark *et al.* (1998).

$$\{T\} = \left[c^E\right]\{S\} - [e]\{E\}, \tag{1}$$

$$\{D\} = [e]^T\{S\} + \left[\varepsilon^S\right]\{E\}, \tag{2}$$

where the superscript $(\ )^S$ means that the values are measured at constant strain and the superscript $(\ )^E$ means that the values are measured at constant electric field, $\{T\}$ is the stress tensor [N/m$^2$], $\{D\}$ is the electric displacement vector [C/m$^2$], $\{S\}$ is the strain tensor [m/m], $\{E\}$ is the electric field [V/m = N/C], $\left[c^E\right]$ is the elasticity tensor at constant electric field [N/m$^2$], $[e]$ is the dielectric permitivity tensor [N m/V m$^2$ = C/m$^2$] and $\left[\varepsilon^S\right]$ is the dielectric tensor at constant mechanical strain (permitivity matrix) [N m/V$^2$ m]. The letters in brackets indicate the units of the variables (in the SI system of units) with N, m, V and C denoting Newton, meter, Volts and Coulomb, respectively.

$$\{T\} = [T_{11}\ T_{22}\ T_{33}\ T_{23}\ T_{13}\ T_{12}]^T,$$

$$\{S\} = [S_{11}\ S_{22}\ S_{33}\ 2S_{23}\ 2S_{13}\ 2S_{12}]^T,$$

$$D = [D_1\ D_2\ D_3]^T, \qquad E = [E_1\ E_2\ E_3]^T,$$

$$\left[\varepsilon^S\right] = \begin{bmatrix} \varepsilon_1^S & 0 & 0 \\ 0 & \varepsilon_1^S & 0 \\ 0 & 0 & \varepsilon_3^S \end{bmatrix}, \qquad [e] = \begin{bmatrix} 0 & 0 & e_{31} \\ 0 & 0 & e_{31} \\ 0 & 0 & e_{33} \\ 0 & 0 & 0 \\ 0 & e_{15} & 0 \\ 0 & e_{15} & 0 \end{bmatrix},$$

$$[c^E] = \begin{bmatrix} c_{11}^E & c_{12}^E & c_{13}^E & 0 & 0 & 0 \\ c_{12}^E & c_{11}^E & c_{13}^E & 0 & 0 & 0 \\ c_{13}^E & c_{13}^E & c_{33}^E & 0 & 0 & 0 \\ 0 & 0 & 0 & c_{55}^E & 0 & 0 \\ 0 & 0 & 0 & 0 & c_{55}^E & 0 \\ 0 & 0 & 0 & 0 & 0 & c_{66}^E \end{bmatrix}.$$

If each element of the matrix of piezoelectric material constant, $[e]$, is designed by $e_{ij}$ where $i$ corresponds to the row and $j$ corresponds to the column of the matrix, then $e_{ij}$ corresponds to the stress developed in the $j$-th direction due to an electric field applied in the $i$-th direction. The piezoelectric strain constants $d_{ij}$, relating the voltage applied in the $i$-th direction to a strain developed in $j$-th direction, are provided more often than the stress constants. However, the piezoelectric stress constants can be obtained from the strain constants since the constitutive equation can also be written as:

$$\{S\} = [s^E]\{T\} + [d]\{E\}, \tag{3}$$

$$\{D\} = [d]^T\{T\} + [\varepsilon^T]\{E\}, \tag{4}$$

where $\varepsilon^T$ is the dielectric tensor at constant stress. The relative dielectric constant, $K^T$, is the ratio of the permitivity of the material, $\varepsilon^T$, to the permitivity of the free space, $\varepsilon_0$. ($\varepsilon_0 = 8.9 \cdot 10^{-12}$ Farads/m or Amp sec/V m). Then,

$$[c^E] = [s^E]^{-1}, \qquad [e] = [c^E][d],$$

$$[\varepsilon^S] = [\varepsilon^T] - [d]^T[c^E][d], \qquad K^T = \frac{\varepsilon^T}{\varepsilon_0},$$

$$[d] = \begin{bmatrix} 0 & 0 & d_{31} \\ 0 & 0 & d_{31} \\ 0 & 0 & d_{33} \\ 0 & 0 & 0 \\ 0 & d_{15} & 0 \\ 0 & 0 & d_{15} \end{bmatrix}.$$

## 3   Finite Element Model

The equations of motion for coupled electromechanical systems are derived using the coordinate system of Figure 1. The Rayleigh–Ritz formulation is used to derive the equations of motion of the electroelastic beam. The assumed displacement field shapes within the elastic body and electric potential field shapes will be combined through the piezoelectric properties to form a set of coupled electromechanical equations of motion.

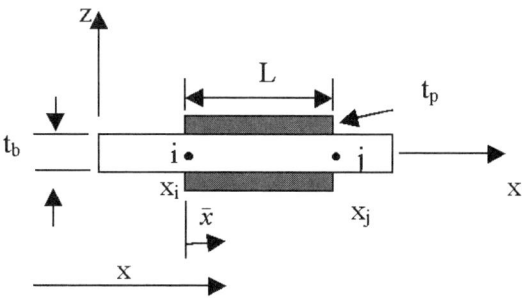

Figure 1 Section of the beam with bonded PZT.

Hagood *et al.* (1990) and Allik and Hughes (1970) applied the generalized form of Hamilton's Principle for a coupled electromechanical system:

$$\int_{t_1}^{t_2} [\delta (T - U + W_e - W_m) + \delta W] dt = 0, \qquad (5)$$

where $t_1$ and $t_2$ are two arbitrary instants, $T$ is the kinetic energy, $U$ is the potential energy, $W_e$ is the work done by electrical energy and $W_m$ is the work done by magnetic energy, which is negligible for piezoceramic material. These values are:

$$T = \int_{V_S} \frac{1}{2} \rho_S \dot{u}^T \dot{u} \, dV + \int_{V_P} \frac{1}{2} \rho_P \dot{u}^T \dot{u} \, dV, \qquad (6)$$

$$U = \int_{V_S} \frac{1}{2} S^T T \, dV + \int_{V_P} \frac{1}{2} S^T T \, dV, \qquad (7)$$

$$W_e = \int_{V_P} \frac{1}{2} E^T D \, dV, \qquad (8)$$

where $\rho$ is the mass density and the subscripts $s$ and $p$ refer to the structure and piezoelectric material, respectively. The virtual work, $\delta W$, done by external forces and the prescribed surface charge, $Q$, is,

$$\delta W = \int_{V_S} \delta u^T P_b \, dV + \int_{S_S} \delta u^T P_S \, ds_s + \delta u^T P_C - \int_{S_P} \delta \phi \, Q \, ds_P, \qquad (9)$$

where $P_b$ is the body force, $P_S$ is the surface force, $P_C$ is the concentrated load and $Q$ is the surface charge. To formulate an electroelastic matrix of a finite element method, the displacement vector, $\{u\}$, and the electric potential, $\phi$, must be expressed in terms of nodal value, $i$, via the interpolation function

$$u(x) = [N_u] \{u_i\}, \qquad (10)$$

$$\phi(x) = [N_\phi] \{\phi_i\}. \qquad (11)$$

Since the constitutive relations for both the structure and the piezoelectric material have been defined, the strain can be linked to the nodal displacement through the derivative of $\{u\}$. Likewise, a gradient operator can link the electrical field with the electrical potential. These relationships are:

$$\{S(x,t)\} = [B_u(x)]\{u_i(t)\}, \tag{12a}$$

$$\{E(x,t)\} = -[B_\phi(x)]\{\phi_i(t)\} \tag{12b}$$

with

$$[B_u(x)] = [L_u][N_u(x)], \tag{13a}$$

$$[B_\phi(x)] = \nabla[N_\phi(x)], \tag{13b}$$

where $[L_u]$ is the differential operator for the particular elasticity problem, and $\nabla$ is the gradient operator. At this point, substituting the generalized stress, strain and energy equations into the variational equation (5) can derive the coupled electromechanical system equation. Allowing arbitrary variations of $\{u_i\}$ and $\{\phi_i\}$, two equilibrium matrix equations, in generalized coordinates, are obtained:

$$([M_S^e] + [M_P^e])\{\ddot{u}_i\} + ([K_S^e] + [K_{uu}^e])\{u_i\} - [K_{u\phi}^e]\{\phi_i\} = \{F^e\}, \tag{14}$$

$$[K_{\phi u}^e]\{u_i\} - [K_{\phi\phi}^e]\{\phi_i\} = \{Q^e\}, \tag{15}$$

where the mass and stiffness matrices for the structure and piezoelectric path elements are defined as:

$$[M_S^e] = \int_{V_S} [N_u]^T \rho_S [N_u]\, dV, \tag{16a}$$

$$[M_P^e] = \int_{V_P} [N_u]^T \rho_P [N_u]\, dV, \tag{16b}$$

$$[K_S^e] = \int_{V_S} [B_u]^T [c_S][B_u]\, dV, \tag{17a}$$

$$[K_{uu}^e] = \int_{V_P} [B_u]^T [c^E][B_u]\, dV. \tag{17b}$$

The piezoelectric capacitance matrix, $K_{\phi\phi}$, and the electromechanical coupling matrix, $K_{u\phi}$, are

$$[K_{\phi\phi}^e] = \int_{V_P} [B_\phi]^T [\varepsilon^S][B_\phi]\, dV, \tag{18a}$$

$$[K_{u\phi}^e] = \int_{V_P} [B_u]^T [e][B_\phi]\, dV \tag{18b}$$

with $\left[K_{\phi u}^e\right] = \left[K_{u\phi}^e\right]^T$.

The force vectors are given by:

$$\{F^e\} = \int_{V_S} [N_u]^T \{P_B\}\ dV + \int_{S_S} [N_u]^T \{P_S\}\ ds_S + [N_u]^T \{P_C\}, \qquad (19)$$

$$\{Q^e\} = -\int_{S_P} [N_\phi]^T Q\ ds_P. \qquad (20)$$

For the entire structure, using the standard assembly technique for the finite element method, we obtain the complete equation for a coupled electromechanical system as

$$\begin{bmatrix} [M_{uu}] & 0 \\ 0 & 0 \end{bmatrix} \begin{Bmatrix} \{\ddot{u}\} \\ \{\ddot{\phi}\} \end{Bmatrix} + \begin{bmatrix} [K_{uu}] & [K_{u\phi}] \\ [K_{\phi u}] & [K_{\phi\phi}] \end{bmatrix} \begin{Bmatrix} \{u\} \\ \{\phi\} \end{Bmatrix} = \begin{Bmatrix} \{F\} \\ \{Q\} \end{Bmatrix}, \qquad (21)$$

where

$$[M_{uu}] = \sum_{i=1}^{ne} [M_S^e]_i + \sum_{j=1}^{np} [M_P^e]_j, \qquad (22a)$$

$$[K_{uu}] = \sum_{i=1}^{ne} [K_S^e]_i + \sum_{j=1}^{np} [K_{uu}^e]_j, \qquad (22b)$$

$$[K_{u\phi}] = \sum_{j=1}^{np} [K_{u\phi}^e]_j, \qquad [K_{\phi\phi}] = \sum_{j=1}^{np} [K_{\phi\phi}^e]_j, \qquad (22c)$$

where $ne$ is the number of structural elements and $np$ is the number of piezoelectric patches in the structure. The summation symbol, in the above equations, implies the finite element assembling. At this point it is important to note that the mass and stiffness matrices for a finite element and therefore for the complete structure are not positive definite. Lopes *et al.* (2000b) give more details about the application of the general equations presented above to given structures.

## 4  Modal Analysis and Control

Among the most popular control systems we can mention the following: output feedback controller, linear quadratic regulator and proportional derivative (PD) controller (Veley and Rao, 1996). In this study a PD controller is used to relate the sensor and actuator degrees of freedom of the finite element model as inputs and outputs, respectively. The vector of electric potential $\{\phi\}$ can be computed from the system outputs measured by the sensors, and for a PD controller is given by

$$\{\phi\} = [G_P]\{u_S\} + [G_D]\{\dot{u}_S\}, \qquad (23)$$

where $\{u_S\}$ is the vector of sensor positions and $[G_P]$ and $[G_D]$ are the matrices of gains of the proportional and derivative controllers, respectively.

The sensor (and actuator) degrees of freedom are a subset of the generalized coordinates and can be related to them by a distribution matrix $[T_S]$ as

$$\{u_s\} = [T_S]\{u\} \qquad \text{and} \qquad \{\dot{u}_s\} = [T_s]\{\dot{u}\}\,. \tag{24}$$

Matrix $[T_S]$ describes the PZT sensor-actuator position using zero-one entries. Zero indicates that no sensor-actuator exists at the corresponding position.

Considering proportional damping, the general equation of motion can be obtained by substituting equation (23) in the first part of equation (21) and taking into account equation (24):

$$[M_{uu}]\{\ddot{u}\} + [D^*]\{\dot{u}\} + [K^*]\{u\} = \{F\}, \tag{25}$$

where

$$[D^*] = ([D] + [K_{u\phi}][G_D][T_S])\,,$$

$$[K^*] = ([K_{uu}] + [K_{u\phi}][G_P][T_S])\,,$$

$[D]$ is the proportional damping matrix. Equation (25) can be integrated using any available numerical procedure. In this paper the numerical integration was conducted using the Newmark scheme corresponding to the constant-average-acceleration method (Bathe and Wilson, 1976).

As $\{\phi\}$ is known from the control law, we consider that the system is voltage controlled, and the second part of equation (21) can be used to determine the surface charge $Q$.

Equation (25) can be written in state space form as follows:

$$\{\dot{z}\} = [A]\{z\} + [B]\{u_f\}, \tag{26}$$

where $\{z\} = [\{u\}\,\{\dot{u}\}]^T$ is the state vector, $[A]$ is the state matrix, $B$ is the input matrix and $\{u_f\}$ is the applied force.

Taking into account simple linear transformations we have

$$[A] = \begin{bmatrix} [0] & [I] \\ -[M_{uu}]^{-1}[K^*] & -[M_{uu}]^{-1}[D^*] \end{bmatrix},$$

$$[B] = \begin{bmatrix} [0] \\ [M_{uu}]^{-1} \end{bmatrix}, \qquad \{u_f\} = \{F\}. \tag{27}$$

Natural frequencies and mode shapes can be obtained from the dynamic matrix (matrix $[A]$).

It is possible to measure the system performance for a particular choice of controller, placement and PZT characteristics by the rate of decay of system states. This is obtained by placing the poles of the system far into the left half of the complex plane. However, in doing so the control effort may exceed the maximum excitation voltage, and the second equation (21) for $\{\phi\}$ can be used as a constraint equation for design purposes.

Another common optimal control problem is called the *linear regulator problem*, which has application in vibration control suppression (see, e.g., Inman, 1989).

The optimal solution is obtained from the minimization of the performance index given by

$$J = \int_{t_0}^{t_1} (u^T Q_1 u + \phi^T Q_2 \phi) dt \,, \tag{28}$$

where $Q_1$ and $Q_2$ are weighting matrices. Optimal control requires choosing $Q_1$ and $Q_2$. Gabbert *et al.* (1997) considers that the mechanical energy is physically restricted and does not take into account the first term $Q_1$ of the performance index for minimization purposes. The resulting cost functional can be interpreted as electrical work if $Q_2 = K_{\phi\phi}$. Here we also use this approach in the optimization strategy in order to determine the optimal positions of the actuators for a given set of control gains.

## 5 Optimization

The determination of optimal positions of actuators and control gains in adaptive mechanical structures requires the formulation of a discrete–continuous optimization problem. In this case, the nonlinear optimization problem is defined as:

$$\text{Minimize } f(x) \tag{29}$$

with the constraints

$$g_i(x) \le 0 \,, \qquad i = 1, \ldots, m \tag{30}$$

and the variables

$$x_j \in [0, 1] \,, \qquad j = 1, ..., n_D \,, \tag{31}$$

$$x_{j,lb} \le x_j \le x_{j,ub} \,, \qquad j = n_D + 1, ..., n \,, \tag{32}$$

where $n = n_D + n_C$, $n_C$ is the number of continuous variables, $n_D$ is the number of discrete variable and $x_{j,lb}$ and $x_{j,ub}$ bounds on the continuous design variables.

In our case, the optimization goal aims at two aspects related to the dynamical behavior of the system. The first one is the minimization of the electrical work, and the second is the maximization of the vibration suppression of the mechatronic structure by placing the poles of the system far into the left half of the complex plane. Therefore the objective function is a combination of the following criteria

$$J = \min \left[ \int_{t_0}^{t_1} (u^T K_{\phi\phi} u) dt \right] \tag{33}$$

and

$$f_{\text{obj}}(x) = \max_k \left[ \text{Re } \lambda_k([A]) \right] \,, \tag{34}$$

where $\lambda_k([A])$ is the $k$-th eigenvalue of $[A]$ and Re stands for the real part of the eigenvalue.

The surface charge is given by the second equation (21) as

$$[K_{\phi u}]\{u\} + [K_{\phi\phi}]\{\phi\} = \{Q\} \,, \tag{35}$$

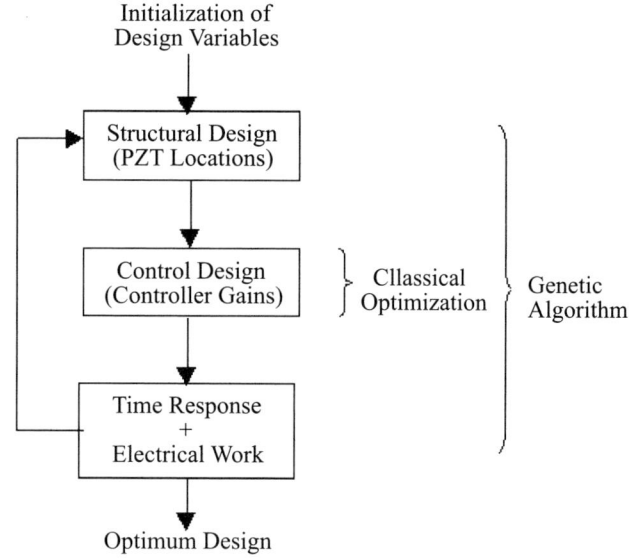

Figure 2  Flowchart for the optimization procedure.

where $\{\phi\}$ is given by equation (23). As the electric charge is limited to a given value depending on the actuator characteristics, from equation (35) it is possible to write a set of $n_p$ inequality constraints given by

$$\{Q\} \leq \{Q_{\max}\}. \tag{36}$$

Equation (36) guarantees that the maximum value for electric charge is not violated while maximizing damping in the structure.

The optimization is performed in a sequence of two interdependent steps:

1. $J$ is minimized with respect to the discrete design variables $x_j$, $j = 1, \ldots, n_D$;

2. $f_{\mathrm{obj}}(x)$ is maximized with respect to the continuous design variables $x_j$, $j = n_D + 1, \ldots, n$.

These steps are repeated until convergence is reached. The minimization over the discrete variables is performed using genetic algorithms and the maximization over the continuous variables is performed using classical optimization techniques, taking into account the constraints given by equation (36). Figure 2 shows the flowchart corresponding to the optimization procedure.

Genetic algorithms are random search techniques based on Darwin's "survival of the fittest" theories, as presented by Goldberg (1989). A basic feature of the method is that an initial population evolves over generations to produce new and hopefully better designs. The elements (or designs) of the initial population are randomly or heuristically generated. A basic genetic algorithm uses four main operators, namely *evaluation, selection, crossover* and *mutation* (Michalewicz, 1996), which are briefly described in the following. *Evaluation* – the genetic algorithms require information about the fitness of each population member (fitness corresponds to the objective function in the classical optimization techniques). The fitness measures the adaptation grade of the individual. An individual is understood as a set of design variables. No gradient or auxiliary information is required, only the fitness function is needed. *Selection* – the operation of choosing members of the current generation to produce the prodigy of the next generation. Better designs, viewed from the fitness function, are more likely to be chosen as parents. *Crossover* – the process in which the design information is transferred from the parents to the prodigy. The results are new individuals created from existing ones, enabling new parts of the solution space to be explored. This way, two new individuals are produced from two existing ones. *Mutation* – a low probability random operation used to perturb the design represented by the prodigy. It alters one individual to produce a single new solution that is copied to the next generation of the population to maintain population diversity. *Crossover* and *mutation* are called genetic operators. Genetic algorithms were originated with a binary representation of the parameters and have been used to solve a variety of discrete optimization problems. In this paper, the program *GAOT – The Genetic Algorithm Optimization Toolbox for Matlab 5* was used (Hook *et al.*, 1995).

The continuous optimization problem is solved using classical sequential unconstrained techniques. For this purpose the Augmented Lagrange-function Method is used to define a pseudo-objective function which takes into account the constraints. The unconstrained minimization is performed using the BFGS (Broyden–Fletcher–Goldfarb–Shanno) method through which an estimate of the Hessian matrix is updated each iteration. The one-dimensional search is performed using polynomial interpolation techniques. In this paper, *The Optimization Toolbox for Matlab* was used (Grace, 1993).

# 6   Numerical Application

To verify the design methodology presented above, an aluminum cantilever beam with 500x30x5 mm of length, width and thickness, respectively was considered. The properties of aluminum are: $E_S$ = 70 GPa, $\rho_S$ = 2710 kg/m$^3$. The total number of structural degrees of freedom used was 42 (20 elements, 2 DOF by node). The number of electrical degrees of freedom changes as a function of the number of PZTs considered in each example (2 DOF by PZT ). The piezoelectric material properties based on material designation PSI-5A-S4 (Piezo Systems, Inc.) are given in Table 1.

Table 1   Material properties of the PZT.

| | |
|---|---|
| $E_P$ = 63 GPa | $\rho_P$ = 7650 kg/m$^3$ |
| $d_{31}$ = $-190$e$-12$ m/V | $d_{33}$ = 390e$-12$ m/V |
| $c_{11}$ = 1.07e11 N/m$^2$ | $K^T$ = 1800 |
| $e_{31}$ = 30.705 N m/V m$^2$ | $\varepsilon_{33}^S$ = 7.33e$-9$ F/m |

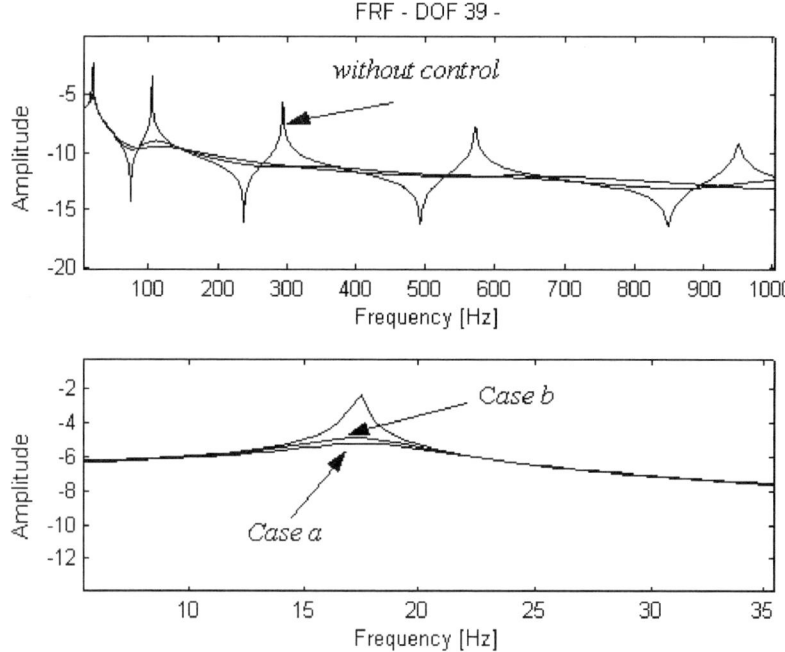

Figure 3  (a) FRF at DOF 39; (b) zoom at the first natural frequency.

Two different cases were considered to find the optimum number of actuators, their position and the corresponding optimum control gains:

**Case a)** Maximization of the vibration suppression for any number of actuators. The PZT size is equal to the discrete finite element size; hence there are $2^{20}$ possible discrete solutions for this case. Usually, some dynamic properties of the system are known, and some constraints can be imposed in order to reduce the computational time. In this specific example, we limited the possible placement of actuators to the 10 first elements based on the physics of a cantilevered beam. The optimal locations for the PZTs were found to be at the element numbers 1, 2, 3, 4, 6, 7, 8, 9 and 10.

**Case b)** Maximization of the vibration suppression for a fixed number of actuators. In practical situations, it is desired to have a small number of actuators. In this case, the total number of actuators is five. The optimal locations for the PZTs were found to be at element numbers 1, 2, 3, 4 and 7. Figure 3 shows the frequency response function (FRF) at the free end of the beam (DOF 39) and a zoom at the first natural frequency for the original and the optimal system configurations for the cases $a$ and $b$, respectively. Figure 4 shows the displacement time response at DOF 39 for cases $a$ and $b$.

In all examples analyzed, the capacitance for each PZT was equal to 200 nF. During the control optimization step, 100 volts was considered as the maximum voltage value, and the surface charge was limited to $2.10^{-5}$ Coulomb. Figure 5 shows the maximum surface charge corresponding to the PZT actuator moment applied at DOF 22 (element 10) for case $a$, and DOF 16 (element 7) for case $b$, respectively.

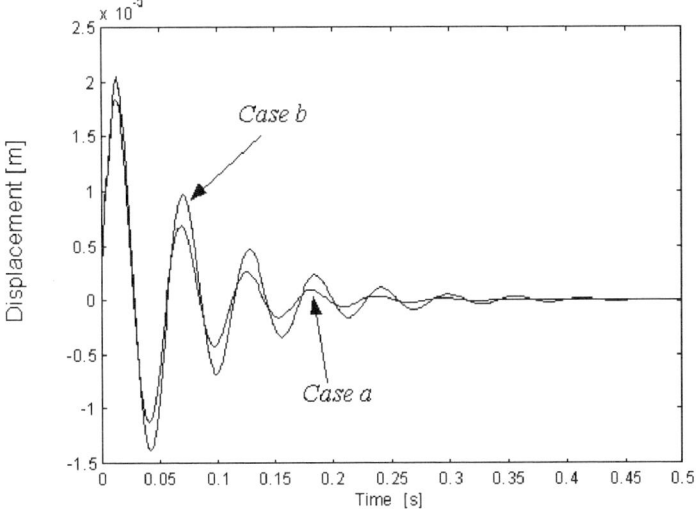

Figure 4  Time response at DOF 39 for cases **a)** and **b)**.

# 7   Conclusions

This paper presented a general methodology for mechatronic structures design using bonded piezoelectrics for vibration control. The actuator placement and control gain determination were characterized as a discrete–continuous problem. A hybrid approach using genetic algorithms and classical techniques was used to obtain the optimal design.

The hybrid technique involving genetic algorithms and classical optimization methods proved to be very effective in solving the discrete–continuous problem.

Two interdependent criteria were used for optimization purposes: the minimal electric work to obtain the optimal actuator placement and the maximum vibration damping to obtain the controller gains.

To avoid heavy calculation and high computation time, it is interesting to restrain the number of actuator candidate positions. In the application presented in this paper, the second half of the cantilever beam was not considered for actuator placement, since the first half of the beam corresponds to higher strain energy locations.

The authors are encouraged by the preliminary results obtained, and it seems that the methodology developed can be extended to practical applications of mechatronic systems.

# Acknowledgments

Dr. Lopes acknowledges the support of the Research Foundation of the State of Sao Paulo (FAPESP – Brazil). Dr. Steffen is thankful to Capes Foundation (Brazil) and to Fulbright Program (USA) for the scholarship awarded during his visit at

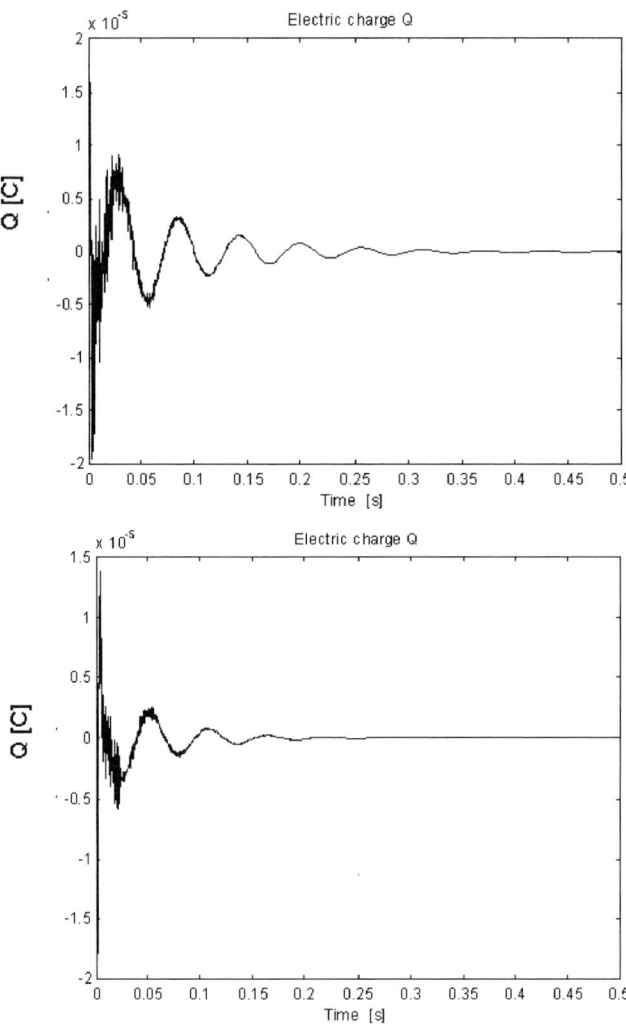

Figure 5  Surface charges for cases **a)** and **b)**.

Virginia Polytechnic Institute and State University in 1999. Dr. Inman gratefully acknowledges the support of National Science Foundation, grant no. CMS-9713453-001.

# References

1. Allik, H. and Hughes, T.J.R. (1970) Finite Element Method for Piezoelectric Vibration. *Int. J. for Numerical Methods in Eng.*, **2**, 151–157.

2. Bathe, K.-J. and Wilson, E.L. (1976) *Numerical Methods in Finite Element Analysis*. Prentice-Hall, Inc., New Jersey.

3. Clark, R.L., Sauders, W.R. and Gibbs, G.P. (1998) *Adaptive Structures: Dynamics and Control*. John Wiley & Sons, Inc., Chichester.

4. Crawley, E.F. and De Luis, J. (1987) The use of Piezoelectric Actuators as Elements of Intelligent Structures. *AIAA Journal*, **25** (10); *AIAA Paper*, N 86-0878.

5. Crawley, E.F. and Anderson, H.H. (1990) Detailed Models of Piezoceramic Actuation of Beams. *Journ. of Intell. Mater. Systems and Structures*, **1** (1), 4–25.

6. Dosch, J.J., Inman, D.J. and Garcia, E. (1992) A Self-Sensing Piezoelectric Actuator for Collocated Control. *Journ. of Intell. Mater. Systems and Structures*, **3** (1), 166–185.

7. Gabbert, U., Schultz, I. and Weber, C.-T. (1997) Actuator Placement in Smart Structures by Discrete-Continuous Optimization. In *ASME Design Eng. Tech. Conferences*, September, Sacramento, CA.

8. Goldberg, D. (1989) *Genetic Algorithms in Search, Optimization, and Machine Learning*. Addison-Wesley, Boston.

9. Grace, A. (1993) *Optimization ToolBox For Use with Matlab*. The Math Works, Inc., Natick, MA.

10. Hagood, N.W., Chung, W.H. and von Flotow, A. (1990) Modeling of Piezoelectric Actuator Dynamics for Active Structural Control. *Journal of Intell. Mater. Syst. and Struct.*, **1**, July, 4–25.

11. Hook, C.R., *et al.* (1995) A Genetic Algorithm for Function Optimization: A Matlab Implementation. NCSU-IE Technical Report 95-09 (North Carolina State University – Raleigh, NC 27695, USA).

12. Inman, D.J. (1989) *Vibration with Control, Measurement and Stability*. Prentice Hall, New Jersey.

13. Kirby III, G.C., Matic, P. and Lindner, D.K. (1994) Optimal Actuator Size and Location using Genetic Algorithms for Multivariable Control. In *AD-Vol. 45/MD-Vol. 54, Adaptive Structures and Composite Materials: Analysis and Application*, ASME, pp. 325–335.

14. Lopes Jr., V., Steffen Jr., V. and Inman, D.J. (2000a) Optimal Design of Smart Structures using Bonded Piezoelectrics for Vibration Control. In *Proc. of the International Conference on Noise and Vibration Engineering – ISMA 25*, Leuven, Belgium, September 13–15, Vol. 1, pp. 95–102.

15. Lopes Jr., V., Pereira, J.A. and Inman, D.J. (2000b) Structural FRF Acquisition via Electric Impedance Measurement Applied to Damage Detection. In *Proc. XVIII International Modal Analysis Conference – IMAC*, San Antonio, Texas, pp. 1549–1555.

16. Michalewicz, Z. (1996) *Genetic Algorithms + Data Structures = Evolution Programs*. Springer Verlag AI Series, New York.

17. Schiehlen, W. and Schonerstedt, H. (1998) Controller Design for the Active Vibration Damping of Beam Structures. In *SAE Conference*, pp. 137–146.

18. Steffen Jr., V. and Inman, D.J. (1999) Using Piezoelectric Materials for Vibration Damping in Mechanical Systems. *Journal of Intell. Mater. Systems and Structures*, **10** (12), 945–955.

19. Rao, S.S., Pan, T.S. and Venkayya, V.B. (1991) Optimal Placement of Actuators in Actively Controlled Structures using Genetic Algorithms. *AIAA Journal*, **29** (6), 942–943.

20. Veley, D.E. and Rao, S.S. (1996) Multiobjective Design of Piezoelectrically Damped Structures. In $6^{th}$ *AIAA/NASA/ISSMO Symposium on Multidisciplinary Optimization*, September, pp. 857–865.

# A Review of New Vibration Issues due to Non-Ideal Energy Sources

J.M. Balthazar,[1] R.M.L.R.F. Brasil,[2] H.I. Weber,[3]
A. Fenili,[4] D. Belato,[4] J.L.P. Felix[4] and F.J. Garzelli[2]

[1] *Dept. of Statistics, Applied Mathematics and Computation*
*Institute of Geosciences and Exact Sciences*
*UNESP, PO Box 178, 13500-200, Rio Claro, SP, Brazil*
[2] *Dept. of Structural and Foundations Engineering*
*Polytechnic School, University of São Paulo*
*PO Box 61548, 05424-930, SP, Brazil*
[3] *Dept. of Mechanical Engineering, Catholic University of Rio de Janeiro*
*Rua Marquês de São Vicente 225, 22453-900 Gávea, RJ, Brazil*
[4] *School of Mechanical Engineering*
*UNICAMP, PO Box 6122, 13800-970, Campinas, SP, Brazil*

We analyze the dynamical coupling between energy sources and structural response that must not be ignored in real engineering problems, since real motors have limited output power. We present models of certain problems that render descriptions that are closer to real situations encountered in practice.

## 1   Introduction

The study of non-ideal vibrating systems, that is, when the excitation is influenced by the response of the system, has been considered a major challenge in theoretical and practical engineering research. When the excitation is not influenced by the response, it is said to be an ideal excitation or an ideal source of energy. On the other hand, when the excitation is influenced by the response of the system, it is said to be non-ideal. Then, depending on the excitation, one refers to vibrating systems as ideal or non-ideal.

The behavior of ideal vibrating systems is well known in the current literature, but there are few results on non-ideal ones. Generally, non-ideal vibrating systems are those for which the power supply is limited. The behavior of the vibrating systems departs from the ideal case, as power supply becomes more limited. For non-ideal dynamical systems, one must add an equation that describes how the energy source "supplies the energy to the equations" that governs the corresponding ideal dynamical systems and the response is unknown. Thus, as a first characteristic, the non-ideal vibrating system has one more degree of freedom than its ideal counterpart.

The first kind of non-ideal problem to arise in the current literature is the so-called Sommerfeld effect, discovered in 1902 (see [1]), commented on in a book by Prof. Kononenko [2], entirely devoted to the subject. An experiment was described which detected interactions between a motor and its elastic foundation (a cantilever beam supporting a non-ideal energy source at its free end). In this case, the system exhibited unstable motions in the regions of resonance. Furthermore, the form of the resonance curve depended on which direction the frequency of excitation was being altered. The form of vibrations was changed, and the character of the transition through resonance was also altered when the frequency of excitation was changed. Further investigation of this behavior led to what is commonly called jump phenomena. As the driving frequency approaches the natural frequency, the vibrating system can suddenly jump from one side of the resonance peak to another. That is, the system operating in a steady state cannot realize certain frequencies near resonance. The jump appears on the frequency response curve as a discontinuity that indicates a region where steady-state conditions do not exist. Thus we see that modeling vibrating systems by ideal sources may be inadequate if the driving frequency lies near a natural frequency of the system, as may be the case. If we consider the region before resonance on a typically frequency–response curve, we note that as the power supplied to the source increases, the speed of rotation of the motor increases accordingly. However, this behavior does not continue indefinitely. The closer the motor speed moves toward the resonant frequency, the more power the source requires to increase the motor speed, as part of the energy is consumed in moving the supporting structure. A large change in the power supplied to the motor results in a small change in its frequency and a large increase in the amplitude of the resulting oscillations. Thus, near resonance, it appears that additional power supplied to the motor only increases the amplitude of the response of the structure while having little effect on the RPM of the motor. Jump phenomena and the increase in power required by a source operating near resonance are manifestations of a non-ideal energy source and are often referred to as the Sommerfeld effect, in honor of the first man who observed it [1]. Sommerfeld suggested the that structural response provided a sort of energy sink. Thus, we pay to vibrate our structure rather than operate the machinery. One of the problems often faced by designers is how to drive a system through resonance and avoid the "energy sink" described by Sommerfeld.

In this paper we will illustrate this problem as a motivation before passing to several other related problems. In order to understand better what happened in the experiment performed in the past [1], some authors investigated the dynamics of an unbalanced direct current motor mounted on an elastically supported table through experimental and numerical simulations. Recently, [3], [4], [5] and [6] analyzed, with interesting details, the governing equations of a cantilever beam supporting a non-ideal energy source at its free end. They presented some experimental results on this problem. Numerical simulations of a simplified model of the same problem, using an unbalanced rotor attached to an elastic nonlinear support with internal and external damping driven by a non-ideal energy source, were analyzed by [4]. In [3], [5] and [6] the response of the beam was monitored by a non-contacting inductance gage or by an accelerometer near the end of the beam. To control the speed of the motor, a power supply was used and the applied voltage was controlled by the operator. The current was left free according to the demand of the motor, but the available power was limited. The response of the beam was monitored by a chain-driven encoder, 560 pulses per revolution. They conclude that: (i) the interactions between the motor and its support induce non-constant rotation speed of a mean value frequency; (ii) gravity can induce even harmonics in the frequency response of the system; (iii) cubic terms in the nonlinear stiffness is responsible for odd harmonics in the frequency response of the system; (iv) the current demanded by the motor varies with the same frequencies of the vibration of the support structure; (v) the quadratic and cubic nonlinearities in the problem are of same order of magnitude; and (vi) in the experiments, sub- and super-harmonic as well as parametric resonances were not detected.

Further contributions to non-ideal problems were presented in books [7], [8] and [9] and papers by Prof. Dimentberg and coworkers [10] and [11]. Profs. Nayfeh and Mook [12] give a comprehensive and complete review of different approaches to the problem up to 1979. Recently, a complete review of different theories on non-ideal vibrating systems was discussed and presented in [13] and [14]. We present, in a further section of this paper, comments on a number of scientific papers published in the period from 1969 to 2000 concerning a number of different non-ideal problem applications.

The goal of this paper is to present an overview of various aspects of non-ideal vibrating systems, such as discussion of the physical phenomena involved and the study of an adequate methodology to deal with them, and give a report of selected papers published recently and in the past. We remark that the present work does not claim literature completeness, since the available literature is dispersed over many distinct sources. It is restricted to the main references on non-ideal vibrating systems and some related papers on this subject. This paper is organized as follows: in Section 2 we discuss the ability to deal with non-ideal vibrating problems; in Section 3 we discuss the modeling of non-ideal vibrating systems; in Section 4 we present the main properties of DC machines; in Section 5 we review current literature; in Section 6 we make some comments on some recent work; in Section 7 we make some concluding remarks on the subject; in Section 8 we acknowledge financial support by research funding agencies; finally we list the references.

# 2    On the Ability to Deal with Non-Ideal Dynamical Systems

The extended Hamilton's Principle [15] plays a very important role in the field of analytical dynamics. Essentially, this variational principle reduces the formulation of a dynamical system to the calculation of variation $\delta$ of only three scalar quantities:

the kinetic energy $T$, the potential energy $V$ (including the work of the conservative forces, $W_C$) and the work of the nonconservative forces, $W_{NC}$. It is invariant under generalized coordinate transformation. It is a variational principle that states that among all possible paths, non-ideal (or ideal) dynamical systems will take the path that makes

$$\delta \int_{t_1}^{t_2} \Gamma dt = 0 \,, \tag{2.1}$$

where $\Gamma = T - (V + W_{NC})$ and $\int_{t_1}^{t_2} \Gamma dt$ will be an extreme for instants time $t_j$ $(j = 1, 2)$ such as $\delta(t_1) = \delta(t_2) = 0$. We will obtain from (2.1) the governing equations of motion of non-ideal dynamical systems and the associated possible boundary. After obtaining these equations of motion, they can be analyzed via a number of different procedures such as, for example, asymptotic methods (in the case of moderately large amplitudes) such as the Multiple Scales Method [16] or Method of Normal Forms [17]. In some other cases, of higher dimension, we can also use the Finite Element Method or other numerical methods.

## 3    On Modeling of Non-Ideal Vibrating Systems

We suppose that the governing equations, for a system driven by a single non-ideal source (using the approach discussed in the last section), have the form:

$$\left\{ \begin{array}{l} \text{First set:} \quad \textit{Dynamic Supporting Equations} \ = \ \textit{Interaction Terms} \\[2mm] \text{Second set:} \quad J\left[\dfrac{d^2\varphi}{dt^2}\right] + L\left(\dfrac{d\varphi}{dt}\right) = \textit{Other Interaction Terms} \end{array} \right\} , \tag{3.1}$$

where $J$, $L$ and $\dfrac{d\varphi}{dt}$ are, respectively, the moment of inertia of the rotating mass, the driving torque of the energy source (DC motor) and the angular velocity of the rotor. Note that it is possible to include damping and the torque resisting the rotation of the motor in the above expressions. The procedure described here is a general one. Details will depend on the particular problem analyzed, the properties of the supported structure and the characteristics of the rotor. Also note that, here, the excitation is always limited in two senses: (i) by the characteristic curves of the particular energy source (DC motor), and (ii) by the dependence of the motion of the dynamical system on the motion of the energy source, that is, the coupling between the governing equations of motion of the dynamical system and the energy source (DC motor). Generally, $\dfrac{d^2\varphi}{dt^2}$ varies slightly around the period of the vibration, that is, it is of $O(\varepsilon)$, where $\varepsilon$ is a small parameter of the problem. Note that the small parameter $\varepsilon$ does not appear explicitly in equation (3.1), because it is an intrinsic term of the interaction terms. Nevertheless, one should have in mind that the parameter $\varepsilon$ is being considered in equation (3.1). If we consider the unperturbed motion, obtained by imposing $\varepsilon = 0$ in the above equations, we have the simplest dynamic support oscillation with a constant rotation speed of the rotor. Therefore, for $\varepsilon \neq 0$, it is natural to expect that the oscillations are approximately the simplest dynamic support oscillations and the rotation speed of the rotor is approximately constant, i.e., it varies slowly.

The non-ideal vibrating system may be studied by determining the effect of changing the physical parameters in the system, that is, the unbalanced masses, mass of the motor, moment of inertia of the motor and other rotating masses, eccentricity of the motor, damping, characteristic of the motor, etc. Supposing that the non-ideal vibrating system is started from some initial conditions, one could increase its speed until it reaches resonance conditions. Then, depending upon these initial conditions imposed on the motion and the range of physical parameters, the angular velocity $\dfrac{d\varphi}{dt}$ will continue to increase beyond the resonance condition (passage through resonance) or it will remain close to the natural frequency of the vibrating system (capture in resonance). Obviously, the time of passage through resonance also depends on the initial conditions. We remark that we may be unable to operate a rotating vibrational system beyond the critical speed due to a limited power supply. If it is possible to pass through the critical speed, it often yields a comparatively large transient dynamic response. To solve these problems, optimum design methods can be applied to the operating curve of a limited power supply. In this way, the operating curve is so optimized as to reduce the transient dynamic responses around the critical speed. We mention the application of two optimization methods in two different non-ideal problems: [18] considered variations of the acceleration rate using a gradient-based optimization method in order to minimize the motion during passage through resonance and [19] analyzed a synthesis of an optimal control through the first resonance peak by use of Tikhonov's method of regularization. We also mention that the topologic properties of parametric and non-parametric non-ideal vibrations, as defined in Kononenko's book [1], were analyzed by [20] and [21]. The use of geometric approaches, such as Poincaré Maps, can be the best way to analyze this kind of phenomenon.

## 4   On the Characteristics of DC Machines

At this point we must consider the characteristics of the motor. For each potential $V$, considered constant, the characteristics of the DC motor are defined by the curves Driving Torque *vs.* Angular Velocity of the motor, as we can see in Fig. 1, where the indexes make reference to the potential $V$. The motor will be assumed to drive a mechanical load corresponding to a mechanical torque. The power supplier permits control of the maximum voltage (the current is left free in this case) or the maximum current (the voltage is left free in this case). We have preferred to control the voltage, because it is more adequate and more common in practical situations.

We obtain

$$L = T_m + J\left[\frac{d\varphi}{dt}\right], \tag{4.1}$$

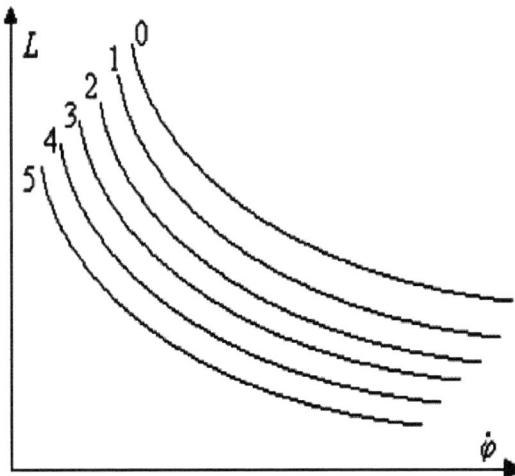

Figure 1  Characteristic curves of the energy source of limited power.

where $T_m$ is the mechanical torque. It can be modeled as [22]:

$$T_m = k_1 + k_2\left[\frac{d\varphi}{dt}\right] + k_3\left[\frac{d^2\varphi}{dt^2}\right]. \tag{4.2}$$

Note that the first term in equation (4.2) corresponds to a constant Coulomb friction due to the friction of the bearing, the second term represents the viscous damping due to eddy currents and the third term is due to the inertia of the rotor. The various coefficients in equation (4.2) are constants.

# 5   Brief Literature Review

The vibrations of linear dynamical systems have been studied exhaustively in recent years and significant contributions have also been made to the theory of vibrations of non-linear dynamical systems. The problem of passage through resonance of vibrating systems has been paid special attention by engineering researchers in recent years but unfortunately little literature on this subject is available. Here, we mention a number of works related to non-ideal vibrating systems. Probably, C. Laval was the first to work with non-ideal problems via an experiment. He built, in 1889, a single-stage turbine and demonstrated that in the case of rapid passage through resonance with enough power, the maximum vibration amplitude may be reduced significantly compared with that obtained in steady-state resonant vibration. Simultaneously, it was known that sometimes the passage through resonance required more input power than the vibrating system driver had available. The consequence is the so-called Sommerfeld effect that the vibrating system cannot pass the resonance or requires an intensive interaction between the dynamical system and the motor to do it. The worst case is where a dynamical system, constructed for an over-critical operation, becomes stuck *just* before resonance conditions are reached. A strong interaction results in fluctuating motor speed and fairly large vibration amplitudes. This phenomenon was studied intensively by [1]. The work of Sommerfeld was the starting point for the research in this field. Prof. Kononenko [2] presented the first detailed study of the non-ideal problem of passage through resonance. Later, [23] examined Kononenko's vibrating systems in his PhD thesis in Germany. Prof. Evan-Iwanoski [8] also contributed work on the non-ideal Duffing oscillator. After these results, the problem of passing through resonance was investigated independently by a number of international scientific groups that produced some essential results. The majority of these studies were based on an ideal power supply with prescribed speed increase or decrease. Only Prof. Dimentberg [10, 11] included a limited power source in a simple one-degree-of-freedom dynamical system. We mention that [24] considered the dynamics of a flexible slider crank mechanism driven by a non-ideal energy source analyzing the steady-state behavior and the transient startup and rundown, using numerical simulations. In [10, 11] the authors considered a one-degree-of-freedom dynamical system composed of a rigid base flexibly attached to the ground and excited by an unbalanced motor. The support stiffness was *switched* from a high value to a low value as the rotational speed of the motor increased. Some experiments were conducted for comparison with the numerical simulations. A non-ideal dynamical system was also studied by [25] using a similar approach. They considered a flexible simply supported vertical shaft with damping, and it was assumed that the bending stiffness of the shaft could be *switched* from one value to another. As the rotational speed of the shaft increased, the bending stiffness was changed at a certain time to avoid passage through resonance. The transient motion induced by this change of stiffness could be large, and hence the decrease in maximum response was not always significant. Suherman and Plaut [26] investigated the transient motions of a flexible, simply supported, horizontal shaft with a central disk (including eccentricity of the disk, gravitational forces and damping) based on PhD thesis of Suherman [27]. In order to suppress vibrations they assumed that a flexible, asymmetric internal support could be activated at a certain position along the shaft to change the stiffness of the system. We note that the possibility of passing through resonance without large oscillation amplitudes by switching the system stiffness has attracted the attention of the mechanical engineering community in recent years, but few authors have considered non-ideal dynamical systems in their work. Finally, we mention a paper by [28], considering variations of the acceleration rate in order to minimize the motion during passage through resonance, and two earlier works [29] and [30] in the non-ideal dynamical systems field.

Next we will comment some recent results on non-ideal problem.

# 6 Comments on Some Recent Results

In this section we present some recent studies on non-ideal vibrating problems. In particular, we present results obtained by the authors of this paper on: (1) non-ideal nonlinear portal frame structures; (2) non-ideal slewing structures; (3) stick-slip non-ideal vibrations; (4) a non-ideal gear rattling problem; and (5) a non-ideal electro-motor-pendulum system.

## 6.1   A portal frame structure

Nonlinear vibrations of frames have been investigated by a number of authors. We note that the nonlinear vibrations of a portal frame under a single ideal harmonic excitation will be found in [31] and multiple scales analysis of nonlinear oscillations of a portal frame foundation for several machines is considered in [32]. Geometric studies of ideal nonlinear vibrations are presented in [33]. Now, the simple portal frame of Fig. 2, of nonlinear behavior and excited by a non-ideal motor, is considered in the analysis. It has two columns clamped at their bases with height $h$ and constant cross-section, of area $A_c$ and moment of inertia $I_c$, with concentrated weights at their tops of mass $m$. The horizontal beam is pinned to the columns at both ends with length $L$ and constant cross section, of area $A_b$ and moment of inertia $I_b$. A linear elastic material is considered whose Young modulus is $E$. The structure will be modeled as a three-degree-of-freedom system. The adopted model is shown in Fig. 2 (portal frame and motor). The generalized coordinate $q_1$ is related to the horizontal displacement of the central section of the beam in sway mode (with natural frequency $\omega_1$) and $q_2$ to the mid-span vertical displacement of the same section of beam in the first symmetrical mode (with natural frequency $\omega_2$). An unbalanced (non-ideal) DC motor is placed at the mid-span of the beam. The angular displacement of its rotor is adopted as generalized coordinate $q_3$. The total mass of the motor is $M$ and its rotor has moment of inertia $I$ and carries an unbalanced mass $m_0$ at a distance $r$ from the axis. The characteristic driving torque of the motor is $\Theta(\dot{q}_3)$ and the resisting torque is $\Psi(\dot{q}_3)$. For each given power level, they are assumed to be known, either from the manufacturer or from previous experiments. The natural frequencies are set in such a way that the motor will excite either the first anti-symmetric mode (sway) or the first symmetrical mode. The physical and geometric properties of the frame are chosen to tune the natural frequencies of these two modes into a 1:2 internal resonance ($\omega_2 \approx 2\omega_1$).

To derive the equations of motion, Hamilton's Principle is used. $V$ is the total potential energy of the system, $T$ is the kinetic energy of the system and $N_i$ are the generalized non-conservative forces applied to the system. The total potential energy is:

$$V = MgLq_2 + (k_c - mgC)h^2q_1^2 + \frac{k_bL^2}{2}q_2^2 + \frac{k_bCh^2L}{2}q_1^2q_2 + m_0rg\sin(q_3)\,. \quad (6.1)$$

The kinetic energy is:

$$T = \frac{1}{2}\{(2m + M - m_0)h^2\dot{q}_1^2 + (M - m_0)L^2\dot{q}_2^2 + I\dot{q}_3^2 \rightarrow$$
$$\rightarrow m_0[(h\dot{q}_1 - r\dot{q}_3 sin(q_3))^2 + (L\dot{q}_2 + r\dot{q}_3\cos(q_3))^2]\}\,. \quad (6.2)$$

The generalized non-conservative forces are:

$$N_1 = -c_1h\dot{q}_1\,, \qquad N_2 = -c_2L\dot{q}_2\,, \qquad N_3 = \Theta(\dot{q}_3) - \Psi(\dot{q}_3)\,. \quad (6.3)$$

In (6.3) $c_1$ and $c_2$ are modal linear viscous damping, $\Theta(\dot{q}_3)$ is the characteristic driving torque of the motor, and $\Psi(\dot{q}_3)$ is the resisting torque. Thus, we obtain the

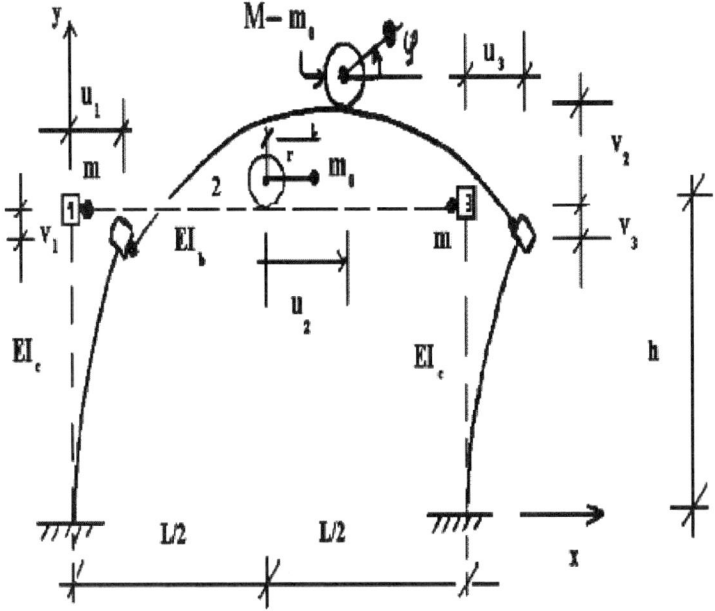

Figure 2  Non-ideal portal frame model.

following equations of motion of this model:

$$
\begin{cases}
\ddot{q}_1 + \omega_1^2 q_1 = \varepsilon[-\alpha_5 q_1 q_2 + \alpha_1(\ddot{q}_3 sin(q_3) + \dot{q}_3^2 \cos(q_3)) - \mu_1 \dot{q}_1], \\[2mm]
\ddot{q}_2 + \omega_2^2 q_2 = \varepsilon[-\alpha_6 q_1^2 + \alpha_2(-\ddot{q}_3 \cos(q_3) + \dot{q}_3^2 sin(q_3)) - \mu_2 \dot{q}_2 - \alpha_8], \\[2mm]
\ddot{q}_3 = \varepsilon[\alpha_3 \ddot{q}_1 sin(q_3) - \alpha_4 \ddot{q}_2 \cos(q_3) - \alpha_7 \cos(q_3) + F(\dot{q}_3)],
\end{cases}
\tag{6.4}
$$

where:

$$
\varepsilon\alpha_1 = \frac{m_0 r}{(2m+M)h}, \quad \varepsilon\alpha_2 = \frac{m_0 r}{ML}, \quad \varepsilon\alpha_3 = \frac{m_0 h r}{I + r^2 m_0}, \quad \varepsilon\alpha_4 = \frac{m_0 L r}{I + r^2 m_0},
$$

$$
\varepsilon\alpha_5 = \frac{k_b C L}{2m+M}, \quad \varepsilon\alpha_6 = \frac{k_b C h^2}{2mL}, \quad \varepsilon\alpha_7 = \frac{m_0 r g}{I + m_0 r^2},
$$

$$
\varepsilon\mu_1 = \frac{c_1}{(2m+M)h}, \quad \varepsilon\mu_2 = \frac{c_2}{ML}, \quad \varepsilon F(\dot{q}_3) = \frac{\Theta(\dot{q}_3) - \Psi(\dot{q}_3)}{I + r^2 m_0},
$$

$$
\omega_1^2 = \frac{2(k_c - mgC)}{2m+M}, \quad \omega_2^2 = \frac{k_b}{M}.
\tag{6.5}
$$

We now consider real, limited power supply, motors. For simplicity, their characteristic curves of the energy source (DC motor) are assumed to be straight lines of form

$$
\Theta(\dot{q}_3) - \Psi(\dot{q}_3) = \hat{a} - \hat{b}\dot{q}_3.
\tag{6.6}
$$

Note that the $\hat{a}$ is related to the voltage, and $\hat{b}$ is a constant for each model of motor considered. The voltage will be the parameter control of the problem. The numerical values adopted by the authors, in [34], [35] and [36], are EI $= 128$ N m$^2$, $h = 0.36$ m, $L = 0.5$ m, $M = 2.0$ kg, $m = 0.5$ kg, $_m0 = 0.1$ kg, $I = 0.000017$ kg m$^2$, $r = 0.01$ m, $c_1 = 2.55$ N s/m and $c_2 = 3.14$ N s/m. A standard RK-4 algorithm is used. Passage through resonance with the second natural frequency, corresponding to the mid-span vertical displacement of the beam in the first symmetrical mode, with frequency $\omega_2 = 157$ rad/s, is presented. These values were also chosen to allow for a 1:2 internal resonance condition for the structure, as the frequency of the first mode is $\omega_1 = 78$ rad/s. They obtained the following results: (i) with slow increase of power levels we have the possibility of occurrence of the Sommerfeld effect (getting stuck in resonance), that is, not enough power to reach higher speed regimes with lower energy consumption; (ii) we have the possibility of occurrence of saturation of high frequency low amplitude mode and transference of energy to low frequency high amplitude mode; and (iii) we show the possibility of occurrence of regular (periodic) motion and irregular (chaotic) behavior, depending on the value of the control parameter (voltage of the DC motor). We also have compared numerical results of the modulation equations, obtained by use of an averaging method [37], and the original governing equation, using the same physical parameters and initial conditions. In [38], we obtained excellent agreement between the two solutions.

## 6.2   On non-ideal slewing structures

The study of the dynamical behavior of slewing flexible structures has in view the production of lighter and faster systems of this type. The study of the dynamics and control of these structures is complex and of continuing interest to researchers and scientists. The applications may be divided into two groups: robotics and space structures. Typically, the modal analysis approach is the most popular model approach to space slewing structures and the Finite Element approach to robotics. The slewing structures were first considered in the literature by [39], applying a truncated modal model on a two-beam and a two-joint system, including discussion of a controlled design with a torque source. Considerable efforts have been made since then by other authors. We mention that in [40] hardware was set up to study slewing control for flexible structures, experimentally. That paper was the motivation of the present section. A long flexible panel is used for laboratory experiments in air and vacuum at the thermovacuum chamber at the NASA Langley Research Center. The test model is cantilevered in a vertical plane, driven by an electric gear-motor. The instrumentation consists of three full-bridge strain gauges to measure bending deformation and an angular potentiometer to measure the angle of rotation at the root. The strain gauges are located near the root and at the mid-span. Signals from all four sensors are amplified and then monitored by an analog data acquisition system. An analog computer closes the control loop, generating a voltage signal for the gear-motor based on an output feedback law. The test apparatus is mathematically modeled as a flexible beam rotating about a vertical axis: the flexible beam is cantilevered from the motor at the root $x = 0$ and free at the tip $x = L$. In [41] the author considered a single link flexible slewing beam, driven by a DC motor at its end. They discussed modal factors, which is

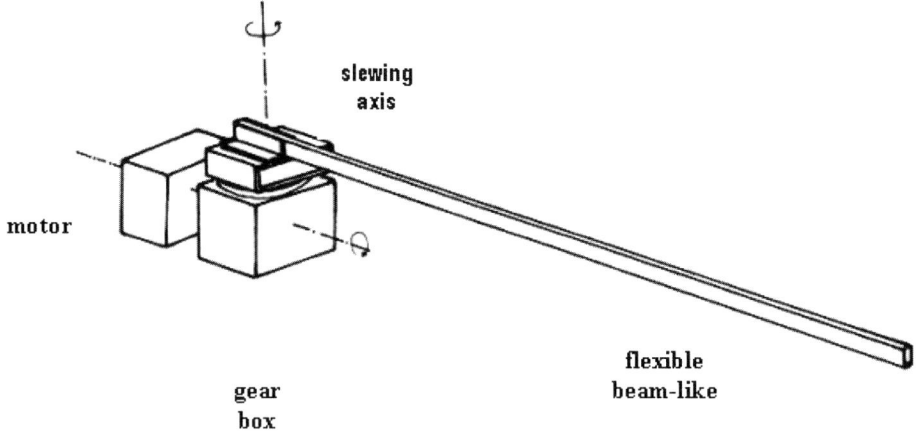

Figure 3  Schematic of a slewing flexible structure system.

indicative of the degree that actuators and flexible structures interact with one another dynamically. In [42] and [43], a Finite Element algorithm was used to consider the dynamic interaction between a DC motor and a slewing beam. They found that systems with an appropriate amount of DC-motor/beam interaction tended to be easy to control and required only a modest amount of actuator effort. Systems with less DC-motor/beam interaction were prone to beam vibrations and required feedback of beam motions for good closed-loop performance. Systems with excessive interaction required stabilization efforts for good transient performance and tended to consume high levels of actuation energy. We remark that little effort has been focused on DC motor–structure interaction, which obviously affects the dynamic properties of the slewing system. The main objective of this section is to deal with this kind of interaction. We adopted the physical and mathematical model defined in [44]. In Fig. 3 we show the physical model.

We remark that we want to change $\theta$ from zero to some desired angle, say, $\theta_f$. Naturally we specify $\theta$ as a pure function of time $t$ in order to produce this change. Note that when $\theta$ changes, the beam deflects (and its deflection can be calculated by the governing equations). A typical function $\theta(t)$ may look like the curve exhibited in Fig. 4 [46]. It's necessary to apply a forcing moment to produce this $\theta(t)$.

Note that using the equation for the DC motor it is possible to find the voltage $V(t)$ that has to be applied to the DC motor to make it move according to $\theta(t)$. This will render the vibrating system non-ideal [2]. Fig. 5 [45] shows the schematic of forces and moments actuating in this non-ideal vibrating system.

In [45] we discuss a derivation of the governing equations and perform a nonlinear analysis of the motion of a flexible rotating cantilever beam, without assuming that the beam is inextensive. We also present the derivation of the equations that govern the weak electric motor used to rotate the base of the beam and their coupling to those that govern the motions of the beam.

We remark that in [47] and [48] the authors obtained that in the study of slewing flexible structures it is important to consider small values of the gear ratio or

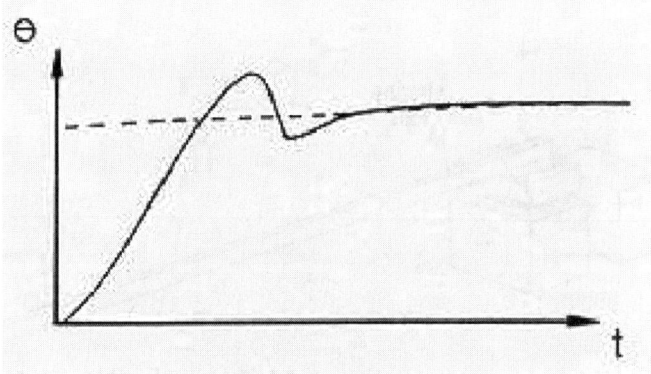

Figure 4  Typical motion.

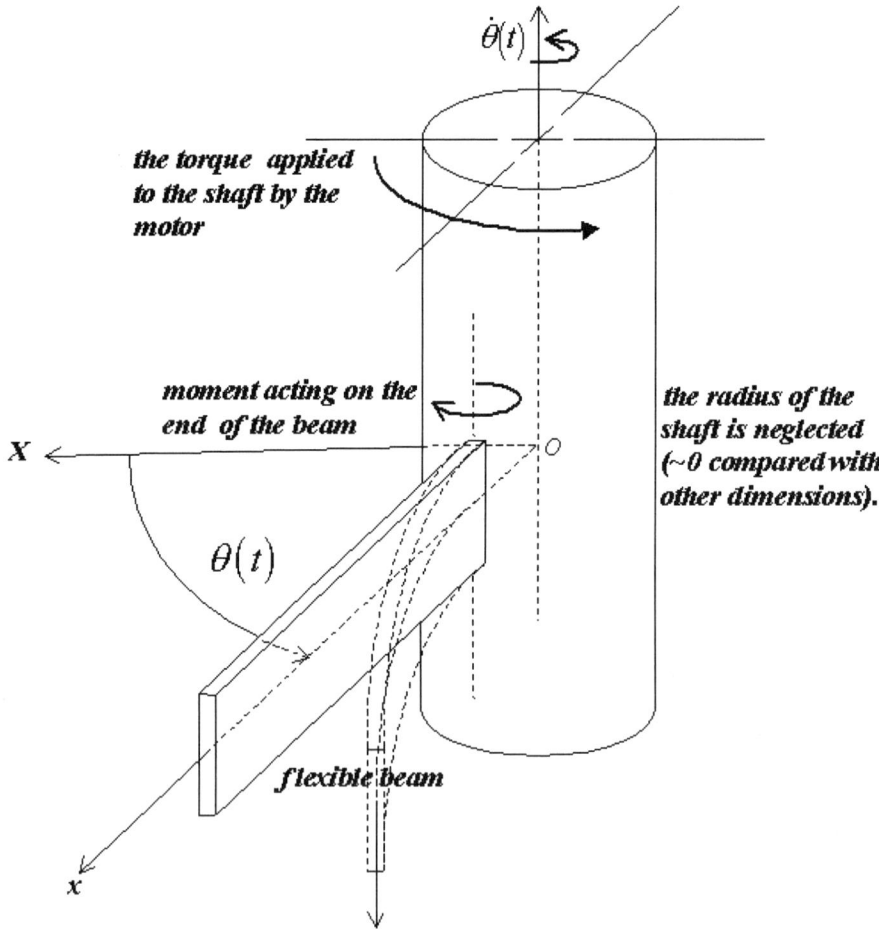

Figure 5  Schematics of forces and moments actuating in non-ideal slewing motions.

great velocities of slewing. The simulations have also shown that by increasing the gear ratio considerably, the differences between the ideal system approach and the non-ideal system approach tend to vanish. A damping effect in the amplitudes of vibration (in the absence of the beam structural damping) due to the actuator–structure interaction in the non-ideal case was also observed. In an application of the Center Manifold Theory [49] to a non-ideal slewing flexible structure, the authors obtained, in [50], the conditions for the occurrence of a Hopf bifurcation in this kind of problem. The analysis of these problems by use of center manifold theory was realized around these conditions. It is known in the current literature that a Hopf bifurcation is one possible route to chaos. Next, we present the governing equations of motion for a slewing flexible structure under large deflections and coupled to a DC motor, written in matrix form [50]:

$$
\left\{ \begin{array}{c} \dot{x}_1 \\ \dot{x}_2 \\ \dot{x}_3 \\ \dot{x}_4 \end{array} \right\} = \left\{ \begin{array}{l} x_3 \\ \\ x_4 \\ \\ \dfrac{D}{AD-BC}\left[-a_6x_3 + a_8U - a_7x_2 + \sigma^2(-a_3^N x_4^2 x_2 - a_5^N x_2 x_3 x_4)\right] - \\ \dfrac{-C}{AD-BC}\left[-a_6x_3 + a_8U - a_7x_2 + \sigma^2(-a_3^N x_4^2 x_2 - a_5^N x_2 x_3 x_4)\right] + \\ \\ \\ -\dfrac{B}{AD-BC}\left[-b_1x_2 + \sigma^2(-\mu_1 x_4 - b_3^N x_3^2 x_2 + b_4^N x_2 x_3 x_4 + b_6^N x_2^3 - b_7^N x_2^3 + \right. \\ \left. +b_8^N x_4^2 x_2)\right] \\ \\ +\dfrac{A}{AD-BC}\left[-b_1x_2 + \sigma^2(-\mu_1 x_4 - b_3^N x_3^2 x_2 + b_4^N x_2 x_3 x_4 + b_6^N x_2^3 - b_7^N x_2^3 + \right. \\ \left. +b_8^N x_4^2 x_2)\right] \end{array} \right.
$$
(6.7)

where $A = 1+\sigma^2(a_2^N x_2^2)$; $B = a_1 - \sigma^2(a_4^N x_2^2)$; $C = b_2 + \sigma^2(b_5^N x_2^2)$; $D = 1-\sigma^2(b_8^N x_2^2)$, the parameters $a_i$, $a_i^N$ and $b_i$, $b_i^N$ are functions of the geometric and material properties of the system, and $\mu_1$ represents the structural damping. The point $(x_1, x_2, x_3, x_4) = (0,0,0,0)$ is a fixed point (or a point of equilibrium) for the vibrating system (6.7). The following analysis will be performed in the neighborhood of this particular point. Expanding (6.7) in Taylor series around $(0,0,0,0)$ and keeping terms to order three, one finds:

$$
\left\{ \begin{array}{c} \dot{x}_1 \\ \dot{x}_2 \\ \dot{x}_3 \\ \dot{x}_4 \end{array} \right\} = \left[ \begin{array}{cccc} 0 & 0 & 1 & 0 \\ 0 & 0 & 0 & 1 \\ 0 & \dfrac{-a_7+a_1b_1}{1-a_1b_2} & \dfrac{-a_6}{1-a_1b_2} & \dfrac{\sigma^2\mu_1 a_1}{1-a_1b_2} \\ 0 & \dfrac{-b_1+a_7b_2}{1-a_1b_2} & \dfrac{a_6b_2}{1-a_1b_2} & \dfrac{-\sigma^2\mu_1}{1-a_1b_2} \end{array} \right] \left\{ \begin{array}{c} x_1 \\ x_2 \\ x_3 \\ x_4 \end{array} \right\} +
$$

$$+ \left\{ \left\{ \begin{array}{c} 0 \\ 0 \\ \left( \dfrac{a_8}{1 - a_1 b_2} \right) U \\ \left( \dfrac{-b_2 a_8}{1 - a_1 b_2} \right) U \end{array} \right\} + \right.$$

$$\left. + \sigma^2 \left\{ \begin{array}{c} 0 \\ 0 \\ a_{400}^N U x_2^2 + a_{500}^N x_2^3 + a_{600}^N x_2^2 x_3 + a_{700}^N x_2 x_3^2 + a_{800}^N x_2 x_3 x_4 + a_{900}^N x_2 x_4^2 \\ a_{450}^N U x_2^2 + a_{550}^N x_2^3 + a_{650}^N x_2^2 x_3 + a_{750}^N x_2 x_3^2 + a_{850}^N x_2 x_3 x_4 + a_{950}^N x_2 x_4^2 \end{array} \right\} \right\} ,$$

$$(6.8)$$

where

$$K_1 = -b_8^N + a_2^N - a_1 b_5^N + b_2 a_4^N , \quad a_{400}^N = \frac{-a_8 b_8^N (1 + a_1 b_2) - a_8 K_1}{2(1 - a_1 b_2)^2} ,$$

$$a_{500}^N = \frac{(a_7 b_8^N - a_1 b_{67}^N - b_1 a_4^N)(1 - a_1 b_2) - 3(-a_7 + a_1 b_1) K_1}{6(1 - a_1 b_2)^2} ,$$

$$a_{600}^N = \frac{a_6 b_8^N}{2(1 - a_1 b_2)} , \quad a_{700}^N = \frac{a_1 b_3^N}{2(1 - a_1 b_2)} , \quad a_{800}^N = \frac{-a_5^N - a_1 b_4^N}{2(1 - a_1 b_2)} ,$$

$$a_{900}^N = \frac{-a_3^N - a_1 b_8^N}{2(1 - a_1 b_2)} , \quad a_{450}^N = \frac{-a_8 b_5^N (1 + a_1 b_2) - a_8 b_2 K_1}{2(1 - a_1 b_2)^2} , \quad (6.9)$$

$$a_{550}^N = \frac{(a_7 b_5^N + b_{67}^N - b_1 a_2^N)(1 - a_1 b_2) - 3(a_7 b_2 - b_1) K_1}{6(1 - a_1 b_2)^2} ,$$

$$a_{650}^N = \frac{a_6 b_5^N}{2(1 - a_1 b_2)} , \quad a_{750}^N = \frac{-b_3^N}{2(1 - a_1 b_2)} ,$$

$$a_{850}^N = \frac{b_2 a_5^N + b_4^N}{2(1 - a_1 b_2)} , \quad a_{950}^N = \frac{b_2 a_3^N + b_8^N}{2(1 - a_1 b_2)} .$$

Excluding the first equation in (6.8), which is responsible for the rigid body motion of the structure and has no influence on the behavior of the structural motion of the system, the Jacobian matrix associated to this system can be given by:

$$J = \begin{bmatrix} 0 & 0 & 1 \\ \dfrac{-a_7 + a_1 b_1}{1 - a_1 b_2} & \dfrac{-a_6}{1 - a_1 b_2} & \dfrac{\sigma^2 \mu_1 a_1}{1 - a_1 b_2} \\ \dfrac{-b_1 + a_7 b_2}{1 - a_1 b_2} & \dfrac{a_6 b_2}{1 - a_1 b_2} & \dfrac{-\sigma^2 \mu_1}{1 - a_1 b_2} \end{bmatrix} . \quad (6.10)$$

The characteristic equation associated to (6.10) is:

$$\lambda^3 + \alpha_1\lambda^2 + \alpha_2\lambda + \alpha_3 = 0 \qquad (6.11)$$

where

$$\alpha_1 = \frac{\sigma^2\mu_1}{1 - a_1b_2} + \frac{a_6}{1 - a_1b_2},$$

$$\alpha_2 = \frac{\sigma^2\mu_1 a_6}{(1 - a_1b_2)^2} - \frac{\sigma^2\mu_1 a_6 a_1 b_2}{(1 - a_1b_2)^2} - \frac{a_7 - a_1b_1}{1 - a_1b_2}, \quad \alpha_3 = \frac{a_6b_1}{1 - a_1b_2}. \qquad (6.12)$$

The Hurwitz criterion for which the Jacobian matrix (6.10) will have a pair of purely imaginary eigenvalues, the condition for the existence of a Hopf bifurcation, is stated as:

$$\alpha_1 \geq 0, \quad \alpha_2 \geq 0, \quad \alpha_3 \geq 0, \quad \alpha_1\alpha_2 = \alpha_3, \quad \alpha_1 > \alpha_2. \qquad (6.13)$$

## 6.3   Stick-slip non-ideal vibrations

Paper [51] predicts stick-slip chaotic motions in a one-degree-of-freedom ideal kind of oscillator using the Melnikov technique and, in [52], chaotic behavior is observed due to the interaction between the system power supply and friction-driven variations.

Recently, papers [53] and [54] have shown that: (i) the occurrence of no sliding (stick) between the belt and the mass block is marked by horizontal lines in the phase portrait. When the belt velocity passes a certain value, the dominant mode is sliding (slip) and the system undergoes a limit cycle; (ii) the dynamical influence of the motor on the vibrating system is evident by the angular velocity time response and, due to the investigation carried out, it was possible to observe power supply influence on the vibrating system along with non-periodic motions with chaotic characteristics. Earlier, in paper [55], the authors considered a limited power supply in a similar problem using a perturbation approach. They found that the dry friction damper has a substantial influence on the nature of the resonance curve. Finally we remark that in [56] a non-ideal parametrically and self-excited vibrating system model was investigated by analytical and numerical methods. Some contributions of the non-ideal energy source are observed by internal stable loops located inside the main parametric resonance present in the amplitude versus angular velocity diagram. In the ideal model the authors also obtained an internal loop but it presents stable and unstable parts. In this particular case the loop appears on a distinct part of the resonance diagram.

## 6.4   A non-ideal gear rattling

Rattling in gear changes in automobiles is an unwanted comfort problem. It is excited by torsional vibrations of the drive train system at the entrance to the gear-box, where these torsional vibrations are generated by unbalances in the machines. All gear wheels not under load rattle due to backlashes in the meshes of the gears. In recent years, general models have been developed to analyze these rattling phenomena, mainly with the goal of finding some means to reduce them by parameter

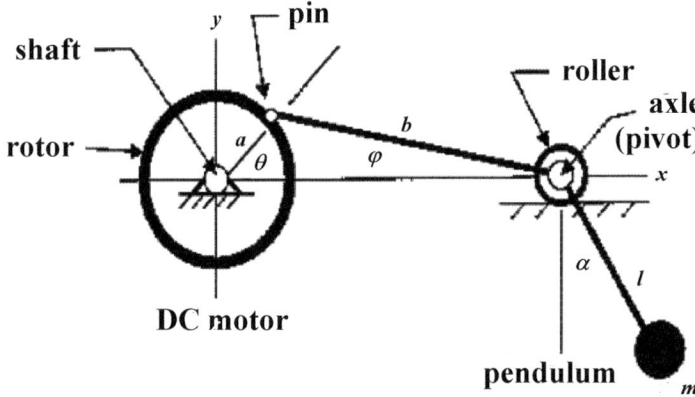

Figure 6  Electro-motor pendulum non-ideal system.

variation. Here, we present a procedure based on impact theory. Rattling vibrations possess typical non-linear behavior, leading to periodic and chaotic regimes. The subject has been analyzed by a number of authors, such as [57] and [58], to detect chaotic behavior. In [59] we confine our considerations to single stage rattling. The authors consider two spur gears with different diameters and gaps between the teeth and suppose that the motion of one gear is given while the motion of the other is governed by its dynamics. In the ideal case, the driving wheel is supposed to undergo a sinusoidal motion with given constant amplitudes and frequency. In the non-ideal approach, one considers this motion to be a function of the system response and a limited energy source is adopted. The numerical investigations confirm that a rich dynamics structure and chaotic behavior are present in the adopted model.

## 6.5  A non-ideal electro-motor pendulum vibrations

Consideration of the steady states in resonant motions of a pendulum when its support is driven by a mechanism with restricted power motivated the work of [60]. Fig. 6 shows the physical model. The shown crank-slider mechanism is coupled via the rod to the support of a pendulum. When the electro-motor shaft and crank $a$ rotates through an angle $\varphi$ the suspension shows a displacement $d(t) = a \cos \varphi$. An OXZ Cartesian coordinate system is used to describe the pendulum oscillations and the mass of the slider is taken to be negligible. Let $\alpha$ denote the angular deflection of the pendulum from the z-axis.

The authors, in [60], [61], [62] and [63], prove that in the resonance curve, for the fundamental resonance, when one increases the value of the frequency the unstable point on the resonance curve corresponds to a bifurcation point where the system loses stability. This point is associated with a jump phenomenon, and it is known as saddle–node bifurcation (discontinuous bifurcation). Next, an increase in the value of the parameter control (voltage) destroys the resonance periodic attractor, leading to the appearance of a chaotic attractor in the phase space. The attractor has the aspect of a Cantor set. In fact, the chaotic attractor is a single potential well, and it is located near the unstable point on the resonance curve. Mathematically, the explosion that generates the chaotic attractor is defined as an interior crisis, and it is a route to chaos. This phenomenon occurs due to the presence of an unstable orbit (saddle) in the potential well. With the variation of the control parameter the amplitude of the periodic attractor collides with the saddle, and the pendulum motion begins to vary between a periodic behavior (laminar phase), observed inside the attraction basin, and an aperiodic behavior (bursts), observed among the saddle and the hetereoclinic orbit (calculated with the unforced pendulum equation). This change in the pendulum behavior reveals the existence of the intermittence phenomenon.

# 7    Concluding Remarks

Recent and older studies of the main properties of non-ideal vibrating systems have been reviewed. In ideal systems, we assume that a motor operating on a structure requires a certain input (power) to produce a certain output (RPM), regardless of the motion of the structure. For non-ideal systems this may not be the case. Hence, we are interested in what happens to the motor (input-output) as the response of the system changes. The excitation of the vibrating system is always limited in two senses: by the characteristic curves of the particular energy source and by the dependence of the motion of the system on the motion of the energy source, that is, the coupling between the governing equations of motion of the system and of the energy source.

Several interesting phenomena were presented, and studies of the corresponding vibrating systems are leading to advances in their comprehension. Jump phenomena and the increase in power required by a source operating near resonance are manifestations of a non-ideal energy source, and they are often referred to as the Sommerfeld effect. These phenomena suggest that the vibrational response provides an energy sink, and thus we pay to vibrate the structure rather than to operate the machinery. One of the problems often faced by designers is how to drive a system through resonance and avoid this kind of energy sink. The appearance of chaos in various vibrating systems with a limited power supply is a new result that has to be taken into account in the design of machinery. Control based on the Tikhonov regularization algorithm in the passage through resonance seems to be adequate for this kind of problem. Two kinds of bifurcations were observed: Hopf bifurcation in the case of slewing structures and saddle node in the case of an electro-motor pendulum.

The results discussed here are the first stage in the study of realistic machinery. The research in progress will involve other engineering aspects of the problem, to be studied in the near future.

# Acknowledgments

The authors thank Fundação de Amparo à Pesquisa do Estado de São Paulo (**FAPESP**), Brazil, and Conselho Nacional de Pesquisas (**CNPq**), Brazil, for their financial support, in several grants, in order to develop scientific research into non-ideal problems and their applications to Engineering Sciences.

# Basic References

1. Sommerfeld, A. (1902) Beiträge zum Dynamischen Ausbau der Festigkeitslehe. *Physikal Zeitschr.*, **3**, pp. 266 and 286.

2. Kononenko, V.O. (1959, 1969) *Vibrating Systems With a Limited Power Supply* [in Russian (1959); English Translation: Illife Books (1969)].

3. Balthazar, J.M., Mook, D.T., Weber, H.I. and de Mattos, M.C. (1997) Some Remarks On the Behavior of Non-Ideal Dynamical Systems. In *Nonlinear Dynamics, Chaos, Control and Their Applications to Engineering Sciences*, J.M. Balthazar, D.T. Mook, J.M. Rosario (eds), Chapter 1, pp. 88–96.

4. Balthazar, J.M., Rente, M.L., Mook, D.T. and Weber H.I. (1997) Some Observations on Numerical Simulations of a Non-Ideal Dynamical Systems. In *Nonlinear Dynamics, Chaos, Control and Their Applications to Engineering Sciences*, J.M. Balthazar, D.T. Mook, J.M. Rosario (eds), Chapter 1, pp. 97–104.

5. de Mattos, M.C., Balthazar, J.M., Wieczork S. and Mook D.T. (1997) An Experimental Study of Vibrations of Non-Ideal Systems. In *Proceeding of DETC'97, ASME Design Engineering Technical Conference*, September 14–17, Sacramento, California, USA, CD-ROM.

6. de Mattos, M.C. and Balthazar, J.M. (1999) The Dynamics of An Armature Controlled DC Motor Mounted On An Elastically Supportable Table. In *15 Brazilian Congress of Mechanical Engineering*, November 22–26, CD ROM.

7. Blekman, I.I. (1953) *Self-Synchronization of Certain Vibratory Devices*, Eng. Trans., vol. 16, ASME PRESS Translations, Wayne State University, USA.

8. Evan-Iwanowski, R.M. (1976) *Resonance Oscillators in Mechanical Systems*. Elsevier, New York.

9. Dimentberg, M.F. (1988) *Statistical Dynamics of Nonlinear and Time Varying Systems*. John Wiley & Sons, New York.

10. Dimentberg, M.F., Chapdelaine, J., Norton, R.L., Harrison, R. and Norton, R.L. (1994) Passage through Critical Speed With Limited Power by Switching System Stiffness. In *Nonlinear and Stochastic Dynamics*, A.K. Bajai, N.S. Namachchivaya, R.I. Ibrahim (eds), AMD, vol. 192, DE, vol. 78, 57–67.

11. Dimentberg, M.F., McGovern, L., Norton, R.L., Chapdelaine, J. and Harrison, R. (1997) Dynamic of An Unbalanced Shaft Interacting with a Limited Power Supply. *Nonlinear Dynamics*, **13**, 171–187.

12. Nayfeh, A.H. and Mook, D.T. (1979) *Nonlinear Oscillations*. John Wiley & Sons, New York.

13. Balthazar, J.M., Mook, D.T., Weber, H.I., Fenili, A., Belato, D., de Mattos, M.C. and Wieczorek, S. (1999) On Vibrating Systems with a Limited Power Supply and Their Applications to Engineering Sciences. In *Brazilian Seminar of Mathematical Analysis*, Campinas, SP, Brazil, pp. 137–277.

14. Balthazar, J.M., Mook, D.T., Weber, H.I., de Mattos, M.C., Brasil, R.M.L.R.F., Fenili, A., Belato, D. and Palacios, J.L. (2000) Review of Unresolved Machine Torsion Vibrations Issues. In *11*$^{th}$ *Workshop on Dynamics and Control*, Rio de Janeiro, Brazil, October 9–11.

15. Meirovitch, L. (1967) *Analytical Methods in Vibrations*. Macmillan Publishing C$^{o}$, Inc., New York.

16. Nayfeh, A.H. (1981) *Introduction to Perturbation Techniques*. John Wiley & Sons, New York.

17. Nayfeh, A.H. (1993) *Normal Forms*. John Wiley & Sons, New York.

18. Yamanaka and Murakami, H.S. (1989) Optimum Designs of Operating Curves for Rotating Shaft Systems with Limited Power Supplier. In *Current Topics in Structural Mechanics*, H. Chung (ed.), PVP, V. 179, ASME, NY, pp. 181–185.

19. Balthazar, J.M., Cheshankov, B.I., Rushev, D.T., Barbanti, L. and Weber, H.I. (2001) Remarks on the Passage Through Resonance of a Vibrating System with Two Degree of Freedom. *Journal of Sound and Vibration*, **239** (5) 1075–1085.

20. Balthazar, J.M., Tonon, O., Weber, H.I. and Mook, D.T. (1999) Some Remarks on Kononenko's Topology of Non Ideal Dynamical Systems. In *Applied Mechanics in the Americas*, Vol. 8, Chapter: Dynamics, pp. 1231–1234.

21. Balthazar, J.M., Campanha, J.R., Mook, D.T. and Weber, H.I. (2001) A Comment on the Geometry Behaviour of a Parametric Vibrating System with a Limited Power Supply Defined on Kononenko's Book (Kononenko 1979), *International Journal: Problems of Nonlinear Analysis in Engineering Systems*, **7**, 1–17.

22. Hindmarsh, J. (1981) *Electrical Machines and their Applications*. Pergamon Press, Oxford.

23. Christ, H. (1966) Stationärer und Instatioärer Betrieb Eines Federnd Gelagerten, Unwuchtigen Motors. Diss., Universität Karlsruhe, Karlsruhe, Germany.

24. Wauer, J. and Bürle, P. (1997) Dynamics of a Flexible Slider-Crank Mechanism Driven by a Non-Ideal Source of Energy. *Nonlinear Dynamics*, **13**, 221–242.

25. Wauer, J. and Suherman, S. (1997) Vibration Suppression of Rotating Shafts Passing Through Resonance's by Switching Shaft Stiffness. *Journal of Vibration and Acoustics*, **120**, 170–180.

26. Suherman, S. and Plaut, R. (1996) Use of Flexible Internal Support to Suppress Vibrations of a Rotating Shaft Passing Through a Critical Speed, In *Sixth Conference on Nonlinear Stability and Dynamics*, Blacksburg, Virginia, USA.

27. Suherman, S. (1996) Transient Analysis and Vibrating Suppression of a Cracked Rotating Shaft with an Ideal and a Non-ideal Motor Passing Through a Critical Speed. Ph.D. Thesis, Virginia Polytechnic Institute and State University, Blacksburg, Virginia, USA.

28. Iwatsubo, T., Kanki, H. and Kawai, R. (1972) Vibration of Asymmetric Rotor Through Critical Speed with Limited Power Supply. *Journal Mechanical Engineering Science*, **14** (3), 184–194.

29. Suzuki, S.H. (1978) Dynamic Behavior of a Beam Subject to a Force of Time-Dependent Frequency (Continued). *Journal Sound and Vibration*, **60** (3), 417–422.

30. Suzuki, S.H. (1978) Dynamic Behavior of a Beam Subject to a Force of Time-Dependent Frequency. *Journal Sound and Vibration*, **57**, 59–64.

31. Mazzili, C.E.N. and Brasil, R.M.L.R.F. (1995) Effect of Static Loading on the Nonlinear Vibrations of a Three–Time Redundant Portal Frame: Analytical and Numerical Results. *Nonlinear Dynamics*, **8**, 347–366.

32. Brasil, R.M.F.L. (1999) Multiple Scales Analysis of Nonlinear Oscillations of a Portal Frame Foundations for Several Machines. RBCM. *Journ. of the Braz. Soc. Mechanical Sciences*, **XXI** (4), 641–654.

33. Brasil, R.M.F.L. and Balthazar, J.M. (2000) On Chaotic Oscillations of a Machine Foundation: Perturbation and Numerical Studies. In *Nonlinear Dynamics, Chaos, Control and Their Applications to Engineering Sciences*, Vol. 3: New Trends in Dynamics and Control, J.M. Balthazar, P.B. Gonçalves, R.M.F.L.R.F. Brasil (eds), pp. 36–52.

34. Brasil, R.M.F.L. and Mook, D.T. (1994) Vibrations of a Portal Frame Excited by a Non-Ideal Motor. In *Fifth Conference on Nonlinear Vibrations, Stability, and Dynamics of Structures*, Blacksburg, Virginia.

35. Brasil, R.M.F.L, Palacios, J.L and Balthazar, J.M. (2000) On the Nonlinear Dynamic Behavior of a Non-ideal Machine Foundation: Numerical Simulations. In *Nonlinear Dynamics, Chaos and Their Applications to Engineering Sciences*, Vol. 4: Recent Developments in Nonlinear Phenomena, J.M. Balthazar, P.B. Gonçalves, R.M.F.L.R.F. Brasil, I.L. Caldas, F.B. Rizatto (eds), Chapter 10, pp. 326–354.

36. Brasil, R.M.F.L. and Balthazar, J. M. (2000) Nonlinear Oscillations of a Portal Frame Structure Excited by a Non-ideal Motor. In *Control of Oscillations and Chaos*, F.L. Chernousko, A.L. Fradkov (eds), Vol. 2, pp. 275–278.

37. Bogolyubov, N.N. and Mitropol'skii, Y.A. (1962) *Asymptotic Methods in the Theory of Non-Linear Oscillations*. Gordon and Breach, Longhorn.

38. Brasil, R.M.F.L, Palacios, J.L and Balthazar, J.M. (2000) Some Comments on Numerical Analysis of Nonlinear Vibrations of a Civil Structure Induced by a Non-ideal Energy Source. In *Computational Methods in Engineering 2000*, L.E. Vaz (ed.) CD ROM.

39. Book, W.J., Maizza-Neto, O. and Whitney, D.E. (1975) Feedback Control of Two Beam, Two Joint Systems with Distributed Flexibility. *Journ. of Dynamic Systems, Measurement and Control*, 429–431.

40. Juang, J.N., Horta, L.G. and Robertshaw, H.H. (1985) A Slewing Control Experiments for Flexible Structures, *Journal of Guidance, Control and Dynamics*, **9**, 599–607.

41. Garcia, E. (1989) On the Modeling and Control of Slewing Flexible Structures. Ph.D. Thesis, State University of New York at Buffalo.

42. Sah, J.J.F. (1990) On the Interaction Between Actuator and Slewing Structure. Ph.D. Thesis, State University of New York at Buffalo.

43. Sah, J.J.F., Lin, J.S., Crassidis, A.L. and Mayne, R.W. (1993) Dynamic Interaction in a Slewing Motor-Beam System: Closed Loop Control. In *ASME Design Technical Conferences – 14$^{th}$ Biennial Conference on Mechanical Vibration and Noise*, Albuquerque, New Mexico, September, DE, Vol. 61, pp. 257–263.

44. Fenili, A. (2000) On Slewing Structure: Modeling and Dynamical Analysis, Ph.D. Thesis presented at State University of Campinas, SP, Brazil (in Portuguese), 2000.

45. Junkins, J.L. and Kim, Y. (1993) *Introduction to Dynamics and Control of Flexible Structures*. AIAA Educational Series, J.S. Przemieniecki, Series Editor-in-Chief.

46. Fenili, A., Balthazar, J.M., Weber, H.I. and Mook, D.T. (submitted) Nonlinear Analysis of the Motion of a Flexible, Rotating, Cantilever Beam. *Vibration and Control*.

47. Fenili, A., Balthazar, J.M. and Mook, D.T. (2000) On the Comparison Between Two Mathematical Models for Flexible Slewing Structures-Linear and Nonlinear Curvature, In *Nonlinear Dynamics, Chaos and Their Applications to Engineering Sciences*, Vol. 4: Recent Developments in Nonlinear Phenomena, J.M. Balthazar, P.B. Gonçalves, R.M.F.L.R.F. Brasil, I.L. Caldas, F.B. Rizatto (eds), Chapter 10, pp. 372–382.

48. Fenili, A., Balthazar, J.M. and Mook, D.T. (2000) On the Comparison Between Ideal and Non-ideal Systems Modeling for Flexible Slewing Structures (Nonlinear Curvatures). In *Computational Methods in Enginering 2000*, CD-ROM.

49. Carr, J. (1981) *Application of Center Manifold Theory*. Springer Verlag, Berlin.

50. Fenili, A., Balthazar, J.M., Mook, D.T. and Weber, H.I. (1999) Application of the Center Manifold Reduction to the Slewing Flexible Non-Ideal Model, In *15$^{th}$ Brazilian Congress of Mechanical Engineering*, Monte Sião, SP, Brazil, December 4–11, CD ROM.

51. Awrejcewicz, J. and Holicke, M.M. (1999) Melnikov Method and Stick-Slip Chaotic Oscillations in Very Weakly Forced Mechanical Systems. *Int. Journ. Bifurcation and Chaos*, **9** (3), 505–518.

52. Balthazar, J.M., Campanha, J.R, Weber, H.I. and Mook, D.T. (1999) Some Remarks on the Numerical Simulations of Ideal and Non-Ideal Self-Excited Vibrations. In *Applications of Mathematics to Engineering' 24*, B.I. Cheshankov and M.D. Todorov (eds), Heron Press, Sofia, pp. 9–15.

53. Pontes, B.R., de Oliveira, V.A. and Balthazar, J.M. (2000) On Friction-Driven Vibrations in a Mass Block-Belt-Motor with Limited Power Supply. *Journal of Sound and Vibration*, **4** (234), 713–723.

54. Pontes, B.R., de Oliveira, V.A., Balthazar, J.M. (2001) On the Dynamics Response of a Mechanical System with Dry Friction and Limited Power Supply. In *Nonlinear Dynamics, Chaos and Their Applications to Engineering Sciences*, Vol. 4: Recent Developments in Nonlinear Phenomena, J.M. Balthazar, P.B. Gonçalves, R.M.F.L.R.F. Brasil, I.L. Caldas, F.B. Rizatto (eds), Chapter 10, pp. 355–371.

55. Alifov, A. and Frolov, K.V. (1977) Investigation of Self-Excited Oscillations with Friction, Under Conditions of Parametric Excitation and Limited Power of Energy Source. *Mekhanika Tverdogo Tela*, **15** (4), 25–33.

56. Warminski, J., Balthazar, J.M. and Brasil, R.M.L.R.F. (2001) Vibrations of Non-Ideal Parametrically and Self-Excited Model. *Journal of Sound and Vibration*, **245** (2), 363–374.

57. Karagiannis, K. and Pfeifer, F. (1991) Theoretical and Experimental Investigations of Gear-Rattling. *Nonlinear Dynamics*, **2**, 367–387.

58. Moon, F.C. (1992) *Chaos and Fractal Dynamics*. John Wiley & Sons, Chichester.

59. De Souza, S.L.T., Caldas, I.L., Balthazar, J.M., Brasil, R.M.L.R.F. (1999) Analysis of Regular and Irregular Dynamics of a Non-Ideal Gear Rattling Problem. In *15$^{th}$ Brazilian Congress of Mechanical Engineering*, Monte Sião, SP, Brazil, December 4-11, CD ROM.

60. Krasnopol'skaya, T.S. and Shevts, A.Y. (1990) Chaotic Interactions in a Pendulum Energy Source System. *Prikladnaya Mekhanika*, **26** (5), 90–96.

61. Belato, D., Balthazar, J.M., Mook, D.T. and Weber, H.I. (in press) On the Appearance of Chaos in a Non Ideal System. In *Sixth Conference On Nonlinear Stability and Dynamics*, A.H. Nayfeh, D.T. Mook (eds).

62. Belato, D., Weber, H.I., Balthazar, J.M. and Mook, D.T. (1999) Chaotic Vibrations of a Non Ideal Electro Mechanical System. In *Applied Mechanics in the Americas*, Vol. 7, Chapter: Stability, Bifurcation's and Chaos, Mechanics of New Materials, Structural Mechanics and Solid Mechanics, pp. 539–542.

63. Belato, D., Weber, H.I., Balthazar, J.M. and Mook, D.T. (2001) Chaotic Vibrations of a Non-Ideal Electro-Mechanical System. *International Journal of Solids and Structures*, **38**, 1699–1706.

# Identification of Flexural Stiffness Parameters of Beams

José João de Espíndola and João Morais da Silva Neto

*Department of Mechanical Engineering*
*Federal University of Santa Catarina, Brazil*
*E-mails: espindol@mbox1.ufsc.br, joaoneto@emc.ufsc.br*

A three degree of freedom model of the dynamic mass at the middle of a test sample, resembling a Stockbridge neutralizer, is introduced. This model is used to identify the hereby called equivalent complex cross-section flexural stiffness (ECFS) of the beam element which is part of the whole test sample. This ECFS, once identified, gives the effective cross-section flexural stiffness of the beam as well as its effective damping, measured as the loss factor of an equivalent viscoelastic beam. The beam element of the test sample may be of any complexity, such as a segment of stranded cable of the ACSR type. These data are important parameters for the design of overhead power transmission lines and other cable structures. A cost function is defined and used in the identification of the ECFS. An experiment, designed to measure the dynamic masses of two test samples, is described. Experimental and identified results are presented and discussed.

## 1 Introduction

The cross-section stiffness of beams of complex constructions, such as sandwich beams and stranded cables of the ACSR type, are difficult to evaluate by analytical and/or numerical means.

Also difficult to assess is the inherent damping of such constructions. In many instances the damping of ACSR cables, for example, is measured by the logarithmic decrement of a segment of the cable, fixed at both ends, under a certain mechanical tension. This practice is conceptually wrong, for the logarithmic decrement is not a measure of the inherent damping of the "material," that is, of the inherent damping of the cable (Lazan, 1968). Rather, it is a measure of the "system" formed by the piece of cable fixed at both ends. For a different span, a different logarithmic decrement is bound to be measured. Logarithmic decrement is a structural parameter, not a material one.

In this paper the beam, no matter if a piece of ACSR cable or a multi-layered sandwich beam, is modelled as an equivalent continuous viscoelastic beam of complex cross-section flexural stiffness (ECFS) equal to $EI = (EI)_r(1 + i\eta) = (EI)_r + i(EI)_i$, where $(EI)_r$ is the effective cross-section flexural stiffness and $\eta$ is the equivalent loss factor, i.e., the loss factor of the viscoelastic material from which the beam model is made. The purpose of this paper is to introduce the concept of equivalent complex cross-section flexural stiffness (ECFS) for beams and to produce an approach to measure, or identify it. To achieve this objective a builtup structure, or test sample, is constructed and tested experimentally. The beam from which the ECFS is desired is part of this builtup structure, or test sample (see Fig. 1). This builtup structure closely resembles a symmetric Stockbridge neutralizer (Teixeira, 1997). It is stressed here, though, that this paper is not at all concerned with, nor about, Stockbridge neutralizers. As a matter of fact, it is not about neutralizers of any sort. The above resemblance is noted "en passant," and is supposed to clarify, not to confound. Note: In this, as well as in all the previous papers of the first author, the name "dynamic vibration neutralizer," or DVN, is used instead of "dynamic vibration absorber," or DVA, simply because this device is not an absorber, or damper at all. The words "absorber" and "damper" are inaccurate (Crede, 1965).

Once the ECFS is known, both the real flexural stiffness $(EI)_r$ and the effective loss factor $\eta$ of the "material" (for instance, a stranded cable) are also known. These data may be precious for many practical applications, such as the design of overhead power transmission lines and other cable structures.

In beams made up of alternate layers of metal and viscoelastic material, both $(EI)_r$ and $\eta$ may be heavily dependent on frequency (Snowdon, 1968). Although this dependence is currently being investigated by the authors, it is not taken into account in this paper. Hereby, both $(EI)_r$ and $\eta$ are not supposed to vary with frequency.

The parameters $(EI)_r$ and $\eta$ (or $(EI)_r$ and $(EI)_i$) are measured, or identified, by fitting a theoretical model of the built-up beam structure (test sample) to a frequency response function (FRF) measured at the root of it (Silva Neto, 1999).

## 2   Three Degrees of Freedom Model of the Test Sample

Fig. 1 shows half the symmetric beam structure, or test sample. The test sample is build up with a rigid body at one end of the beam and is excited at a central mass by a force $f(t)$. It is assumed that there is no rotation at the fixed section at the central mass, thanks to symmetry. In actual practice, some lack of symmetry is unavoidable. The experiment must then allow for some means of preventing such rotations.

The beam itself is assumed to have no mass.

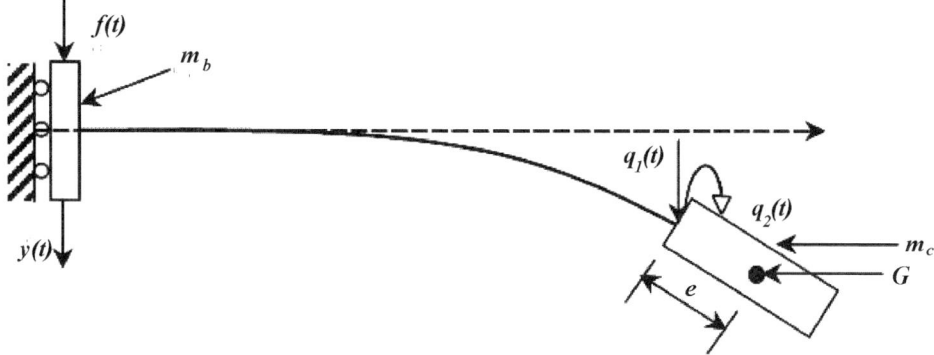

Figure 1  Half the symmetric beam structure.

The stiffness matrix in the co-ordinates $q_1(t)$ and $q_2(t)$ can easily be shown to be (Espíndola, 1987):

$$K = EI \begin{bmatrix} \dfrac{12}{L^3} & -\dfrac{6}{L^2} \\[2mm] -\dfrac{6}{L^2} & \dfrac{4}{L} \end{bmatrix}. \tag{1}$$

This is a complex stiffness matrix, since $EI = (EI)_r(1 + i\eta) = (EI)_r + i(EI)_i$ (the ECFS) is a complex number. $\eta$ is the loss factor of the viscoelastic material of the beam model.

The quantities shown in Fig. 1 are:

$m_b$ — half the mass of basis of the test sample;

$m_c$ — mass of the rigid body of the test sample;

$J_c$ — moment of inertia of the above rigid body in relation to a baricentric axis, normal to the plane of the paper;

$e$ — distance from the free end of the beam element to the mass centre of the rigid body;

$q_1(t)$ — displacement co-ordinate, relative to the root of the test sample, at the free end of the beam element;

$q_2(t)$ — rotational co-ordinate, measuring the small rotation of the rigid body;

$q(t) = [q_1(t) \quad q_2(t)]^T$;

$y(t)$ — displacement of the root of the beam structure;

$f(t)$ — exciting force.

The kinetic and potential energies for half the test sample model are:

$$T = \frac{1}{2}m_c(\dot{y}(t) + \dot{q}_1(t) + e\dot{q}_2(t))^2 + \frac{1}{2}J_c\dot{q}_2^2(t) + \frac{1}{2}m_b\dot{y}^2(t) \tag{2}$$

and

$$V = \frac{1}{2}q^T K q - f(t)y(t),\qquad(3)$$

where the dot over quantities means derivative in relation to time.

Substitution of (2) and (3) into the Lagrange equations (Meirovitch, 1970, 1990) results in:

$$
\begin{bmatrix}
m_c & e\,m_c & m_c \\
e\,m_c & e^2 m_c + J_c & e\,m_c \\
m_c & e\,m_c & (m_b + m_c)
\end{bmatrix}
\begin{Bmatrix}
\ddot{q}_1 \\
\ddot{q}_2 \\
\ddot{y}
\end{Bmatrix}
+
\begin{bmatrix}
K_{11} & K_{12} & 0 \\
K_{21} & K_{22} & 0 \\
0 & 0 & 0
\end{bmatrix}
\begin{Bmatrix}
q_1 \\
q_2 \\
y
\end{Bmatrix}
=
\begin{Bmatrix}
0 \\
0 \\
f(t)
\end{Bmatrix}.
$$

$$(4)$$

This result may be written in the following form:

$$
\begin{cases}
\begin{bmatrix}
m_c & e\,m_c \\
e\,m_c & e^2 m_c + J_c
\end{bmatrix}
\begin{Bmatrix}
\ddot{q}_1 \\
\ddot{q}_2
\end{Bmatrix}
+
\begin{bmatrix}
m_c \\
e\,m_c
\end{bmatrix}
\ddot{y}
+
\begin{bmatrix}
K_{11} & K_{12} \\
K_{21} & K_{22}
\end{bmatrix}
\begin{Bmatrix}
q_1 \\
q_2
\end{Bmatrix}
=
\begin{Bmatrix}
0 \\
0
\end{Bmatrix} \\[4mm]
[m_c \ \ e\,m_c]^T
\begin{Bmatrix}
\ddot{q}_1 \\
\ddot{q}_2
\end{Bmatrix}
+ (m_b + m_c)\,\ddot{y} = f(t).
\end{cases}
$$

$$(5)$$

Fourier transforming both sides of (5) and after a few algebraic manipulations, one gets:

$$M(\Omega) = \frac{F(\Omega)}{-\Omega^2 Y(\Omega)} = (m_b + m_c) + \Omega^2 m_L^T \left[ -\Omega^2 M + K \right]^{-1} m_L.\qquad(6)$$

In the above expression, one has $M = \begin{bmatrix} m_c & e\,m_c \\ e\,m_c & e^2 m_c + J_c \end{bmatrix}$ as the mass matrix in the co-ordinates $q_1(t)$ and $q_2(t)$ and $m_L = [m_c \ \ e\,m_c]^T$. $M(\Omega)$ is the dynamic mass at the root of the test sample.

An alternative expression that has proved to be more convenient to work with is given below, easily derived from (6):

$$M(\Omega) = (m_b + m_c) + \Omega^2 m_L^T \Phi \left[ -\Omega^2 I + \Lambda \right]^{-1} \Phi^T m_L,\qquad(7)$$

where $\Phi$ is modal matrix (real) and $\Lambda$, the spectral matrix (complex) of the eigenvalue problem $K\phi = \lambda M\phi$. The modal matrix is constructed in such a way that its columns are the eigenvectors of $K\phi = \lambda M\phi$. Note that $K$ is also complex.

The spectral matrix is a diagonal one containing the eigenvalues of the above eigenvalue problem. Expression (7) assumes that the eigenvectors are orthonormal in relation to the mass matrix, that is, that they obey the relation

$$\Phi^T M \Phi = I.\qquad(8)$$

In such a condition one has:

$$\Phi^T K \Phi = \Lambda.\qquad(9)$$

The eigenvalues in the diagonal of $\Lambda$ may be written in the form:

$$\lambda_j = \Omega_j^2 (1 + i\eta_j).\qquad(10)$$

In this expression $\Omega_j$, $j = 1, 2$, is the $j$-th undamped natural frequency, and $\eta_j$ is the corresponding modal loss factor of the test sample. The modal loss factors are structural parameters of the whole test sample, not material parameters of the beam segment. So they must not be confused with $\eta$ in the expression $EI = (EI)_r(1+i\eta)$.

# 3   The Cost Function

At this stage a cost function must be defined so that the FRF of expression (7) is made as close as possible to its measured counterpart. This is accomplished by varying the parameters $(EI)_r$ and $(EI)_i$ contained in (7). This approximation of mathematical model (7) to its measured counterpart may be made through an appropriate nonlinear optimization algorithm or a genetic or even a hybrid algorithm (Espíndola and Bavastri, 1997, 1998). The error function is defined by taking the difference, for each value of $\Omega$, between mathematical model (7) and its measured counterpart:

$$E(\Omega) = 2M(\Omega) - \overline{M}(\Omega)\,, \tag{11}$$

where $M(\Omega)$ is given by (7), and $\overline{M}(\Omega)$ is the measured FRF.

Note that $E(\Omega)$ is a complex-valued function defined on the real number set.

The cost function may be defined as follows:

$$f(x, \Omega) = E(\Omega)E^*(\Omega)\,, \tag{12}$$

where the asterisk * stands for complex conjugate, and $x$ is a design vector given by:

$$x = [(EI)_r \quad (EI)_i \quad e \quad J_c \quad R_r \quad R_i]^T\,. \tag{13}$$

In the above expression, $(EI)_r$ is the real part and $(EI)_i$ is the imaginary part of $EI$, so that $\eta = (EI)_r/(EI)_i$. $R_r$ and $R_i$ are the real and imaginary parts of a residual term to be added to the mathematical model (7) to take into account effectively existing modes above the measurement base-band (Ewins, 1984; Natke and Cottin, 1988; Maia and Silva, 1997).

In vector (13), $e$ and $J_c$ are taken as design variables to allow them to account for some inertial effects due to actual mass of the beam. Also, in the case of a Stockbridge neutralizer, $e$ and $J_c$ are difficult to measure accurately due to the shape of the end masses.

Finally, a complementary equation is necessary so that $M(\Omega)$, in equation (7), can be effectively computed. This equation is precisely the eigenvalue problem $K\phi = \lambda M\phi$.

# 4   Measurement Setup

Figure 2 shows a Stockbridge neutralizer (made by Wetzel, SC, Brazil, model AS-2008) which was effectively taken as one particular test sample. This neutralizer was taken as a test sample only because it was readily available. The purpose of this particular experiment was then to find the ECFS of the stranded cable, part of this Stockbridge neutralizer. In the end both $(EI)_r$ and $\eta$ for this cable would be available. The value of this $\eta$ is an important parameter to assess the efficiency

Figure 2   A Stockbridge neutralizer used as a test sample.

(or lack of) of this sort of neutralizer in reducing overhead transmission line vibrations over a wide frequency band (Espíndola and Bavastri, 1997 and Espíndola
*et al.*, 1998).

The Stockbridge test sample was suspended by thin flexible cables at the central
mass and excited horizontally with an electrical shaker. The acceleration response
was measured at the same mass. Both signals were fed into a Dynamic Signal Anal-
yser (DSA) and the FRF computed (see Figure 3). The force signal was a swept-sine
function. Frequency resolution was 0.3Hz. The measurement band ranged from zero
up to three hundred Hertz (base-band). Mathematical model (7) was then fitted
to the measured FRF and the cost function minimized by the DFP (Davidon–
Fletcher–Powell) algorithm (Rao, 1996). After the minimization is accomplished,
the parameters in (13) are available.

## 5   Experimental Results

Table 1 gives the results of such an identification. In Table 1, $f_1 = \Omega_1/2\pi$ and
$f_2 = \Omega_2/2\pi$.

Figure 4 shows a comparison between the measured FRF and the regenerated
one. The agreement seems to be excellent. This is best shown by zooming both
curves around the resonant picks as shown in Figure 5a,b.

To further boost the confidence in the above approach, another test sample was
designed and constructed, with a cylindrical beam element of diameter $d$ made out
of hot rolled carbon steel, with $(EI)_r$ known in advance. Figure 6 is a photo of that
test sample. Figure 7 shows details of the end masses.

The data for this particular test sample are shown in Table 2. The beam was
fixed at the end masses by shrink fit.

Table 3 shows the numerical results after identification.

If a comparison is made of the value of $(EI)_r$ in Table 3 with EI in Table 2 a
conclusion is drawn that they differ by only 1.7 per cent. The loss factor of the

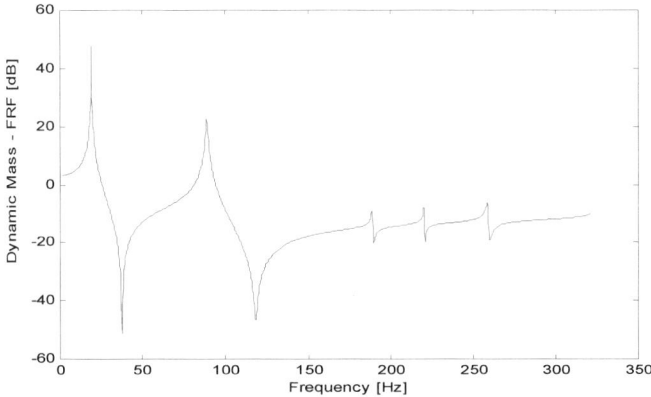

Figure 3   Measured frequency response function for the Stockbridge test sample.

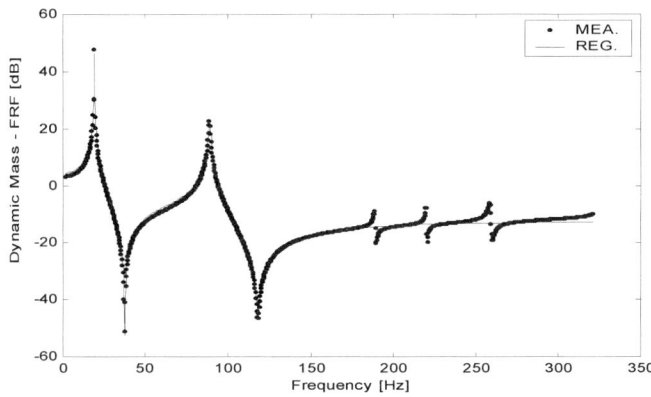

Figure 4   Comparison of the measured FRF and the regenerated one.

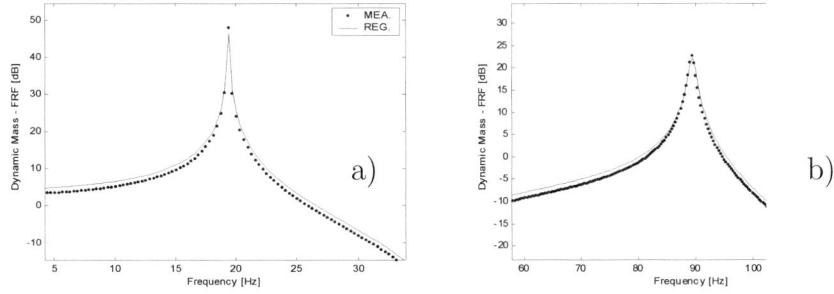

Figure 5   Zoomed FRF around the picks. Dimensions in millimetres. Material: Carbon steel.

Table 1  Parameters for Stockbridge test sample.

| Parameter | Value |
|---|---|
| $(EI)_r$ (N m$^2$) | 3.75 |
| $(EI)_i$ ($10^{-2}$ N m$^2$) | 4.15 |
| $e$ ($10^{-3}$ m) | $-11.49$ |
| $J_c$ ($10^{-3}$ kg m$^2$) | 0.29 |
| $\eta$ ($10^{-3}$) | 11.07 |
| $f_1$ (Hz) | 19.41 |
| $f_2$ (Hz) | 89.40 |
| $\eta_1$ ($10^{-3}$) | 6.16 |
| $\eta_2$ ($10^{-3}$) | 12.66 |
| $R_r$ ($10^{-3}$ N s$^2$/m) | 1.38 |
| $R_i$ ($10^{-3}$ N s$^2$/m) | 1.97 |

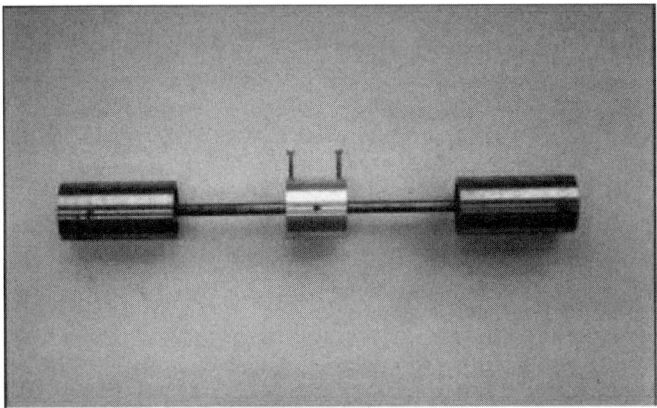

Figure 6  A test sample with a steel beam element

material (low carbon steel) is computed from $(EI)_r$ and $(EI)_i$, giving $2.97 \times 10^{-4}$, which is within the expected range of $2 \times 10^{-4}$ to $6 \times 10^{-4}$ for steels (Rao, 1995).

One can see that the three degree of freedom model of the test sample presented here is well adequate for the purpose of identifying the ECFS of beams and, as a consequence, its effective flexural stiffness and inherent damping.

A comment, "en passant," on the inherent damping of Stockbridge neutralizers is in order, in spite of the fact that this paper is not at all about those devices. The loss factor of 0.011, Table 1, is very low indeed. This explains why such a neutralizer is so efficient in reducing vibrations in a very narrow band of frequencies around a natural frequency, but so poor in mitigating vibrations outside that band. Since the band of frequencies exciting overhead lines is actually much wider, far more damping is needed in a neutralizer than that provided by the Stockbridge one. The first author has designed a neutralizer with a great amount of viscoelastic damping and excellent performance over a wide band of frequencies and patented it (Espíndola *et al.*, 1998).

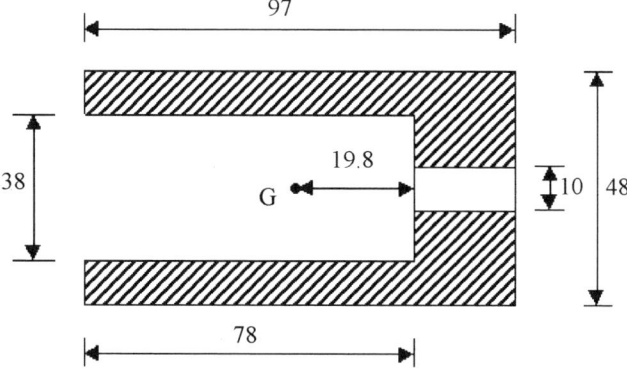

Figure 7  A diametrical cut of one of the end masses

Table 2  Data for the constructed test sample.

| Test Sample Detail | Value |
|---|---|
| $m_b$ (Kg) | 0.142 |
| $m_c$ (Kg) | 0.686 |
| beam element diameter (mm) | 10 |
| beam element length (mm) | 170 |
| $I = \pi d^4/64$ (m$^4$) | $4.91 \times 10^{-10}$ |
| modulus of Young (GPa) | 203 |
| $EI$ (N m$^2$) | 99.65 |
| $e$ (m) | $19.8 \times 10^{-3}$ |
| $J_c$ (Kg m$^2$) | $0.77 \times 10^{-3}$ |

Table 3  Parameters for the constructed test sample.

| Parameter | Value |
|---|---|
| $(EI)_r$ (N m$^2$) | 97.94 |
| $(EI)_i$ ($10^{-2}$ N m$^2$) | $29.16 \times 10^{-3}$ |
| $e$ ($10^{-3}$ m) | $-22.08$ |
| $J_c$ ($10^{-3}$ kg m$^2$) | 0.61 |
| $\eta$ ($10^{-4}$) | 2.97 |
| $f_1$ (Hz) | 44.31 |
| $f_2$ (Hz) | 202.60 |
| $\eta_1$ ($10^{-3}$) | 7.96 |
| $\eta_2$ ($10^{-3}$) | 9.35 |
| $R_r$ ($10^{-3}$ N s$^2$/m) | 4.64 |
| $R_i$ ($10^{-3}$ N s$^2$/m) | 19.30 |

# 6    Final Remarks and Conclusions

A three degree of freedom model of a test sample resembling (but not intended to be) a Stockbridge neutralizer was presented. The primary objective of it is the identification of the ECFS of the segment of beam, part of it, thus permitting the computation of its actual bending stiffness $(EI)_r$ and equivalent loss factor.

One immediate application is the identification of these two parameters for stranded cables, such as those used in overhead transmission lines and other cable structures.

The results from two test samples, one Stockbridge neutralizer and one specially designed and built to check the approach, have given very encouraging results.

This technique is now being extended to sandwich beams, where the core is a very damped viscoelastic material. This poses an additional challenge, for very damped viscoelastic materials have their dynamic properties largely dependent on frequencies. This means that the ECFS for such beams must be identified in a band of frequencies. This technique, once developed, can be used to identify the dynamic properties of the viscoelastic core material over a band of frequencies.

# References

1. Crede, C.E. (1965) *Shock and Vibration Concepts in Engineering Design.* Prentice-Hall, Inc., Englewood Cliffs, N.J.

2. Espíndola, J.J. (1987) *Notes on Vibration Control.* Graduate Program in Vibration and Acoustics, UFSC, Federal University of Santa Catarina (in Portuguese).

3. Espíndola, J.J. and Bavastri, C.A. (1997) Reduction of Vibration in Complex Structures with Viscoelastic Neutralizers: A Generalised Approach and a Physical Realization. In *Proceedings ASME Design Engineering Technical Conferences*, Sacramento, Paper DETC97/VIB-4187.

4. Espíndola, J.J., *et al.* (1998) A Hybrid Algorithm to Compute the Optimal Parameters of a System of Viscoelastic Vibration Neutralizers in a Frequency Band. In *Proceedings MOVIC'98*, Vol. 2, Zurich, Switzerland, pp. 577–582.

5. Ewins, D.J. (1984) *Modal Testing: Theory and Practice.* Research Studies Press LTD.

6. Lazan, B.J. (1968) *Damping of Materials and Members in Structural Mechanics.* Pergamon Press, Oxford.

7. Natke, H.G. and Cottin, N. (1988) Introduction to System Identification: Fundamentals and Survey. In *CISM Courses and Lectures*, No. 296: Application of System Identification in Engineering, edited by H.G.Natke, Springer-Verlag, Udine, Italy.

8. Maia, N.M.M. and Silva, J.M.M. (eds) (1997) *Theoretical and Experimental Modal Analysis.* John Wiley & Sons, Inc., New York, USA.

9. Meirovitch, L. (1990) *Dynamics and Control of Structures.* John Wiley & Sons, Inc., New York, USA.

10. Meirovitch, L. (1970), *Methods of Analytical Dynamics.* John Wiley & Sons, Inc., New York, USA.

11. Rao, S.S. (1995) *Mechanical Vibrations*, Third Edition. Addison-Wesley Publishing Company, New York, USA.

12. Rao, S.S. (1996) *Engineering Optimization: Theory and Practice*, Third Edition. John Wiley & Sons, Inc., New York, USA.

13. Silva Neto, J.M. (1999) Identification of Material and Structures Parameters in the Frequency Domain. M.Sc. Thesis, Federal University of Santa Catarina – UFSC (in Portuguese).

14. Snowdon, J.C. (1968) *Vibration and Shock in Damped Mechanical Systems*. John Wiley & Sons, New York, USA.

15. Teixeira, P.H. (1997) Control of Vibration in Single Cables of Overhead Electric Transmission Line by Means of Viscoelastic Neutralizers. M.Sc. Thesis, Federal University of Santa Catarina – UFSC (in Portuguese).

# Active Noise Control Caused by Airflow through a Rectangular Duct

Seyyed Said Dana, Naor Moraes Melo
and Simplicio Arnaud da Silva

*Graduate Studies in Mechanical Engineering*
*Mechanical Engineering Department, Federal University of Paraiba, Campus I*
*58059-900 Joao Pessoa, Paraiba, Brazil*
*E-mails: dana@ct.ufpb.br, naor@lycos.com*

Controlling the noise coming from outside is a very important factor in bringing about a healthy indoor environment. If an individual is exposed to a noise level beyond a certain limit, serious damage can be done to his hearing system, which is irreversible in most cases. In this paper the possible conditions of cancellation of a noise generated by a single frequency source in a duct with variable rectangular section are verified. The attenuation of the noise level is done on resonance frequency of the duct using piezoelectric sheet as an actuator. This study brings the possibility of using active control principles for noise control to reduce the noise level generated from a specific source and carried by airflow. A finite element model adjusted with experimental results serves as a basis for further research.

## 1 Introduction

Among the various problems that afflict human society, noise stands out as a fundamental factor. Noise is mainly generated by equipment and machines, either industrial or domestic.

After the industrial revolution, a great many machines and equipment that facilitate work and increase productivity were introduced; however, they cause humans to be exposed to very high noise levels that could provoke partial or total deafness. Therefore it is necessary to control the noise generated by equipment, and this can be accomplished by several techniques of acoustic engineering.

The noise cancellation technique is being extensively researched today because it offers operational advantages in relation to other acoustic control techniques. Noise cancellation was developed in the 1930s by Lueg, although its practical use has been developing from the 1980s to the present day with the coming of microcomputers and the development of new material.

Also, the cancellation technique has been tried using well-designed speakers. However, the volume and space required for their installation pose rather a problem for implementation of the cancellation technique in compact areas. The high cost of speakers also makes it infeasible for equipment in daily use.

The objective of this work is to attenuate the treatment of the noise generated inside a metallic duct with a single frequency content, using a feedforward control by introducing a secondary source at an appropriate place in the duct. The source frequency at this point is 347 Hz, which has a wavelength of approximately 1 meter. If one may use passive noise control techniques with specific acoustic materials, it will be necessary to use a material for its isolation of the order of $^1/_4$ of the wavelength, a very high thickness which may be impracticable from the design point of view.

## 2 Approximation Method

The original idea for noise cancellation is a very simple and ingenuous one. However, its implementation needs instrumental adjustment for synchronizing the secondary source with the primary source, generating a noise at a known frequency. Here, contrary to most of the research methods, a rather complex experimental setup is developed and constructed to verify the decisive factors which lead in tuning the secondary source with the generated noise, which may allow attenuation of noise at a considerable level. After discovering the critical factors in a physical setup with a certain complexity and the most practical configuration, a well-concepted finite element model adjusted with suitable parameters approaches the experimental results. With a numerical well fitted model, the simplest and most efficient duct configuration is determined theoretically, for implementation of feedback control as the continuation of the research.

## 3 Experimental Setup

A piezoelectric patch pasted to the surface of the metallic duct is used as a secondary source to control the noise. As the primary source, a simple speaker generates a known frequency noise, as shown in Figure 1.

The secondary source will excite the metallic duct through vibration, at the same time it is transforming the electric signal in mechanical displacement. The vibration field then generates an acoustic field in the interior of the duct whose sound wave undertakes an 180° phase change in relation to the primary source.

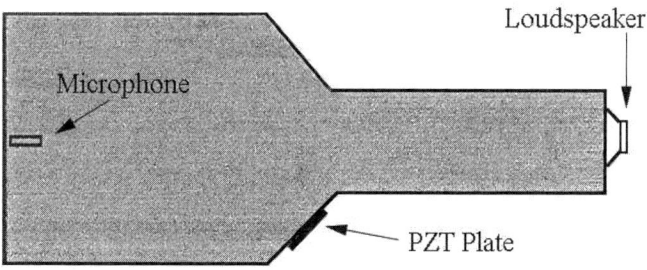

Figure 1  The excitation and measurement configuration.

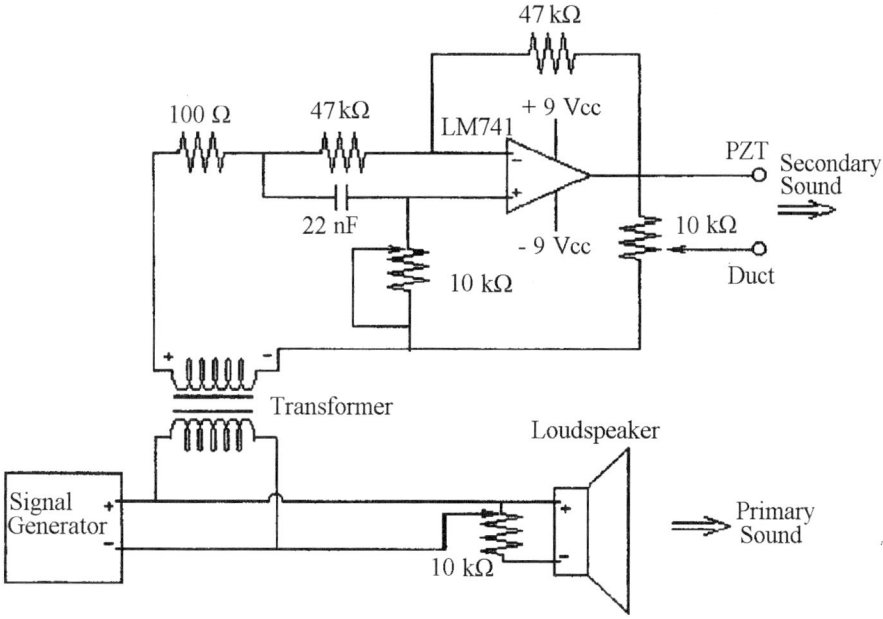

Figure 2  The phase change system and amplitude adjustment electronic outline.

The signal phase change takes place through an electronic device shown in Figure 2, which is capable of adjusting the phase change and the width of the excitation source signals.

In this work only one function generator is used to generate two excitation sources of sinus form. The use of two independent excitation generators does not produce successful results, because of difficulties in synchronizing the two sources.

A microphone is used for measuring the global noise level generated in the duct and an oscilloscope for monitoring the electric signals.

The control system is purely of feedforward previous adjustment, as can be seen in Figure 3, without the use of any data acquisition system for later use in the computer.

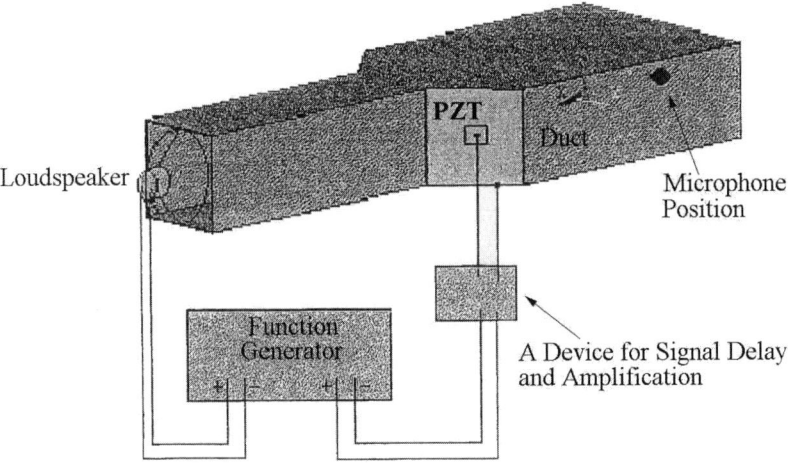

Figure 3 The equipment connections configuration and the devices used in the experiment.

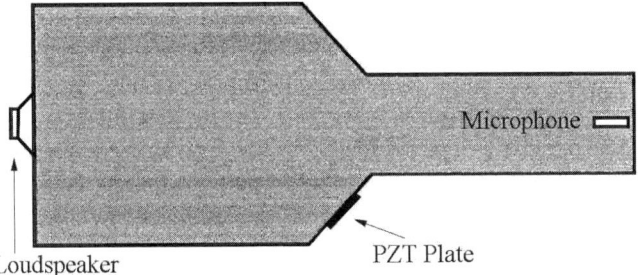

Figure 4 The microphone placement in relation to the piezoelectric sheet – frontal view.

Keeping the same calibration and device adjustments of the experiment, the sound level is measured while the duct is excited at a frequency of 347 Hz, and the microphone and the speaker position are changed as shown in Figure 4.

The equipment employed is:

- A metallic zinc duct with a total length (68 cm), width (larger section = 24 cm, smaller section = 10 cm), height (10 cm)
- Speaker (8 Ω, 1.5 W)
- Piezoelectric ceramic patch (PZT G1 195, dimensions: 19 × 19 mm)
- Electronics components
- Potentiometer 10 kΩ, 100 Ω
- Resistance 47 Ω
- Operational amplifier LM741
- Capacitor 22 nF

Table 1  Noise level values measured with and without the control system.

| Frequency (Hz) | Medium level (dB(A)) | | |
| | (1) PZT (Off) Speaker (on) | (2) PZT (On) Speaker (on) | Difference between (1) and (2) |
| --- | --- | --- | --- |
| 269 | 68.8 | 63.8 | − 4.8 |
| 341 | 75.6 | 73.8 | − 1.8 |
| 347 | 75.2 | 62.6 | −12.6 |
| 442 | 89.5 | 83.9 | − 5.6 |

Table 2  Theoretical resonance frequencies of the acoustic field.

| Modes | Frequencies (Hz) |
| --- | --- |
| 1 | 84.346 |
| 2 | 366.20 |
| 3 | 544.53 |
| 4 | 749.79 |
| 5 | 798.08 |

Table 3  Theoretical resonance frequencies of the metallic duct.

| Frequencies (Hz) |
| --- |
| 246.25 |
| 285.51 |
| 303.74 |
| 307.29 |
| 316.62 |
| 336.77 |
| 347.92 |
| 358.73 |
| 363.62 |
| 376.76 |
| 407.65 |

- Multi-variable transformer

- Signal generator (ETB-511, Entelbra brand)

- Connecting cables (BNC plug cables, alligator clips, test tip)

- Microphone with signal pre-amplifier

- Oscilloscope (MO-1360, Minipa brand).

The results identified for attenuation of the noise are shown in Table 1. The resonance frequencies of the metallic duct, and also the interior volume of the air, are of fundamental importance, because the acoustic energy generated at these frequencies is very strong. The frequencies are listed in Tables 2 and 3.

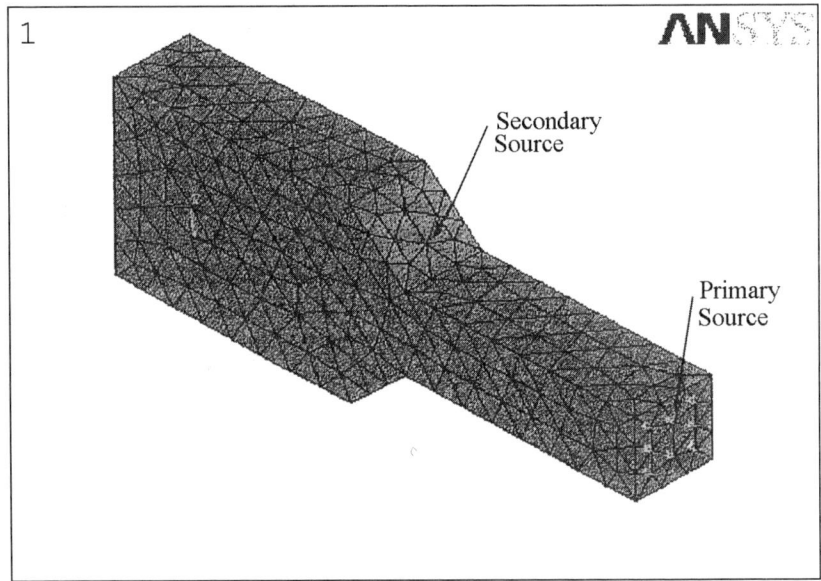

Figure 5  The Finite Element Model showing the regions of the primary and secondary excitation.

# 4   Theoretical Model

Simulating the effect of various parameters for noise cancellation is done by the finite element method which is a rather powerful tool for mathematical discretization of continuous systems. The advantages of using a finite element model are quite obvious, therefore it is not necessary to describe them.

In this paper the finite element model represents the full physical dimensions and the properties of the experimental system. Here the three-dimensional shell element "SHELL43" is used for modeling the metallic duct. The volume element "FLUID30" is applied to model the air volume in the interior of the duct. Moreover, the mesh is done automatically as offered by the software using a solid model method. For studying the acoustic element the mechanical module option is employed as required. For further use, the commands for materials properties are shown in Appendix 1.

The results for modal and harmonic analysis are carried through a frequency band of zero to 1 kHz. In harmonic analysis, where the acoustic behavior verifies the sound pressure level and its distribution through the model, the primary source is exciting the continuum medium with a single frequency. A secondary source used as feedforward control has the same frequency of excitation with a phase change of 180 degrees. The system configuration is shown in Figure 5.

The required phase delay for excitation signal of the secondary source is done by attributing appropriate values for the real and imaginary parts of pressure. The coupling of the behavior of air fluid with solid structures is performed by the FSI (fluid structure iteration) command.

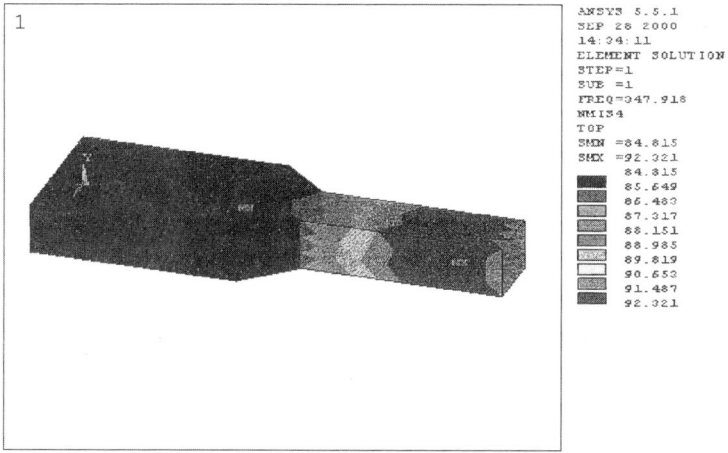

Figure 6  The Finite Element Model without the presence of the control source.

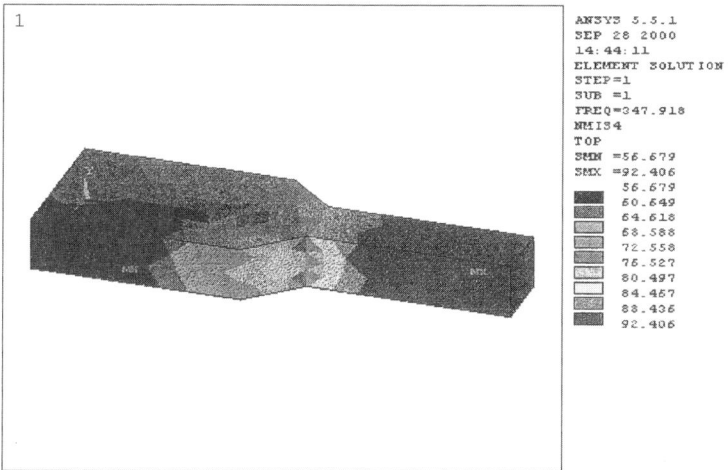

Figure 7  The Finite Element Model with the presence of the control source (amplitude of 1 Pa).

The acoustic impedance command (IMPD) is used to furnish the necessary boundary condition which permits simulation of the opening of the end of the duct. This boundary condition is simulated by the presence of a non-rigid surface. Here a non-rigid surface is defined as a surface which does not reflect any portion of air reflection.

The results obtained from the finite element model are shown in Figures 6 through 10. The resonance frequencies of the metallic duct as much as the interior volume of the air are of fundamental importance, because the acoustic energy generated at these frequencies is very strong.

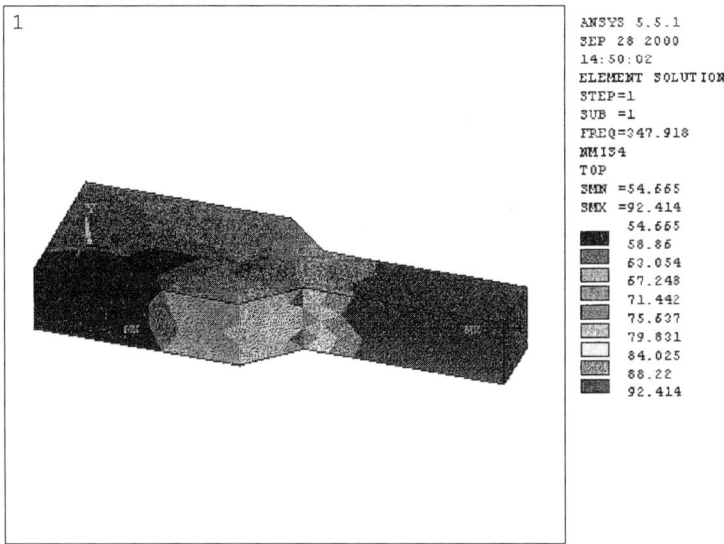

Figure 8  The Finite Element Model with the presence of the control source (amplitude of 0.98 Pa).

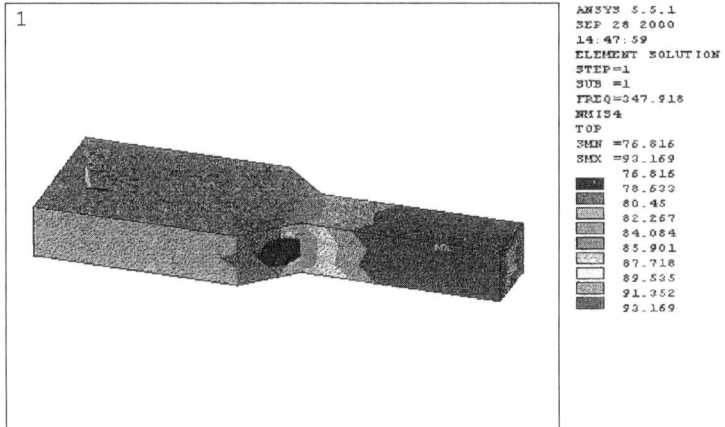

Figure 9  The Finite Element Model with the presence of the control source (amplitude of 0.086 Pa).

The frequencies are listed in Tables 2 and 3. The results obtained for attenuation of the noise are shown in Table 1. The highest reduction of noise level, 12.6 dB, at the frequency 347 Hz is observed, which is also one of the natural frequencies of the duct.

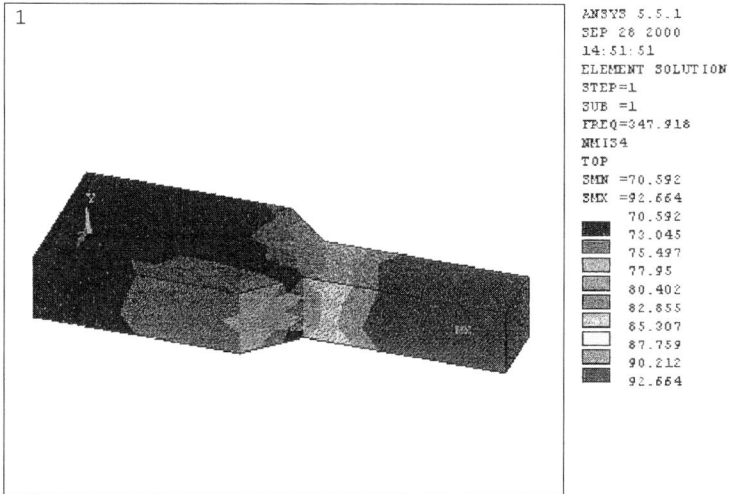

Figure 10  The Finite Element Model with the presence of the control source (amplitude of 0.51 Pa).

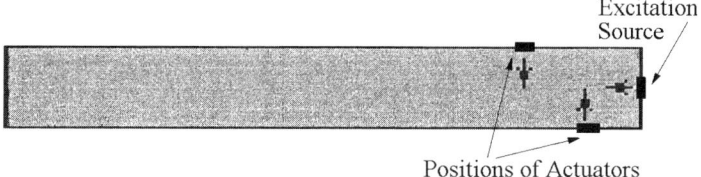

Figure 11  The Finite Element Model showing the regions of the primary and secondary excitation in the rectangular constant area duct.

# 5 A Simplified Duct Configuration

The practical application in small air conditioners mostly used in small indoor spaces demands a simplified duct configuration with a constant duct area. Using a constant duct area lowers the cost of its manufacture, which allows its application in more locations. Since the piezoelectric patches can be placed simply inside of the duct parallel to its surface with a minimum contact area, the application of this simple duct configuration would be of great use if it could bring the same attenuation level of noise as in the experiment with the complex duct. A finite element model of a duct with a constant area with two sources of feedforward control shows the possibility of reducing the noise level by 9 dB. The finite element model is shown in Figure 11 and the results in Figures 12 and 13.

# 6 Conclusion

Obtaining good experimental results demands having equipment capable of very fine adjustment. It is verified in this work that a minimum variation in terms of

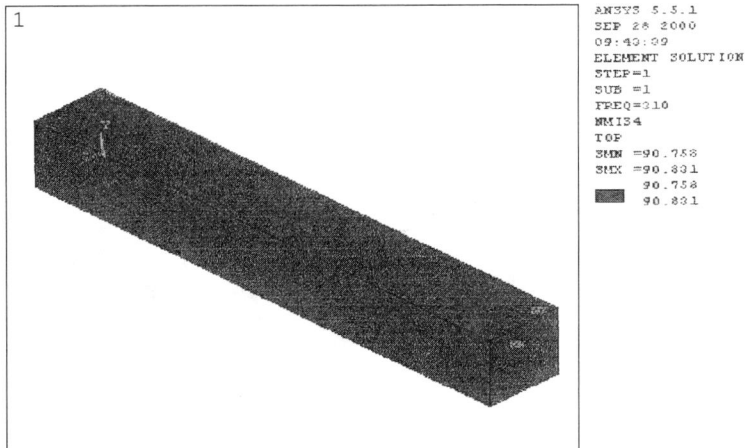

Figure 12  The Finite Element Model without the presence of the control source.

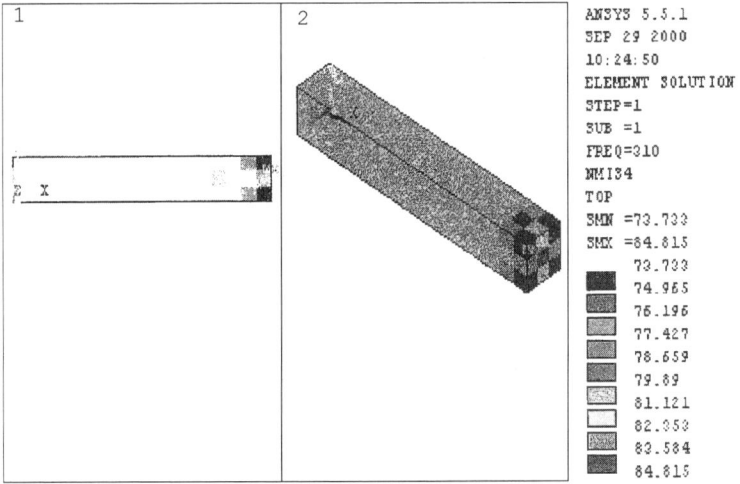

Figure 13  The Finite Element Model with the presence of two control sources (amplitude of each is 0.5 Pa).

connections and variation of values of electronic components may provoke some interference in the piezoelectric feeding signal.

The microphone placement position is a factor of fundamental importance, because it helps to determine the area where the noise cancellation happens. The experimental setup has been tested for other frequencies, as shown in Table 1, but the results are more expressive in terms of noise attenuation at the frequency of 347 Hz. The background noise is another item of great importance because its possible high level can mask measurements. Using the appropriate equipment, attenuation of the order of 12.6 dB(A) is obtained, which is a satisfactory level for the technique used, opening the way for new studies.

The primary source position is also inverted in relation to the duct, changing the direction of primary wave propagation, as can be seen in Figure 4; however, the result of −4.5 dB(A) of attenuation is not satisfactory when compared to the best result obtained at a frequency of 347 Hz.

The theoretical model reproduces the experimental measurement with a certain coherency, and this study shows that the noise attenuation occurs in special regions. Also, it becomes clear that different amplitudes at the secondary source will bring different attenuation levels in the acoustic field. The amplitude excitation of the secondary source should be of a specific value so that it does not mask or is not masked by the primary source. These results can be seen in Figures 6–10.

It is important to mention that as a result of this research it is possible to design a simple duct with two piezoelectric actuators to attenuate the noise generated by air conditioners.

# References

1. Bai, M.R. and Lin, Z. (1998) Active Noise Cancellation for Three Dimensional Enclosure by using Multiple-Channel Adaptive Control and H$\omega$ control. In *Transactions of the ASME*, vol. 120, Department of Mechanical Engineering, National Chiao-Tung University, Republic of China.

2. da Silva, S.A. (1998) Controle Ativo de Vibração e Ruido em Estruturas Flexíveis Utilizando Atuadores Piezoéletricos. Tese Work, Federal University of Paraiba.

3. Howard, C. (2000) Coupled Structural-Acoustic Analysis using ANSYS. Internal Report, Department of Mechanical Engineering, University of Adelaide.

# Appendices

## Appendix 1. The list of commands used to define the materials properties

```
ET,1,SHELL43
ET,2,FLUID30
ET,3,FLUID30,,1

! Zinc
MP,EX,l,8.2e10,     →    Elasticity Module of Zinc
MP.DENS, 1,7100,    →    Density of Zinc
MP,NUXY,1,.25,      →    Coefficient of Poisson
```

```
! Air Volume
MP,DENS,2,1.21,      →   Air Density
MP,SONC,2,343,       →   Sound Velocity
MP,MU,2,1

R,l,.0005, , , , ,    →   Real Constante – Espessure Plate of Zinc
R,2,20E-6            →   Pression of Reference
```

## Appendix 2. The list of commands used for the boundary conditions

```
SFA,8,,FSI                    →   Fluid Structure Iteration
SFA,10,,IMPD,1

FLST,2,4,1,ORDE,4
FITEM,2,120
FITEM,2,135
FITEM,2,142
FITEM,2,157
D,P51X, ,0,, , ,UY            →   Constraint Points from the Duct

D,337,PRES,.707,.707,344,1    →   Sound Pressure from the Primary Source
                                  (1 Pa = 91 dB)
D,310,PRES,−.207,−.207        →   Sound Pressure from the Secondary Source
                                  (0.086 Pa = 80 dB)
```

# Dynamical Features of an Autonomous Two-Body Floating System

Helio Mitio Morishita and Jessé Rebello de Souza Junior

*University of São Paulo*
*Department of Naval Architecture and Ocean Engineering*
*Av. Prof. Mello Moraes, 2231, Cidade Universitária 05508-900*
*São Paulo, SP, Brazil*
*Tel: +55 11 3818-5184, Fax: +55 11 3818-5717, E-mail: hmmorish@usp.br*

Dynamical aspects of a system composed of two interconnected ships floating at sea under the combined action of wind and current are presented here. Realistic, simulation-type, mathematical models are used throughout to represent the hydrodynamic and aerodynamic interactions of the vessels with their environment, as well as to incorporate the restoring forces from mooring lines and connecting hawser. A numerical study reveals that the system can have several theoretical equilibrium positions (up to fourteen), which can be stable (attracting) or unstable, feasible or geometrically impossible (implying superposition of the hulls). The system is also shown to exhibit attracting limit cycles. It is demonstrated that the number of solutions and their stability properties depend on various system parameters such as the wind to current speed ratio, wind and current angles of incidence, and shuttle vessel's draft. The role of the wind to current speed ratio and of wind angles of incidence is examined in detail, allowing the construction of a series of bifurcation diagrams in which stability properties as well as physical feasibility are identified.

## 1 Introduction

One of the most frequently adopted solutions for the exploitation of oil fields in Brazilian offshore basins is the Floating Production Storage and Offloading system (FPSO). This is a conventional tanker converted to operate as a platform to receive the risers (ducts bringing hydrocarbons from the reservoir) and store the production temporarily. Environmental forces tend to displace the vessel from the desired position, and therefore station-keeping devices are required. A common solution is an anchoring system called the Differentiated Compliant Anchoring System (DICAS), in which mooring lines connect the vessel to anchors on the seabed. Pre-tensions of the lines are arranged in such way that the vessel has some freedom to adjust itself to the predominant direction of the incoming weather.

The main vessel has to transfer its cargo periodically to a shuttle vessel that takes the oil to onshore installations. During such offloading operations the distance between the two vessels has to be kept within certain limits. The vessels are then connected to each other through a polyester cable (a hawser) so that a flexible pipe (hose) can be used to transfer the products.

One of the main concerns in the design and operation of these large systems is clearly their dynamical behavior under the action of wind, currents and waves. The dynamics of these and other anchoring systems – taking the FPSO unit alone – have been investigated elsewhere, revealing a variety of regimes of response that, depending on the mathematical model employed, may include bifurcations of static equilibria, limit cycles, and chaotic motions, [1], [2]. Building upon recent studies on the modeling and dynamics both of single-body and two-body FPSO systems under wind and current, [3], [4], [5] this work further investigates a two-body problem where the main ship, whilst moored in a DICAS arrangement, is attached to a shuttle vessel for the offloading operation. This six-degree-of-freedom coupled system is shown to exhibit complex behavior that starts with a multiplicity of static equilibrium solutions whose number and stability properties vary according to wind and current relative speeds and angles.

A preliminary investigation of the dynamical features of a real FPSO–Shuttle vessel system in a tandem configuration under the action of wind and current is presented here. Realistic mathematical models are used throughout this study to represent as accurately as possible the dynamics of the vessels in the horizontal plane. The Newtonian six-degrees-of-freedom model (two linear, one angular displacement for each vessel) results in a complex system of twelve first-order nonlinearly coupled autonomous differential equations. As a first step in characterizing the most relevant aspects of the problem, equilibria are calculated along with their stability properties for a range of values of the main parameters defining each configuration: wind and current speeds, relative angles of incidence, and hawser length. The influence of the draught of the shuttle vessel is also assessed.

Due to the great complexity of the mathematical models involved, solutions are obtained numerically, and their stability is studied mainly through time-domain simulations. The system displays a variety of different regimes of solutions in which both their number and their stability may change as one or more parameters are varied. These results are summarized in a series of bifurcation diagrams in which stability properties as well as physical feasibility are identified.

# 2    Mathematical Model

The dynamical system under investigation here represents the rigid-body motions of two sea vessels in the horizontal plane. The complex motions of ships at sea are usually split into high frequency and low frequency motions. The former motions

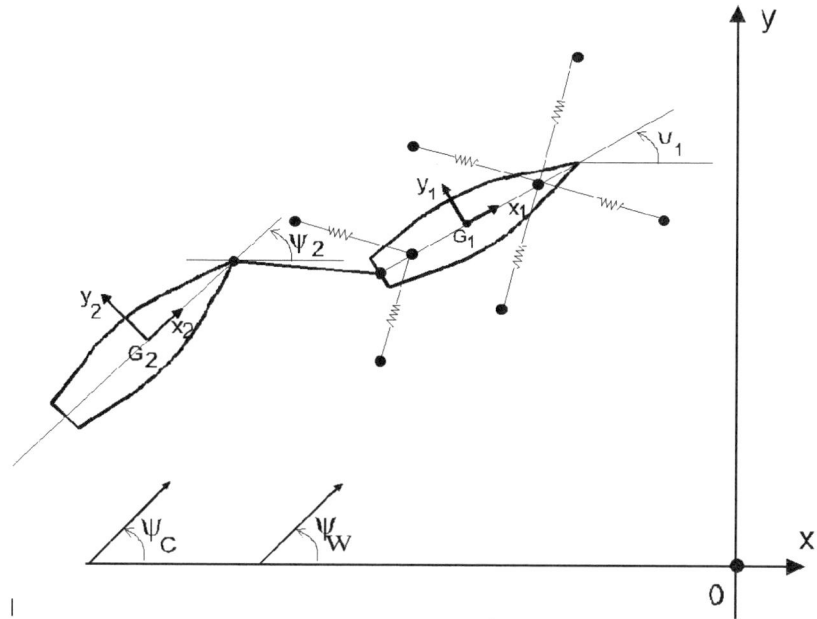

Figure 1  Reference frame.

result mainly from the action of the so-called first-order wave forces, whereas the latter motions are due to a combination of current, wind, and second-order wave forces [6]. In this paper, only low frequency motions due to current and wind forces are taken into account. In addition to those environmental forces, other terms are included related to hydrodynamic added inertia, yaw damping, and the restoring forces due to the action of hawser and mooring lines.

The motions of the vessels are traditionally expressed in two separate co-ordinate systems (see Fig. 1): one is the inertial system fixed to the Earth, OXYZ, and the other is a vessel-fixed non-inertial reference frame. The origin for this system is in the intersection of the mid-ship section with the ship's longitudinal plane of symmetry; its vertical position plays no role in the problem. The axes for this system coincide with the principal axes of inertia of the vessel with respect to the origin. In fact, there are two of these non-inertial systems, one for each vessel: $o_1 x_1 y_1 z_1$, and $o_2 x_2 y_2 z_2$ fixed to the FPSO and shuttle vessel, respectively.

The resulting equations of motions for each vessel can be given by:

$$\dot{u} = \frac{1}{D_x} \left[ (m - m_{22}) \, vr - (mx_g - m_{26}) \, r^2 - (m_{11} - m_{12}) \, v_c r + X \right], \qquad (1)$$

$$\dot{v} = \frac{1}{D_y} \left\{ \left[ (m_{11} - m) + \frac{(mx_g - m_{26})^2}{I_z - m_{66}} \right] ru - (m_{11} - m_{12}) \, u_c r + Y - \right.$$

$$\left. - \frac{mx_g - m_{26}}{I_z - m_{66}} N \right\}, \qquad (2)$$

$$\dot{r} = \frac{1}{D_z} \left\{ \frac{mx_g - m_{26}}{m - m_{22}} \left[ -\left(m_{11} - m_{22}\right) ru + \left(m_{11} - m_{12}\right) u_c r - Y + \right. \right.$$

$$\left. \left. + \frac{m - m_{22}}{mx_g - m_{26}} N \right] \right\} , \tag{3}$$

where $m$ is the mass of the vehicle; $m_{i,j}$, $i,j = 1, 2, 6$ are the added mass in surge, sway and yaw, respectively; $u$ and $v$ are the surge and sway velocities of the vehicle expressed in the body-fixed reference frame, respectively; $u_c$ and $v_c$ are current speeds related to $ox$ and $oy$ directions, respectively; $r$ is the yaw rate; $I_z$ is the moment of inertia about the $oz$ axis; $X$, $Y$, and $N$ represent the total external forces and moments in surge, sway, and yaw directions, respectively (see Appendix A for details of these forces as well as for expressions of $D_x$, $D_y$, and $D_z$); $x_g$ is the co-ordinate of the vessel's center of gravity along the $ox$ axis, and the dot means time derivative of the variable. The position and heading of each vessel related to the Earth-fixed co-ordinate system are obtained from the following equations:

$$\dot{x}_0 = u \cos \psi - v \sin \psi , \tag{4}$$

$$\dot{y}_0 = u \sin \psi + v \cos \psi , \tag{5}$$

$$\dot{\psi} = r , \tag{6}$$

where $\dot{x}_0$ and $\dot{y}_0$ are the components of the vessel's speed in the $ox$ and $oy$ axes, respectively, and $\psi$ is the vehicle heading. The components $u_c$ and $v_c$ of the current are calculated as:

$$u_c = V_c \cos(\psi_c - \psi) , \tag{7}$$

$$v_c = V_c \sin(\psi_c - \psi) , \tag{8}$$

where $V_c$ and $\psi_c$, are the velocity and direction of the current, respectively.

Equations (1) to (6) applied to each vessel yield a nonlinear autonomous dynamical system that can be written concisely as:

$$\dot{\boldsymbol{X}} = f(\boldsymbol{X}, \mu) , \tag{9}$$

where $\boldsymbol{X} = \begin{bmatrix} \boldsymbol{X}_1^T & \boldsymbol{X}_2^T \end{bmatrix}^T$, $\boldsymbol{X}_i = \begin{bmatrix} u_i & v_i & r_i & x_{0_{Bi}} & y_{0_{Bi}} & \psi_i \end{bmatrix}^T$, $i = 1, 2$, $\mu = \begin{bmatrix} \psi_c & V_c & \psi_w & V_w & \Delta \end{bmatrix}$ is the control parameter vector, $\psi_w$ is the angle of incidence of the wind, $V_w$ is the wind speed, and $\Delta$ is the ship displacement.

## 3   Solution of Equilibrium Equations

A comprehensive view of the dynamics of this system can be achieved by determining the equilibrium solutions and their stability properties. Therefore the first step of the study should be the calculation of the static solutions of Eq. (9), imposing $\dot{\boldsymbol{X}} = 0$. Clearly, in this condition velocity terms vanish, i.e., $u = v = r = 0$, as well as the net external forces and moments on the vessel. In fact, as shown in Appendix B, individual external resultant forces and moments $X$, $Y$, and $N$ for each vessel have to be zero. The latter condition establishes the equations to determine the static solutions:

$$X_{c,1}(\psi_c, V_c, \psi_1) + X_{w,1}(\psi_w, V_w, \psi_1) + X_{h,1}(x_{0,1}, y_{0,1}, x_{0,2}, y_{0,2}, \psi_1) + X_m(\psi_1) = 0 \,, \tag{10}$$

$$Y_{c,1}(\psi_c, V_c, \psi_1) + Y_{w,1}(\psi_w, V_w, \psi_1) + Y_{h,1}(x_{0,1}, y_{0,1}, x_{0,2}, y_{0,2}, \psi_1) + Y_m(\psi_1) = 0 \,, \tag{11}$$

$$N_{c,1}(\psi_c, V_c, \psi_1) + N_{w,1}(\psi_w, V_w, \psi_1) - Y_{h,1}(x_{0,1}, y_{0,1}, x_{0,2}, y_{0,2}, \psi_1)e_{h,1} + N_m(\psi_1) = 0 \,, \tag{12}$$

$$X_{c,2}(\psi_c, V_c, \psi_2) + X_{w,2}(\psi_w, V_w, \psi_2) + X_{h,2}(x_{0,1}, y_{0,1}, x_{0,2}, y_{0,2}, \psi_2) = 0 \,, \tag{13}$$

$$Y_{c,2}(\psi_c, V_c, \psi_2) + Y_{w,2}(\psi_w, V_w, \psi_2) + Y_{h,2}(x_{0,1}, y_{0,1}, x_{0,2}, y_{0,2}, \psi_2) = 0 \,, \tag{14}$$

$$N_{c,2}(\psi_c, V_c, \psi_2) + N_{w,2}(\psi_w, V_w, \psi_2) + Y_{h,2}(x_{0,1}, y_{0,1}, x_{0,2}, y_{0,2}, \psi_2)e_{h,2} = 0 \,, \tag{15}$$

where the subscripts $c$, $w$, $h$, and $m$ refer to current, wind, hawser, and mooring lines, respectively, and $e_h$ is the distance between the origin of the ship-fixed coordinate system and the hawser connection point, and it is typically half the length of the ship.

The solutions of Eqs. (10) to (15) are of the form:

$$\boldsymbol{X}^* = \begin{bmatrix} 0 & 0 & 0 & x_{0,1}^* & y_{0,1}^* & \psi_1^* & 0 & 0 & 0 & x_{0,2}^* & y_{0,2}^* & \psi_2^* \end{bmatrix}^T \,.$$

However, these values cannot be obtained straightforwardly since the set of equations involves nonlinear terms and empirical coefficients such as those of the wind that are based on OCIMF data [7]. Therefore a numerical procedure was employed, and a brief description of the algorithm is given since it is useful to understand the influence of some of the parameters on the solution of the problem. The overall procedure involves determining the components of $\boldsymbol{X}^*$ in a three-step sequence, as follows:

a) Shuttle vessel heading ($\psi_2^*$).

The heading of the shuttle vessel is determined considering Eqs. (14), (15) and those related to calculation of the current and wind force and moment shown in Appendix A, which give:

$$\frac{\rho_a A_L}{\rho T L}\sigma^2 \left[ C_{6w}(\psi_w - \psi_2) - \frac{1}{2}C_{2w}(\psi_w - \psi_2) \right] + C_{6c}(\psi_c - \psi_2) - \frac{1}{2}C_{2c}(\psi_c - \psi_2) = 0 \,, \tag{16}$$

where $\sigma = \dfrac{V_w}{V_c}$ is the ratio of wind to current speed, $\rho_a$ and $\rho$ are the densities of the air and seawater respectively, $A_L$ is the projected lateral area, and $T$ and $L$ are the shuttle vessel draft and length, respectively.

Equation (16) shows that, for a given shuttle vessel, once $\psi_c$ and $\psi_w$ are chosen the equilibrium heading $\psi_2^*$ depends only on the ratio of wind to current speed $\sigma$.

With $\psi_2^*$ obtained above, the forces $X_{h,2}$ and $Y_{h,2}$ of the hawser on the shuttle vessel are then calculated through Eqs. (13) and (14). These values allow the determination of the total hawser force $H$, and also its direction $\theta$ referred to the Earth-based co-ordinate system.

$$H = \sqrt{X_{h,2}^2 + Y_{h,2}^2} \qquad \text{and} \qquad \theta = \tan^{-1}\left(\frac{Y_{h,2}}{X_{h,2}}\right) + \psi_2^*,$$

where $H$ is the total force of the hawser, and $\theta$ is the direction of the hawser.

b) FPSO position $(x_{0,1}^* \quad y_{0,1}^* \quad \psi_1^*)$.

The position $(x_{0,1}^* \quad y_{0,1}^* \quad \psi_1^*)$ of the main vessel is determined by numerically solving Eqs. (10) to (12) simultaneously because of the nonlinear coupling imposed by the mooring lines catenary equations.

c) Shuttle vessel linear position $(x_{0,2}^* \quad y_{0,2}^*)$.

The linear position of the shuttle vessel in the horizontal plane is determined from the position of the FPSO obtained in (b) together with the hawser equation and $\psi_2^*$ as determined in (a).

The complexity of the nonlinear equations governing the motions of the vessels precludes any *a priori* conclusion concerning the number of equilibrium solutions. Thus, step (a) of the procedure above can and does indeed find more than one equilibrium solution $\psi_2^*$. For each value of $\psi_2^*$, step (b) could in principle produce any number of corresponding equilibrium values for the remaining phase variables. The numerical investigation carried out here showed, however, that for any given value of the control parameters in $\mu$, step (b) always generates a unique solution for the position of the main vessel. Such a result is consistent with the physical characteristics of the DICAS mooring system. For a different mooring system (for instance, the 'turret') other results could be expected, [8]. Therefore, for the system under study here, the value of $\psi_2^*$ uniquely determines any fixed point, and can be justifiably used to characterize the whole solution.

## 4  Stability and Bifurcation Analysis

The equilibrium solutions for the system clearly depend on the value of the control parameter vector, which has five components: $\mu = [\psi_c \quad V_c \quad \psi_w \quad V_w \quad \Delta]$. The number of possible combinations of system parameters can therefore be quite large. Some facts should, however, be considered. Firstly, it can be seen from Eq. (16) that the values of $\psi_2^*$ do not depend on $V_w$ and $V_c$ individually, but only on their ratio $\sigma$. Secondly, if $\psi_w$ and $\psi_c$ are rotated by a same angle, say $\Delta\psi$, a solution $\psi_2^*$ will obviously be rotated by a corresponding value $-\Delta\psi$. Therefore, it is reasonable to choose a value for one of the angles $\psi_w$ or $\psi_c$ and vary the other. Current speed and incidence vary more slowly than the wind's, making it interesting to keep them constant in the analysis. Moreover, the DICAS system is designed to operate in conditions of near-head current, so that keeping $\psi_c = \pi$ is perhaps a more meaningful choice for the study.

A typical portrait for this system can therefore be obtained by fixing current speed and angle of incidence and varying wind speed and incidence so that the whole range of relative angles and speeds is covered. For the present work the speed of current was fixed at $V_c = 1$ m/s. Representative values for other parameters such as the hawser nominal length were chosen and kept constant throughout the present analysis (see Appendix C for a more complete list of parameters). The FPSO and the shuttle vessel are 330,000 ton dwt and 130,000 ton dwt vessels, respectively. The influence of the displacement of the shuttle vessel was assessed by considering two drafts: full draft, and 40% of full draft. The FPSO was taken at her full displacement. Hawser nominal length was 170m (around half the FPSO's length or 64% of the shuttle vessel's length).

Seven wind incidences were chosen here: $\psi_w = 0°$, $10°$, $20°$, $30°$, $40°$, $50°$, and $60°$. Figures 2 and 3 display the main results for the equilibria of these equations using the ratio of wind to current speed $\sigma = V_w/V_c$ as an independent parameter. Figure 2 is for the full draft condition, whereas Fig. 3 is for partial draft. Each of these figures shows two different representations of the same data: Figs. 2a and 3a are 3D plots while Figs. 2b and 3b are 2D plots where curves for the various values of $\psi_w$ have been superimposed on one another.

Before the stability and bifurcations of equilibria are considered, a few general comments can be made regarding the existence and number of fixed points.

- There are always two or more fixed points for any given condition.

- Small angles of incidence of the wind $\psi_w$ produce more complicated results, with a larger number of fixed points; as $\psi_w$ grows beyond, say, $40°$ the picture becomes significantly simpler.

- There is an intermediate range of $\sigma$-values for which the number of fixed points can be considerably larger than two. In partial draft such range goes approximately from 6 to 16, whereas for full draft conditions the range moves up to the 15 to 50 interval. This difference was to be expected because the relative effect of wind forces are larger for the partial draft condition.

- In broad terms, it can be noticed that the full draft scenario tends to be less complex than that obtained for the partial draft.

Once the static solutions are determined, the next natural step is the analysis of their stability properties. That can be performed by various means: dynamic simulation, eigenvalues (Jacobian matrix), or pseudo potential wells as shown in [5]. For this paper the first procedure was adopted and two angles of incidence of the wind were selected for this more detailed investigation.

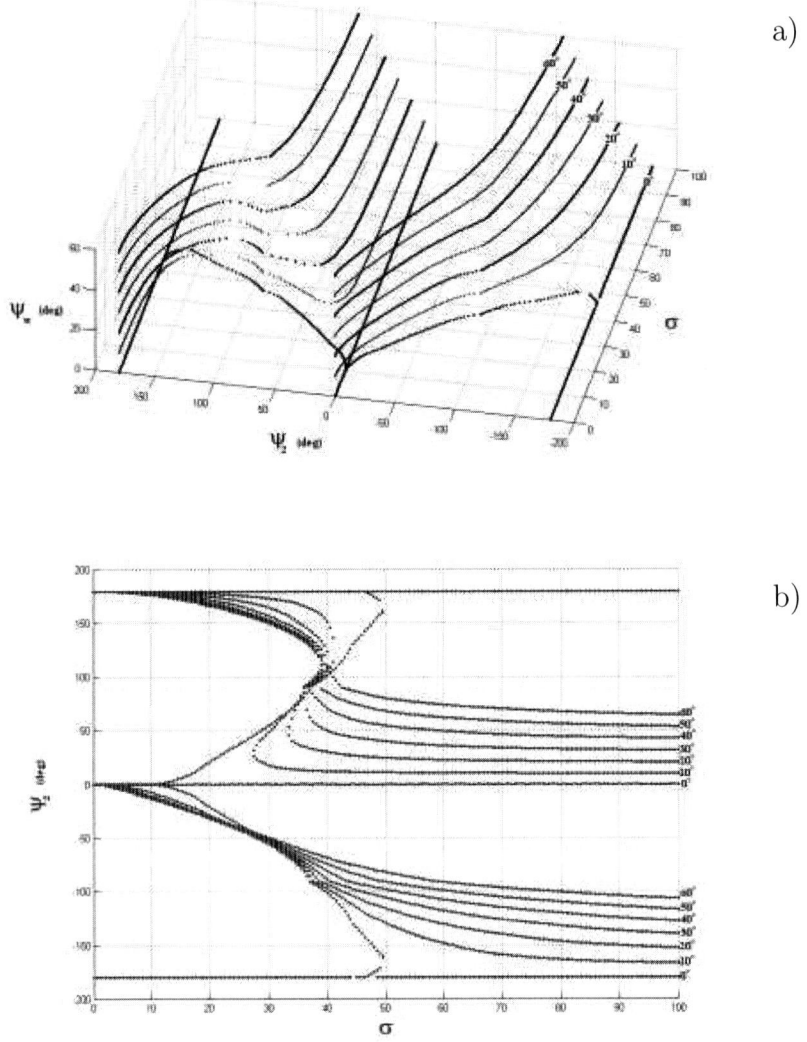

Figure 2 Bifurcation diagrams: overview for full draft condition. (a) 3D plot; (b) Consolidated 2D plot.

Figures 4 and 5 display the results for $\psi_w = 0°$ and $\psi_w = 10°$, respectively. For a more detailed analysis both diagrams can be split into three main regions denoted in Figs. 4 and 5 as regions I, II, and III, whose limits depend on the angle of incidence of the wind. Region I is for relatively low $\sigma$ values, ranging ap proximately from 0.0 to 5.5 for $\psi_w = 0°$ and from 0.0 to 10.0 for $\psi_w = 10°$. Region II is for intermediate values of $\sigma$, ranging approximately from 5.5 to 15.5 for $\psi_w - 0°$ and from 10.0 to 14.0 for $\psi_w = 10°$. Region III is for high $\sigma$, above 15.5 for $\psi_w = 0°$ and above 14.0 for $\psi_w = 10°$.

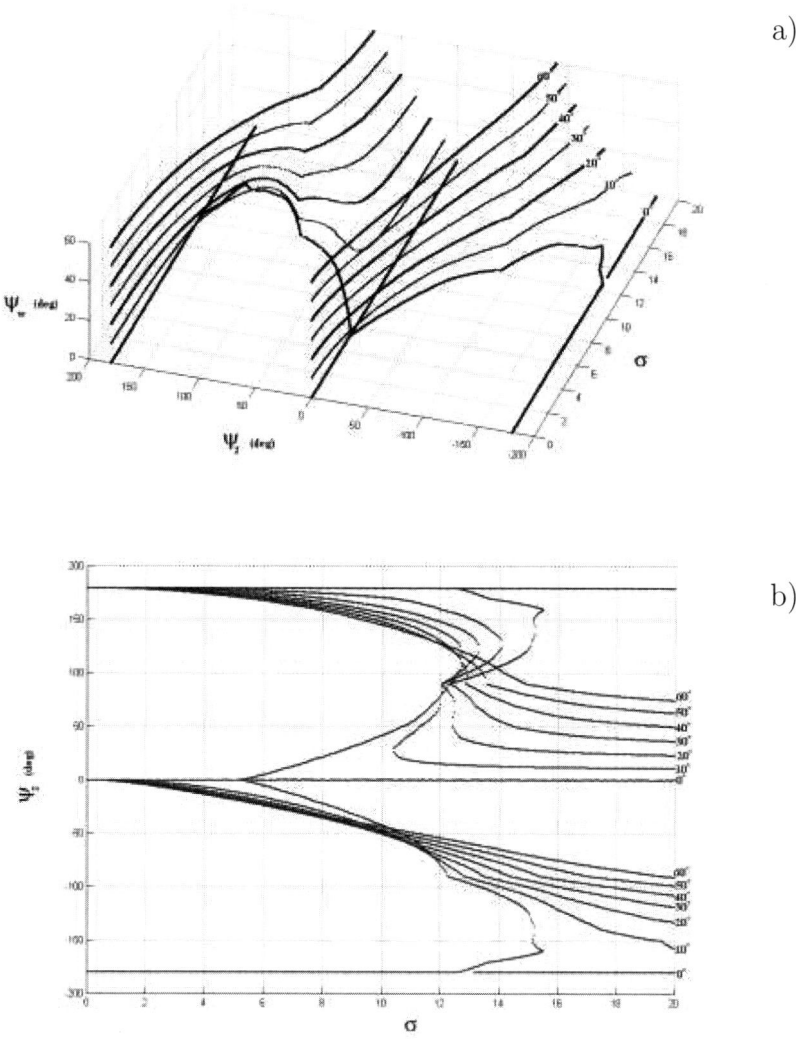

Figure 3 Bifurcation diagrams: overview for partial draft condition. (a) 3D plot; (b) Consolidated 2D plot.

The first result to be observed is that the system always displays either one or more attracting equilibria or one or more attracting limit cycles. In either case the system will also display at least one unstable static equilibrium with no limit cycle around it.

Regions I and III in both diagrams exhibit a similar pattern of behavior where only two solutions are found, one of them unstable and the other being either statically stable (attracting) or a stable (attracting) limit cycle. The attracting periodic solution is found in all of these regions except for region III in Fig. 5. In region I of Fig. 5 a limit cycle is the only attracting solution for values of $\sigma$

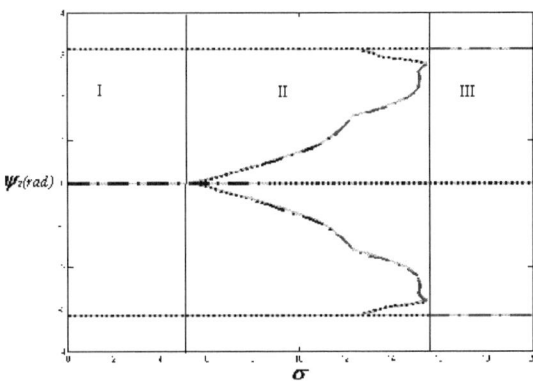

Figure 4  Bifurcation diagrams: detailed plot for $\psi_w = 0°$, partial draft.

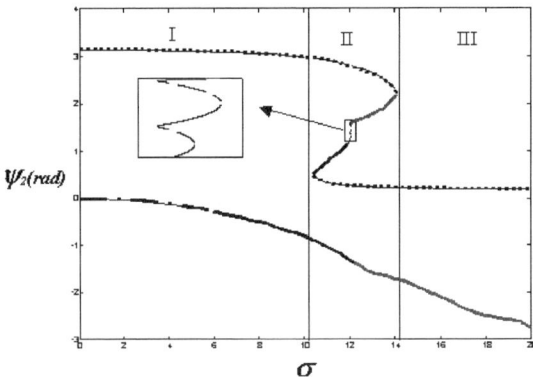

Figure 5  Bifurcation diagrams: detailed plot for $\psi_w = 10°$, partial draft.

of up to 6.0 at which value a supercritical Hopf bifurcation takes place leaving a stable focus as the attracting solution. In the intermediate region II more complex scenarios can be seen with a larger number of solutions (up to 14 for $\psi_w = 0°$). The intricate bifurcational structures that give rise to these solutions originate in a series of fold (saddle-node) bifurcations, and are shown in boxes in Fig. 5. It should be noted that these more complex situations only exist within a narrow range of values of $\sigma$. With respect to the influence of the wind incidence angle it has been found that for angles larger than $40° \sim 50°$ the complexity seen in region II of the diagrams above disappears, resulting in a relatively simple two-solution scenario similar to the ones seen in regions I and III.

# 5 Comments on the Robustness and Feasibility of Solutions

To enhance the usability of these results, two other aspects might deserve attention: the robustness of attracting solutions in terms of their catchment regions (basins of attraction), and their physical viability (feasibility). In other words, from an applied point of view an attracting solution is only important if, in addition to being stable, it will capture a considerable fraction of the phase space: the larger the basin of attraction the more relevant the solution. The second aspect is of obvious importance, particularly if an attracting solution (perhaps even with a large basin of attraction) happens to be unfeasible, that is, if it involves unacceptable proximity or collision between vessels. In a practical setting feasibility will be defined as the compliance with a number of geometric or otherwise operational parameters related primarily to safety issues of the offloading procedure. In the present work conditions were regarded as unacceptable when there was collision between vessels.

The analysis of the robustness of attracting solutions for this system has been performed and reported elsewhere [5]. Basins of attraction were constructed for conditions where four attractors exist, and dominant solutions were identified. These results were complemented by comparison with associated pseudo potential wells.

As far as the feasibility of solutions is concerned, results have indicated that for the angles of incidence of the wind under consideration the stable undesirable solutions occur when $\sigma$ is from moderate to high. For instance in the case of $\psi_w = 10°$ all stable solutions for $\sigma$ larger than 12.2 must be discarded, as indicated in Fig. 5. The stable limit cycles appearing for moderate $\sigma$ (up to $\sigma = 6.0$) are all acceptable. These results can be explained by remembering that wind forces tend here to throw the shuttle vessel against the FPSO. Another unexpected finding of relevance to the engineering of this system is the existence of conditions with more than one acceptable, attracting solution (see region II in Figs. 4 and 5).

# 6 Conclusions

The dynamics of a complex two-body floating system in tandem configuration under the action of wind and current was investigated with the use of concepts and tools of nonlinear dynamics. The system studied consisted of sophisticated simulation-type equations of motion that include detailed modeling of the hydrodynamic and aerodynamic forces.

Studies assessed the influence of several parameters such as wind and current speeds and angles of incidence. Equilibrium solutions for the system were investigated numerically, revealing a complex pattern of behavior, which include folds (saddle-node bifurcations), and Hopf bifurcations to stable limit cycles. Solutions were summarized in bifurcation diagrams where stable, unstable, feasible and unfeasible solutions were depicted. An interesting feature uncovered by this analysis is the existence of sizable regions of parameters where multiple attracting equilibrium solutions coexist.

# Appendix A

This Appendix contains the main formulae detailing the mathematical models employed throughout this study.

## Current

The forces and moment due to current are given by the following equations, [1]:

$$F_c(\beta, V) = \frac{1}{2}\rho T L^p C_{ic}(\beta)|V_c|^2\,, \qquad i = 1, 2, 6\,, \tag{A1}$$

$$p = 1 \quad \text{for} \quad i = 1, 2\,, \qquad p = 2 \quad \text{for} \quad i = 6\,,$$

where the hydrodynamic coefficients are given by:

$$C_{1c}(\beta) = \left[\frac{0.09375}{(\log(\mathrm{Re}) - 2)^2}\frac{S}{TL}\right]\cos(\beta) + \frac{1}{8}\frac{\pi T}{L}(\cos(3\beta) - \cos(\beta))\,, \tag{A2}$$

$$C_{2c}(\beta) = \left[C_Y - \frac{\pi T}{2L}\right]\sin(\beta)|\sin(\beta)| + \frac{\pi T}{2L}\sin^3(\beta) + \frac{\pi T}{L}\left[1 + 0.4\frac{C_B B}{T}\right]\sin(\beta)|\cos(\beta)|\,, \tag{A3}$$

$$C_{6c}(\beta) = \frac{-l_g}{L}\left[C_Y - \frac{\pi T}{2L}\right]\sin(\beta)|\sin(\beta)| - \frac{\pi T}{L}\sin(\beta)\cos(\beta) -$$

$$-\left[\frac{1 + |\cos(\beta)|}{2}\right]^2\frac{\pi T}{L}\left[\frac{1}{2} - 2.4\frac{T}{L}\right]\sin(\beta)|\cos(\beta)|\,, \tag{A4}$$

where $B$ and $T$ are the breadth and draft of the ship respectively; $C_B$ is the block coefficient; $C_Y$ is the lateral force coefficient in transversally steady current; Re is the Reynolds's number (based on the length $L$); $l_g$ measures the longitudinal distance between the hull's centre of mass and the midship section; $\beta$ is the angle of attack defined as:

$$\beta = \tan^{-1}\left(\frac{v - v_c}{u - u_c}\right)\,. \tag{A5}$$

## Damping Due to Yaw

The damping due to yaw is also calculated based on low aspect ratio wing theory and is given by:

$$X_D = -\frac{1}{4}\rho\pi T^2 L v_r r - \frac{1}{16}\rho\pi T^2 L^2\frac{u_r}{|u_r|}r^2\,, \tag{A6}$$

$$Y_D = \frac{1}{2}\rho T L^2 C_{D,2}u_r r - 0.035\rho T L^2 v_r r - 0.007\rho T L^3|r|r\,, \tag{A7}$$

$$N_D = -\frac{1}{2}\rho T L^3 C_{D,6}|u_r|r - \frac{3}{20}\rho T L^3 C_\gamma|v_r|r - \frac{1}{32}\rho T L^4 C_\gamma|r|r\,, \tag{A8}$$

$$u_r = u - u_c\,, \tag{A9}$$

$$v_r = v - v_c\,, \tag{A10}$$

$$C_{D,2} = \frac{\pi T}{2L} \left( 1 - 4.4 \frac{B}{L} + 0.16 \frac{B}{T} \right) , \tag{A11}$$

$$C_{D,6} = \frac{\pi T}{4L} \left( 1 + 0.16 \frac{B}{T} - 2.2 \frac{B}{L} \right) . \tag{A12}$$

## Wind

The wind forces are determined by the following equations:

$$F_w = \frac{1}{2} C_{iw}(\psi_{rw}) \rho_w V_w^2 A L_{BP}^p , \qquad i = 1, 2, 6$$
$$p = 0 \quad \text{for} \quad i = 1, 2 , \qquad p = 1 \quad \text{for} \quad i = 6 , \tag{A13}$$

$$\psi_{rw} = \psi_w - \psi , \tag{A14}$$

where the $C_{iv}$ are coefficients determined experimentally, [7]; $V_w$ is the wind speed; $A$ is the corresponding projected area of the vessel; and $\psi_w$ is the direction of the wind.

## Mooring Lines and Hawser

The forces due to mooring lines and hawser are modeled considering conventional catenary equations.

## $D_x$, $D_y$, $D_z$

These are defined as:

$$D_x = m - m_{11} , \tag{A15}$$

$$D_y = (m - m_{22}) + \frac{(mx_g - m_{26})^2}{I_z - m_{66}} , \tag{A16}$$

$$D_z = (I_z - m_{66}) - \frac{(mx_g - m_{22})^2}{m - m_{22}} . \tag{A17}$$

# Appendix B

The condition $\dot{X} = 0$ implies $r = 0$ (see Eq. (6)), and taking into account this result Eqs. (1) to (3) give the following matrix identity:

$$\begin{bmatrix} \dfrac{1}{D_x} & 0 & 0 \\[2ex] 0 & \dfrac{1}{D_y} & \dfrac{1}{D_y} \dfrac{mx_g - m_{26}}{I_z - m_{66}} \\[2ex] 0 & \dfrac{-1}{D_z} \dfrac{mx_g - m_{26}}{m - m_{22}} & \dfrac{1}{D_z} \end{bmatrix} \begin{bmatrix} X \\ Y \\ N \end{bmatrix} = \begin{bmatrix} 0 \\ 0 \\ 0 \end{bmatrix} . \tag{B1}$$

As the determinant of the square matrix is:

$$\frac{1}{D_x D_y D_z}\left[1 + \frac{(mx_g - m_{26})^2}{(I_z - m_{66})(m - m_{22})}\right] \neq 0,$$

the only possible solution of the linear system (B1) is $X = Y = N = 0$.

# Appendix C

This Appendix contains the main parameters defining the vessels employed in this study.

VESSELS

|  | FPSO | Shuttle Vessel |
|---|---|---|
| Length (m) | 327.0 | 260.0 |
| Beam (m) | 54.5 | 44.5 |
| Draft | 21.6 | 6.5 |
| Block coefficient | 0.83 | 0.77 |
| Wetted surface (m$^2$) | 27500.0 | 11745.0 |
| Mass (kg) | 312.8E6 | 57.3E6 |
| Moment of inertia (kg m$^2$) | 4.12E12 | 5.22E11 |
| Wind transversal area (m$^2$) | 1304.0 | 1339.0 |
| Wind lateral area (m$^2$) | 3893.0 | 4819.0 |

HAWSER

| Length (m) | 170.0 |
|---|---|
| EA (N) | 1.0E7 |
| Linear density (N/m) | 60.0 |

# References

1. Leite, A.J.P., Aranha, J.A.P., Umeda, C. and de Conti, M.B. (1998) Current Forces in Tankers and Bifurcation of Equilibrium of Turret System: Hydrodynamic Model and Experiments. *Applied Ocean Research*, **20** (3), 145–156.

2. Bernitsas, M.M., Garza-Rios, L.O. and Kim, B. (1999) Mooring Design Base on Catastrophes of Slow Dynamics. *Transactions of the SNAME*.

3. Morishita, H.M. and Cornet, B.J.J. (1998) Dynamics of a Turret-FPSO and Shuttle Vessel due to Current. In *IFAC Conference*, Fukuoka. CAMS'98: Control Applications in Marine Systems, Kyushu, Japan, pp. 101–106.

4. Souza Jr., J.R., Morishita, H.M., Fernandes, C.G. and Cornet, B.J.J. (2000) Nonlinear Dynamics and Control of a Shuttle Tanker. In *Nonlinear Dynamics, Chaos, Control and Their Applications*, Vol. 5, J.M. Balthazar, P.B. Gonçalves, R.M.F.L.R.F Brasil, I.L. Caldas, F.B. Rizatto (eds), Chapter 2, pp. 137–146.

5. Morishita, H.M., Souza Jr., J.R. and Fernandes, C.G. (2001) Nonlinear Dynamics of a FPSO and Shuttle Vessel in Tandem Configuration. In *11th Int. Offshore and Polar Engineering Conf. (ISOPE'2001)*, Stavanger, Norway, June 17–22, 2001, vol. 1, pp. 336–342.

6. Faltinsen, O.M. (1990) *Sea Loads on Ships and Offshore Structures*. Cambridge University Press.

7. OCIMF (1994) Predictions of Wind and Current Loads on VLCCs. Oil Companies International Marine Forum.

8. Morishita, H.M., Souza Jr., J.R. and Cornet, B.J.J. (2001) Systematic Investigation of the Dynamics of a Turret FPSO Unit in Single and Tandem Configuration. In *20th Int. Conf. on Offshore Mechanics and Arctic Engineering (OMAE'2001)*, Rio de Janeiro, Brasil, June 3–8, 2001 (published on CD-ROM).

# Dynamics and Control of a Flexible Rotating Arm Through the Movement of a Sliding Mass

Agenor de Toledo Fleury [1,2]
and Frederico Ricardo Ferreira de Oliveira [2]

[1] *Control Systems Group/Mechanical & Electrical Engineering Division*
*IPT/ São Paulo State Institute for Technological Research*
*P.O. Box 0141, 01064-970, São Paulo, SP, Brazil*
[2] *Mechanical Engineering Department/Escola Politécnica*
*USP – University of São Paulo*
*P.O. Box 61548, 05508-900, São Paulo, SP, Brazil*

Several mechanical devices in the real world utilize multiple elements and their corresponding movements to accomplish a series of given tasks. Cranes and robots are just a few examples of such devices. Generally speaking, composition of element movements is not explored to any advantage. If one intends to improve the design of multiple body devices, reducing weights and getting faster dynamic responses, one has to deal with flexibility and as a result with vibration problems. Active control of structural vibrations is addressed in this work where, motivated by the above reasoning, the problem of a flexible slewing arm carrying a sliding mass with independent motion is discussed. A dynamical model is deduced using the Extended Hamilton's Principle. The resulting model is a coupled integro-differential system of nonlinear equations, time and space variant due to changes in the inertia terms. In order to circumvent mathematical difficulties, substructuring techniques are employed, thus allowing us to consider separately the flexible arm and slider motions. Assuming synchronous motions, the response of the system can be expanded in products of spatial and time functions, resulting in a simpler model, suitable for control design. This model is generic in the sense that one can include any number of eigenfunctions. Two approaches are then considered for the mass movement when the rotating flexible arm performs a typical maneuver. The first assumes that the mass positioning over the arm is prescribed or restrained to a given length interval and the external torque on the arm is used both to generate the maneuver and to control the elastic motion. Simulation results are presented for some cases. Although they are very helpful in understanding how the composed torque-mass position control can contribute to the vibration control of the rotating arm, the second approach, where an external force actuating over the mass is a second control variable, allows more significant results. An optimal control problem, where mass initial position and velocity are design parameters, is then proposed and numerical solutions are investigated.

## List of Symbols

| | |
|---|---|
| $e(x,t)$, $e(x)$: | local elastic transverse displacement of the arm; |
| $\tau(t)$, $\tau$: | torque applied to the rotating hub; |
| $l(t)$, $l$: | slider position regarding a moving referential M(Oxyz); |
| $F(l,t)$, $F(l)$: | external force applied to the slider, tangent to the arm at point $l$; |
| $F_N(l,t)$: | normal component of the force applied to the mass by the arm; |
| $F_T(l,t)$: | tangent component of the force applied to the mass by the arm; |
| $\eta_r(t)$, $\eta_r$: | $r$-th generalized coordinate; |
| $\phi_r(x)$, $\phi_r$: | $r$-th admissible function; |
| $\omega_r$: | $r$-th natural frequency; |
| $\theta(t)$, $\theta$: | angle of rotation of the arm regarding a inertial referential I(OXYZ); |
| $M$: | mass of the slider; |
| $J_C$: | mass moment of inertia of the hub; |
| $J_B$: | mass moment of inertia of the arm; |
| $I_B$: | area moment of inertia of the arm around an axis normal to the plane of rotation; |
| $\rho$: | mass per unity length of the arm; |
| $E$: | Young's modulus of elasticity; |
| $h$: | thickness of the arm; |
| $b$: | height of the arm; |
| $L$: | total length of the arm. |

## 1   Introduction

The dynamics and control of flexible structures have merited increasing attention in recent years, because of the large potential to generate solutions to problems in important areas such as aeronautical and space engineering, precision mechanics and robotics. State-of-the-art in this field can be found in papers such as Book (1990, 1993), and books as Meirovitch (1990). A quite interesting question is related to the attenuation and control of vibrations induced in flexible devices during structure large maneuvers, which are the target of several papers such as Warren *et al.* (1995), Juang *et al.* (1986), Gildin (1998) and Lozano and Brogliato (1992), to give a few examples. Up to now, research has not tackled most of the theoretical difficulties concerning this field, especially in robotics, where the flexible motions are avoided

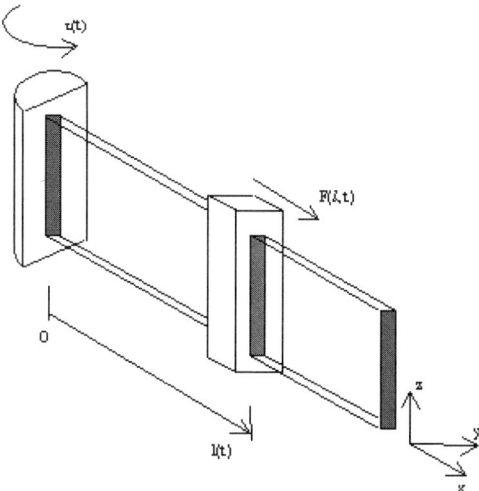

Figure 1   Flexible rotating arm and slider system.

– and not properly faced – by employing maneuvers of high rigidity. Nowadays, the main limitations for practical usage of flexible structures are: first, the complex modeling involved in the treatment of this kind of device, if one intends to generate a simple and satisfactory model for controller design; second, the limitations of the actuators available at the present time. Concerning the first problem, one can refer to the works of Liu and Skelton (1993) and Kajiwara and Nagamatsu (1994), who discuss some aspects involved in simultaneous modeling and control design of flexible systems. This topic has to be considered as a fundamental one: it is quite difficult, if not impossible, to control movements of a structure which has not been designed for that. Even simple aspects in the design of flexible structures must be seen from a different approach. Simo and Vu-Quoc (1986) propose the formulation of a dynamic response of a flexible beam relative to an inertial frame, which is an alternative approach to the traditional formulations, referred to as a floating frame called a "shadow beam." Özguner and Barbieri (1988) compare constrained and unconstrained mode expansions for a flexible slewing arm, remarking that the constrained mode expansion yields good results only if the flexible arm and the rigid hub moments of inertia ratio is small. About the second difficulty, several methods of control of vibrations have been proposed recently, such as the extensive usage of piezoelectric materials in the construction of sensors and actuators or the employment of variable positioning weights.

From the other side, several mechanical devices in the real world utilize multiple elements and their corresponding movements to accomplish a series of given tasks. Cranes and robots are examples of such devices. Generally speaking, composition of element movements is not explored to any advantage, except by the main purpose for which the mechanism is employed. If one intends to improve the design of multiple body devices, reducing weights and getting faster dynamic responses, one has to deal with flexibility and, as a consequence, with vibration problems.

Motivated by this reasoning, this work addresses the active control of structural vibrations of a flexible slewing arm carrying a sliding mass with independent motion, as shown in Figure 1. The objectives are the achievement of a good system model and the synthesis of an optimal controller using two control variables: the torque applied to the rotating hub in which the arm is fixed; and the motions of a slider that moves over the arm.

The dynamic model of this distributed system can be deduced by means of the Extended Hamilton's Principle, resulting in a coupled integro-differential system of nonlinear equations, as typical for this kind of system (Meirovitch, 1997). Besides this, in the particular case of the flexible arm with a slider, the system matrices are time and space variants due to changes in the inertia terms. The resulting system is too complicated for practical use. Substructuring techniques were then used to separate flexible arm and slider movements, and the system response was expanded in products of spatial and time variables. The control model is still a time variant one, although linear. Friedland *et al.* (1987) present the "adiabatic approximation" for the design of linear quadratic (LQ) control laws for linear, time varying systems. The authors prove that the Lyapunov Second Method assures system response stability and that the control strategy, although not optimal, also assures good performance in many cases.

Based on this approach, several cases where the slider moves on prescribed trajectories while the flexible arm rotates have been simulated and analyzed (Oliveira and Fleury, 1999 and Oliveira, 2000). Although these cases have been very helpful in understanding how the composed torque-mass position control contributes to control the elastic arm vibrations, many questions about adequate design parameters remained inconclusive. A second approach, where the slider movement is a control variable, led to an optimal control problem. The control model is now a nonlinear one, and the optimal arm and slider trajectories have to be investigated through the use of a good numerical technique (Citron, 1969). The technique of Schwartz and Polak (1996), which is based on the consistent approximations method for the numerical solution of optimal control problems (Polak, 1993), has been chosen and implemented. RIOTS_95 (Schwartz, 1997), a computational package for use with MATLAB, has assumed a special importance to this research.

# 2   Dynamic Modeling of the System

Figure 2 shows a top view diagram of the flexible arm motion, during a rotation maneuver, which is developed in the horizontal plane Oxy, in such a way that the weight force can be disregarded. Variables in Figure 2 are defined in the text.

## 2.1   Elastic potential and kinetic energy of the arm

Carrying out one by one the terms involved in the Extended Hamilton's Principle, one can express the elastic potential energy of an Euler–Bernoulli beam by

$$U(t) = \frac{1}{2} \int\limits_0^L E \cdot I_B \cdot \left( \frac{\partial^2 e}{\partial x^2} \right)^2 \, dx \,, \tag{1}$$

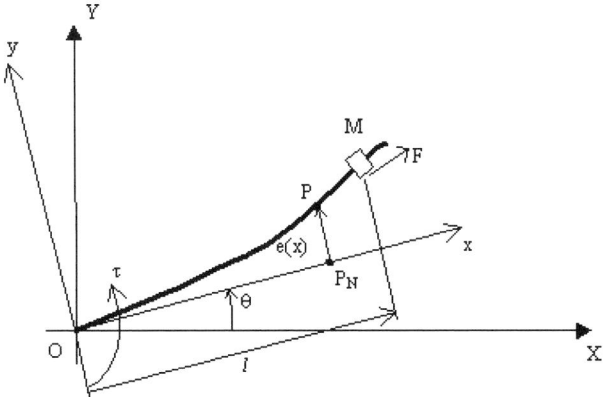

Figure 2  Displacement of the arm performing a rotation maneuver (top view).

$E$ denotes the Young's modulus of elasticity; $I_B$ is the area moment of inertia of the arm around the Oz axis, normal to the plane Oxy of rotation; and $e = e(x, t)$ is the transverse displacement of the beam. $\dfrac{\partial^2 e}{\partial x^2}$ denotes the second partial derivative of $e(x, t)$ with respect to $x$.

The total kinetic energy of the arm is

$$T_B(t) = \frac{1}{2} \int_0^L \rho \cdot \left[ \left( \frac{\partial e}{\partial t} \right)^2 + 2 \cdot \frac{\partial e}{\partial t} \cdot x \cdot \frac{d\theta}{dt} + (x^2 + e^2) \cdot \left( \frac{d\theta}{dt} \right)^2 \right] dx, \qquad (2)$$

where $L$ and $\rho$ denote the total length and the linear mass density of the flexible beam. $\theta(t)$ is the arm angle of rotation, with respect to the inertial reference system Oxyz, as shown in Figure 2.

## 2.2  Kinetic energy of slider

The kinetic energy of the slider is

$$T_M(t) = \frac{1}{2} \cdot M \cdot \left[ \left( \frac{\partial e}{\partial t} \right)_{x=l}^2 + 2 \cdot \left( \frac{\partial e}{\partial t} \right)_{x=l} \cdot l \cdot \frac{d\theta}{dt} + [l^2 + e^2(l)] \cdot \left( \frac{d\theta}{dt} \right)^2 + \right.$$

$$\left. + \left( \frac{dl}{dt} \right)^2 - 2 \cdot \frac{dl}{dt} \cdot e(l) \cdot \frac{d\theta}{dt} \right], \qquad (3)$$

$M$ is the slider mass and $l$ its position in the moving reference system Oxyz, which is rotating with the arm, as in Figure 2.

## 2.3   Kinetic energy of the hub

If $J_C$ is the mass moment of inertia of the rigid hub where the arm is fixed, the kinetic energy stored by the hub is

$$T_C(t) = \frac{1}{2} \cdot J_C \cdot \left(\frac{d\theta}{dt}\right)^2.$$

(4)

## 2.4   Complete dynamic model

The Extended Hamilton's Principle states that, between two arbitrary instants $t_1$ and $t_2$,

$$\int_{t_1}^{t_2} \delta L \, dt + \int_{t_1}^{t_2} \delta W \, dt = 0,$$

(5)

where $L$ expresses the Lagrangian of the system, a scalar function defined as

$$L = T - U = T_V + T_M + T_C - U.$$

(6)

The virtual work of the non-conservative forces is given by

$$\delta W = \tau \cdot \delta\theta + F \cdot \delta l,$$

(7)

where $\tau$ is the torque applied to the hub by a motor (not included in the modeling); and $F$, a force actuating over the slider, tangent to the arm in its position.

Thus, the Hamilton's Principle reduces to

$$\int_0^L \left\{ E \cdot I_B \cdot \frac{\partial^4 e}{\partial x^4} + \rho \cdot \left[ \frac{\partial^2 e}{\partial t^2} + x \cdot \frac{d^2\theta}{dt^2} - e \cdot \left(\frac{d\theta}{dt}\right)^2 \right] + \right.$$

$$\left. + M \cdot \left[ \frac{\partial^2 e}{\partial t^2} + 2 \cdot \frac{dl}{dt} \cdot \frac{d\theta}{dt} + l \cdot \frac{d^2\theta}{dt^2} - e \cdot \left(\frac{d\theta}{dt}\right)^2 \right] \cdot \Delta_l \right\} dx = 0,$$

(8)

$$\int_0^L \left[ \rho \cdot \left( e^2 \cdot \frac{d^2\theta}{dt^2} + 2 \cdot e \cdot \frac{\partial e}{\partial t} \cdot \frac{d\theta}{dt} + x \cdot \frac{\partial^2 e}{\partial t^2} + x^2 \cdot \frac{d^2\theta}{dt^2} \right) + \right.$$

$$\left. + M \cdot \left( \frac{\partial^2 e}{\partial t^2} \cdot l - e \cdot \frac{d^2 l}{dt^2} + e^2 \cdot \frac{d^2\theta}{dt^2} + 2 \cdot e \cdot \frac{\partial e}{\partial t} \cdot \frac{d\theta}{dt} \right) \cdot \Delta_l \right] dx +$$

$$+ M \cdot \left( l^2 \cdot \frac{d^2\theta}{dt^2} + 2 \cdot l \cdot \frac{\partial l}{\partial t} \cdot \frac{d\theta}{dt} \right) + J_C \cdot \frac{d^2\theta}{dt^2} = \tau,$$

(9)

$$M \cdot \left[ \frac{d^2 l}{dt^2} - l \cdot \left(\frac{d\theta}{dt}\right)^2 \right] - \int_0^L M \cdot \left( e \cdot \frac{d^2\theta}{dt^2} + 2 \cdot \frac{\partial e}{\partial t} \cdot \frac{d\theta}{dt} \right) \cdot \Delta_l \, dx = F.$$

(10)

Here $\Delta_l$ is the Dirac delta function, introduced to simplify the notation. Equations (8) to (10) represent the dynamic model of the system, with boundary conditions given by

$$e(0,t) = 0,, \qquad \left(\frac{\partial e}{\partial x}\right)_{x=0,\,t} = 0, \qquad \left(\frac{\partial^2 e}{\partial x^2}\right)_{x=L,\,t} = 0, \qquad \left(\frac{\partial^3 e}{\partial x^3}\right)_{x=L,\,t} = 0.$$

(11)

One can observe in Equation (11) that displacement and rotation are zero at the clamped end, and that, when the slider is not at the beam tip, the bending moment and the shearing force are zero at this point.

# 3 Substructure Synthesis

The model deduced in the last section consists of a boundary value problem, composed by a coupled integro-differential set of partial differential equations. The mathematical difficulty inherent in the direct solution of this kind of problem leads one to look for approximate methods of solution, like Rayleigh–Ritz techniques. Meirovitch (1997) presents a set of information concerning the vibration of distributed parameter systems, like flexible rotating arms. Component mode synthesis has been employed in the modeling of flexible systems since the 1960s, after the works published by Hurty (1965) and Craig and Bampton (1968). Meirovitch and Kwak (1991) discuss substructure synthesis as an extension of the Rayleigh–Ritz method to flexible multibody systems, in which the full structure is treated as an assemblage of distinct substructures. The structure dynamic behavior is based on an expansion of response of the individual substructures, described in terms of admissible functions (and not proper modes of vibrations).

Since a direct solution to the present problem does not seem feasible, substructuration – as substructure synthesis is also called – represents an alternative way to model the structure. Interaction between the flexible arm and the slider, analyzed as two distinct substructures, can be resumed to a force $F_E(l)$, external with respect to the beam and punctual, applied to it in the mass position $l$.

## 3.1 First subsystem: the slider

For small oscillations, one can assume that the force $F_E(l)$ applied to the arm by the slider is

$$F_E(l) = -M \cdot \left(\left.\frac{\partial^2 e}{\partial t^2}\right|_{x=l} + l \cdot \frac{d^2\theta}{dt^2} + 2 \cdot \frac{dl}{dt} \cdot \frac{d\theta}{dt}\right).$$

(12)

On the other hand, supposing the quadratic terms to be small, the external force which moves the slider results in

$$F = M \cdot \left[\frac{d^2 l}{dt^2} - l \cdot \left(\frac{d\theta}{dt}\right)^2\right].$$

(13)

## 3.2 Second subsystem: flexible arm

In the sequence, one has to consider the second subsystem composed by the flexible beam and the rotating rigid hub to which the beam is fixed. Two external forces act on this subsystem: the torque $\tau$ employed to generate the rotation maneuver; and the force $F_E(l)$, in the normal direction to the arm at the mass position $l$.

The virtual work of non-conservative forces in the second subsystem can be written as

$$\delta W = \tau \cdot \delta\theta + F_E(l) \cdot \delta e \,. \tag{14}$$

Use of the Extended Hamilton's Principle results in

$$\int_0^L E \cdot I_B \cdot \frac{\partial^4 e}{\partial x^4} + \rho \cdot \left[ \frac{\partial^2 e}{\partial t^2} + x \cdot \frac{d^2\theta}{dt^2} - e \cdot \left( \frac{d\theta}{dt} \right)^2 \right] dx = F_E(l) \,, \tag{15}$$

$$\int_0^L \rho \cdot \left( e^2 \cdot \frac{d^2\theta}{dt^2} + 2 \cdot e \cdot \frac{\partial e}{\partial t} \cdot \frac{d\theta}{dt} + x \cdot \frac{\partial^2 e}{\partial t^2} + x^2 \cdot \frac{d^2\theta}{dt^2} \right) dx + J_C \cdot \frac{d^2\theta}{dt^2} = \tau \,, \tag{16}$$

with the same boundary conditions given by Equation (11). In Equation (16), $J_B$ denotes the moment of inertia of the flexible arm. Equations (15) and (16) compose the model obtained via substructure synthesis, where the quadratic terms are ignored.

It is convenient to define a new variable $z(x,t)$ through

$$z(x,t) = e(x,t) + x \cdot \theta(t) \,. \tag{17}$$

Then, from Equations (15) and (16),

$$E \cdot I_B \cdot \frac{\partial^4 z}{\partial x^4} + \rho \cdot \frac{\partial^2 z}{\partial t^2} = F_E(l) \,, \tag{18}$$

$$\int_0^L \rho \cdot x \cdot \frac{\partial^2 z}{\partial t^2} \, dx + J_C \cdot \frac{d^2\theta}{dt^2} = \tau \,. \tag{19}$$

Exploring the assumption of synchronous motions, the system response can be approximated through the expansion of independent spatial and time functions, that is,

$$z(x,t) = \sum_{r=1}^{\infty} \phi_r(x) \cdot \eta_r(t) \,, \tag{20}$$

where $\phi_r(x)$ are admissible functions and $\eta_r(t)$, generalized coordinates. Therefore, for free vibration,

$$E \cdot I_B \cdot \frac{d^4\phi_r}{dx^4} - \omega_r^2 \cdot \rho \cdot \phi_r = 0 \,. \tag{21}$$

By inspection, one can verify that the general form $\phi_r(x)$

$$\phi_r(x) = a_r \cdot \sin(\beta_r \cdot x) + b_r \cdot \cos(\beta_r \cdot x) + c_r \cdot \sinh(\beta_r \cdot x) + d_r \cdot \cosh(\beta_r \cdot x) \tag{22}$$

satisfies Equation (21) and leads to the problem eigenfunctions. Coefficients $a_r$, $b_r$, $c_r$, $d_r$, and $\beta_r$ in Equation (22) are dependent on the system parameters (Meirovitch, 1997).

Thus, for normalized modes, the displacements due to system flexibility are simply expressed by

$$\frac{d^2\eta_r}{dt^2} + \omega_r^2 \cdot \eta_r = \int_0^L F_E(l) \cdot \phi_r \, dx + \tau \cdot \left. \frac{d\phi_r}{dx} \right|_{x=0}. \tag{23}$$

Equation (23) describes the dynamic behavior of the system in terms of generalized coordinates.

## 3.3 Substructure synthesis model

Combining Equations (12) and (23) of the force and time response of the arm, one can conclude that

$$\frac{d^2\eta_r}{dt^2} = \frac{1}{M} \cdot \sum_{s=1}^{p} \left[ S_{rs} \cdot \left( -\omega_s^2 \cdot \eta_s - 2 \cdot M \cdot \frac{dl}{dt} \cdot \frac{d\theta}{dt} \cdot \int_0^L \phi_s \, dx + \left. \frac{d\phi_s}{dx} \right|_{x=0} \cdot \tau \right) \right], \tag{24}$$

where the matrix elements $S_{rs}$ are dependent on the physical parameters of the system and of the slider position $l$, and given by

$$S_{rs} = \begin{cases} \dfrac{M + M \cdot \displaystyle\sum_{\substack{i=1, \\ i\neq r}}^{p} \left[ M \cdot \phi_i(l) \cdot \displaystyle\int_0^L \phi_i \, dx \right]}{1 + \displaystyle\sum_{i=1}^{p} \left[ M \cdot \phi_i(l) \cdot \displaystyle\int_0^L \phi_i \, dx \right]}, & \text{for} \quad r = s, \\[4em] \dfrac{-M^2 \cdot \displaystyle\int_0^L \phi_r \, dx \cdot \phi_s(l)}{1 + \displaystyle\sum_{i=1}^{p} \left[ M \cdot \phi_i(l) \cdot \displaystyle\int_0^L \phi_i \, dx \right]}, & \text{for} \quad r \neq s. \end{cases} \tag{25}$$

Using the same steps, the dynamics of the arm angular movement, Equation (16), may be transformed to

$$\frac{d^2\theta}{dt^2} = \frac{1}{(J_B + J_C + M \cdot l^2)} \cdot \left\{ l \cdot \sum_{r=1}^{p}\sum_{s=1}^{p} \left[ \phi_r(l) \cdot S_{rs} \cdot \omega_s^2 \cdot \eta_s \right] + \right.$$

$$+ 2 \cdot M \cdot l \cdot \frac{dl}{dt} \cdot \frac{d\theta}{dt} \cdot \left\{ \sum_{r=1}^{p}\sum_{s=1}^{p} \left[ \phi_r(l) \cdot S_{rs} \cdot \int_0^L \phi_s \, dx \right] - 1 \right\} +$$

$$\left. + \left\{ 1 - l \cdot \sum_{r=1}^{p}\sum_{s=1}^{p} \left[ \phi_r(l) \cdot S_{rs} \cdot \left. \frac{d\phi_s}{dx} \right|_{x=0} \right] \right\} \cdot \tau \right\}. \tag{26}$$

Equations (13), (24), and (26) compose the system model by means of substructure synthesis. Those terms describe, respectively, the translational dynamics of the slider, the transverse oscillations due to flexibility and rotational dynamics of the arm.

## 4　Pre-Specified Maneuver of the Slider

As stated before, a first approach to the vibration control problem employing the slider movement considers its position $l$ and velocity $\dfrac{dl}{dt}$ as prescribed trajectories. In this case, Equation (13), which describes the longitudinal positioning of the sliding mass, is rendered unnecessary, since the position $l$ and velocity $\dfrac{dl}{dt}$ are treated from an exclusive kinematic point of view.

Equations (24) and (26), which describe respectively oscillations and rotation of the flexible arm, allow one to write, for a $p$-order model (which means a model whose response is expanded by $p$ eigenfunctions), the following equations:

$$
\begin{cases}
\dfrac{dx_{(2r-1)}}{dt} = x_{(2r)}, \\[2mm]
\dfrac{dx_{(2r)}}{dt} = \dfrac{1}{M} \cdot \displaystyle\sum_{s=1}^{p} \left[ S_{rs} \cdot \left( -\omega_s^2 \cdot x_{(2s-1)} - 2 \cdot M \cdot \dfrac{dl}{dt} \cdot \int_0^L \phi_s \, dx \cdot x_{(2p+2)} + \right.\right. \\[4mm]
\qquad\qquad\qquad \left.\left. + \left.\dfrac{d\phi_s}{dx}\right|_{x=0} \cdot u \right) \right], \\[4mm]
\dfrac{dx_{(2p+1)}}{dt} = x_{(2p+2)}, \\[4mm]
\dfrac{dx_{(2p+2)}}{dt} = \dfrac{1}{J_B + J_C + M \cdot l^2} \cdot \left\{ l \cdot \displaystyle\sum_{r=1}^{p}\sum_{s=1}^{p} \left[ \phi_r(l) \cdot S_{rs} \cdot \omega_s^2 \cdot x_{(2s-1)} \right] + \right. \\[4mm]
\qquad + 2 \cdot M \cdot l \cdot \dfrac{dl}{dt} \cdot \left\{ \displaystyle\sum_{r=1}^{p}\sum_{s=1}^{p} \left[ \phi_r(l) \cdot S_{rs} \cdot \int_0^L \phi_s \, dx \right] - 1 \right\} \cdot x_{(2p+2)} + \\[4mm]
\qquad \left. + \left\{ 1 - l \cdot \displaystyle\sum_{r=1}^{p}\sum_{s=1}^{p} \left[ \phi_r(l) \cdot S_{rs} \cdot \left.\dfrac{d\phi_s}{dx}\right|_{x=0} \right] \right\} \cdot u \right\},
\end{cases}
\tag{27}
$$

where the state, $x$, and control, $u$, variables are defined by

$$
x_{(1)} = \eta_1, \qquad x_{(2)} = \dfrac{d\eta_1}{dt}, \qquad x_{(3)} = \eta_2, \qquad x_{(4)} = \dfrac{d\eta_2}{dt},
$$

$$
x_{(5)} = \eta_3, \qquad x_{(6)} = \dfrac{d\eta_3}{dt}, \qquad \cdots \qquad x_{(2p-1)} = \eta_p, \qquad x_{(2p)} = \dfrac{d\eta_p}{dt}, \tag{28}
$$

$$
x_{(2p+1)} = \theta, \qquad x_{(2p+2)} = \dfrac{d\theta}{dt}, \qquad u = \tau.
$$

One can observe that the slider position $l$ arises implicitly on the state model of the system. Only one control variable is present, the torque $\tau$ applied to the hub.

Friedland *et al.* (1987) show, through the Lyapunov Theory, that a sub-optimal LQ solution to the control problem of a time varying system can be obtained by solving the algebraic Riccati equation for the optimum control law at discrete instants of time. In this way, a time discretization of the control problem is performed, with the advantage of resulting asymptotically stable controlled systems with good performance. Since the state model given by Equation (27) is linear and time varying, such a method can be used.

# 5   Optimal Control of the Slider Movement

A second approach to the problem is to make use of the proper movement of the slider in the controller project. A force $F$ applied to the moving mass, as illustrated in Figure 2, is used as second control variable, and the position and velocity of the slider become state variables. The dynamics of arm oscillations and rotation can be deduced from Equations (24) and (26), and reduces to

$$
\left\{
\begin{aligned}
\frac{dx_{(2r-1)}}{dt} &= x_{(2r)}\,, \\[2mm]
\frac{dx_{(2r)}}{dt} &= \frac{1}{M} \cdot \sum_{s=1}^{p} \Bigg[ S_{(2r-1)s} \cdot \Bigg( -\omega_s^2 \cdot x_{(2s-1)} - \\
&\qquad - 2 \cdot M \cdot \int_0^L \phi_s\, dx \cdot x_{(2p+2)} \cdot x_{(2p+4)} + \left.\frac{d\phi_s}{dx}\right|_{x=0} \cdot u_1 \Bigg) \Bigg]\,, \\[2mm]
\frac{dx_{(2p+1)}}{dt} &= x_{(2p+2)}\,, \\[2mm]
\frac{dx_{(2p+2)}}{dt} &= \frac{1}{J_B + J_C + M \cdot x_{(2p+3)}^2} \cdot \Bigg\{ \sum_{r=1}^{p}\sum_{s=1}^{p} \left[ \phi_r(x_{(2p+3)}) \cdot S_{rs} \cdot \omega_s^2 \cdot x_{(2s-1)} \right] \times \\
&\qquad \times x_{(2p+3)} + 2 \cdot M \cdot \Bigg\{ \sum_{r=1}^{p}\sum_{s=1}^{p} \left[ \phi_r(x_{(2p+3)}) \cdot S_{rs} \cdot \int_0^L \phi_s\, dx \right] - 1 \Bigg\} \times \\
&\qquad \times x_{(2p+2)} \cdot x_{(2p+3)} \cdot x_{(2p+4)} + \\
&\qquad + \Bigg\{ 1 - \sum_{r=1}^{p}\sum_{s=1}^{p} \left[ \phi_r(x_{(2p+3)}) \cdot S_{rs} \cdot \left.\frac{d\phi_s}{dx}\right|_{x=0} \right] \cdot x_{(2p+3)} \Bigg\} \cdot u_1 \Bigg\}\,, \\[2mm]
\frac{dx_{(2p+3)}}{dt} &= x_{(2p+4)}\,, \\[2mm]
\frac{dx_{(2p+4)}}{dt} &= \frac{u_2}{M} + x_{(2p+2)}^2 \cdot x_{(2p+3)}\,,
\end{aligned}
\right.
\tag{29}
$$

where the state and control variables are defined as

$$x_{(1)} = \eta_1 , \qquad x_{(2)} = \frac{d\eta_1}{dt} , \qquad x_{(3)} = \eta_2 , \qquad x_{(4)} = \frac{d\eta_2}{dt} ,$$

$$x_{(5)} = \eta_3 , \qquad x_{(6)} = \frac{d\eta_3}{dt} , \qquad \cdots \qquad x_{(2p-1)} = \eta_p , \qquad x_{(2p)} = \frac{d\eta_p}{dt} , \qquad (30)$$

$$x_{(2p+1)} = \theta , \qquad x_{(2p+2)} = \frac{d\theta}{dt} , \qquad x_{(2p+3)} = l , \qquad x_{(2p+4)} = \frac{dl}{dt} ,$$

$$u_1 = \tau , \qquad u_2 = F .$$

An optimal control problem (OCP) to the system can now be established, aiming at the minimization of an index of performance in the form of

$$J = \int_0^{t_f} \left( \sum_{i=1}^{2p+4} Q_{ii} \cdot x_i^2 + \sum_{j=1}^{2} R_{jj} \cdot u_j^2 \right) dt \qquad (31)$$

and subject to the dynamic constraints given by Equation (29) and to the state variable inequality constraints

$$0 \leq l \leq L . \qquad (32)$$

# 6   Numerical Simulation

Numerical simulation was implemented using MATLAB and RIOTS_95, a computer program for numerical solution of optimal control problems, developed by Schwartz and Polak (1996). Cases considering the two approaches described in preceding sections have been analyzed, taking arm rotation from $-45°$ to $0°$, simulation time of 4 seconds and 514 time discretization points. Values of the physical parameters assumed for simulation are given in Table 1.

## 6.1   Pre-specified maneuver of the slider

Some of the cases selected for simulation considered the response of the structure described in terms of three autofunctions. These autofunctions correspond to the three first natural frequencies of the arm-hub device, in the absence of the sliding mass. An LQ controller with variable gains was implemented, through solving the algebraic Riccati equation at 514 time points of discretization.

Results achieved during the analysis of preliminary cases suggest that the slider movement should be performed in such a way that the Coriolis forces applied to the arm by the sliding mass should be in opposite phase with the oscillations of $\eta_1$, generalized coordinate related to the first natural frequency. Figure 3 shows the system response to such a case. From left to right and top to bottom, one can see: the pre-specified slider position and velocity; angular position and velocity of the rotating arm; displacement of the free end of the arm; and finally, the applied torque $\tau$ coordinate related to the first natural frequency. The slider moves according to a sinusoidal trajectory, with central point at $x = 0.2552$ m (which corresponds to the

Table 1  Values of physical parameters employed in simulations.

| Physical parameter | Nomenclature | Value | Unit (SI) |
|---|---|---|---|
| Young's modulus of elasticity | $E$ | $7.1 \cdot 10^{10}$ | Pa |
| thickness of the arm | $h$ | 0.001 | m |
| height of the arm | $b$ | 0.0254 | m |
| length of the arm | $L$ | 0.7 | m |
| mass density of aluminum | $\rho_B$ | 2710 | kg/m$^3$ |
| total mass of the arm | $m$ | $\rho_B \cdot L \cdot b \cdot h$ | kg |
| linear mass density of the arm | $\rho$ | $m/L$ | kg/m$^2$ |
| area moment of inertia of the arm | $I_B$ | $b \cdot h^3/12$ | m$^4$ |
| mass of the slider | $M$ | 0.01 | kg |

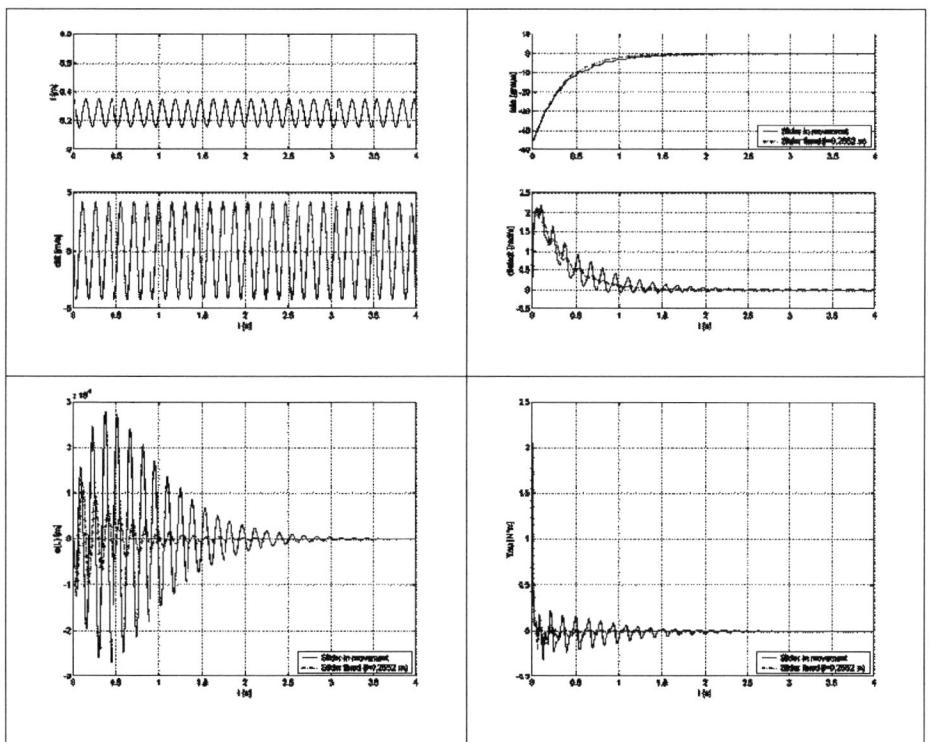

Figure 3  Slider sinusoidal motion.

point of maximum displacement of the first mode between the hub and the node) and frequency of oscillation equal to the first natural frequency of the arm. This trajectory is chosen aiming for the Coriolis forces to be in opposite phase to $\eta_1$.

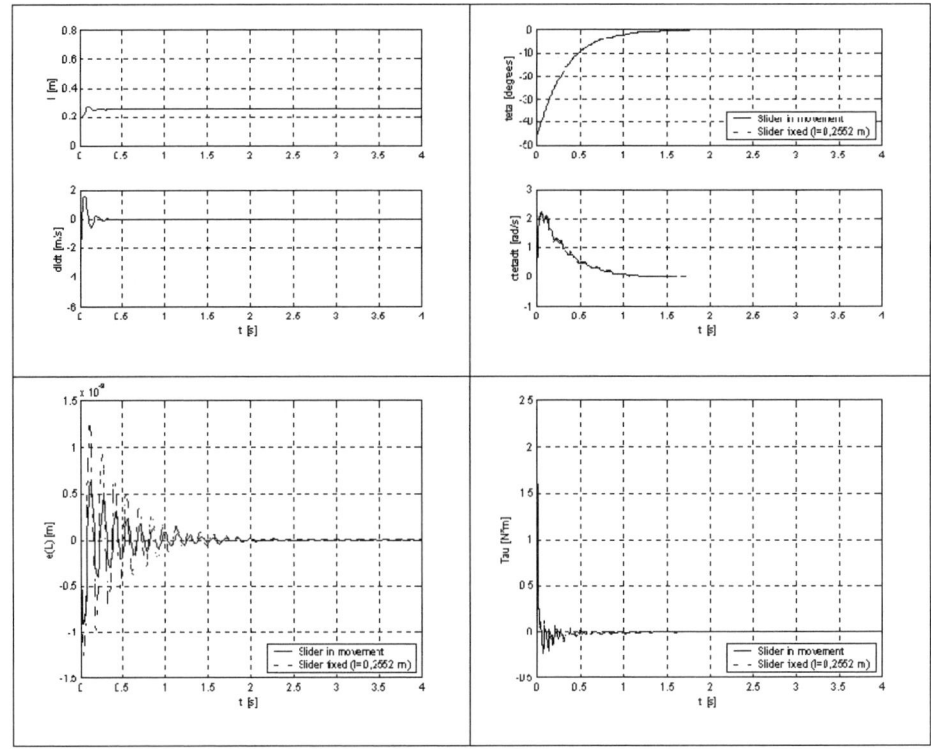

Figure 4  Slider damped motion.

The $Q_{rr}$ weights of $\eta_r$ and $\dfrac{d\eta_r}{dt}$ states in the $Q$ matrix of the index of performance defined for the LQ control are about a thousand times greater than those associated to $\theta$, $\dfrac{d\theta}{dt}$ and to the $R_{rr}$ weight of the control variable $\tau$. All these weights are time invariable.

One can observe an attenuation of the first peak of the arm tip oscillation, and, from then on, an increased motion amplitude.

Figure 4 shows a different case where the slider performs an attenuated periodic trajectory, also with central point at $x = 0.2552$ m. Again, the slider trajectory is chosen so that the Coriolis forces will be in opposite phase to $\eta_1$. In this particular case, one can observe a good attenuation of the vibration motions. Despite this, some maneuvers, like the sinusoidal oscillation of the slider shown in Figure 3, led to a worse response, with the growth of oscillation amplitudes. This fact and many others concerning the system parameters to be defined, such as slider/beam mass ratio, initial slider position, type of slider maneuver and so on, cannot be easily analyzed with this approach, since only a trial-and-error setup and simulation scheme can be used. An optimal control approach may give more direct answers to these questions.

## 6.2 Optimal control of the slider movement

The cases simulated so far using the second approach took into account only one autofunction, associated to the first natural frequency of the arm-hub set. Employing RIOTS_95, system responses were analyzed for two distinct cases: in the first, the slider moves without any state variable inequality constraints; in the second, the sliding mass trajectory is subjected to the constraints pointed out in Equation (32), which mean that the slider is restricted to moving between the hub and the arm tip. The objective function corresponds to the minimization of the arm tip vibration integrated along the time interval, that is

$$J = \int_0^4 x_i^2 dt \,. \tag{33}$$

Figure 5 shows the resulting optimal slider trajectory for the case where no state variable inequality constraints (or trajectory constraints) were considered. Position and velocity of the slider and applied force $F$, the second control variable, are illustrated. The sliding mass trajectory is smooth, nevertheless the hub position, indicated by $x = 0$, is exceeded. The initial control guess is also indicated for comparison purposes. Figure 6 shows that the oscillations induced at the free end of the arm are almost totally attenuated by the slider movement. A zoom view is provided in order to show the final result. Figure 7 shows the slider trajectory in case of state-variable inequality constraints, given by Equation (32). Figure 8 indicates the arm free end displacement and the torque applied to the hub. Once more, one can verify very good vibration attenuation. The slider trajectory is also smooth and satisfying the inequality constraints, given by the physical limits of the arm, $x = 0$ and $x = L$.

Tables 2 and 3 show the peak, mean values and standard deviation of amplitude oscillations at the arm tip, in the absence and presence of trajectory constraints. The resulting values of the objective function for initial and optimal control actions are also given, and one can verify a great reduction in these values.

# 7 Concluding remarks

The results allow one to conclude that the coupling of movements of different elements of a multibody system can be used for attenuation of the induced vibrations in a mechanical device. Even in the case of simple flexible maneuvers, the dynamical effects generated by a slider motion can constitute an effective vibration control method.

Both approaches to the vibration control problem treated in this work have led to good performance systems. The first strategy has shown that good results can be obtained even when the slider trajectory is simple, like a damped periodic motion. The large number of parameters that could be modified (such as the slider/arm mass ratio, slider initial position, types of trajectory, etc.) makes it difficult to propose a more general solution, and just a trial-and-error scheme, based on heavy numerical simulations, should be used. Other concluding remarks, therefore, would demand a deeper and more systematic analysis of the system dynamic behavior,

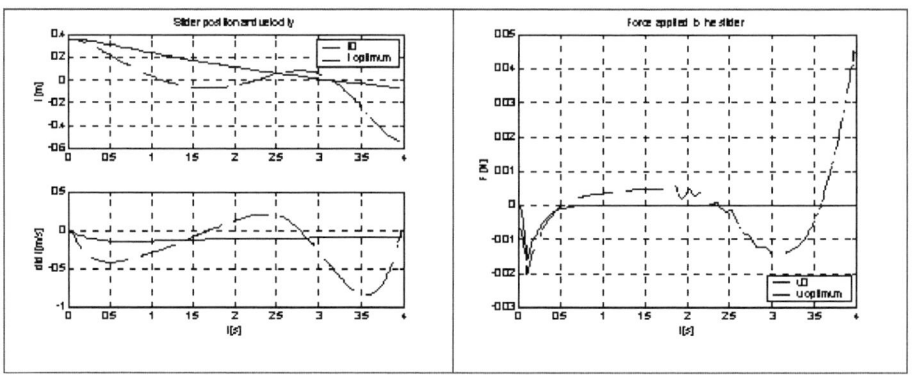

Figure 5  Slider motion and corresponding force: no trajectory constraints.

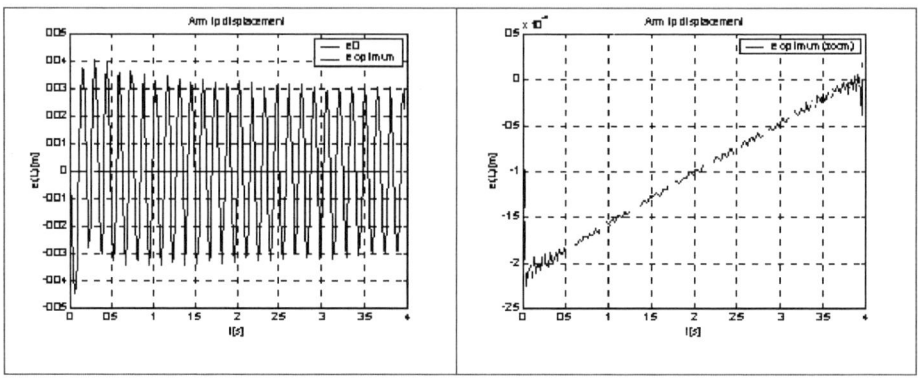

Figure 6  Arm tip displacements: no trajectory constraints.

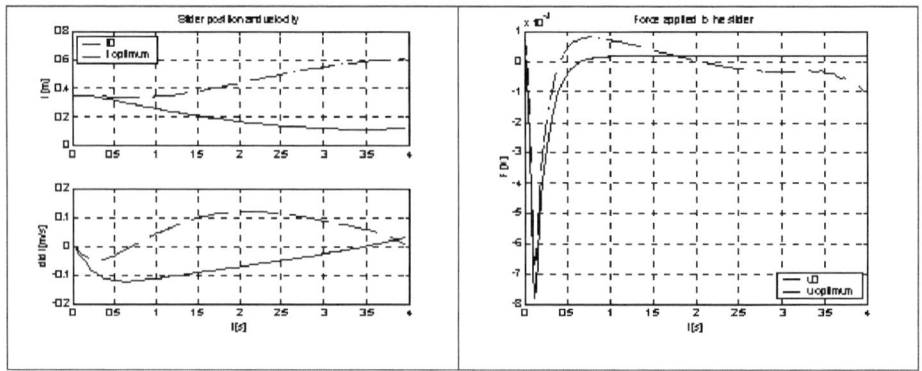

Figure 7  Slider motion and corresponding force: active trajectory constraints.

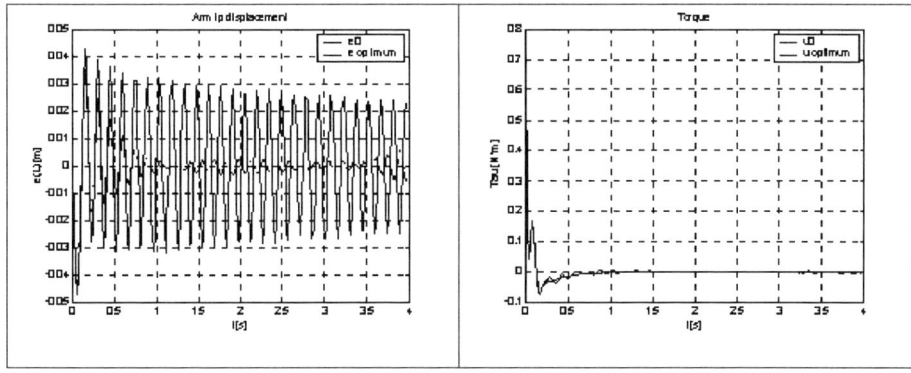

Figure 8  Arm tip displacement: active trajectory constraints.

Table 2  Peak and mean values, standard deviation and objective function for the initial and optimal control case in the absence of trajectory constraints.

| | torque $\tau$ [N m] | | | force $F$ [N] | | | $J$ |
|---|---|---|---|---|---|---|---|
| | peak | mean | SD | peak | mean | SD | |
| initial control | $7.85 \times 10^{-1}$ | $1.88 \times 10^{-3}$ | $5.70 \times 10^{-2}$ | $-1.59 \times 10^{-2}$ | $-8.15 \times 10^{-4}$ | $0.26 \times 10^{-2}$ | $3.1 \times 10^{-5}$ |
| optimal control | $4.59 \times 10^{-1}$ | $2.20 \times 10^{-3}$ | $3.05 \times 10^{-2}$ | $4.63 \times 10^{-2}$ | $2.75 \times 10^{-4}$ | $1.12 \times 10^{-2}$ | $8.0 \times 10^{-10}$ |

Table 3  Peak and mean values, standard deviation and objective function for the initial and optimal control case in the presence of trajectory constraints.

| | torque $\tau$ [N m] | | | force $F$ [N] | | | $J$ |
|---|---|---|---|---|---|---|---|
| | peak | mean | SD | peak | mean | SD | |
| initial control | $7.85 \times 10^{-1}$ | $1.95 \times 10^{-3}$ | $5.61 \times 10^{-2}$ | $8.00 \times 10^{-3}$ | $-4.03 \times 10^{-4}$ | $1.28 \times 10^{-3}$ | $2.5 \times 10^{-5}$ |
| optimal control | $7.96 \times 10^{-1}$ | $3.12 \times 10^{-3}$ | $5.88 \times 10^{-2}$ | $6.86 \times 10^{-3}$ | $-2.30 \times 10^{-4}$ | $1.12 \times 10^{-3}$ | $3.3 \times 10^{-6}$ |

but these preliminary results are fundamental to understanding the very nature of arm-slider interaction.

In the general approach corresponding to the optimal control problem employing the slider motion as a second control variable, besides the torque applied to the hub, just the slider optimal trajectories have been analyzed. On the other hand, the theoretical and numerical base for the optimization of any other physical parameters and for the study of different system configurations is ready. The preliminary results shown above for the case of the absence of state-variables inequality constraints are original and point to the need for careful compensation weight design when one is dealing with flexible rotating structures.

# Acknowledgments

The authors gratefully acknowledge CNPq – Conselho Nacional de Desenvolvimento Científico e Tecnológico – for scholarships granted to the author.

# References

1. Book, W.J. (1990) Modeling, Design and Control of Flexible Manipulator Arms: a Tutorial Review. In *Proceedings of the $29^{th}$ Conference on Decision and Control*, Honolulu, pp. 500–506.

2. Book, W.J. (1993) Controlled Motion in an Elastic World. *ASME Journal of Dynamic Systems, Measurement and Control*, **115** (2B), June, 252–261.

3. Citron, S.J. (1969) *Elements of Optimal Control*. Holt, Rinehart and Winston Inc., New York, USA.

4. Craig, R.R., Jr. and Bampton, M.C.C. (1968) Coupling of Substructures for Dynamic Analyses. *AIAA Journal*, **6** (7), April, 1313–1319.

5. Friedland, B., Richman, J. and Williams, D.E. (1987) On the "Adiabatic Approximation" for Design of Control Laws for Linear, Time-Varying Systems. *IEEE Transactions on Automatic Control*, **AC-32** (1), 62–63.

6. Gildin, E. (1998) Desenvolvimento de um Controlador Adaptivo para Manipuladores Flexiveis com Incerteza de Cargas. MSc Dissertation, Escola Politecnica, Universidade de Sao Paulo, Brasil.

7. Hurty, W.C. (1965) Dynamic Analysis of Structural Systems using Component Modes. *AIAA Journal*, **3** (4), April, 678–685.

8. Kajiwara, I. and Nagamatsu, A. (1994) Approach for Simultaneous Optimization of a Structure and Control System. *AIAA Journal*, **32** (4).

9. Juang, J., Horta, L.G. and Robertshaw, H.H. (1986) A Slewing Control Experiment for Flexible Structures. *AIAA Journal of Guidance and Control*, **9** (5), 599–607.

10. Liu, K. and Skelton, R. (1993) Integrated Modeling and Controller Design with Application to Flexible Structure Control. *Automatica*, **29** (5), 1291–1314.

11. Lozano, R. and Brogliato, B. (1992) Adaptive Control of Robot Manipulators with Flexible Joints. *IEEE Transactions on Automatic Control*, **37** (2), 174–181.

12. Meirovitch, L. (1980) *Computational Methods in Structural Dynamics*. Alphen aan den Rijin Sijthoff, Noorthoof.

13. Meirovitch, L.(1990) *Dynamics and Control of Structures*. John Wiley & Sons, New York, USA.

14. Meirovitch, L. and Kwak, M.K. (1991) Rayleigh–Ritz Based Substructure Synthesis of Flexible Multibody Systems. *AIAA Journal*, **29** (10), 1709–1719.

15. Meirovitch, L. (1997) *Principles and Techniques of Vibrations*. Prentice-Hall, New York, USA.

16. Oliveira, F.R.F. (2000) Controle de um Braço Rotativo Flexível pelo Movimento de uma Massa sobre o Braço. Dissertação de Mestrado, Universidade de São Paulo, Brasil.

17. Oliveira, F.R.F. and Fleury, A.T. (1999) Controle de um Braço Rotativo Flexível pelo Movimento de uma Massa sobre o Braço. In *Anais do XV COBEM – Congresso Brasileiro de Engenharia Mecânica*, Águas de Lindóia, São Paulo, Brasil.

18. Özguner, Ü. and Barbieri, E. (1988) Unconstrained and Constrained Mode Expansions for a Flexible Slewing Link. *Journal of Dynamic Systems, Measurement and Control*, **110** (4), 416–421.

19. Polak, E. (1993) On the Use of Consistent Approximations in the Solutions of Semi-Infinite Optimization and Optimal Control Problems. *Mathematical Programming*, **62**, 385–415.

20. Schwartz, A. and Polak, E. (1996) Consistent Approximations for Optimal Control Problems Based on Runge–Kutta Integration. *SIAM Journal of Control Optimization*, **34** (4).

21. Simo, J.C. and Vu-Quoc, L. (1986) On the Dynamics of Flexible Beams under Large Overall Motions – The Plane Case: Parts I and II. *Journal of Applied Mechanics, Transactions of the ASME*, **53**, 849–854.

22. Warren, S., Voulgaris, P. and Bergman, L. (1995) Robust Control of a Slewing Beam System. *Journal of Vibrations and Control*, **1**, 251–271.

# Measuring Chaos
# in Gravitational Waves

Humberto Piccoli [1] and Fernando Kokubun [2]

[1] *Department of Materials Science, Federal University of Rio Grande*
*Rio Grande, RS, Brazil*
[2] *Department of Physics, Federal University of Rio Grande*
*Rio Grande, RS, Brazil*

The present work is devoted to the investigation of chaotic behavior of gravitational waves from the Hénon–Heiles systems describing galactic dynamics, considered as measured quantities. Numerical simulation was applied to produce five time series with different energy levels and initial conditions. Spectral analysis were applied in all time series. State space reconstruction with delay coordinates was adopted for embedding the dynamics and its parameters were computed (time delay and embedding dimension). The search for chaotic behavior was performed by observation of Poincaré diagrams, observation of visualizations tools, and invariant computations as Lyapunov exponents and correlation dimensions for all series. The results will be used in future studies of measured emissions and are an important basis for identification of characteristics of bursts behaviors in galactic systems.

## 1    Introduction

Cosmic gravitational waves were first reported in the literature by Einstein, a few months after his formulation of general relativity. The theoretical basis of the theory was linear but restricted to weak waves emitted by bodies with very low self-gravity. A few years later a nonlinear approach was developed to include sources with significant self-gravity. Emissions of gravitational waves can be originated by collapsing stellar cores or black holes (coherent bulk motions of matter or coherent vibrations of spacetime curvature) in regions of spacetime where gravity is relativistic and the motions of matter are near the speed of light. Knowledge about the characteristics of the emissions of gravitational waves is very important in physics because they can provide new information about the universe, and they could be helpful in yielding experimental tests of fundamental laws of physics which cannot be tested in any other way (coalescence of black-hole binaries, for instance) [1].

The main difficulty lies in the measurement of gravitational waves. A lot of effort has been dedicated to experimental detection of the gravitational waves emission. In the sixties and seventies a lot of effort was dedicated in trying to build a gravitational wave detector based on resonant bars. The results were not satisfactory. In the eighties new detectors based on laser interferometry were developed and are in use. The new generation of gravitational wave detectors will provide observations of more systems and make the discovery of new possible sources of these cosmic emissions [1,2]. In Brazil, a detector is under construction at INPE (Graviton Project), designed to operate at a high frequency range. Under these circumstances, there is a strong need to provide tools for identification of gravitational waves characteristics in order to permit faithful analysis of measurements.

The Hénon–Heiles system, used to study the dynamics of galactic systems, represents the emission of gravitational waves from a system that can exhibit chaotic behavior. This happens when the emitted power exceeds some threshold value. Below this value for the emitted power the behavior is regular. An important characteristic of chaotic systems is their sensitivity to initial conditions which can generate a complex behavior with a high number of periodic components. Besides Hénon–Heiles models of galactic dynamics, chaotic behavior can be found in the three body problem, the irregular motion of Saturn's satellite Hyperion, geodesic motion around black holes; these and others, are examples of gravitational systems exhibiting chaotic behavior [2].

When the dynamics are known it is possible to search for chaotic behavior in a theoretical exact approach. However, when only measurements of the variables are available and, consequently, the dynamics are not well understood it is rather a difficult task to make any assertion about chaos occurrence. The delay coordinates technique, based on reconstruction of a state space with topological characteristics equivalent to the theoretical one, is a powerful tool to permit the investigation of signals generated by chaotic systems [3].

In this work a search is made, trying to identify chaotic behavior directly from the energy emitted by gravitational waves from Hénon–Heiles systems, here considered as a measured variable obtained by simulation, by means of a delay coordinates reconstruction of the dynamics. Spectral densities, Poincaré diagrams, Lyapunov exponents and correlation dimensions were computed from four simulations with different energy levels and/or initial conditions. The main goal is to build a report on the basic characteristics to be expected in future measurements of gravitational waves.

# 2   Theoretical Backgrounds

## 2.1   The Hénon–Heiles system

The Hénon–Heiles system is described by the potential

$$V\left(x,y\right) = m\omega^2\left[\frac{x^2+y^2}{2} + \frac{1}{a}\left(x^2y - \frac{y^3}{3}\right)\right], \tag{1}$$

where $m$ is the mass of the system, $\omega$ is the frequency, $a$ is a scale parameter, $x$ and $y$ are the coordinates.

The power emitted by a gravitational wave can be computed using

$$-\frac{dE}{dt} = \frac{G}{45c^2}\overset{\cdots}{Q}{}^2_{\alpha\beta}\,,$$  (2)

where $G$ is the gravitational constant, $c$ the light speed and, in the present case

$$Q_{xx} = 2x^2 - y^2\,,$$
$$Q_{yy} = 2y^2 - x^2\,,$$  (3)
$$Q_{xy} = 3xy\,,$$

where $x$ and $y$ are obtained by solving Eq. (1).

## 2.2  Chaotic investigation of series

The first step in this investigation is the state space reconstruction by delay coordinates [3,4]. If the measured time series is represented by $s(n)$, $n = (d-1)\tau+1,\ldots,N$, where $N$ is the total number of sampled points, then the vector in the reconstructed state space is given by

$$y(n) = \{s(n), s(n-\tau), \cdots, s(n-(d-1)\tau)\}^T\,,$$  (4)

where the superscript $T$ denotes the transpose of the matrix, $\tau$ and $d$ are the embedding parameters.

The time delay $\tau$ is taken as an integer multiple of the sampling time and was computed using the average mutual information [5], given by

$$I(\tau) = \sum_{n=1}^{N} P(s(n), s(n+\tau)) \log_2 \left| \frac{P(s(n), s(n+\tau))}{P(s(n))P(s(n+\tau))} \right|\,,$$  (5)

where $P(x)$ is the probability of a measurement $x$, and $P(x,y)$ is the joint probability of measurements $x$ and $y$. The average mutual information is plotted *versus* the time delay and the chosen value is the corresponding first local minimum. In general, first local minimum is well established. When the first local minimum is not well determined we observe the topology for values near the first local minimum and choose the time delay, which produces a Poincaré diagram with a better-defined attractor.

The embedding dimension was obtained by the method of false neighbors, based on searching for a $d$-dimensional state space in which points are neighbors due only to their proximity and not due to an attractor observation by a "window" with a non-sufficient dimension. In other words: there are no false crossings of the trajectory in the reconstructed state space. This search for false neighbors was performed with an algorithm developed by Kennel *et al.* [6].

After the state space reconstruction a topological analysis is performed. Poincaré diagrams are observed to search for evidence of chaotic behavior.

The chaotic behavior can be quantified by computing Lyapunov exponents and correlation dimensions of attractors in the reconstructed state space. A system with chaotic behavior has at least one positive Lyapunov exponent and a non-integer correlation dimension.

The correlation dimension was presented by Grassberger and Procaccia [7], as

$$d_{cor} = \lim_{\varepsilon \to 0} \frac{\log C(\varepsilon)}{\log \varepsilon}, \tag{6}$$

where $C(\varepsilon) = \lim_{N \to \infty} \dfrac{N_P}{N^2}$, $N$ is the number of points in the time series, and $N_P$ is the number of pairs $(s_i, s_j)$ with $|s_i - s_j| < \varepsilon$.

The usual procedure to estimate the correlation dimension is to plot $C(\varepsilon)$ *versus* $\varepsilon$ in log-log scales. In the limit, the slope of this curve is an estimate of the correlation dimension.

Lyapunov exponents are the characteristic exponents associated to an invariant measure. They can be computed directly from experimental data, and they are of easy interpretation. The origin of Lyapunov exponents is on the multiplicative ergodic theorem of Oseledec [8,9]. This theorem establishes that if an evolution in state space given by the map $y(n+1) = F(y(n))$ is known, and if some mathematical requirements are fulfilled, it is possible to construct a matrix

$$\Lambda = \lim_{n \to 0} \left( DF^n \left( y(0) \right) DF^n \left( y(0) \right)^T \right)^{1/2n} \tag{7}$$

known as the *Oseledec matrix* ($D$ is a differential operator). If $\alpha_1, \alpha_2, \ldots, \alpha_n$ are the eigenvalues of $\Lambda$, the associated Lyapunov exponents are

$$\lambda_1 = \log |\alpha_1|, \quad \lambda_2 = \log |\alpha_2|, \quad \ldots, \quad \lambda_n = \log |\alpha_n|. \tag{8}$$

# 3   Analysis

There were generated five long-term series by numerical simulation of the Hénon–Heiles system ($s_1$, $s_2$, $s_3$, $s_4$, $s_5$). Four time series were generated with an energy level in the limit state between regular and irregular behavior, with different initial conditions, and the fifth lies in a chaotic regime, and with an energy level near the escape value. The trajectories in the potential field can be observed in Figure 1. The series $s_1$, $s_2$, $s_3$, and $s_4$, showed a behavior strongly influenced by initial conditions, as can be seen in Figure 1, and series $s_5$ is representative of a chaotic system. The computations were performed with the package TISEAN [10], from Universities of Dresden and Wuppertal.

There are some beautiful pictures built with the trajectory plots for regular behaviors. When chaoticity begins, the trajectories frequent all regions in the potential field with the same proportion.

## 3.1   Spectral analysis

The first step in the investigation is the spectral analysis. In spite of the complexity of these spectral densities (Figure 2), it is possible to conclude that Figures 2a to 2c are computed from time series with regular quasi-periodic behavior. In Figures 2b and 2c we can observe weak evidence of chaotic behavior but, even for the most

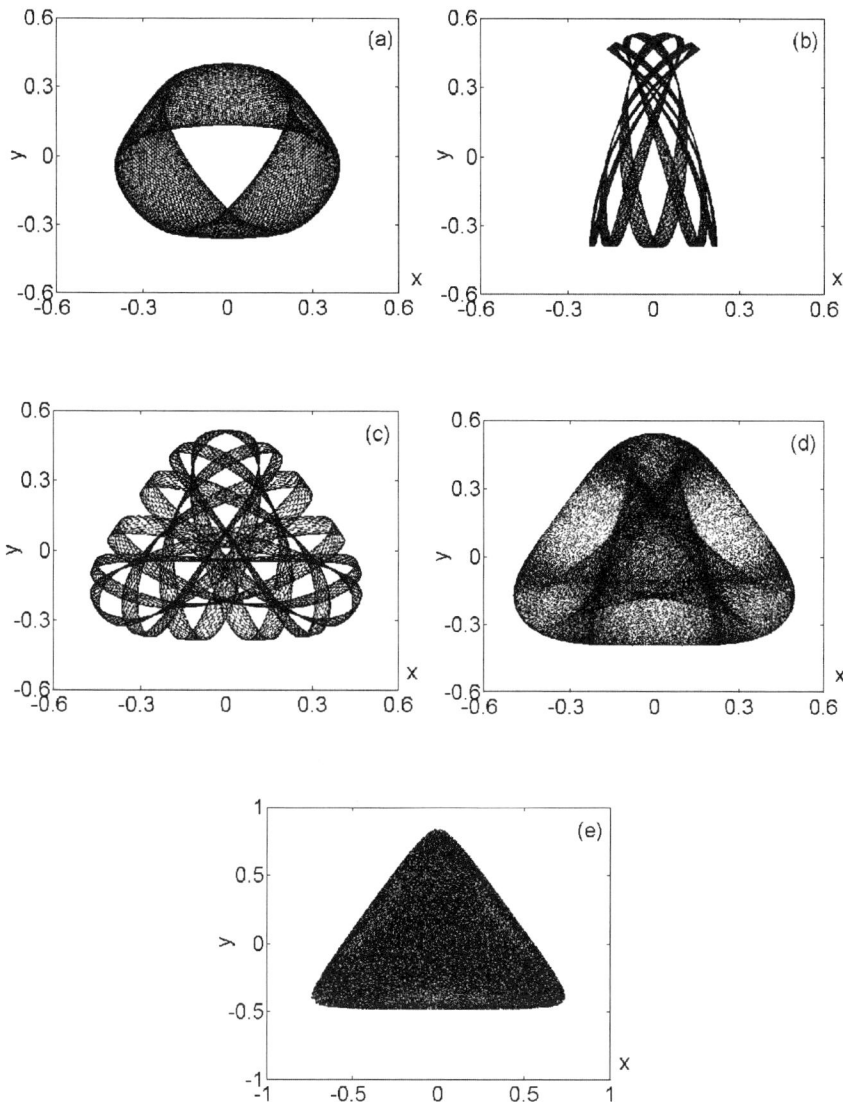

Figure 1  Trajectories in potential field.

experienced and skilled analyst, it is impossible to draw any conjecture about that. Figure 2d shows a more complex behavior, and we can suspect chaotic behavior but the evidence is not conclusive yet. The spectrum presented in Figure 2e is more representative of a chaotic behavior, and this possibility must be investigated. One remarkable observation in these plots is the spreading of the frequency contents when the series becomes more chaotic. There is also a growing of the low frequency components, especially in Figure 2e.

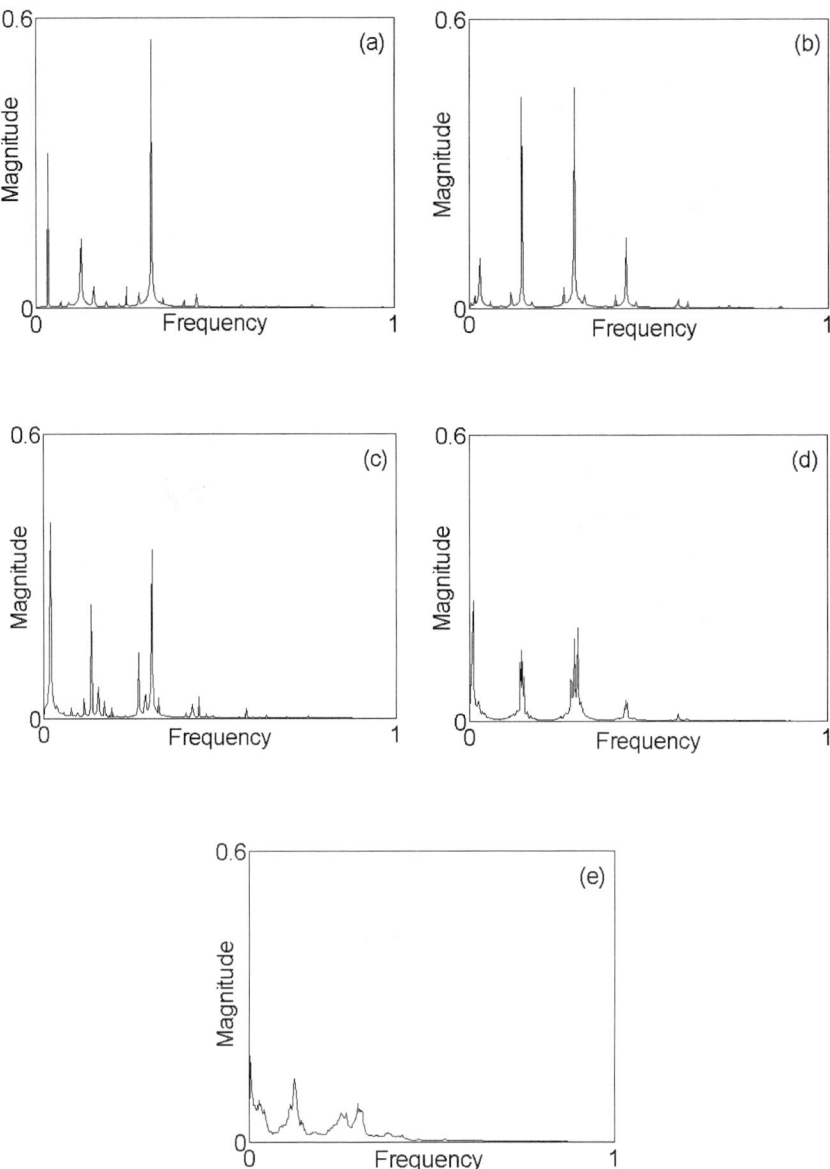

Figure 2  Fourier transforms: (a) series $s_1$, quasi-periodic behavior; (b) series $s_2$, predominantly periodic behavior; (c) series $s_3$, predominantly quasi-periodic behavior; (d) series $s_4$, periodic behavior with irregularities growing; (e) series $s_5$, irregular behavior.

## 3.2   State space reconstruction

Following the investigation, the state space reconstruction by delay coordinates is performed. The embedding dimension and time delay are chosen by using some different approaches. For the choice of the time delay the adopted approaches are

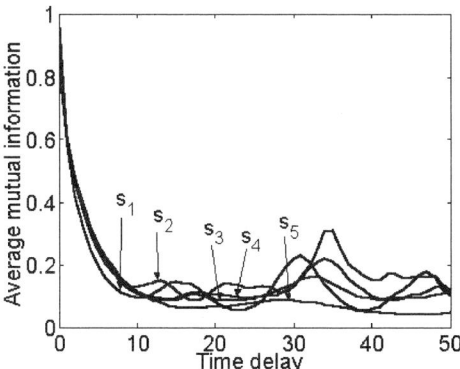

Figure 3  Average mutual information for determination of optimum time delay.

the first minimum in the average mutual information, the first zero crossing of the autocorrelation function and the worst linear predictor error. The plots of average mutual information are presented in Figure 3.

In Figure 3 it is observed that the first minimum of the average mutual information is well defined for the quasi-periodic series ($s_1$, $s_2$, $s_3$, $s_4$) and the chaotic series presents a broad range of minimum values. It was decided to take the first numerical minimum of each curve in spite of its definition difficulty. These values are shown in the first row of Table 1 (the units are number of time steps).

Table 1

| Time series | $s_1$ | $s_2$ | $s_3$ | $s_4$ | $s_5$ |
|---|---|---|---|---|---|
| Time delay | 10 | 10 | 14 | 14 | 17 |
| Embedding dimension | 3 | 3 | 3 | 4 | 4 |

Next the embedding dimension can be selected using the false nearest neighbors method. A good embedding is obtained with dimension 3 for the first three time series and 4 for the other two, as observed from Figure 4. These results are shown in the second row of Table 1.

## 3.3  Poincaré diagrams

Considering that we could obtain Poincaré diagrams directly from the solution process of describing differential equations we have here a first opportunity to compare the topology from a real state space and a reconstructed state space. Figure 5 shows the real Poincaré diagrams, and Figure 6 presents the Poincaré diagrams obtained from the evolution in the reconstructed state space (delay coordinates were used with time delays and embedding dimensions from Table 1). In spite of the strong differences between related figures when observed as pictures, there is a complete relationship when the topological structures are compared. Figures 5a and 6a show

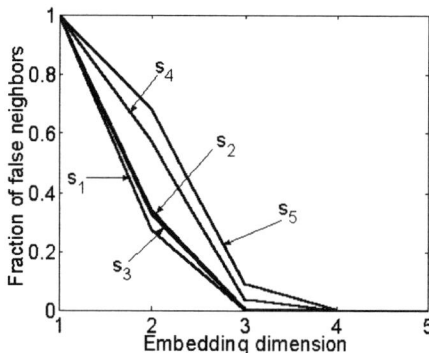

Figure 4   Computation of false nearest neighbors for embedding dimension estimation: time series $s_1$, $s_2$, and $s_3$ have near zero false neighbors at dimension 3, while for time series $s_4$ and $s_5$ the related value is 4.


biperiodic behavior. In Figures 5b-6b and 5c-6c a finite number of periodic components can be observed. In Figures 5d-6d there is a clear formation of a strange attractor, and in Figures 5e-6e a chaotic behavior can be identified.

## 3.4   Lyapunov exponents

To quantify the chaotic (or non-chaotic) behavior Lyapunov exponents must be evaluated for the adopted embedding. Figures 7 and 8 present Lyapunov exponents computed with two different algorithms: Kantz algorithm and Rosenstein algorithm [9]. In both cases it is clear that there is a more pronounced slope (dominant Lyapunov exponent) of curve $S$ in series $s_5$.

An analysis of Figures 7 and 8 shows that both algorithms were efficient in classifying the chaotic intensity for the present series. Clearly the slope grows with growing complexity of the time series.

## 3.5   Correlation dimension

Correlation dimension computations depends highly on time series length. It is possible here to generate time series with any desirable length. In practical situations, however, this is not feasible because we cannot control the duration of a burst in galactic dynamics, and the time scales could be so long that the analysis would become impractical. Under these limitations we can conclude from observation of Figure 9 that correlation dimension is not a very good tool for identifying chaos in measurements of power emissions of gravitational waves in Hénon-Heiles systems.

## 3.6   Visualization tools – recurrence plots

The qualitative structure of complex time series can be observed with the help of recurrence plots [10,11,12]. The method consists in placing a dot for a pair of close points (distance less than some neighborhood limit value). Line segments parallel to the main diagonal indicates points close successively in time, and occur

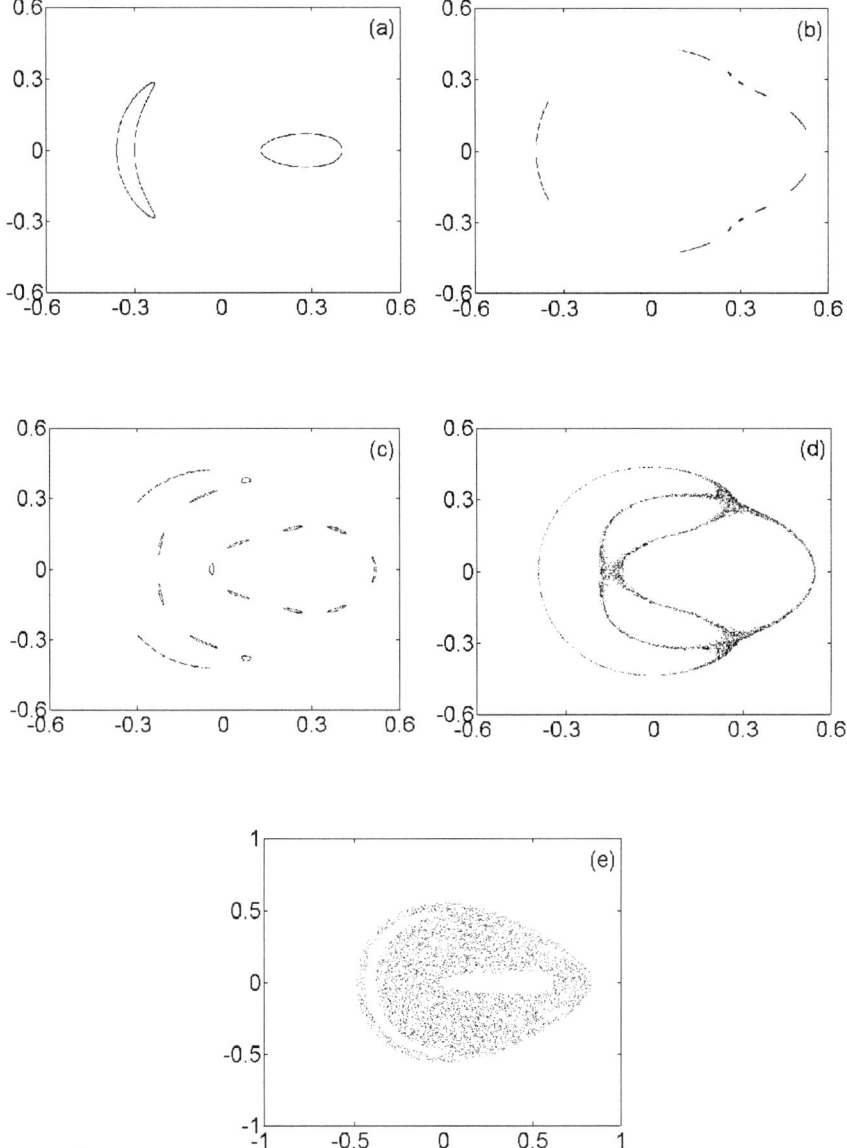

Figure 5  Theoretical Poincaré diagrams: (a) series $s_1$, regular attractor; (b) series $s_2$, regular attractor; (c) series $s_3$, regular attractor; (d) series $s_4$, strange attractor with multiperiodic behavior; (e) series $s_5$, strange attractor with chaotic behavior.

in a deterministic process. If ergodicity exists the points cover the plane uniformly. Recently a method of quantifying regularity in recurrence plots has been proposed, combining it with principal component analysis [12]. Figure 10 shows recurrence plots for the time series under investigation. We can see lines parallel to main diagonal in plots for series $s_1$, $s_2$, and $s_3$. Series $s_4$ shows some regions of the plane

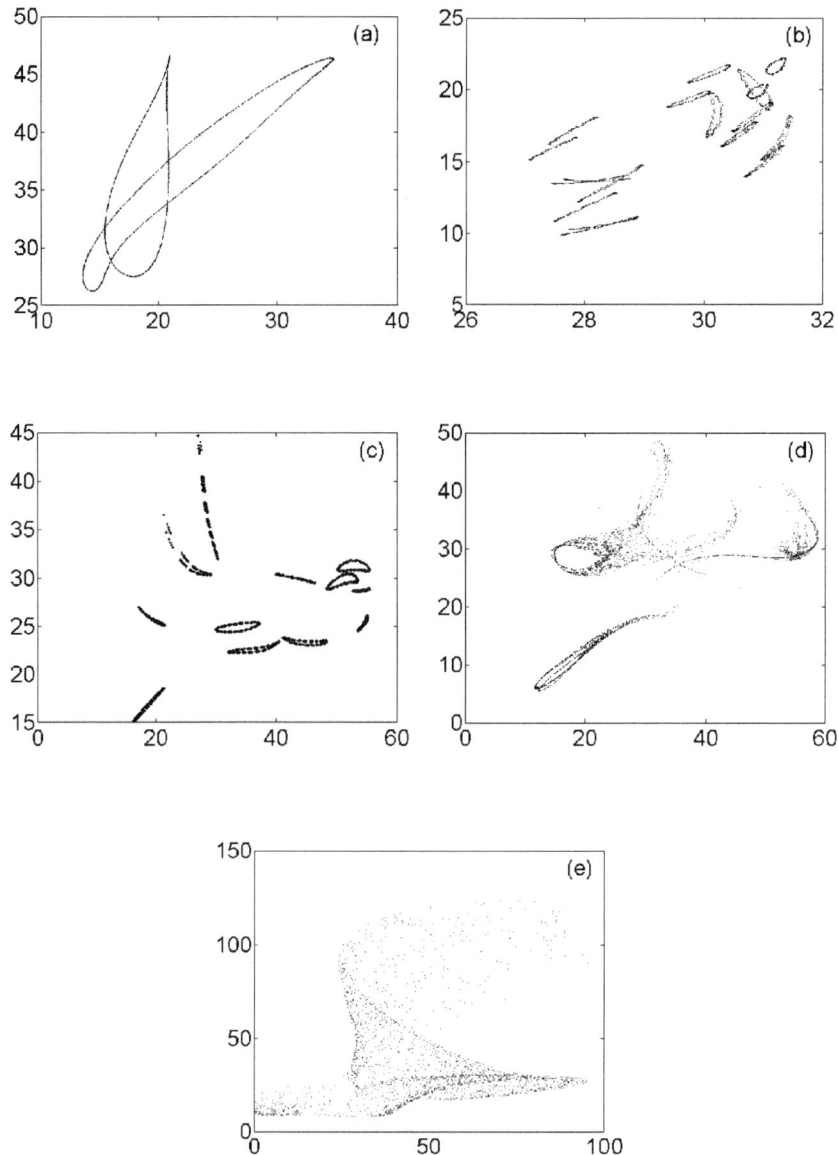

Figure 6 Poincaré diagrams obtained from state space reconstruction by time delay co-ordinates: (a) series $s_1$, regular attractor; (b) series $s_2$, regular attractor; (c) series $s_3$, regular attractor; (d) series $s_4$, strange attractor with multiperiodic behavior; (e) series $s_5$, strange attractor with chaotic behavior.

with more density of points than others with some characteristics of regularity. For series $s_5$ there is an almost uniform filling of the plane with a conclusion of irregular behavior.

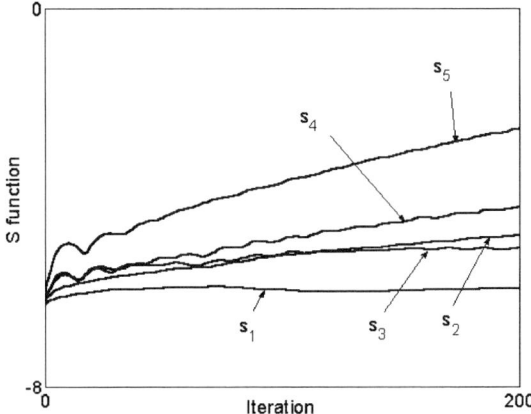

Figure 7 Lyapunov exponents computation by Kantz algorithm: (a) series $s_1$, flat curve, null dominant exponent; (b) series $s_2$, slightly growing curve, positive dominant exponent but very close to zero; (c) series $s_3$, slightly growing curve, positive dominant exponent but very close to zero; (d) series $s_4$, slightly growing curve, positive dominant exponent close to zero yet; (e) series $s_5$, clearly growing curve, positive dominant exponent.

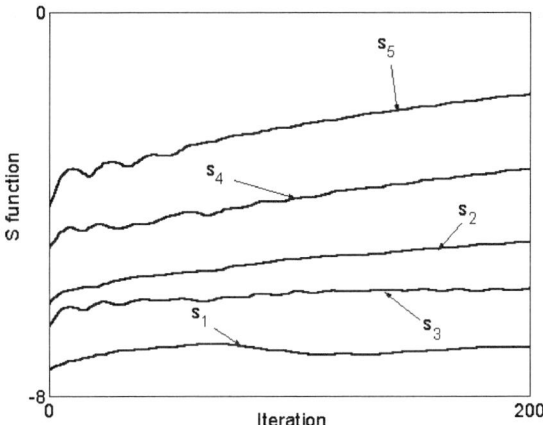

Figure 8 Lyapunov exponents computation by Rosenstein algorithm: (a) series $s_1$, flat curve, null dominant exponent; (b) series $s_2$, slightly growing curve, positive dominant exponent but very close to zero; (c) series $s_3$, almost flat curve, null dominant exponent; (d) series $s_4$, slightly growing curve, positive dominant exponent close to zero yet; (e) series $s_5$, clearly growing curve, positive dominant exponent.

## 3.7 Visualization tools – space time separation plots

Space time separation plots are an evolution of recurrence plots [10,13]. They consist in drawing lines of constant probability per unit time of a point to be a neighbor of a reference point for a determined distance in time. The time correlations inside the time series are then identified. The evolution of these combined evolution

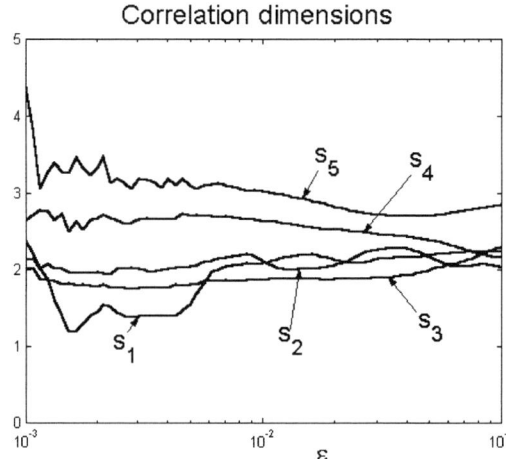

Figure 9  Correlation dimensions: series $s_2$ and $s_3$ have a reasonable constant region of dimension near 2; series $s_1$ and $s_5$ have very irregular curves with no clear constant region, and series $s_4$, shows a short flat region at approximately 2.7.

space distance versus time distance can be helpful in identifying when the time distance is sufficiently big that points are independent. Figure 11 presents space time separation plots for the present time series. We choose curves constructed for a short space distance sufficiently small to preserve the neighborhood characteristics. There are well-defined relative times in series $s_1$, and $s_2$. Besides the separation times observed in the previous series, $s_3$ shows another larger time, indicating an increasing in low frequency content, a characteristic of chaotic behavior. Series $s_4$ still shows some regularity yet but with an increase of the large separation time. For series $s_5$ there is much more irregularity in the separation times, with another conclusion of irregular behavior.

These two visualization tools have a special ability to deal with relatively short time series. This is an important characteristic in the present investigation because of an intrinsic difficulty in obtaining large time series in observing emissions of gravitational waves.

## 4   Conclusions

This investigation of chaotic behavior in power emissions from gravitational waves was performed to make possible the identification of this kind of behavior in measurements of this variable. This identification would be useful in leading to a better understanding of cosmic events and consequently a better understanding of the universe itself. The possibility of chaos identification from measured variables with reconstruction of state space with delay coordinates by computation of Poincaré diagrams, Lyapunov exponents, correlation dimensions, and visualization tools showed different features. Lyapunov exponents and correlation dimension computations seem to be very sensitive to shortness of time series. Besides, correlation dimension

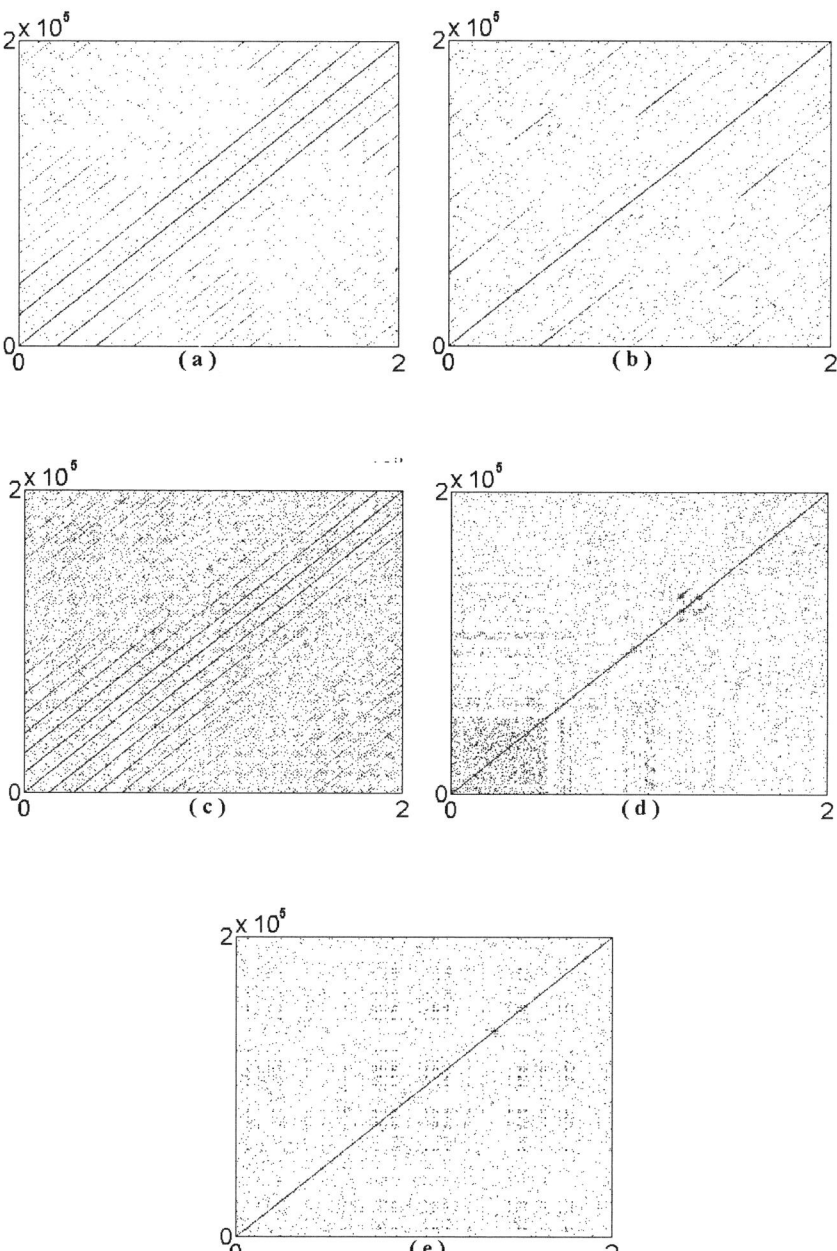

Figure 10   Recurrence plots: (a) series $s_1$, straight lines parallel to main diagonal; (b) series $s_2$, slight straight lines parallel to main diagonal, with few recurrence points; (c) series $s_3$, straight lines parallel to main diagonal with some spreading of recurrence points; (d) series $s_4$, some dominant regions of recurrence but a scarce number of points; (e) series $s_5$, a scarce number of spreading recurrence points.

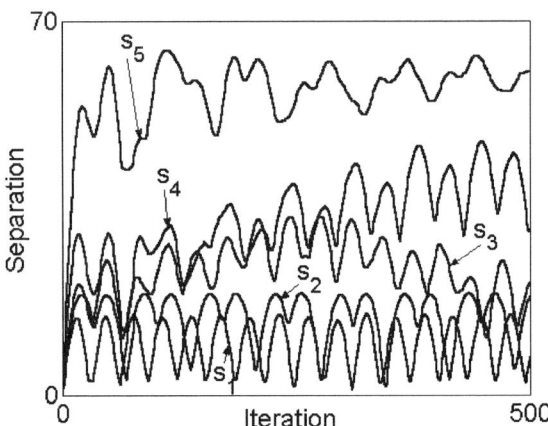

Figure 11  Space time separation plots: (a) series $s_1$, regular return time separations; (b) series $s_2$, regular return time separation; (c) series $s_3$, two orders of return time separations; (d) series $s_4$, irregular return time separation; (e) series $s_5$, irregular return time separation.

has a reasonable computation cost. In spite of these, all tools were able to give a safe conclusion about chaotic (or, at least, irregular) behavior. More investigation has been undertaken to improve the ability to forecast chaotic situations from measurement of power emissions of gravitational waves in Hénon–Heiles systems, particularly trying to deal with short time series.

# References

1. Thorne, K.S. (1987) Gravitational Radiation. In *300 Years of Gravitation*, S.W. Hawking, W. Israel (eds), Cambridge University Press, 330–458.

2. Kokubun, F. (1998) Gravitational Waves from the Hénon–Heiles System. *Physical Review D*, **57**, 2610–2612.

3. Takens, F. (1981) Detecting Strange Attractors in Turbulence. *Lecture Notes in Mathematics*, **898**, 366–381.

4. Packard, N.H., Crutchfield, J.P., Farmer, J.D. and Shaw, R.S. (1980) Geometry from a Time Series. *Physical Review Letters*, **45**, 712–716.

5. Fraser, A.M. (1989) Reconstructing Attractors from Scalar Time Series: a Comparison of Singular System and Redundancy Criteria. *Physica D*, **34**, 391–404.

6. Kennel, M.B., Brown, R. and Abarbanel, H.D.I. (1992) Determining Embedding Dimension for Phase-Space Reconstruction using a Geometrical Construction. *Physical Review A*, **45**, 3403–3411.

7. Grassberger, P. and Procaccia, I. (1983) Measuring the Strangeness of Strange Attractors. *Physica D*, **9**, 189–208.

8. Oseledec, V.I. (1968) A Multiplicative Ergodic Theorem. Lyapunov Characteristic Numbers for Dynamical Systems. *Transactions of Moscow Mathematical Society*, **19**, 197–231.

9. Eckmann, J.-P. and Ruelle, D. (1985) Ergodic Theory of Chaos and Strange Attractors. *Review of Modern Physics I*, **57**, 617–656.

10. Hegger, R., Kantz, H. and Schreiber, T. (1999) Practical Implementation of Nonlinear Time Series Methods: The TISEAN Package. *Chaos*, **9**, 413–435.

11. Eckmann, J.-P., Oliffson Kamphorst, S. and Ruelle, D. (1987) Recurrence Plots on Dynamical Systems. *Europhysics Letters*, **4**, 973.

12. Zbilut, J.P., Giuliani, A. and Webber, C.L., Jr. (1998) Recurrence Quantification Analysis and Principal Components in the Detection of Short Complex Signals. *Phys. Lett. A*, **237**, 131–135.

13. Provenzale, A., Smith, L.A., Vio, R. and Murante, G. (1992) Distinguishing Between Low-Dimensional Dynamics and Randomness in Measured Time Series. *Physica D*, **58**, 31.

# PART III

# Estimation of the Attractor for an Uncertain Epidemic Model

E. Crück,[1] N. Seube[2] and G. Leitmann[3]

[1] *Laboratoire de Recherches Balistiques et Aérodynamiques*
*BP 914, 27207 Vernon Cedex, France*
*Tel: +33 02 27 24 44 26, Fax: +33 02 27 24 44 96*
*E-mail: e.cruck@mageos.com*
[2] *Ecole Nationale Supérieure des Ingénieurs des Etudes et Techniques d'Armement*
*29806 BREST Cedex, France*
[3] *College of Engineering, University of California*
*Berkeley CA 94720, USA*

We study an uncertain model of communicable disease with the assumption of constant screening rate. For a given screening rate, we wish to determine the residual ratio of infected population. We prove that this ratio can be characterized by the universal attractor of an uncertain dynamical system. Then, we provide upper and lower estimates of this set which can be computed by using numerical tools devoted to viability kernel approximation.

## 1 Introduction

Many diseases, such as gonorrhea, syphilis, and tuberculosis, cannot be totally eradicated, due to several limitations of public health policies. The aim of this paper is to study an uncertain model of communicable disease, with the assumption of constant screening rate. For a given screening rate, we characterize the residual ratio of infected population. We use two population models of dynamic evolution of the disease under parameter uncertainty, similar to the one used in [1]. In particular, it takes into account temporal variation of disease communication.

We shall estimate the attractor of the uncertain dynamics modelling gonorrhea evolution by using tools from viability theory. We shall prove that the endemic stage of gonorrhea can be characterized by the universal attractor of an uncertain dynamical system. Furthermore, we shall provide upper and lower estimates of this set, which we can compute by using numerical tools devoted to viability kernel approximation.

## 2   Problem Statement

### 2.1   Modeling gonorrhea transmission

The model used in this paper is similar to the one used in [1]. We consider the spread of gonorrhea among the sexually mature male and female populations.

A model without intervention is:

$$\begin{cases} I_1' = a_1 I_2 \left(1 - I_1\right) - d_1 I_1 \,, \\ I_2' = a_2 I_1 \left(1 - I_2\right) - d_2 I_2 \,, \end{cases} \tag{1}$$

where $I_1$ and $I_2$ denote the fraction of infected. Here $a_1$, $a_2$, $d_1$, $d_2$ are uncertain parameters; that is

$$\begin{aligned} a_1(\cdot) \in \mathcal{D}\left([0,\overline{a_1}]\right) \quad &\text{and} \quad a_2(\cdot) \in \mathcal{D}\left([0,\overline{a_2}]\right) \,, \\ d_1(\cdot) \in \mathcal{D}\left([\underline{d_1},\overline{d_1}]\right) \quad &\text{and} \quad d_2(\cdot) \in \mathcal{D}\left([\underline{d_2},\overline{d_2}]\right) \,, \end{aligned} \tag{2}$$

where $\overline{a_1} > 0$, $\overline{a_2} > 0$, $0 < \underline{d_1} < \overline{d_1}$, and $0 < \underline{d_2} < \overline{d_2}$ are known constants, and $\mathcal{D}\left([\alpha,\beta]\right)$ is the set of measurable functions ranging the interval $[\alpha,\beta]$.

With intervention, the model becomes

$$\begin{cases} I_1'(t) = a_1(t)I_2(t)\left(1 - I_1(t)\right) - d_1(t)I_1(t) - s_1(t)b_1(t)I_1(t) \,, \\ I_2'(t) = a_2(t)I_1(t)\left(1 - I_2(t)\right) - d_2(t)I_2(t) - s_2(t)b_2(t)I_2(t) \,, \end{cases} \tag{3}$$

where $s_1(t)$ and $s_2(t)$ are controls representing screening rates for population 1 and 2. Denoting by $N_1(t)$ and $N_2(t)$ the sizes of the populations, $N_1(t)s_1(t)$ and $N_2(t)s_2(t)$ are the numbers of, respectively, individuals of each sex to be screened at time $t$. Limitations in facilities and medical staff induce a control constraint; that is

$$0 \leq s_1(t) \leq \overline{s_1} \quad \text{and} \quad 0 \leq s_2(t) \leq \overline{s_2} \,, \tag{4}$$

where $\overline{s_1}$ and $\overline{s_2}$ are known bounds. The screening efficiency coefficients $b_1(t)$ and $b_2(t)$ are also allowed to be uncertain, that is:

$$b_1(\cdot) \in \mathcal{D}\left([\underline{b_1},\overline{b_1}]\right) \quad \text{and} \quad b_2(\cdot) \in \mathcal{D}\left([\underline{b_2},\overline{b_2}]\right) \,, \tag{5}$$

where $0 < \underline{b_1} < \overline{b_1}$ and $0 < \underline{b_2} < \overline{b_2}$ are known constants.

The problem we address in this paper is the design of a realistic control strategy which guarantees convergence to a prescribed region of the state space. This strategy takes into account some constraints: The state vector $I = (I_1(t), I_2(t))^T$ cannot be measured reliably. Therefore, *the control must be designed in open loop.* Furthermore, the implementation of a time-varying screening policy $(s_1(t), s_2(t))$ requires infrastructure and a staff that can be adjusted throughout time. This is clearly not realistic. Therefore, following [1], we shall investigate the effects of constant screening rates.

Set $s_1(t) = s_1$ and $s_2(t) = s_2$ satisfying (4). Then the controlled system can be written as [1]

$$\begin{cases} I_1' &= a_1 I_2 (1 - I_1) - c_1 I_1 &:= f_1(I_1, I_2, a_1, c_1), \\ I_2' &= a_2 I_1 (1 - I_2) - c_2 I_2 &:= f_2(I_1, I_2, a_2, c_2), \end{cases} \tag{6}$$

where $c_i(t) = d_i(t) + s_i b_i(t)$. Hence,

$$c_1(\cdot) \in \mathcal{D}\left(\left[\underline{c_1}, \overline{c_1}\right]\right) \quad \text{and} \quad c_2(\cdot) \in \mathcal{D}\left(\left[\underline{c_2}, \overline{c_2}\right]\right) \tag{7}$$

with

$$\begin{cases} \underline{c_1} = \underline{d_1} + s_1 \underline{b_1} \quad \text{and} \quad \underline{c_2} = \underline{d_2} + s_2 \underline{b_2}, \\ \overline{c_1} = \overline{d_1} + s_1 \overline{b_1} \quad \text{and} \quad \overline{c_2} = \overline{d_2} + s_2 \overline{b_2}, \end{cases} \tag{8}$$

and, as before,

$$a_1(\cdot) \in \mathcal{D}\left([0, \overline{a_1}]\right) \quad \text{and} \quad a_2(\cdot) \in \mathcal{D}\left([0, \overline{a_2}]\right). \tag{9}$$

System (6) can be written as a differential inclusion, that is:

$$I' \in F(I) := F_1(I) \times F_2(I), \tag{10}$$

where $I = (I_1, I_2)^T$ is the state vector, and the set-valued dynamics are given by

$$F_1(I_1, I_2) = \bigcup_{\substack{a_1 \in [0, \overline{a_1}] \\ c_1 \in [\underline{c_1}, \overline{c_1}]}} f_1(I_1, I_2, a_1, c_1),$$

$$F_2(I_1, I_2) = \bigcup_{\substack{a_2 \in [0, \overline{a_2}] \\ c_2 \in [\underline{c_2}, \overline{c_2}]}} f_2(I_1, I_2, a_2, c_2).$$

We shall denote by $S_F(I_0)$ the set of all the absolutely continuous solutions to differential inclusion (10) starting at $I_0$. Notice that $I(\cdot) \in S_F(I_0)$ means that there exists a set of admissible measurable functions $a_1(\cdot)$, $a_2(\cdot)$, $c_1(\cdot)$, and $c_2(\cdot)$ such that $I(\cdot)$ is the solution to (6) starting at $I_0$.

It is well known that absolutely continuous solutions of (1) coincide with absolutely continuous solutions of (10).

As the sets of admissible parameters are closed and convex, we know that the dynamics $F$ is Marchaud.[2] Therefore the differential inclusion (10) admits absolutely continuous solutions.

In the following, we shall assume that the control rates $s_1$ and $s_2$ are fixed.

Let $\mathcal{T}$ be the region of admissible states, that is

$$\mathcal{T} = \{(I_1, I_2) : 0 \leq I_1 \leq 1 \text{ and } 0 \leq I_2 \leq 1\}.$$

As stated in [1], $\mathcal{T}$ is invariant for system (10) in the following sense:

---

[1]Notice that system with constant control (6) is similar to system without control (1).

[2]A set-valued map is said to be Marchaud if it is upper semi-continuous, with convex and compact images, and with linear growth (see [2]).

**Definition 1 (Invariant set and invariance kernel)** *Let $K$ be a closed subset of $\mathbf{R}^n$. We shall say that the set $K$ is* invariant *under $F$ if for all initial conditions $I_0 \in K$, all solutions $I(\cdot) \in S_F(I_0)$ to (10), verify*

$$\forall t \geq 0, \quad I(t) \in K. \tag{11}$$

*If the set $K$ does not enjoy the invariance property, one can consider the set of all initial conditions $I_0 \in K$ that satisfy (11). This set is called the* invariance kernel *of $K$ under $F$, and is denoted by $\mathrm{Inv}_F(K)$.*

From [3] we know that the stability properties of the nominal system of (6) with constant parameters

$$\begin{cases} I_1'(t) &=& a_1 I_2(t)\,(1 - I_1(t)) - c_1 I_1(t), \\ I_2'(t) &=& a_2 I_1(t)\,(1 - I_2(t)) - c_2 I_2(t) \end{cases} \tag{12}$$

are characterized by the ratio $c_1 c_2 / a_1 a_2$:

- If $c_1 c_2 \geq a_1 a_2$, then $I = 0$ is a globally asymptotically stable (GAS) equilibrium point in $\mathcal{T}$.

- Else, $I = 0$ is an unstable equilibrium point, and there exists a GAS equilibrium point $I^* \in \mathcal{T} \setminus \{0\}$, with

$$I_1^* = \frac{a_1 a_2 - c_1 c_2}{a_2(a_1 + c_1)} \quad \text{and} \quad I_2^* = \frac{a_1 a_2 - c_1 c_2}{a_1(a_2 + c_2)}. \tag{13}$$

From [1] we know that if $\underline{c_1 c_2} \geq \overline{a_1 a_2}$, then $I = 0$ is a GAS equilibrium point in $\mathcal{T}$ for system (6). Otherwise, $I = 0$ is an unstable equilibrium point, and all solutions to (10) converge towards a bounded set in $\mathcal{T}$.

The aim of this paper is to characterize a non-conservative estimate of the attractor for system (10). In Section 2 we characterize analytically the set of equilibria of the differential inclusion (10) and derive some of its properties. It appears that this subset of $\mathcal{T}$ will play a crucial role for the determination of the attractor. In Section 3 we provide an estimate of the attractor of the differential inclusion (10). This estimation is based on the invariance envelope of an auxiliary subset that we can compute by using numerical tools from viability theory.

## 3 The Set of Equilibria

From the analysis of the nominal system (i.e., with constant parameters), we know that we can associate with each set of constant parameters $(a_1^0, a_2^0, c_1^0, c_2^0)$ a GAS equilibrium point $I^0(a_1^0, a_2^0, c_1^0, c_2^0)$ satisfying

$$f_1(I_1^0, I_2^0, a_1^0, c_1^0) = f_2(I_1^0, I_2^0, a_2^0, c_2^0) = 0.$$

Obviously, this point does not need to be a universal equilibrium point for system (6) in the sense that it does not satisfy

$$\begin{cases} f_1(I_1^0, I_2^0, a_1, c_1) &=& 0, \\ f_2(I_1^0, I_2^0, a_2, c_2) &=& 0 \end{cases}$$

for all constant and admissible parameters $(a_1, a_2, c_1, c_2)$. However, each $I^0(a_1^0, a_2^0, c_1^0, c_2^0)$ can be viewed as the limit of at least one trajectory originating in $\mathcal{T}$.

**Lemma 1** *Let be $I^0 = (I_1^0, I_2^0)$ a GAS equilibrium point associated with a set of admissible parameters $(a_1^0, a_2^0, c_1^0, c_2^0)$. From each $I^1 \in \mathcal{T}$, there exists a solution $I(\cdot) \in S_F(I^1)$ such that $\lim\limits_{t \to +\infty} I(t) = I^0$.*

**Proof:** The set of uncertain parameters $a_1(t)$, $a_2(t)$, $c_1(t)$, and $c_2(t)$ may take on the constant values

$$\begin{cases} a_1(t) \equiv a_1^0, & c_1(t) \equiv c_1^0, \\ a_2(t) \equiv a_2^0, & c_2(t) \equiv c_2^0. \end{cases}$$

Since the parameters are constant, $I^0$ is a GAS equilibrium point for this realization of the uncertain system (6). $\qquad\square$

Let us denote by $E$ the closed set of equilibrium points of system (6); that is

$$E := \bigcup_{(a_1, a_2, c_1, c_2) \in P} \left\{ \left( \frac{a_1 a_2 - c_1 c_2}{a_2(a_1 + c_1)}, \frac{a_1 a_2 - c_1 c_2}{a_1(a_2 + c_2)} \right) \right\}, \tag{14}$$

where $P$ denotes the subset of $[0, \overline{a_1}] \times [0, \overline{a_2}] \times [\underline{c_1}, \overline{a_1}] \times [\underline{c_2}, \overline{a_2}]$ such that $c_1 c_2 \leq a_1 a_2$.

It is possible to find an analytical characterization of the boundary of $E$. Indeed, $E = \{I : 0 \in F(I)\}$, and we observe that

$$\min_{a_1, c_1 \in P} F_1(I_1, I_2) = -\overline{c_1} I_1 \quad \text{and} \quad \max_{a_1, c_1 \in P} F_1(I_1, I_2) = \overline{a_1} I_2 (1 - I_1) - \underline{c_1} I_1$$

$$\min_{a_2, c_2 \in P} F_2(I_1, I_2) = -\overline{c_2} I_2, \quad \text{and} \quad \max_{a_2, c_2 \in P} F_2(I_1, I_2) = \overline{a_2} I_1 (1 - I_2) - \underline{c_2} I_2.$$

Since $F$ has convex values, we deduce that

$$0 \in F_1(I_1, I_2) \iff \overline{a_1} I_2 (1 - I_1) - \underline{c_1} I_1 \geq 0,$$

$$0 \in F_2(I_1, I_2) \iff \overline{a_2} I_1 (1 - I_2) - \underline{c_2} I_2 \geq 0.$$

Hence,

$$E = \left\{ (I_1, I_2) \mid I_1 \geq \frac{\underline{c_2}}{\overline{a_2}} \frac{I_2}{1 - I_2} \text{ and } I_2 \geq \frac{\underline{c_1}}{\overline{a_1}} \frac{I_1}{1 - I_1} \right\}. \tag{15}$$

**Proposition 1** *Let $K$ be a closed subset of $\mathcal{T}$. Suppose that $K$ is invariant under the differential inclusion (10). Then $E \subset K$.*

**Proof:** Let $I_* \in \text{Int}(E)$ and $I_0 \in K$. From Lemma (1), we know that there exists $I(\cdot) \in S_F(I_0)$ such that $\lim\limits_{t \to +\infty} I(t) = I_*$. As $K$ is invariant under $F$, $\forall t > 0$, $I(t) \in K$. Therefore, $I_* \in \overline{K} = K$. We deduce that $\text{Int}(E) \subset K$. Since $E$ and $K$ are closed, we have $E \subset K$. $\qquad\square$

By using the same argument we used in the proof of Lemma (1), we can characterize the invariance kernel of the set of equilibria.

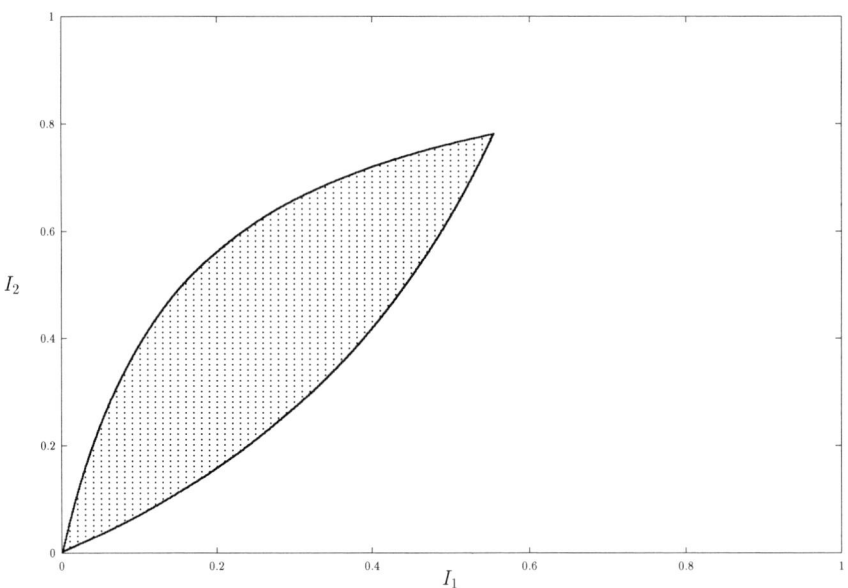

Figure 1  The set of equilibria $E$ ($\overline{a_1} = 0.04$, $\underline{c_1} = 0.025$, $\overline{a_2} = 0.08$, $\underline{c_2} = 0.0125$).

**Proposition 2** *The invariance kernel of $E$ under the differential inclusion (10) satisfies*

$$\mathrm{Inv}_F(E) = E \qquad \text{or} \qquad \mathrm{Inv}_F(E) = \emptyset.$$

**Proof:** Assume that $\mathrm{Inv}_F(E) \neq \emptyset$. As $\mathrm{Inv}_F(E)$ is invariant, we know, by proposition (1) that $E \subset \mathrm{Inv}_F(E)$. Since $\mathrm{Inv}_F(E) \subset E$, we deduce that $\mathrm{Inv}_F(E) = E$. $\square$

From proposition (2), we deduce a criterion for characterizing the invariance kernel of $E$. This criterion is straightforward; namely if there exists a trajectory which leaves $E$, then $\mathrm{Inv}_F(E) = \emptyset$; otherwise, $\mathrm{Inv}_F(E) = E$.

Actually, we can exhibit a trajectory leaving the set $E$. For this purpose, let

$$\partial E_1 := \left\{ I_1, I_2 \mid \quad I_1 = \frac{c_2 I_2}{\overline{a_2}(1 - I_2)} \text{ and } I_2 \geq \frac{\underline{c_1}}{\overline{a_1}} \frac{I_1}{1 - I_1} \right\}, \tag{16}$$

$$\partial E_2 := \left\{ I_1, I_2 \mid \quad I_2 = \frac{\underline{c_1} I_1}{\overline{a_1}(1 - I_1)} \text{ and } I_1 \geq \frac{\underline{c_2}}{\overline{a_2}} \frac{I_2}{1 - I_2} \right\}. \tag{17}$$

Then $\partial E = \partial E_1 \cup \partial E_2$. Set $(I_1, I_2) \in (\partial E_1 \setminus \{0\})$; then $\displaystyle\sup_{y_2 \in F_2(I_1, I_2)} y_2 = 0$. The outward normal vector on $\partial E_1$ is $\nu = \left( -1, \dfrac{\underline{c_2}}{\overline{a_2}} \dfrac{1}{(1 - I_2)^2} \right)^T$. Hence,

$$\sup_{(y_1, y_2) \in F(I_1, I_2)} \{\nu.(y_1, y_2)^T\} = \sup_{y_1 \in F_1(I_1, I_2)} \{-y_1\} + \frac{\underline{c_2}}{\overline{a_2}} \frac{1}{(1 - I_2)^2} \sup_{y_2 \in F_2(I_1, I_2)} y_2 =$$

$$= \underline{c_1 I_1} > 0. \tag{18}$$

Therefore, there exists at least one trajectory leaving $E$ from $(I_1, I_2)$. We deduce the following:

**Proposition 3** *The invariance kernel of the set of equilibria of system (10) is empty.*

**Remark:** In deriving proposition (3), we actually proved that at every point of $\partial E_1 \setminus \{0\}$, there exists a trajectory leaving $E$. The same result holds true for $\partial E_2 \setminus \{0\}$.

We now state a finite time convergence result, that will be very useful for characterizing the attractor of system (10). For the sake of concision, its proof is postponed to the Appendix.

**Proposition 4** *Let $\varepsilon > 0$, and define $B(E, \varepsilon) := \{I \in \mathcal{T} \mid d(I, E) \leq \varepsilon\}$. Then, there exists a finite time $T$ such that for all solutions $I(\cdot)$ of (10), originating from $\mathcal{T}$, $I(T) \in B(E, \varepsilon)$.*

For a given $\varepsilon > 0$, $\mathcal{B}(E, \varepsilon)$ does not enjoy the invariance property, but any invariant set containing $\mathcal{B}(E, \varepsilon)$ entraps all solutions to (10) in the sense that every trajectory reaches this set in finite time and remains inside hereafter. Furthermore, if we can find a "small" invariant set containing $\mathcal{B}(E, \varepsilon)$ for some sufficiently small $\varepsilon$, it will be a good estimate of the attractor.

## 4    Estimation of the Attractor

**Definition 2** *Let $x(\cdot)$ be a function from $[0, +\infty[$ to $\boldsymbol{R}^n$. We say that the subset*

$$\omega(x(\cdot)) := \bigcap_{T \geq 0} \mathrm{cl}(x([T, +\infty[))$$

*of its cluster points when $t \to +\infty$ is the limit set of $x(\cdot)$. We denote by $\omega_F(K)$ the set of cluster points of all solutions of (10) starting from $K$.*

*The universal attractor of $\mathcal{T}$ under $F$ is the closed subset $\mathcal{A}_F(\mathcal{T}) := \overline{\omega_F(\mathcal{T})}$.*

*Let $\mathcal{C} \in \mathcal{T}$ be a closed set. Then $\mathcal{C}$ attracts the point $I_0 \in \mathcal{T}$ if and only if*

$$\forall I(\cdot) \in S_F(I_0), \quad \lim_{t \to +\infty} d(I(t), \mathcal{C}) = 0.$$

*The set of all points attracted by $\mathcal{C}$ is called the basin of attraction of $\mathcal{C}$. It will be denoted by $\mathrm{Attr}_F(\mathcal{C})$.*

**Proposition 5** *$\mathcal{A}_F(\mathcal{T})$ is the smallest closed set $K$ verifying $\mathrm{Attr}_F(K) = \mathcal{T}$.*

**Proof:** Clearly, $\mathrm{Attr}_F(\mathcal{A}_F(\mathcal{T})) = \mathcal{T}$. Let us show that any closed subset $K$ such that $\mathrm{Attr}_F(K) = \mathcal{T}$ is a superset of $\mathcal{A}_F(\mathcal{T})$.

Let $I_0 \in \mathcal{A}_F(\mathcal{T})$. Then $I_0$ is a cluster point of at least one trajectory, say $I(\cdot)$. Hence, there exists a sequence $(t_n)_{n \in N}$, such that $\lim_{n \to +\infty} I(t_n) = I_0$. As $K$ attracts $\mathcal{T}$, $\lim_{n \to +\infty} d_K(I(t_n)) = 0$.

Then, for all $\varepsilon > 0$, there exists $N > 0$ such that

$$\forall n > N, \quad I(t_n) \in \mathcal{B}(I_0, \varepsilon) \quad \text{and} \quad I(t_n) \in \mathcal{B}(K, \varepsilon).$$

Hence, for all $\varepsilon > 0$, $\mathcal{B}(I_0, \varepsilon) \cap \mathcal{B}(K, \varepsilon) \neq \emptyset$. Hence, since $K$ is a closed set, we conclude that $I_0 \in K$. Therefore, $\mathcal{A}_F(\mathcal{T}) \subset K$. $\qquad\square$

The existence of a universal attractor provides a good characterization of the asymptotic behavior of an uncertain system. Actually, it characterizes the asymptotic behavior of *all trajectories*. We can prove that system (10) admits a universal attractor of $\mathcal{T}$.

As in [1] we define

$$\mathcal{R}(\eta) := \{(I_1, I_2) \in \mathcal{T} \mid V(I_1, I_2) \leq \eta\},$$

where $V(I_1, I_2) = I_1^2 + \dfrac{\overline{a_1}}{\underline{a_2}} I_2^2$.

It is proved in [1] that there exist a strictly positive $\eta^*$ such that all trajectories of system (6) reach the subset $\mathcal{R}(\eta^*)$ and remain inside hereafter.

Let $\varepsilon > 0$. Since $\dot{V}$ is negative definite outside of $\mathcal{R}(\eta^*)$, $\text{Int}(\mathcal{R}(\eta^* + \varepsilon))$ is invariant. Furthermore, every trajectory of system (6) enters $\mathcal{R}(\eta^* + \varepsilon)$ in finite time. We know that $F$ is Marchaud, and we have found an open and bounded subset of $\mathcal{T}$ which is invariant under $F$ and such that every trajectory of system (10) reaches it in finite time. Theorem (6.2.16) in [4, p. 239] enables us to conclude that the universal attractor of $\mathcal{T}$ under $F$ exists and is equal to $\omega_F(\mathcal{R}(\eta^* + \varepsilon))$. Furthermore, since all solutions to system (10) converge to $\mathcal{R}(\eta^*)$, their cluster points lie within $\mathcal{R}(\eta^*)$. We deduce that $\mathcal{A}_F(\mathcal{T}) \subset \mathcal{R}(\eta^*)$.

We have now a characterization of $\mathcal{A}_F(\mathcal{T})$: it is the set $\omega_F(\mathcal{R}(\eta^* + \varepsilon))$ for a given $\varepsilon > 0$. Its approximation would require the computation of all the trajectories of system (10) originating in $\mathcal{R}(\eta^* + \varepsilon)$. Here, we propose another approach based on proposition (4) and the notion of *invariance envelope* introduced in [5].

**Definition 3** *Let $K$ be a subset of the domain of a set-valued map $F$. The smallest closed subset invariant under $F$ containing $K$ is the invariance envelope of $K$ under $F$. It is denoted by $\text{Env}_F(K)$.*

Using the results proved in [5], we know that, if $F$ is Lipschitz with non-empty closed values and if $K \subset \mathbf{R}^2$ is a closed subset such that $K = \overline{\text{Int}(K)}$, then

$$\text{Env}_F(K) = \overline{\mathbf{R}^2 \setminus \text{Inv}_{-F}\left(\hat{K}\right)} \quad \text{where} \quad \hat{K} := \overline{\mathbf{R}^2 \setminus K}, \tag{20}$$

$$\text{and} \quad \text{Env}_F(K) = \overline{R_F(K)}, \tag{21}$$

where $\overline{R_F(K)}$ denotes the set of points reachable from $K$.

Thanks to (20), it is possible to compute the invariance envelope of a set by using the invariance kernel algorithm [2]. In Figure 2, a numerical approximation of the invariance envelope of the set of equilibria is displayed.

Characterization (21) provides information about trajectories of (10). Indeed, every trajectory reaches $\mathcal{B}(E, \varepsilon)$ in finite time; and $\text{Env}_F(\mathcal{B}(E, \varepsilon))$ contains the

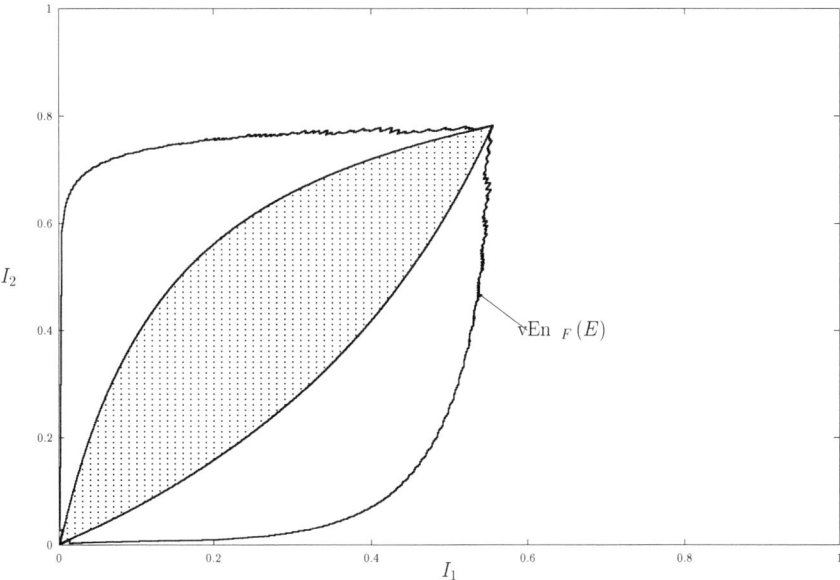

**Figure 2**  The set of equilibria $E$ and its invariance envelope ($\overline{a_1} = 0.04$, $\underline{c_1} = 0.025$, $\overline{a_2} = 0.08$, $\underline{c_2} = 0.0125$).

points reachable from $\mathcal{B}(E, \varepsilon)$. Therefore, $\mathrm{Env}_F\,(\mathcal{B}(E, \varepsilon))$ is a superset of the attractor. Actually, we can derive an upper estimate and a lower estimate of the universal attractor of system (10).

*Let us consider an uncertain system defined on a compact set $\mathcal{T}$ and described by the differential inclusion*

$$x' \in F(x) := \bigcup_{d \in \mathcal{D}} f(x, d) \qquad (22)$$

*where $d$ denotes a disturbance ranging in the compact set $D$. We assume that $F$ is Marchaud. Furthermore, we assume that*

*(i)  For any constant disturbance $d \in \mathcal{D}$, system $x' = f(x, d)$ admits a globally asymptotically stable equilibrium point.*

*(ii)  For any $\varepsilon > 0$, for any initial condition $x_0 \in \mathcal{T}$ and for all solutions $x(\cdot) \in S_F(x_0)$, there exists a finite time $T$ such that $x(T) \in \mathcal{B}\,(E, \varepsilon)$, where $E$ denotes the set of equilibrium points, $E := \{x \mid \{0\} \in F(x)\}$.*

*Then the universal attractor of $\mathcal{T}$ under $F$ exists and can be estimated by*

$$\mathrm{Env}_F\,(E) \subset \mathcal{A}_F\,(\mathcal{T}) \subset \bigcap_{\varepsilon > 0} \mathrm{Env}_F\,(\mathcal{B}(E, \varepsilon))\ .$$

**Proof:** Let us prove the right-hand-side inclusion. Let $\varepsilon > 0$ and $x^* \in \mathcal{A}_F(\mathcal{T})$. Then, there exist $x(\cdot)$ and a time sequence $(t_n)_n$ such that $\lim\limits_{t \to +\infty} x(t_n) = x^*$. There-fore, there exists a time $T$ such that $x(T) \in \text{Env}_F(\mathcal{B}(E, \varepsilon))$. We deduce that $\forall t \geq T$, $x(t) \in \text{Env}_F(\mathcal{B}(E, \varepsilon))$. Therefore, every cluster point of $x(\cdot)$ lies within $\text{Env}_F(\mathcal{B}(E, \varepsilon))$. Hence, for all $\varepsilon > 0$, $\mathcal{A}_F(\mathcal{T}) \subset \text{Env}_F(\mathcal{B}(E, \varepsilon))$.

We now prove the left-hand-side inclusion. We denote by $k$ the Lipschitz constant of $F$. From the Filippov theorem [7, Corollary 1, p. 121] we deduce that

$$\forall x_0 \in \mathcal{T}, \quad \forall \delta > 0, \quad \forall y_0 \in \mathcal{B}(x_0, \delta), \quad \forall y \in S_F(y_0),$$

$$\exists x(\cdot) \in S_F(x_0) \quad \text{such that} \quad \forall t \geq 0, \quad \|x(t) - y(t)\| \leq e^{kt}\delta. \tag{23}$$

Let $x_0 \in \text{Env}_F(E)$. Let us prove that $x_0$ is a cluster point of a solution to (22). We define four sequences $(x_n)_n$, $(x_n(\cdot))_n$, $(y_n)_n$, $(y_n(\cdot))_n$ and $T_n$, and the time $T$ by:

- $T_0 = 0$. Thanks to (21), we can define $y_0 \in E$, $y_0(\cdot) \in S_F(y_0)$ and $T > 0$ such that $x_0 = y_0(T)$.

- For $n \in \mathbf{N}^*$,

    - thanks to Lemma (1), we can define a solution to (22) $x_n(\cdot) \in S_F(x_{n-1})$, a point $y_n \in \mathcal{B}(y_0, \exp(-kT)/n)$ and a finite time $T_n > 0$, such that $x_n(T_n) = y_n$;
    - thanks to (23), we can define a solution of (22) $y_n(\cdot) \in S_F(y_n)$, and a point $x_n \in \mathcal{B}(x_0, 1/n)$ such that $y_n(T) = x_n$.

Let us consider the trajectory of system (22) starting at $x_0$ and such that:

$$
\begin{array}{rcccl}
0 & \leq & t & \leq & T_1 \\
T_1 & \leq & t & \leq & T_1 + T \\
T_1 + T & \leq & t & \leq & T_1 + T_2 + T \\
T_1 + T_2 + T & \leq & t & \leq & T_1 + T_2 + 2T
\end{array}
\qquad
\begin{array}{l}
x(t) = x_1(t) \\
x(t) = y_1(t - T_1) \\
x(t) = x_2(t - (T_1 + T)) \\
x(t) = y_2(t - (T_1 + T_2 + T))
\end{array}
$$

$$\vdots \qquad\qquad\qquad\qquad \vdots$$

$$\sum_{i=1}^{n-1} T_i + (n-1)T \leq t \leq \sum_{i=1}^{n} T_i + (n-1)T$$

$$x(t) = x_n(t - (\sum_{i=1}^{n-1} T_i + (n-1)T))$$

$$\sum_{i=1}^{n} T_i + (n-1)T \leq t \leq \sum_{i=1}^{n} T_i + nT$$

$$x(t) = y_n(t - (\sum_{i=1}^{n} T_i + (n-1)T))$$

For integer $n$ set $t_n = \sum_{i=1}^{n} T_i + nT$. Then $\forall n \in \mathbf{N}, x(t_n) = x_{n+1} \in \mathcal{B}(x_0, 1/(n+1))$.

Hence, $\lim\limits_{n \to \infty} x(t_n) = x_0$. Which means that $x_0$ is a cluster point of $x(\cdot)$. $\qquad\square$

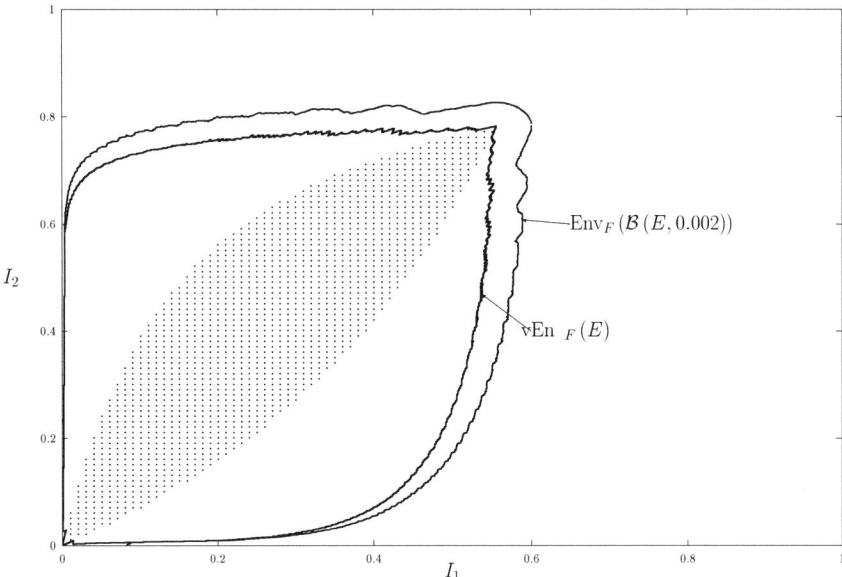

**Figure 3** Numerical approximation of $\mathcal{A}_F\left(\mathcal{T}\right)$ ($\overline{a_1} = 0.04$, $\underline{c_1} = 0.025$, $\overline{a_2} = 0.08$, $\underline{c_2} = 0.0125$).

Thanks to theorem (1), we approximate numerically the universal attractor of system (10). This has been done by using a numerical approximation scheme of invariance envelopes based on the viability kernel algorithm. In Figure 3, the universal attractor of system (10) is plotted for a given level of uncertainty on the system.

# 5 Conclusion

We have proved that the attractor of the uncertain system modeling a disease evolution law can be estimated by using the invariance envelope of the set of equilibria. Our approach does not require knowledge of a Lyapunov function. By using numerical algorithms from viability theory we can approximate estimates of the attractor for a given screening rate. One could therefore determine what screening policy is desirable, in examining the attractor graph versus screening rate.

# Appendix: Proof of Proposition 4

Let us set $\phi_1(I_1, I_2) := \overline{a_1}I_2\left(1 - I_1\right) - \underline{c_1}I_1$ and $\phi_2(I_1, I_2) := \overline{a_2}I_1\left(1 - I_2\right) - \underline{c_2}I_2$. Let $\alpha_1 > 0$ and $\alpha_2 > 0$. Define (see Figure 4)

$$\mathcal{C}_1 = \left\{(I_1, I_2), \quad \phi_1(I_1, I_2) = -\alpha_1\right\},$$

$$\mathcal{C}_2 = \left\{(I_1, I_2), \quad \phi_2(I_1, I_2) = -\alpha_2\right\}.$$

$\mathcal{C}_1$ and $\mathcal{C}_2$ divide $\mathcal{T}$ into four domains:

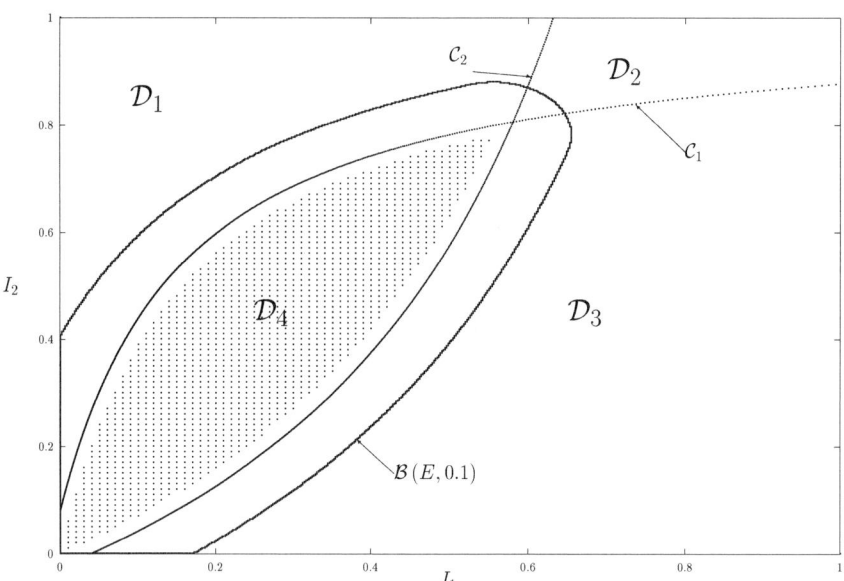

Figure 4  Definition of subsets $\mathcal{D}_1$, $\mathcal{D}_2$, $\mathcal{D}_3$, and $\mathcal{D}_4$ for proof of Proposition 4 ($\overline{a_1} = 0.04$, $\underline{c_1} = 0.025$, $\overline{a_2} = 0.08$, $\underline{c_2} = 0.0125$).

$$\mathcal{D}_1 := \{(I_1, I_2), \quad \phi_1(I_1, I_2) \geq -\alpha_1 \text{ and } \phi_2(I_1, I_2) \leq -\alpha_2\}$$

$$\mathcal{D}_2 := \{(I_1, I_2), \quad \phi_1(I_1, I_2) \leq -\alpha_1 \text{ and } \phi_2(I_1, I_2) \leq -\alpha_2\}$$

$$\mathcal{D}_3 := \{(I_1, I_2), \quad \phi_1(I_1, I_2) \leq -\alpha_1 \text{ and } \phi_2(I_1, I_2) \geq -\alpha_2\}$$

$$\mathcal{D}_4 := \{(I_1, I_2), \quad \phi_1(I_1, I_2) \geq -\alpha_1 \text{ and } \phi_2(I_1, I_2) \geq -\alpha_2\}$$

Assume that $\alpha_1 > 0$ and $\alpha_2 > 0$ are such that $\mathcal{D}_4 \subset B(E, \varepsilon)$. We observe that $\forall (I_1, I_2) \in \mathcal{D}_1 \cup \mathcal{D}_2$, $I_2' \leq -\alpha_2$ and that $\forall (I_1, I_2) \in \mathcal{D}_2 \cup \mathcal{D}_3$, $I_1' \leq -\alpha_2$.

Let $I^0 \in \mathcal{D}_1 \cup \mathcal{D}_2$ and $I(\cdot) \in S_F(I_0)$, a solution to (10). Assume that there exists $T > 0$ such that $\forall t \leq T, I(t) \in \mathcal{D}_1 \cup \mathcal{D}_2$. Now, $\forall t \leq T, I_2'(t) \leq -\alpha_2$, and then $\forall t \leq T, I_2(t) \leq I_0 - \alpha_2 t$. As $I'$ is bounded and $\mathcal{T}$ is invariant under $F$, $I(t)$ has to reach $\mathcal{C}_2$ in a finite time $\tau$. Either $I(\tau) \in \mathcal{D}_4$ and then $I(\tau) \in B(E, \varepsilon)$, or $I(\tau) \in \mathcal{D}_3$.

Similarly, we can prove that every trajectory originating in $\mathcal{D}_2 \cup \mathcal{D}_3$ reaches $\mathcal{C}_1$ in finite time. We deduce that if a solution $I(\cdot)$ to (10) never reaches $\mathcal{D}_4$, then it has to cross $\mathcal{C}_1$ and $\mathcal{C}_2$ an infinite number of times. That is, there exists a time sequence $(t_i)_{i=1,N}$ such that,

$$\forall p \in \mathbf{N}, \quad \left\{ \begin{array}{l} I(t_{2p}) \in \mathcal{C}_1 \cap \mathcal{D}_3 \\ I(t_{2p+1}) \in \mathcal{C}_2 \cap \mathcal{D}_1 \end{array} \right. \text{ and } \left\{ \begin{array}{l} \forall t \in [t_{2p}, t_{2p+1}], \quad I(t) \in \mathcal{D}_2 \cup \mathcal{D}_3, \\ \forall t \in [t_{2p+1}, t_{2p+2}], \quad I(t) \in \mathcal{D}_2 \cup \mathcal{D}_1. \end{array} \right.$$

Now we prove that this trajectory cannot lie outside of $B(E, \varepsilon)$ forever.

Let $p \in \mathbf{N}$. Let $t_p^* := \sup_{t \in [t_{2p}, t_{2p+1}]} \{t : I(t) \in \mathcal{D}_3\}$. Because $I(\cdot)$ is continuous, we know that $I(t^*) \in \mathcal{C}_2$, and $\forall \in [t_p^*, t_{2p+2}]$, $I(t) \in \mathcal{D}_2 \cup \mathcal{D}_1$. Notice that $I_2(t^*) < I_2(t_{2p})$.

Indeed, $I_1(t_p^*) \leq I_1(t_{2p}) - \alpha_1(t_p^* - t_{2p})$, and a point of $\mathcal{C}_2$ is given by $\left(I_1, \dfrac{\alpha_2 + \overline{a_2}I_1}{\overline{a_2}I_1 + c_2}\right)$, which means that $I_2$ decreases if $I_1$ decreases. Furthermore, $I_2(t_{2p+2}) \leq I_2(t_p^*) - \alpha_2(t_{2p+2} - t_p^*)$. We deduce that $I_2(t_{2p+2}) \leq I_2(t_{2p}) - \alpha_2(t_{2p+2} - t_p^*)$ and that $I_1(t_{2p+2}) < I_1(t_{2p})$. Hence, the sequence $(I_2(t_{2p})_{p \in \mathbb{N}})$ is decreasing strictly. Moreover, it admits a limit only if $t_{2p+2} - t_p^* \to 0$. But $t_{2p+2} - t_p^*$ is the reaching time from $\mathcal{C}_1$ to $\mathcal{C}_2$. This means that $I(t) \to \mathcal{C}_1 \cap \mathcal{C}_1 \in \text{Int}\,(B(E, \varepsilon))$. Hence, there exists a finite time $T$ such that $I(T) \in B(E, \varepsilon)$. $\qquad\qquad\square$

# References

1. Corless, M. and Leitmann, G. (2000) Analysis and Control of a Communicable Disease. *Nonlinear Analysis*, **40** (1), 145–172.

2. Aubin, J.-P. (1991) *Viability Theory*. Birkhauser, Boston.

3. Lajmanovich, A. and Yorke, J.A. (1976) A Deterministic Model for Gonorrhea in a Nonhomogeneous Population. *Mathematical Biosciences*, **28**, 221–236.

4. Aubin, J.-P. (1998) *Dynamic Economy Theory – A Viability Approach*. Springer-Verlag, Berlin.

5. Quincampoix, M. (1992) Enveloppes d'Invariance pour des Inclusions Différentielles Lipschitziennes: Applications aux Problèmes de Cibles [Invariance Envelope for Lipschitzean Differential Inclusions]. *Comptes-Rendus de l'Academie des Sciences I*, vol. 318, pp. 343–347, Paris.

6. Quincampoix, M. and Saint-Pierre, P. (1998) An Algorithm for Viability Kernels in Holderian Case: Approximation by Discrete Dynamical Systems. *Journal of Mathematical Systems, Estimation and Control*, **8** (1), 17–29.

7. Aubin, J.-P. and Cellina, A. (1984) *Differential Inclusions*. Springer-Verlag, Berlin.

8. Hethcote, H.W. and Yorke, J.A. (1985) Gonorrhea Dynamics Transmission and Control. *Lecture Notes in Biomathematiques*, No. 56, Springer-Verlag, Berlin.

9. Yorke, J.A., Hethcote, H.W. and Nold, A. (1978) Dynamics and Control of the Transmission of Gonorrhea. *Sexually Transmissible Diseases*, **5** (2), 51–56.

# Liar Paradox Viewed by the Fuzzy Logic Theory

Ye-Hwa Chen

*The George W. Woodruff School of Mechanical Engineering*
*Georgia Institute of Technology, Atlanta, Georgia 30332, USA*
*Tel: +1 404 894 3210, Fax: +1 404 894 9342*
*E-mail: yehwa.chen@me.gatech.edu*

The liar paradox problem is studied within the fuzzy context. By tracing the argument which really explains why the liar paradox is a paradox, we suggest a framework to analyze it. This involves the analysis of the truth values of fuzzy implications. The result is compared with Zadeh's proposal, which is the only past work on the same subject.

## 1  Introduction

The modern version of the classic liar paradox centers around the following (self-referential) sentence: *This sentence is false.* If the sentence is true, then it is false. If the sentence is false, then *this sentence is false* is false. This must mean the sentence is true. The effort of invoking the classical two-valued logic and later three-valued logic in resolving this paradox is well summarized in [1] and [2, Chap. 9] and their bibliographies.

Zadeh [3] is the first in casting the liar paradox into the fuzzy framework. With his unique translation that the liar paradox is a proposition $p$ such that $p := p \ is \ false$, Zadeh proposed the truth-qualification principle. The principle in turn provides a mechanism for the computation of the possibility distribution of the truth of $p$.

It is interesting to note that, despite significant progress in the fuzzy community on many other subjects in latest years, Zadeh's proposal in resolving the liar paradox via fuzzy logic still remains the only work.

The paper endeavors to retrace the origin of the paradox. By examining the argument which really explains why the liar paradox is a paradox, a new framework which invokes the truth values of fuzzy implications is suggested. Within the framework, we reevaluate the status of the liar paradox and its legitimacy. A comparison with Zadeh's result is made. The framework is applicable to both self-referential and non-self-referential forms of the liar paradox.

## 2   Preliminaries

Let $X$ be a variable taking values in a universe of discourse $U \subseteq \mathbf{R}$. Let $F$ be a normal fuzzy set in $U$ which is characterized by the membership function $\mu_F : U \to [0,1]$. If $u$ is a point in $U$, $\mu_F(u)$ is the grade of membership of $u$. The possibility that $X$ may take the value $u$ is a number in $[0,1]$, denoted by $\pi_X(u)$, which is assigned numerically equal to $\mu_F(u)$. That is,

$$Possibility\,(X = u) = \pi_X(u) = \mu_F(u) \quad \forall u \in U\,. \tag{2.1}$$

The mapping $\pi_X : U \to [0,1]$ is called the possibility distribution function.

Let $p$ be a proposition of the form

$$X \quad is \quad F\,. \tag{2.2}$$

We write

$$\pi_X(u) = \mu_F(u) \tag{2.3}$$

as a translation of (2.2).

The simplest form of the liar paradox can be represented by

$$p \quad is \quad (p \ is \ \tau) \tag{2.4}$$

where $\tau$ is a truth qualifier (such as *true, false, very true, not quite true, more or less true*, etc., each of which corresponds to a fuzzy set on $[0,1]$). This is called the liar paradox in the self-referential form.

Let $\pi : U \to [0,1]$ denote a possibility distribution function. Let also $\pi_\tau : [0,1] \to [0,1]$ denote the possibility distribution function of a truth qualifier $\tau$ on $[0,1]$. The possibility distribution of $p$ *is* $\tau$ is given by $\pi_\tau(\pi_p(u))$, $u \in U$.

We say that the proposition $p$ *is* $\tau$ is truth-qualified. The truth value (or the degree of truth) of the proposition for each $u \in U$ is given by $\pi_\tau(\pi_p(u))$.

As an example, $X$ may stand for *Susan's age*, $F$ may stand for *young*, and $\tau$ may stand for *very true*. Thus $p$ *is* $\tau$ may stand for *Susan's age is young is very true* or in short, *Susan is young is very true*.

A fuzzy complement is a mapping $c(\cdot) : [0,1] \to [0,1]$ such that $c(0) = 1$ and $c(1) = 0$. This is often used to negate a fuzzy set. A fuzzy implication is a mapping $I : [0,1] \times [0,1] \to [0,1]$ such that $I(1,1) = I(0,1) = I(0,0) = 1$, $I(1,0) = 0$. This is often used to assign the truth value to a fuzzy conditional proposition. A summary of various forms of fuzzy complement and fuzzy implication can be found in, e.g., [4, p. 309]. We note that a number of axioms are often used to legitimize the implications (and different implications are based on different axioms). However, they are not needed in the following work. Therefore as far as the analysis of the current paper is concerned, none of the axioms is addressed. They are, however, important in understanding the basis of fuzzy implications.

Consider a mapping $f : D \to D$ where $D \subset \mathbf{R}^n$. A point $x \in D$ is called a *fixed point* if $f(x) = x$.

# 3 Zadeh's Interpretation and the Truth-Qualification Principle

In [3], the self-referential form of the liar paradox (2.4) is interpreted as

$$p := p \quad is \quad \tau, \tag{3.1}$$

where := stands for "is defined to be." This interpretation is the basis for any further development. Based on this, Zadeh proposed the truth-qualification principle which asserts that for (3.1)

$$\pi_p(u) = \pi_\tau(\pi_p(u)), \qquad u \in U. \tag{3.2}$$

This in turn implies that $\pi_p(u)$, which denotes the value of the mapping $\pi_p(\cdot)$ at a given $u \in U$, is a fixed point of the mapping $\pi_\tau(\cdot)$. For example, suppose $\pi_\tau(v) = v$ (and hence $\tau$ denotes *true*), then (3.2) is

$$\pi_p(u) = \pi_p(u) \tag{3.3}$$

which holds for all $u \in U$. This means that the proposition *p is true* holds for any $u$ such that the proposition $p$ holds.

For another example, suppose that $\tau$ denotes *false* with $\pi_\tau(v) = 1 - v \ \forall v \in U$ (and hence a special case of the fuzzy complement $c(\cdot)$), then (3.2) is

$$\pi_p(u) = 1 - \pi_p(u). \tag{3.4}$$

This, of course, only holds for the $u \in U$ (if it exists) such that $\pi_p(u) = {}^1/_2$.

# 4 An Alternative Interpretation

Zadeh's interpretation is based on the direct translation of (2.4) to (3.1). In this section, we shall endeavor to explore an alternative interpretation.

Our starting point is the root of the paradox. Consider the special case that $\tau$ denotes *false*. Eq. (2.4) is

$$p \quad is \quad (p \ is \ false). \tag{4.1}$$

The paradox occurs when one tries to decide if $p$ is *true* or *false* by the "IF...THEN" proposition (i.e., the conditional proposition).

If $p$ is true, then by (4.1), *p is false* is true. This in turn means $p$ is false. If $p$ is false, then by (4.1), *p is false* is false. This in turn means $p$ is true. In summary, we have the following two conditional propositions:

$$If \ p \ is \ true, \quad then \ p \ is \ false; \tag{4.2}$$

$$If \ p \ is \ false, \quad then \ p \ is \ true. \tag{4.3}$$

On the surface, there may appear to be two layers of paradox in (4.2) and (4.3). The first appears at each of the two conditional propositions: Both *p is true* and *p is false* are in one proposition. The second appears as one relates (4.2) to (4.3). That is, the existence of two (seemingly) inconsistent conditional propositions.

We now propose a new aspect to interpret (4.2) and (4.3). The possibility distribution of the proposition $p$ *is true* is $\pi_p(u)$, $u \in U$. Also, by adopting the fuzzy complement: $\pi_\tau(v) = c(v)$, $v \in U$, we propose the possibility distribution of the proposition $p$ *is false* to be $\pi_\tau(\pi_p(u)) = c(\pi_p(u))$, $u \in U$.

Next, for a fuzzy implication $I : [0,1] \times [0,1] \to [0,1]$, the truth value of the conditional proposition (4.2) is given by $I(\pi_p(u), c(\pi_p(u)))$ [4]. Similarly, the truth value of the conditional proposition (4.3) is given by $I(c(\pi_p(u)), \pi_p(u))$. We now claim that (4.2) and (4.3), which are interpretations of (4.1), must be equally true in the sense that their truth values are identical. That is,

$$I(\pi_p(u), c(\pi_p(u))) = I(c(\pi_p(u)), \pi_p(u)), \qquad \forall u \in U. \tag{4.4}$$

This is justified by the fact that both (4.2) and (4.3) are derived from (4.1). Their status is symmetrical in the sense that one provides as much truth as the other. Our problem is to solve (4.4) for $\pi_p(u)$. The solution $\pi_p(u)$ then serves as the truth value of $p$ that justifies the paradox (4.1).

**Theorem 1** *Suppose that there is a $u \in U$ such that*

$$c(\pi_p(u)) = \pi_p(u). \tag{4.5}$$

*Then the $\pi_p(u)$ solves (4.4).*

**Proof:** The proof follows directly from (4.4) that as

$$c(\pi_p(u)) = \pi_p(u),$$

$$I(\pi_p(u), c(\pi_p(u))) = I(\pi_p(u), \pi_p(u)) = I(c(\pi_p(u)), \pi_p(u)). \qquad \square$$

**Remark:** This result is independent of the use of any particular form of the fuzzy implication $I(\cdot)$.

*Example.* Consider the Sugeno class of fuzzy complements [4]:

$$c(a) = \frac{1-a}{1+\lambda a}, \qquad \lambda \in (-1, \infty). \tag{4.6}$$

We can express (4.5) as

$$\pi_p(u) = \frac{1 - \pi_p(u)}{1 + \lambda \pi_p(u)}. \tag{4.7}$$

To solve for $\pi_p(u)$, suppose first that $\lambda \neq 0$, then

$$\pi_p(u) + \lambda \pi_p^2(u) = 1 - \pi_p(u) \tag{4.8}$$

or

$$\lambda \pi_p^2(u) + 2\pi_p(u) - 1 = 0. \tag{4.9}$$

This leads to

$$\pi_p(u) = \frac{1}{\lambda}(-1 \pm \sqrt{1 + \lambda}). \tag{4.10}$$

Since $\pi_p(u) \geq 0$, we neglect the minus sign in front of the square root and hence

$$\pi_p(u) = \frac{1}{\lambda}\left(-1 + \sqrt{1 + \lambda}\right). \tag{4.11}$$

The value of $\pi_p(u)$ is dependent on $\lambda$. For example, as $\lambda = 1$, then $\pi_p(u) = -1 + \sqrt{2} \approx 0.41$.

If $\lambda = 0$, then (4.7) leads to

$$\pi_p(u) = \frac{1}{2}. \tag{4.12}$$

This is the same as [3].

Suppose, instead, we adopt the Yager class of fuzzy complement [4]:

$$c(a) = (1 - a^w)^{1/w}, \qquad w \in (0, \infty). \tag{4.13}$$

We can express (4.5) as

$$\pi_p(u) = (1 - \pi_p^w(u))^{1/w} \tag{4.14}$$

or

$$\pi_p(u) = 2^{-1/w}. \tag{4.15}$$

The value of $\pi_p(u)$ is dependent on $w$. For example, if $w = 2$, then $\pi_u(u) = 2^{-1/2} \approx 0.71$. The result in [3] corresponds to $w = 1$ which leads to $\pi_p(u) = 2^{-1} = 0.5$.

# 5  The Formal Setting: The $\alpha$–$\beta$–$\tau$ Legitimacy

We now proceed to the formal setting. Consider truth qualifiers $\alpha$, $\beta$, and $\tau$. The self-referential form of the liar paradox is given by

$$p \quad is \quad (p \ is \ \tau). \tag{5.1}$$

One intends to decide if $p$ is $\alpha$ or $\beta$.

This setting subsumes the previous section as a special case. For example, if $\tau$ and $\alpha$ both denote *false* and $\beta$ denotes *true*, then this setting returns to the previous section. Notice that the paradox has a strong connection with what one wants to decide (i.e., if $p$ is $\alpha$ or $\beta$).

We proceed with the analysis of (5.1) by a similar procedure. If $p$ is $\alpha$, then $p$ *is $\tau$ is $\alpha$*. This in turn means the truth value of this fuzzy conditional proposition is given by $I(\pi_\alpha(\pi_p(u)), \pi_\alpha(\pi_\tau(\pi_p(u))))$. If $p$ is $\beta$, then $p$ *is $\tau$ is $\beta$*. This in turn means the truth value of this fuzzy proposition is given by $I(\pi_\beta(\pi_p(u)), \pi_\beta(\pi_\tau(\pi_p(u))))$. Upon arguing that the truth values of both conditional propositions are equal, we have

$$I(\pi_\alpha(\pi_p(u)), \pi_\alpha(\pi_\tau(\pi_p(u)))) = I(\pi_\beta(\pi_p(u)), \pi_\beta(\pi_\tau(\pi_p(u)))). \tag{5.2}$$

**Definition 1**    *(i) The liar paradox (5.1) is $\alpha$–$\beta$–$\tau$ legitimate if there is at least one solution $\pi_p(u)$, where $u \in U$, to (5.2).*

*(ii) Suppose that the liar paradox (5.1) is $\alpha$–$\beta$–$\tau$ legitimate. The value(s) of $I$ in (5.2) is the $I$-truth value(s) of the liar paradox.*

**Remark:** The legitimacy of (5.1) is subject to which $\tau$, $\alpha$, and $\beta$ one wishes to consider. It is possible, that for given $\tau$, (5.1) is legitimate under some $\alpha$ and $\beta$ but not under some others. Under this definition, the liar paradox problem addressed in Section 4 is called *true–false–false* legitimate. Also, the truth value of the liar paradox depends on the specific form of $I(\cdot)$ that is used.

**Lemma 1**    *(i) The liar paradox (5.1) is always $\alpha$–$\alpha$–$\tau$ legitimate.*

*(ii) If the liar paradox (5.1) is $\alpha$–$\beta$–$\tau$ legitimate, then it is also $\beta$–$\alpha$–$\tau$ legitimate.*

**Proof:** (i): Obvious. (ii): Only $\alpha$ and $\beta$ are reversed in (5.2).    □

To be more specific about the existence of the solution, we now turn to a few special choices of $\alpha$, $\beta$, and $\tau$. Let us first consider that $\alpha$ denotes *true* (and hence $\pi_\alpha(v) = v$) and $\beta$ denotes $\tau$ (and hence $\pi_\beta(v) = \pi_\tau(v)$). Equation (5.2) is reduced to

$$I(\pi_p(u), \pi_\tau(\pi_p(u))) = I(\pi_\tau(\pi_p(u)), \pi_\tau(\pi_\tau(\pi_p(u)))). \qquad (5.3)$$

The liar paradox problem considered in the previous section is in fact a special case of the current consideration by specializing $\tau$ to be *false*. For the solution of (5.3), the following theorem (see, e.g., [5]) is instrumental.

**Theorem 2** *Any continuous mapping $\pi_\tau : [0,1] \to [0,1]$ has at least one fixed point.*

**Remark:** This is in fact a special case of the well-known Brouwer fixed-point theorem (see, e.g., [5]). Although the theorem does not provide any explicit expression for the fixed point, it is a standard numerical task to solve for it. Methods such as bisection and Newton–Raphson can be used.

**Theorem 3** *Suppose that the mapping $\pi_p : U \to [0,1]$ is surjective, i.e., onto (hence for any $k \in [0,1]$, there is at least a $u \in U$ such that $\pi_p(u) = k$). If $\pi_\tau(\cdot) : [0,1] \to [0,1]$ is continuous, then Eq. (5.3) has at least one solution, which is the $\pi_p(u)$ such that*

$$\pi_\tau(\pi_p(u)) = \pi_p(u). \qquad (5.4)$$

*This in turn means that (5.1) is true–$\tau$–$\tau$ legitimate.*

**Proof:** By Theorem 2, there is at least a fixed point, say $\epsilon$, of the mapping $\pi_\tau : [0,1] \to [0,1]$. This means that, since $\pi_p(\cdot)$ is surjective, for the $u \in U$ such that $\pi_p(u) = \epsilon$, we have

$$I(\pi_p(u), \pi_\tau(\pi_p(u))) = I(\epsilon, \pi_\tau(\epsilon)) = I(\epsilon, \epsilon), \qquad (5.5)$$

$$I(\pi_\tau(\pi_p(u)), \pi_\tau(\pi_\tau(\pi_p(u)))) = I(\pi_\tau(\epsilon), \pi_\tau(\pi_\tau(\epsilon))) = I(\epsilon, \pi_\tau(\epsilon)) = I(\epsilon, \epsilon). \qquad (5.6)$$

Hence (5.3) holds. $\qquad\qquad\qquad\qquad\qquad\qquad\qquad\qquad\qquad\qquad\square$

**Remark:** In Zadeh's work in [3], the fixed point(s) of the mapping $\pi_\tau(\cdot)$, which is the sufficient condition for (5.3), was suggested to be the solution (to the truth-qualification principle). The current result coincides with Zadeh's in this sense. We note, however, that Zadeh's reasoning started with (3.1) (instead of (5.2)), which was in fact his rather unique interpretation of the liar paradox. One cannot be more amazed by Zadeh's insight!

The fixed point, which is a *sufficient* condition for the legitimacy, does not need to be *necessary*. This is illustrated as follows.

*Example.* Consider that $\pi_p(u) = \sin u$, $u \in [0, \pi]$ and

$$\pi_\tau(\xi) = \begin{cases} \xi^2, & 0 \le \xi \le 0.1\,, \\ 0.01 + 0.9(\xi - 0.1)\,, & 0.1 < \xi \le 1\,. \end{cases} \tag{5.7}$$

The mapping $\pi_p : [0, \pi] \to [0, 1]$ is surjective. The mapping $\pi_\tau : [0, 1] \to [0, 1]$ is continuous. Consider the Yager implication [4, p. 309]

$$I(a, b) = \begin{cases} 1 & if \ a = b = 0\,, \\ b^a & otherwise\,. \end{cases} \tag{5.8}$$

The fixed point is $\pi_p(u) = 0$ (note that $\pi_\tau(0) = 0$), which is a solution to (5.3) since

$$I(0, 0) = I(0, 0)(= 1)\,. \tag{5.9}$$

In addition, the point $\pi_p(u) = 0.2$ (and hence $\pi_\tau(0.2) = 0.1$, $\pi_\tau(\pi_\tau(0.2)) = 0.01$) also solves (5.3). This is easily shown, since at this point, (5.3) is

$$I(0.2, 0.1) = I(0.1, 0.01) = 0.63\,. \tag{5.10}$$

We next turn to another special choice of $\alpha$–$\beta$–$\tau$: let $\alpha$ be $\tau$ and $\beta$ be $c(\tau)$. Here $c(\tau)$ stands for the fuzzy complement [4] of $\tau$. For example, if $\tau$ denotes *more or less true*, then $c(\tau)$ denotes *not more or less true*. In the special case that $\tau$ denotes *true* and $c(v) = 1 - v$ (which is, of course, a way to denote *false*), the interpretation of if $p$ in (5.1) is $\tau$ or $c(\tau)$ is the same as if $p$ in (4.1) is *true* or *false*.

Equation (5.2) is reduced to

$$I(\pi_\tau(\pi_p(u)), \pi_\tau(\pi_\tau(\pi_p(u)))) = I(c(\pi_\tau(\pi_p(u))), c(\pi_\tau(\pi_\tau(\pi_p(u)))))\,. \tag{5.11}$$

**Theorem 4** *Suppose that $\pi_p : U \to [0, 1]$ is surjective and $\pi_\tau : [0, 1] \to [0, 1]$ is continuous. If $I : [0, 1] \times [0, 1] \to [0, 1]$ is such that*

$$I(l_1, l_1) = I(l_2, l_2) = \cdots, \qquad \forall l_1, l_2 \cdots \in [0, 1]\,, \tag{5.12}$$

*then the equation (5.11) has at least one solution, which is the fixed point of $\pi_\tau(\cdot)$. This in turn means that (5.1) is $\tau$–$c(\tau)$–$\tau$ legitimate.*

**Proof:** Let the fixed point of $\pi_\tau(\cdot)$ be denoted by $\epsilon$. Under this fixed point, Eq. (5.11) is given as (note that $\pi_\tau(\epsilon) = \epsilon$, $\pi_\tau(\pi_\tau(\epsilon)) = \pi_\tau(\epsilon) = \epsilon$)

$$I(\epsilon, \epsilon) = I(c(\epsilon), c(\epsilon)),\tag{5.13}$$

which always holds by (5.12).                                                    □

**Remark:** It is possible in general that $c(\epsilon) \neq \epsilon$.

*Example.* Suppose that $\pi_\tau(v) = v^2$, $v \in [0, 1]$ (and hence $\tau$ is *very true*), $\pi_p(u) = \sin u$, $u \in U = [0, \pi]$. The mapping $\pi_\tau(\cdot)$ is continuous, and the mapping $\pi_p(\cdot)$ is surjective. There is a fixed point of the mapping $\pi_\tau(\cdot)$, which is $v = 1$. This occurs as $u = \pi/2$. Consider the Gödel implication [4, p. 305]

$$I(a, b) = \begin{cases} 1 & if \ a \leq b, \\ b & if \ a > b. \end{cases}\tag{5.14}$$

It is obvious that $1 = I(l_1, l_1) = I(l_2, l_2) = \cdots$ for all $l_1, l_2 \cdots \in [0, 1]$. Hence the liar paradox is $\tau - c(\tau) - \tau$ legitimate and its $I$-truth value is 1. Note that the result is independent of any specific form of $c(\cdot)$.

# 6   Non-Self-Referential Liar Paradox

The previous framework was devoted to the self-referential form of the liar paradox. There are other forms of the liar paradox that contain no self-reference. Consider, for example, the following form:

$$\begin{aligned} X \quad is \quad & (Y \ is \ false), \\ Y \quad is \quad & (X \ is \ true). \end{aligned}\tag{6.1}$$

As was mentioned in the last section, the paradox is closely related to what one wants to decide about $X$ and $Y$. Let us say we wish to consider if $X$ and $Y$ are *true* or *false*. First, consider that $X$ is true. This means that $Y$ *is false* is true. That is, $Y$ is false. However, what $Y$ says is that $X$ *is true*. Therefore, we conclude that, since $Y$ is false, $X$ *is true* is false. That is, $X$ is false.

Second, consider that $X$ is false. This means that $Y$ *is false* is false. That is, $Y$ is true. However, what $Y$ says is that $X$ *is true*. Therefore, we conclude that, since $Y$ is true, $X$ *is true* is true. That is, $X$ is true. In summary, we have the following:

$$\begin{aligned} If \ X \ is \ true, \quad & then \ X \ is \ false; \\ If \ X \ is \ false, \quad & then \ X \ is \ true. \end{aligned}\tag{6.2}$$

Let the possibility distribution of $X$ be denoted by $\pi_X(u)$ (and hence $X$ is false is $c(\pi_X(u))$). We can interpret (6.2) as

$$I(\pi_X(u), c(\pi_X(u))) = I(c(\pi_X(u)), \pi_X(u)).\tag{6.3}$$

The reasoning of the true or false of $Y$ can be carried out in a similar way. We can also conclude that

$$If\ X\ is\ true\,,\quad then\ Y\ is\ false\,;$$
$$If\ Y\ is\ false\,,\quad then\ X\ is\ true\,,$$

(6.4)

and hence

$$I(\pi_Y(u), c(\pi_Y(u))) = I(c(\pi_Y(u)), \pi_Y(u))\,.$$

(6.5)

Here $\pi_Y(u)$ denotes the possibility distribution of $Y$ (and hence $X$ is false is $c(\pi_X(u))$).

**Theorem 5** *Suppose that there is a $u \in U$ such that*

$$c(\pi_X(u)) = \pi_X(u), \qquad c(\pi_Y(u)) = \pi_Y(u)\,.$$

(6.6)

*Then $\pi_X(u)$ and $\pi_Y(u)$ solve (6.3) and (6.5).*

**Proof:** Similar to Theorem 1 and is omitted. ☐

In the most general setting, the non-self-referential form of the liar paradox can be represented by

$$X \quad is \quad (X\ is\ \gamma)\,,$$
$$Y \quad is \quad (Y\ is\ \delta)\,.$$

(6.7)

Here both $\gamma$ and $\delta$ are truth values. One intends to decide if $X$ and $Y$ are $\alpha$ or $\beta$. If $X$ is $\alpha$, i.e., $Y\ is\ \gamma$ is $\alpha$, then $X\ is\ \delta$ is $\gamma$ is $\alpha$. The truth value of this fuzzy conditional proposition is given by $I(\pi_\alpha(\pi_X(u)), \pi_\alpha(\pi_\gamma(\pi_\delta(\pi_X(u)))))$. If $X$ is $\beta$, i.e., $Y\ is\ \gamma$ is $\beta$, then $X\ is\ \delta$ is $\gamma$ is $\beta$. The truth value of this fuzzy conditional proposition is given by $I(\pi_\beta(\pi_X(u)), \pi_\beta(\pi_\gamma(\pi_\delta(\pi_X(u)))))$. Again, the truth values of both fuzzy conditional propositions are argued to be identical:

$$I(\pi_\alpha(\pi_X(u)), \pi_\alpha(\pi_\gamma(\pi_\delta(\pi_X(u))))) = I(\pi_\beta(\pi_X(u)), \pi_\beta(\pi_\gamma(\pi_\delta(\pi_X(u)))))\,.$$

(6.8)

Similarly, by deciding on whether $Y$ is $\alpha$ or $\beta$, we also arrive at the following:

$$I(\pi_\alpha(\pi_Y(u)), \pi_\alpha(\pi_\delta(\pi_\gamma(\pi_Y(u))))) = I(\pi_\beta(\pi_Y(u)), \pi_\beta(\pi_\delta(\pi_\gamma(\pi_Y(u)))))\,.$$

(6.9)

**Remark:** Note that the order of $\pi_\gamma$ and $\pi_\delta$ is reversed from (6.8) to (6.9).

**Definition 2** *The liar paradox (6.7) is $\alpha$–$\beta$–$\gamma$–$\delta$ legitimate if there is at least one solution pair $(\pi_X(u),\ \pi_Y(u))$, where $u \in U$, to (6.8) and (6.9).*

**Remark:** The liar paradox (6.1) addressed in Theorem 5 is *true–false–false–true legitimate.*

To explore further on the existence of solutions, we consider the special case that both $\alpha$ and $\delta$ denote *true* and both $\beta$ and $\gamma$ denote $\tau$. This consideration renders (6.1) a special case since the latter is nothing but $\tau$ denotes *false*. Equations (6.8) and (6.9) are reduced to

$$I(\pi_X(u), \pi_\tau(\pi_X(u))) = I(\pi_\tau(\pi_X(u)), \pi_\tau(\pi_\tau(\pi_X(u))))\,,$$

(6.10)

$$I(\pi_Y(u), \pi_\tau(\pi_Y(u))) = I(\pi_\tau(\pi_Y(u)), \pi_\tau(\pi_\tau(\pi_Y(u))))\,.$$

(6.11)

**Theorem 6** *Suppose that $\pi_X(u) = \pi_Y(u) \; \forall u \in U$ and the mapping $\pi_X(\cdot)$ is surjective. Suppose also that $\pi_\tau(\cdot)$ is continuous. The fixed point of $\pi_\tau(\cdot)$ is a solution to (6.10) and (6.11). This in turn shows that (6.7) is true–$\tau$–$\tau$–true legitimate.*

**Proof:** Similar to that of Theorem 3 and is omitted. $\qquad\square$

Next we consider another special case, that both $\alpha$ and $\gamma$ denote $\tau$, $\beta$ denotes $c(\tau)$, and $\delta$ denotes *true*. Equations (6.8) and (6.9) are reduced to

$$I(\pi_\tau(\pi_X(u)), \pi_\tau(\pi_\tau(\pi_X(u)))) = I(c(\pi_\tau(\pi_X(u))), c(\pi_\tau(\pi_\tau(\pi_X(u))))), \qquad (6.12)$$

$$I(\pi_\tau(\pi_Y(u)), \pi_\tau(\pi_\tau(\pi_Y(u)))) = I(c(\pi_\tau(\pi_Y(u))), c(\pi_\tau(\pi_\tau(\pi_Y(u))))). \qquad (6.13)$$

**Theorem 7** *Suppose that $\pi_X(u) = \pi_Y(u) \; \forall u \in U$ and the mapping $\pi_X(\cdot)$ is surjective. Suppose also that $\pi_\tau(\cdot)$ is continuous. If $I(\cdot)$ is such that*

$$I(l_1, l_1) = I(l_2, l_2) = \cdots, \qquad \forall l_1, l_2 \cdots \in [0, 1], \qquad (6.14)$$

*then the fixed point of $\pi_\tau(\cdot)$ is a solution to (6.12) and (6.13). This in turn shows that (6.7) is $\tau$–$c(\tau)$–$\tau$–true legitimate.*

**Proof:** Similar to that of Theorem 4 and is omitted. $\qquad\square$

**Remark:** The fixed point(s) in Theorem 7 and 8 only provides a *sufficient* (but not *necessary*) condition for the solution. There may be other solutions to the liar paradox.

**Theorem 8** *(i) If the liar paradox (6.7) is $\alpha$–$\beta$–$\gamma$–$\delta$ legitimate, then it is also $\beta$–$\alpha$–$\gamma$–$\delta$ legitimate.*

*(ii) The liar paradox (6.7) is always $\alpha$–$\alpha$–$\gamma$–$\delta$ legitimate.*

*(iii) Suppose that $\pi_\gamma(\pi_\delta(v)) = v$ and $\pi_\delta(\pi_\gamma(v)) = v$. This means that $\pi_\gamma(\cdot)$ and $\pi_\delta(\cdot)$ are the inverse of each other. If $I(\cdot)$ meets (6.14), then the liar paradox (6.7) is $\alpha$–$\beta$–$\gamma$–$\delta$ legitimate.*

**Proof:** (i): Only reverse $\alpha$ and $\beta$ in (6.8) and (6.9). (ii): Equations (6.8) and (6.9) are reduced to

$$I(\pi_\alpha(\pi_X(u)), \pi_\alpha(\pi_\gamma(\pi_\delta(\pi_X(u))))) = I(\pi_\alpha(\pi_X(u)), \pi_\alpha(\pi_\gamma(\pi_\delta(\pi_X(u))))), \qquad (6.15)$$

$$I(\pi_\alpha(\pi_Y(u)), \pi_\alpha(\pi_\delta(\pi_\gamma(\pi_Y(u))))) = I(\pi_\alpha(\pi_Y(u)), \pi_\alpha(\pi_\delta(\pi_\gamma(\pi_Y(u))))). \qquad (6.16)$$

(iii): Equations (6.8) and (6.9) are reduced to

$$I(\pi_\alpha(\pi_X(u)), \pi_\alpha(\pi_X(u))) = I(\pi_\beta(\pi_X(u)), \pi_\beta(\pi_X(u))), \qquad (6.17)$$

$$I(\pi_\alpha(\pi_Y(u)), \pi_\alpha(\pi_Y(u))) = I(\pi_\beta(\pi_Y(u)), \pi_\beta(\pi_Y(u))). \qquad (6.18)$$

$$\qquad\square$$

**Remark:** The analysis in this paper is generic. It does not depend on any specific choice of fuzzy implications $I(\cdot)$. One might wonder what is then the "best" choice of $I(\cdot)$. The issue, however, is not defined until the associated "cost," which is to be minimized, is given. In the current framework, all fuzzy implications are valid for analysis.

# 7 Conclusion

By tracing the argument which really explains why the liar paradox is a paradox, we suggest a new framework for its analysis. The framework is formulated by invoking the truth values of a fuzzy implication. The problem is to decide the $\alpha$–$\beta$–$\tau$ legitimacy (for the self-referential form) or the $\alpha$–$\beta$–$\gamma$–$\delta$ legitimacy (for the non-self-referential form). In the analysis of the self-referential form, Zadeh's result coincides with the sufficient condition of the *true*–$\tau$–$\tau$ legitimacy. This reaffirms Zadeh's accuracy (and of course insight) in his interpretation of the liar paradox. The framework is general enough that it is also applicable to the non-self-referential form. The main contribution of the paper is, judging from this, a proposal for a new framework. Some other logic issues, such as Russell's Paradox, Santa Sentences, Antistrephon, Size Paradoxes, Game Paradoxes, and Cantor's Paradox (see., e.g., [6]), can be further studied in this framework. This may help to provide new perspectives on some long-lasting debates in this area.

# References

1. Martin, R.L. (1970) *The Paradox of the Liar*. Yale University Press, New Haven and London.

2. Kirkham, R.L. (1995) *Theories of Truth*. The MIT Press, Cambridge and London.

3. Zadeh, L.A. (1979) Liar's Paradox and Truth-Qualification Principle. Electronics Research Laboratory Memorandum M79/34, University of California, Berkeley; also in G.J. Klir, B. Yuan (1996, eds) *Fuzzy Sets, Fuzzy Logic, and Fuzzy Systems: Selected Papers by Lotfi A. Zadeh*. World Scientific Publishing, Singapore, pp. 449–463.

4. Klir, G.J. and Yuan, B. (1995) *Fuzzy Sets and Fuzzy Logic: Theory and Applications*. Prentice-Hall, Upper Saddle River, New Jersey.

5. Marsden, J.E. and Hoffman, M.J. (1993) *Elementary Classical Analysis*, Second Edition. W.H. Freeman and Company, New York.

6. Falletta, N. (1983) *The Paradoxicon*. Doubleday and Company Inc., Garden City, New York.

# Pareto-Improving Cheating
# in an Economic Policy Game

Christophe Deissenberg [1][†]
and Francisco Alvarez Gonzalez [2]

[1] *CEFI, UMR CNRS 6126, Université de la Méditerranée (Aix-Marseille II)*
*Château La Farge, Route des Milles, 13290 Les Milles, France*
*Tel: +(33) 4 42 93 59 93, Fax: +(33) 4 42 38 95 85*
*E-mail: deissenb@univ-aix.fr*
[2] *Dpto. Economia Cuantitativa, Universidad Complutense, Madrid, Spain*

This paper presents a simple repeated-game model of interaction between the government and the private sector where, at each repetition, the government first makes a non-binding announcement about its future actions. The private sector, unsure whether or not this announcement will be respected, either acts (with probability $\pi$) as if it trusted the announcement, or disregards it in its decision-making. After observing the reaction of the private sector, the government implements the actual policy measures. Finally, the private sector updates $\pi$ as a function of the payoff it received. We show that, although they are never respected, the government's announcements may allow reaching an outcome that improves the situation of both players compared to the standard equilibrium solutions. This result is in stark contrast to the conclusions usually presented in the related economic literature.

## 1   Introduction

One of the most ubiquitous and stable characteristics of policy-making appears to be that decision-makers repeatedly make announcements and promises that they later do not respect. Based on previous experience, private agents are fully aware that the promises they hear are unlikely to be kept. Yet, the governmental announcements are not neutral. They have an impact on the private agents' decisions.

---

[†]Corresponding author.

One may wonder why announcements that are suspected from the onset not to be respected are not totally disregarded. In this paper, we analyze a situation where these announcements are taken into account because, although it is known that they will be violated, they contain useful information that may allow the agents to increase their welfare. Specifically, they may help as a device to coordinate a superior outcome that would not otherwise result from the interplay between government and private sector. Clearly, there are presumably many other mechanisms that may explain the real impact of deceitful announcements. These additional or alternative explanations will not be reviewed here.

The approach we suggest – introducing reinforcement learning in a repeated reversed Stackelberg game – can be and has been applied to many other socio-economic problems. We present it here in the context of a modified version of the so-called Kydland–Prescott model – a model that is famed in the economic literature for having popularized a diametrically opposed point of view, namely, that the inability of a government to commit to its announcements necessarily implies a very poor economic outcome.

The paper is organized as follows. We first present the version of the Kydland-Prescott model used in this paper and the main conclusions usually drawn thereof in the related literature. Based on perceived ambiguities in the standard analysis, we then reinterpret the underlying static game between the government and the private sector, and present the concepts of optimal and pareto-improving cheating strategies, that are central to the further analysis. In the next section, the static game is used to define a repeated game between a government that makes optimally false announcements, and a single private agent that learns whether or not to disregard these announcements. The last section concludes.

## 2   The Motivating Model

Numerous variants of the basic Kydland-Prescott model have been proposed, that do not crucially modify its basic message. This paper uses the [2, 3] formulation, as presented in the recent work by [4], to which we refer at later places. In the Stokey formulation, the Kydland–Prescott model is as follows. A government, $L$, is engaged in a one-shot game with a continuum of private agents. There is perfect information. The player's objective functions and constraints are common knowledge. The government attempts to maximize through its choice of $y$ its payoff function:

$$J^L = -\frac{1}{2}\left(U^2 + y^2\right),\tag{1}$$

where $U$ is the unemployment rate and $y$ the inflation rate. That is, the government goal is to minimize the sum of (squared) unemployment $U$ and inflation $y$.

Each private agent tries to maximize through its choice of $\xi$ the payoff function:

$$J^{Fi} = -\frac{1}{2}\left[(y - \xi)^2 + y^2\right],\tag{2}$$

where $\xi$ is the agent's expectation about $y$. All private agents being identical, the average expectation, $x$, is equal to $\xi$. The game between the government $L$ and the private agents reduce then to a two-player game between $L$ and the aggregate private sector $F$, whereby the private sector uses $x$ as instrument and has the payoff function:

$$J^F = -\frac{1}{2}\left[(y-x)^2 + y^2\right] . \tag{3}$$

Thus, the goal of the private sector $F$ is to minimize the sum of (squared) inflation $y$ and of the prediction error $|y-x|$ on inflation.

The unemployment is supposed to depend upon inflation through an expectations augmented Phillips curve:

$$U = U^* - \theta(y-x) , \tag{4}$$

where $U^* > 0$ is the natural rate of unemployment, and $\theta > 0$ is a parameter. This last equation asserts that unemployment $U$ deviates from its natural rate only when the private sector anticipates incorrectly the inflation rate, $x \neq y$. Following [4] and with inconsequential loss of generality, we assume in this paper that $\theta = 1$. Thus, taking into account (4), $J^L$ can be rewritten as:

$$J^L = -\frac{1}{2}\left[(U^* + x - y)^2 + y^2\right] . \tag{5}$$

The optimal reaction functions of $L$ and $F$, that is, $T^L : x \to y$ and $T^F : y \to x$, indicate the best response (in terms of one's own objective function) of $L$ or $F$ for any given action of its opponent. They are given by:

$$T^L \triangleq \arg\max_y J^L = \frac{1}{2}(U^* + x) , \tag{6}$$

$$T^F \triangleq \arg\max_x J^F = y. \tag{7}$$

That is, the best the government can do if the private sector chooses $x$ is to choose $y = \frac{1}{2}(U^* + x)$. Likewise, the best the private sector can do is to predict $y$ if the government chooses $y$, that is, to make a perfect prediction.

There are several standard solution concepts associated with the so-defined game. Two of them have been traditionally studied in the context of the Kydland–Prescott model, namely, the Nash equilibrium ($N$) and the Stackelberg equilibrium with $L$ as a leader ($SL$).

The Nash equilibrium $N$ is given when each player chooses a best response to the action of his opponent, that is, by a pair of mutual best responses. In other words, the Nash equilibrium is defined by the intersection of the best response functions $T^L$ and $T^F$. It is given by:

$$x^N = y^N = U^*, \qquad J^{L,N} \triangleq J^L\left(y^N, x^N\right) = -U^{*2} ,$$

$$J^{F,N} \triangleq J^F\left(y^N, x^N\right) = -\frac{U^{*2}}{2} . \tag{8}$$

The Stackelberg equilibrium with $L$ as a leader $SL$, or Ramsey equilibrium, describes the outcome of a hierarchical situation where $L$ chooses his action $y$ knowing that $F$ will react to this choice with his best response $T^F(y)$, that is, where

$y = \arg\min_y J^L\left(y, T^L\left(y\right)\right)$. Using from now on the notation introduced in (8), this equilibrium is given by:

$$x^{SL} = y^{SL} = 0, \quad J^{L,SL} = -\frac{U^{*2}}{2}, \quad J^{F,SL} = 0. \tag{9}$$

The Ramsey equilibrium can be interpreted as a solution where the government manipulates the private sector's expectations on inflation, knowing that this sector will choose $x = y$ if it chooses an inflation level of $y$. Clearly, this equilibrium strictly pareto-dominates the Nash equilibrium: both players are better off under Ramsey than under Nash. Assume, however, that the government had initially played Ramsey, and that the private sector accordingly expects $x = x^{SL} = 0$. The best answer of the government to $x = 0$ is $y = {}^1\!/_2 U^*$. That is, the government has ex post an incentive to deviate from Ramsey – a property known in the economic literature under the generic heading of "time inconsistency."[1] If the government does deviate, the private sector will revise its expectations, leading to a new best answer by the government, and so on. It is easy matter to show that this iterative process converges towards the Nash equilibrium.

This simple observation led to the pessimistic conclusion that, in the absence of binding commitments that force it to play Ramsey, a government would tend to renege on its previous engagements and act in a way that leads to the inferior Nash solution. An important and influential trend of research used this finding as a justification for advocating strict restrictions on government discretion in economic policy-making.

The story, however, is less clear-cut than the above presentation may have led the reader to believe. In particular, note that in the course of our presentation we departed from the original static, one-shot game description of the problem to argue within a dynamic framework. However, in a dynamic game, the curse of time inconsistency is by no means inevitable, see e.g., [5]. In particular, the Ramsey outcome can be supported by the need for the government to maintain a favorable reputation, see e.g., [6]. Similarly, it can be supported if the private sector has the ability to punish the government whenever the latter deviates, see among others [7]. The time inconsistency problem can also be mitigated by incentive contracts or by delegation, [8, 9], for example. More fundamentally, in a dynamic framework, almost any outcome can be supported as an equilibrium. This includes many outcomes that pareto-dominate both the Nash and the Ramsey equilibria – a point that has not been previously studied in the literature.

Furthermore, note that it is not conceptually straightforward to interpret $y$ as an actual economic policy decision if one defines $x$ as an expectation on this decision. Consider for example the game sequence supporting the Ramsey outcome: $L$ plays first by choosing $y$; $F$, who knows $y$, plays second by making an expectation $x$ on $y$. The expectation is made after the action is realized, clearly a rather contrived scenario. This last observation (and more elaborate versions of it) led diverse authors to argue very early that the original Kydland–Prescott problem is not well-defined, see among others [10–12]. To avoid possible logical contradictions, it is necessary

---

[1]This property is generic for Stackelberg equilibria and reflects the fact that, contrary to a Nash equilibrium, a Stackelberg equilibrium does not correspond to a fixed point in the space of strategies.

to make a clear differentiation between the announcement of a policy measure, $y^a$, and the measure actually implemented, $y$,

As a final remark, notice that with both the Ramsey and the Nash solutions, the private sector makes perfect anticipations, $x = y$. This is not accidental, but captures the assumption that economic agents are sufficiently rational and knowledgeable always to anticipate the future correctly if there is no exogenous uncertainty. Moreover, the Ramsey solution coincides with the unconstrained optimum of $F$. The equality $x = y$, however, is no longer given in the Stackelberg game with $F$ as a leader. That is, the assumption of perfect prediction is not compatible with the natural sequence of the game where anticipations on a variable are made on the basis of the currently available information prior to the realization of this variable. Furthermore, the equality $x = y$ is no longer given either at the Nash nor at the Stackelberg game with $F$ as a leader if one modifies even slightly the players' objectives – as can be seen e.g., by assigning to $F$ the (more) plausible payoff function $J^F = -\frac{1}{2}\left[(y-x)^2 + y^2 + U^2\right]$. Such a modification also destroys the coincidence between the Ramsey solution and the unconstrained optimum of $F$. On the other hand, removing the term $y^2$ from the original payoff-function $J^F$ has no impact on the game outcome. Thus, one may argue that Kydland–Prescott's private sector does not care about the real economic outcome: their only preoccupation is with not being fooled upon, a strong and not innocuous assumption. By contrast, our private agents will be exclusively or at least primarily interested in the real consequences of the game.

One can wonder why, in a repeated game, fully rational and perfectly informed agents like $L$ and $F$ in the Kydland–Prescott model would coordinate on $N$ rather than on some of these superior outcomes. In this paper, we will show that even less ideally rational agents can easily coordinate on superior solutions in a set-up that appears both natural and robust.

We now present the framework used in the remainder of the paper for analyzing the interplay between the government and the private sector. Unless otherwise specified, all assumptions underlying the Kydland–Prescott model presented in the previous section are valid for the modified model.

# 3   The Single-Shot Game

Assume the following sequence of play in the single-shot game between $L$ and $F$:

1. The government announces that it will choose some inflation rate $y^a$. The announcement is cheap talk, in the sense that it is not binding and does not enter as an argument of the payoff functions $J^L$ and $J^F$.

2. After hearing this announcement, the private sector forms an anticipation $x$ on the inflation rate $y$ that will be effectively implemented.

3. Given this anticipation, the government chooses $y$.

In this so-called *reversed Stackelberg game*, $L$ has not one, but two instruments at his disposal, $y^a$ and $y$. The outcome of the game will depend on the way in which $y^a$ influences $F$'s anticipation $x$. We single out two important benchmarks:

- $F$ fully disregards the announcement, $dx/dy^a = 0$. In that case, the announcement has no impact at all, neither on the actions nor on the payoffs. Thus, the natural outcome of the single-shot game is the Stackelberg equilibrium with $F$ as a leader ($SF$):

$$x^{SF} = 0, \qquad y^{SF} = \frac{U^*}{2}, \qquad J^{L,SF} = J^{F,SF} = -\frac{U^{*2}}{4}. \qquad (10)$$

- $F$ believes (or: acts as if he believed) that $L$ will realize $y$ as announced, that is, that $y = y^a$ holds. In that case, $x = T^F(y^a)$. The best $L$ can do in the context of a single-shot game is to maximize, with respect to $y^a$ and $y$, (5) with $x$ replaced by $T^F(y^a)$. The resulting *optimal cheating solution* ($OC$) is given by:

$$y^{a,OC} = -U^*, \qquad x^{OC} = -U^*, \qquad y^{OC} = 0,$$

$$J^{L,OC} = 0, \qquad J^{F,OC} = -\frac{U^{*2}}{2}. \qquad (11)$$

The $OC$ solution was first introduced in the literature by [13]. It was derived and analyzed in the case of linear-quadratic dynamic games in [14, 15], and other papers by the same authors. The term "cheating" refers to the fact that under this solution generically one has $y \neq y^a$, that is, the announcement will not be respected.

The optimal cheating solution $OC$ is extremely attractive for $L$: there is no feasible pair $(x, y)$ under which $L$ fares better than under $OC$. However, $OC$ is not a reasonable candidate for the coordination in a repeated interaction between $L$ and $F$, since playing $x^{SF}$ guarantees $F$ the payoff $J^{F,SF} > J^{F,OC}$. Natural choices are strategies where both $L$ and $F$ have higher payoffs than under $SF$.[2] We are going to show that there are other cheating strategies with this property. Before characterizing these *pareto-improving cheating strategies*, however, a look at the geometry of the one-shot game may be useful.

In Figure 1, thick lines refer to $L$ and thin ones to $F$. The straight lines represent the optimal reaction functions, the curved lines indifference curves (that is, level curves of their respective payoff function). The maximum possible payoffs of the two players are represented by large dots – moving away from these dots in any direction implies decreasing payoffs. The Nash equilibrium $N$ lies at the intersection of the reaction functions. The Stackelberg equilibrium $SL$ ($SF$) is situated at the tangency point of $T^F$ ($T^L$) with an indifference curve of $F$ ($L$). That is, the Stackelberg solution allows the leader ($L$ or $F$) to attain his best point on the reaction curve of his opponent. By contrast, the optimal cheating solution $OC$ allows $L$ to attain his best point on his own reaction function – that is, for the class of problems we are considering here, his unconstrained optimum. The cheating mechanism allowing this outcome is apparent on the figure. An announcement $y^a$ by $L$ incites $F$ to reply with $x = T^F(y^a)$ – see point $A$ on $F$'s reaction curve. Given $x$, $L$ replies with $y = T^L(x)$ – see point $B$ on $L$'s own reaction curve. Since $L$'s reaction curve goes

---

[2]Notice that, since $y$ is now always realized after $x$, $SL$ no longer plays any role in the argumentation.

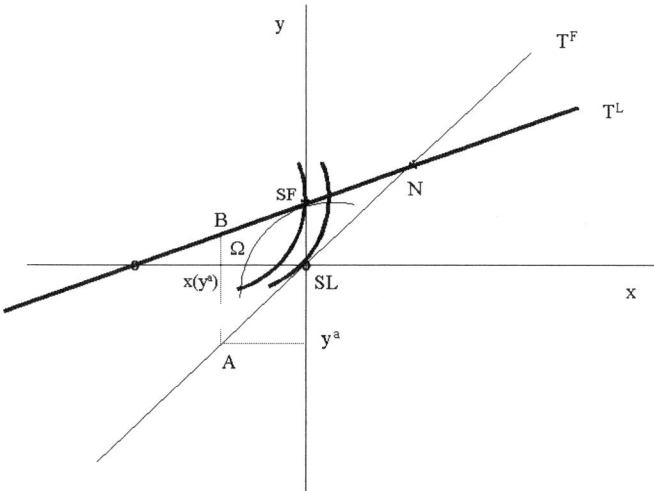

Figure 1 The geometry of the Kydland–Prescott model, $\theta = 1$.

by construction through $L$'s unconstrained maximum, $L$ can insure that the final outcome $B$ coincides with this maximum trough, a proper choice of $y^a$.

Now, neither one of the two equilibria $N$ and $SF$ is pareto-efficient. In particular, there exist feasible points $(x, y)$ that pareto-dominate $SF$. Those are the points situated within the convex lens $\Omega$ delimited by $L$'s and $F$'s indifference curves through $SF$. Any point in $\Omega$ is a candidate for the outcome of coordination over time. Ideally, one might want to restrict the set of candidates to those points within $\Omega$ that are pareto-efficient, that is, to points where indifference curves of $L$ and $F$ are tangent. Without apologies, let us investigate coordination on a simpler set of points $\Gamma$. To characterize $\Gamma$, define:

$$ J^\alpha \triangleq \alpha J^L \left( x^\alpha \left( y^\alpha \right), y \right) + (1 - \alpha) J^F \left( x^\alpha \left( y^\alpha \right), y \right), \quad x^\alpha = T^F \left( y^\alpha \right), \quad (12) $$

and consider the set $O$ given by:

$$ O \triangleq \left\{ (y^{a,\alpha}, x^\alpha, y^\alpha) \ : \ (y^{a,\alpha}, y^\alpha) \triangleq \arg\max_{y^a, y} J^\alpha, \alpha \in [0,1] \right\}. \quad (13) $$

That is, $O$ is the set of the optimal cheating solutions $OC^\alpha$ that would be generated if $L$ based his decisions not on his original payoff function $J^L$, but on a convex combination $J^\alpha$ of $J^L$ and of his opponent's payoff function $J^F$. As $\alpha$ decreases, this injects an increasing dose of altruism into $L$'s actions. For $\alpha = 1$, $OC^\alpha$ coincides with the optimal cheating solution (11). For $\alpha = 0$, it coincides with the unconstrained optimum for $F$, that is here, with the origin.

It is straightforward to prove the following:

**Proposition 1** *There exists a non-empty interval $[\alpha_1, \alpha_2] \subset [0,1]$ such that $(y^\alpha, y^{a,\alpha}, x^\alpha)$ lies within $\Omega$ and thus pareto-dominates $SL$ iff $\alpha \in [\alpha_1, \alpha_2]$, with strict dominance whenever $\alpha_1 < \alpha < \alpha_2$.*

The set $\Gamma$ of pareto-improving cheating solutions is defined as $\Gamma \triangleq \{(y^{a,\alpha}, x^{\alpha}, y^{\alpha}),$ $\alpha \in [\alpha_1, \alpha_2]\}$. As one would expect intuitively, $\alpha_1$ is the unique value of $\alpha$ for which $L$ is indifferent between playing $SL$ or $OC^{\alpha}$, that is for which $J^{L,\alpha} \triangleq J^L(y^{a,\alpha}, x^{\alpha}, y^{\alpha}) = J^{L,SL}$. Similarly, $\alpha_2$ is the unique value of $\alpha$ for which $F$ is indifferent between $SL$ and $OC^{\alpha}$. The values of $\alpha_1$ and $\alpha_2$ do not depend upon $U^*$. They will not be given here. The proposition holds for any game with payoff functions strictly concave in the (scalar) actions.

# 4  The Repeated Game

From the perspective of a single play of the game, the $OC^{\alpha}$ solutions suffer from not being equilibria. Once $F$ has played $x^{\alpha}$ in response to an announcement $y^{a,\alpha}$, $L$ can fare better by playing $T^L(x^{\alpha})$ than $y^{a,\alpha}$. Playing $T^L(x^{\alpha})$, however, leads to a very bad outcome for $F$. Being gullible is very costly. Moreover, in a repeated perspective, the $OC^{\alpha}$ solutions are associated with an apparent irrationality of the private sector: $F$ repeatedly believes, or acts as if he believed, the announcement, although he is regularly cheated.

We shall argue that this is not necessarily the case. For ease of argumentation, and as in the original Kydland–Prescott model, we make the strong assumption that each player knows exactly the objective functions $J^L$ and $J^F$ and the structure of the economy defined by the Phillips curve (4). In particular, $F$ knows $x^{FS}$ and $J^{F,SF}$. However, the private sector is not sure of the relationship between the announced and the realized inflation, $y^a$ and $y$. Will $L$ indeed choose $y^a = y$? Or, at least, will the realized $y$ be such that it is rational for $F$ to choose $x = T^F(y^a)$? In other words, the only incertitude lies in the fact that $F$ does not know for sure the type of government he is facing.

## 4.1  The behavior of the private sector

Notice that, after observing an announcement $y^a$, $F$ has fundamentally two alternatives, denoted $OCA$ (for: Optimal Cheating Accommodate) and $PAL$ (for: Play as A Leader):

1. $OCA$ : He can believe (or act as if he believed) the announcement, and thus choose $x = x^{OCA} = T^F(y^a)$.

2. $PAL$ : He can choose any other value of $x$. In particular, he can choose his best action in the one-shot game, $x^{FS}$. We assume in the following that this is the case.

The $PAL$ alternative so defined essentially mimics dynamics that leads to the inferior Nash equilibrium in [4]. Other answers of $F$ are clearly possible. Later we will give some additional arguments to justify the fact that we are not considering them in this paper.

Consider now the following repeated game: At each repetition

1. $L$ chooses an $\alpha \in [0, 1]$ and announces $y^{a,\alpha}$.

2. $F$ plays $OCA$ with probability $\pi \in (0, 1)$, and $PAL$ with probability $1 - \pi$.

3. $L$ observes the action $x$ of $F$. (i) If $x = x^\alpha$, $L$ plays $y^\alpha$. (ii) If not, he plays his best answer $T^L(x)$.

4. $F$ revises his probability of playing $OCA$ or $PAL$ at the next repetition according to whether or not his realized payoff is higher or lower than $J^{SF}$.

Notice that, under the very simplifying full information assumption made in this paper, the payoffs under 3. (i) will depend uniquely on the value of $\alpha$ chosen by $L$. The payoffs under 3. (ii) are always equal to $J^{L,SF}$ and $J^{F,SF}$.

Specifically, we assume that $\pi$ is updated according to:

$$\pi_{t+1} = \phi\left(\pi_t, \rho_t, \delta_t\right), \qquad \pi_1 = \pi_2 \quad \text{given}, \tag{14}$$

where $\rho_t \triangleq \max_{\tau < t} J_\tau^F - J_t^F$ ($J_t^F$ being the payoff realized by $F$ at the $t$-th repetition), and where $\delta_t = 1$ if $OCA$ is selected at repetition $t$, and $\delta_t = 0$ otherwise. The function $\phi$ is supposed to satisfy:

i) $\phi(\pi, 0, \cdot) = \pi$;

ii) $\phi(\pi, \rho, 1) \lesseqgtr \pi \Leftrightarrow \rho \gtreqless 0$ and $\phi(\pi, \rho, 0) \gtreqless \pi \Leftrightarrow \rho \gtreqless 0$;

iii) $\phi : (0, 1) \to (0, 1)$.

Equation (14) defines a so-called *reinforcement rule*, see e.g., [16–18] for a discussion of reinforcement learning in the context of repeated games. Roughly speaking, this rule implies that a positive experience with $OCA$ ($PAL$) at the current repetition increases (and that a negative experience decreases) the probability with which $F$ will play $OCA$ ($PAL$) in later repetitions. Here, making a positive experience means: obtaining a higher payoff at the current repetition than ever in the past. Thus, the private sector's changes in behavior depend exclusively on the economic outcome, that is, on the values taken by $U$ and $y$. In contrast to the original Kydland–Prescott model and to most of the related literature, cheating, that is, the occurrence of a discrepancy between $y^a$ and $y$, is inconsequential – see our corresponding remarks towards the end of Section 2. Of course, we do not exclude that cheating causes per se real or psychological costs for the private agents. This can easily be taken into account by introducing a corresponding cost term in $F$'s payoff function. As long as this cost term is not overwhelming compared to the other real payoffs, the results we are going to present remain qualitatively valid.

The assumptions i) and ii) on $\phi$ are self-explanatory: favorable outcomes lead the private sector to increase its confidence in the government, whether or not the latter actually cheats (and similarly for unfavorable or neutral outcomes). The assumption iii) ensures that $L$ will never get locked into playing exclusively one of the two alternatives $OCA$ or $PAL$. The assumption is crucial, since it is necessary to insure that the private sector always (a) can be incited to move away from $SF$ to a better equilibrium; and (b) cannot be lastingly lured into an unfavorable solution. In that sense, giving the government credit for favorable outcomes but never fully trusting it is a very strong element of rationality.

This completes the description of $F$'s behavior as an agent with (a) perfect information on all elements of the problem save on the actual choice of $y$, (b) perfect memory, but (c) simple, reinforcement-based adaptive behavior. The assumptions (a) and (b) can easily be weakened to imperfect information and limited memory.

## 4.2 The behavior of the government

In contrast to $F$, the government is supposed to be a maximizing agent. Specifically, we assume that $L$ maximizes at any repetition $t$, through his choice of $\alpha_t$, $\alpha_{t+1}$, ..., $\alpha_{t+h}$, his expected cumulated payoff over a revolving horizon $t + h$, $h \geq 0$ a constant:

$$\sum_{\tau=t}^{t+h} E\left[J_\tau^L\right] \longrightarrow \max_{\{\alpha_\tau\}} \qquad (15)$$

subject to (14).

For $h = 0$, that is, in the case of a myopic government, the problem is trivial. The perfectly myopic government is exclusively interested in maximizing its payoff in the current repetition. Since the payoff under $PAL$ is a constant, and since the probability that $F$ plays $OCA$ does not depend on $L$'s choice of $\alpha$, $L$'s best choice is the value of $\alpha$ that maximizes the $OCA$ payoff, that is, $\alpha = 1$. A myopic government is not benevolent. As a consequence, $\pi$ decreases each time it is updated. In the long run, the private sector tends to systematically disregard the government's announcements.

For $h = 1$, $L$'s problem is to find an optimal compromise between (a) obtaining a good (expected) payoff at the current repetition $t$, which implies a high value of $\alpha_t$; and (b) obtaining a good payoff at the next repetition $t + 1$, assuming that $\alpha_{t+1}$ will be set equal to one. To obtain a good payoff under (b), $\pi_{t+1}$ must be high, which implies a low value of $\alpha_t$. The optimal $\alpha_t$ is given by:

$$\alpha^* = \arg\max_{\alpha \in (0,1]} H(\alpha) - \frac{1}{4}\frac{\left(J_{t-1}^{F^*} - K(\alpha)\right)U^{*2}}{1 + \left|J_{t-1}^{F^*} - K(\alpha)\right|} \times \begin{array}{ll} \pi_t & \text{if } \ J_{t-1}^{F^*} - K(\alpha) \geq 0 \\ 1 - \pi_t & \text{otherwise} \end{array} \qquad (16)$$

where $J_{t-1}^{F^*} \triangleq \max_{\tau < t} J_\tau^F$, $H(\alpha) = -\frac{1}{2}(1-\alpha)^2 U^{*2}$, and $K(\alpha) = -\frac{1}{2}\alpha^2 U^{*2}$. The corresponding expression for $h = 2$ is very messy and will not be reproduced here. For $h > 2$, the expressions for $\alpha^*$ are no longer practically amendable, analytically and numerically.

The dependency of $\alpha^*$ upon $U^*$, $\pi$, and $J_{t-1}^{F^*}$ in the cases $h = 1$ and $h = 2$ is not monotonic, as one may recognize from Figures 2 and 3. In these figures, we plot $\alpha^*$ on the vertical axis against $\pi$ (Figure 2) and $J^{F^*}$ (Figure 3) on the horizontal axis when $h = 1$ (circles) and $h = 2$ (squares). In Figure 2, $J^{F^*}$ is given the value $-7.5625$, in Figure 3 we set $\pi = 0.5$. Due to computational constraints, the indicated values of $\alpha^*$ are somewhat coarse approximations of the true values. They are restricted to multiples of 0.05, which explains the flat sections and the occasional exact coincidence of the value of $\alpha^*$ for $h = 1$ and $h = 2$.

Two features of the results are worth emphasizing. First, with increasing $J^{F^*}$, $\alpha^*$ first decreases, and then increases until it takes the value $\alpha^* = 1$. When the private sector has very high expectations (when $J^{F^*}$ is close to 0), it will be disappointed under $OCA$ (that is, $\rho$ will be a relatively large positive number) independently of the value of $\alpha$ chosen by the government. It turns out that under these conditions the government is better off by shortsightedly maximizing its payoff under $OCA$ than by trying to build up a reputation, that is, to increase $\pi$ by choosing a low $\alpha$.

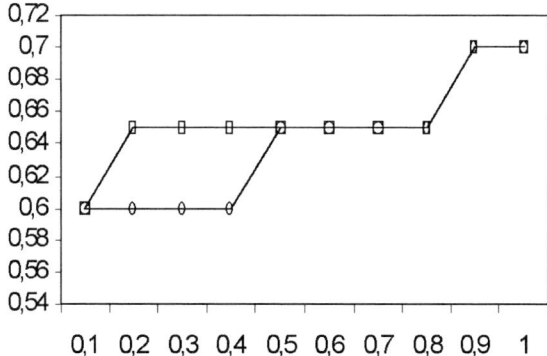

Figure 2  $\alpha^*$ as a function of $\pi$, $J^{F^*} = -7.5625$, $h = 1$ and $h = 2$.

Since we assumed in the paper that $F$ has infinite memory and uses the best payoff he obtained in the past as benchmark $J^{F^*}$, independently of how far in the past it was realized, one cannot exclude a lock-in where $F$ will play $\alpha = 1$ in any future repetition. This problem disappears if one more realistically assumes that $F$ has a finite memory, and/or revise downwards the benchmark $J^{F^*}$ if it is not attained or surpassed over several repetitions. Second, contrary to what one might had expected, a government that has a longer planning horizon ($h = 2$ instead of $h = 1$) will not necessarily choose a lower value of $\alpha$. The complexity of the trade-offs involved precludes giving a clear-cut explanation of this last result.

## 5   Simulation Results

In this section, we present some illustrative simulation results when $h = 1$. The graphs pertain to the instantaneous averages for $t = 1, \ldots, 100$ over 50 Monte Carlo runs. The thick lines correspond to the averages themselves, the dotted ones to the averages $\pm$ two standard deviations. The natural rate of unemployment $U^*$ was set equal to 5.5. Thus, $J^{L,N} = -30.25$ and $J^{F,N} = -15.125$, $J^{L,SF} = J^{F,SF} = -7.5625$, and $J^{F^*} \in [-15.125, 0]$.

The runs were initialized by conducting two repetitions assuming a constant value for $\pi$, $\pi_{-2} = \pi_{-1} = 0.5$. The best payoff for $F$ obtained over these two repetitions was used as initial best past payoff, $J_0^{F^*}$.

Figure 2 shows the evolution of the value of $\alpha$ chosen by $L$ at each repetition. This value is slowly decreasing from about 0.64 to an almost constant value of about 0.587. It is worth noting that this value is considerably lower than the value that makes $F$ indifferent between playing $OCA$ and $PAL$, $\alpha = 0.88$. This reflects the fact that the fear that $F$ might turn to playing $OCA$, which is associated with a very poor payoff for $F$, forces the latter to very much take into account the former's interest.

Figure 3 shows the frequency with which $F$ plays $OCA$. This frequency increases very rapidly, until stabilizing around a value only slightly lower than 1. That is, the private agents quickly learn to play almost always $OCA$.

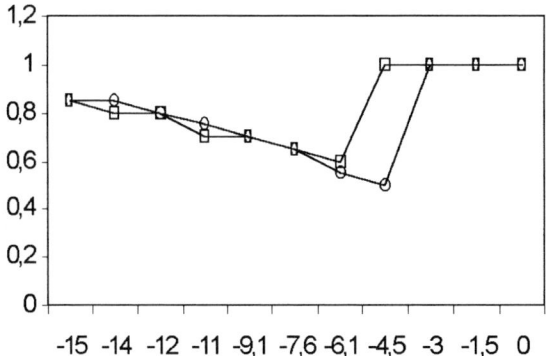

Figure 3  $\alpha^*$ as a function of $J^{F^*}$, $\pi = 0.5$, $h = 1$ and $h = 2$.

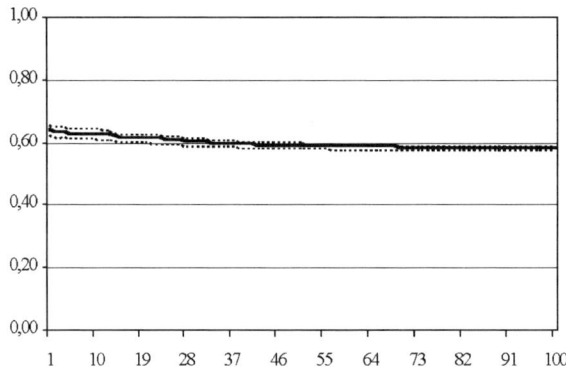

Figure 4  The time evolution of $\alpha$.

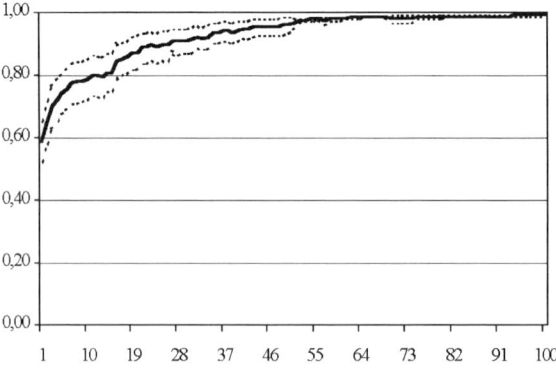

Figure 5  The time evolution of the probability of playing $OCA$.

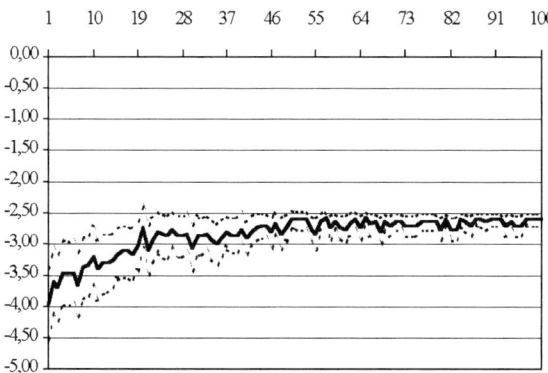

Figure 6  The time evolution of $J_t^L$.

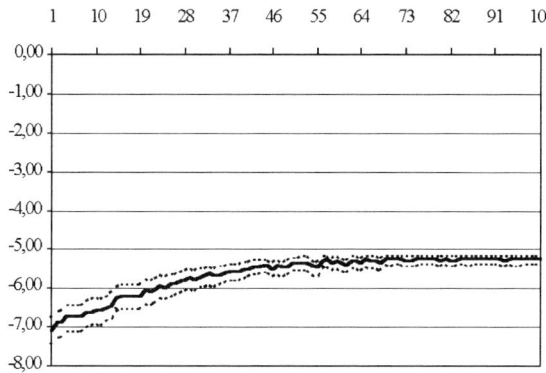

Figure 7  The time evolution of $J_t^F$.

Figures 4 and 5 show the evolution over time of $J^L$ and $J^F$. Both payoffs increase over time, until practically stabilizing about some almost constant value. The payoff improvement over time is significantly more important for $F$ than for $L$. From the onset, both players fare much better than under either $N$ or $SF$. The payoff volatility decreases over time for both players. Statistical tests, not presented here, show that the volatility is similar for both players, contrary to what the graphs may suggest due to the use of different scales. The volatility of all the variables presented in the figures is fairly small, so that the mean appears a good indicator of what might happen in a given historical situation.

# 6   Conclusions

The model presented here does not pretend to be a more realistic description of a real economy than the original Kydland–Prescott model. Nonetheless, it brings some strong messages. Compared to Kydland–Prescott, much less sophisticated economic agents arrive at a much better result. Being cheated upon is no longer a sufficient reason for rejecting a priori governmental announcements. On the contrary, lies can be accepted by the agents as long as they are not detrimental to them. Thus, a government that is neither too myopic nor too stupid may be able to use knowingly wrong announcements to better its payoffs, and the payoffs of the private agents as well.

As usual, numerous extensions of the model can be suggested. In particular, one might want to introduce uncertainty, both on the state of the economy following a choice of actions by the players, and on the preferences of and behavioral rules followed by the players. One extension, however, appears much more crucial. The results presented here cannot be directly interpreted as pertaining to an economy with many independent private agents. Indeed, if there are many agents that independently choose $OCA$ or $PAL$ at each repetition, these agents will have different individual histories and different probabilities of playing $OCA$. This profoundly modifies the problem faced by the government, and introduces questions about possible mechanisms for information exchanges and coordination among private agents. The field of investigation appears vast and constitutes the reserved domain of future research.

# Acknowledgments

The authors would like to thank many friends and colleagues for kind and constructive comments, and most particularly, Willi Semmler and the participants of the January 2000 Bielefeld workshop in honor of Edmond Malinvaud.

# References

1. Barro, R.J. and Gordon, D.B. (1983) Rules, Discretion, and Reputation in a Model of Monetary Policy. *Journal of Monetary Economics*, **12** (1), 101–121.

2. Stokey, N.L. (1989) Reputation and Time Inconsistency. *American Economic Review, Papers and Proceedings*, **79** (2), 134–139.

3. Stokey, N.L. (1990) Credible Public Policy. *Journal of Economic Dynamics and Control*, **15** (5), 627–656.

4. Sargent, T.J. (1999) *The Conquest of American Inflation*. Princeton University Press, Princeton, N.J.

5. McCallum, B. (1997) Crucial Issues Concerning Central Bank Independence. *Journal of Monetary Economics*, **39**, 99–112.

6. Backus, D. and Driffill, J. (1985) Inflation and Reputation. *American Economic Review*, **75**, 530–538.

7. Rogoff, K. (1987) Reputational Constraints on Monetary Policy. *Carnegie-Rochester Conference Series on Public Policy*, **26**, 141–182.

8. Personn, T. and Tabelinni, G. (1993) Designing Institutions for Monetary Stability, *Carnegie-Rochester Conference Series on Public Policy*, **39**, 53–84.

9. Rogoff, K. (1985) The Optimal Degree of Commitment to an Intermediate Monetary Target, *Quarterly Journal of Economics*, **100**, 1169–1189.

10. Basar, T. (1989) Time Inconsistency and Robustness of Equilibria in Non-Cooperative Dynamic Games. In *Dynamic Policy Games in Economics*, F. van der Ploeg, A.J. de Zeeuw (eds), North-Holland, Amsterdam.

11. Hall, S. and Henry, S. (1989) *Macroeconomic Modelling*. North-Holland, Amsterdam.

12. Miller, M. and Salmon, M. (1982) Policy Coordination and the Time Inconsistency of Optimal Policy in Open Economies. *Economic Journal*, **7** (2), 124–137.

13. Hämäläinen, R. (1981) On the Cheating Problem in Stackelberg Games. *International Journal of Systems Science*, **12**, 753–770.

14. Vallée, Th., Deissenberg, Ch. and Basar, T. (1999) Optimal Open Loop Cheating in Dynamic Reversed Linear-Quadratic Stackelberg Games. *Annals of Operations Research*, **88**, 247–266.

15. Vallée, Th. and Deissenberg, Ch. (1998) Pareto Efficient Cheating in Dynamic Reversed Linear-Quadratic Stackelberg Games: the Open-Loop Linear-Quadratic case. In *Issues in Computational Economics*, S. Holly, S. Greenblatt (eds), Elsevier, pp. 127–140.

16. Fudenberg, D. and Levine, K. (1998) *The Theory of Learning in Games*. MIT Press, Cambridge, MA.

17. Börgers, T. and Sarin, R. (1997) Learning Through Reinforcement and Replicator Dynamics. *Journal of Economic Theory*, **77**, 1–14.

18. Brenner, T. (1999) *Modelling Learning in Economics*. Edward Elgar, Cheltenham.

# Dynamic Investment Behavior Taking into Account Ageing of the Capital Goods

Gustav Feichtinger,[1][†] Richard F. Hartl,[2]
Peter Kort[3] and Vladimir Veliov[1]

[1] *Institute for Econometrics, OR and Systems Theory, University of Technology*
*Argentinierstrasse 8, A-1040 Vienna, Austria*
*Tel: +43 1 58801 / 11927, Fax: +43 1 58801 / 11999*
*E-mail: or@eos.tuwien.ac.at*
[2] *Institute of Management, University of Vienna, Vienna, Austria*
[3] *Department of Econometrics and Operations Research and CentER*
*Tilburg University, Tilburg, The Netherlands*

In standard capital accumulation models all capital goods are equally productive and produce goods of the same quality. However, due to ageing, in reality it holds most of the time that newer capital goods are more productive. Implications of this feature for the firm's investment policies are investigated in an optimal control problem with distributed parameters. It turns out that investing in capital goods of different age is done such that the net present value of marginal investment equals zero. Comparing the returns of investment in capital goods of different age, the higher productivity of younger capital goods has to be weighed against the lower costs of depreciation, discounting and acquisition of older capital goods. In the steady state it holds that, in the most reasonable scenario, the firm should invest at the highest rate in new capital goods, and disinvestment can only be optimal when costs of acquisition are large and machines are old.

## 1   Introduction

One of the driving forces in a market economy is the growth of firms and industries. In the literature the analysis of firm growth started out in the sixties with Eisner and Strotz [1]. In the framework they considered the firm owns a stock of capital goods

---

[†]Corresponding author.

that is needed to produce goods, which are sold on the market to obtain revenue. The firm is able to increase capital stock by investing. This profit maximization problem thus involves the choice of investments to expand the stock of capital goods. After this first contribution by Eisner and Strotz [1], many others have followed (e.g., Lucas [2], Davidson and Harris [3], Barucci [4]), and they mostly differ in the specifications of revenue and investment cost functions. All these contributions have in common that capital stock is homogeneous. Hence, its features do not change over the years, so that it can be concluded that matters like ageing and technological progress are not taken into account.

The aim of this paper is to analyze a model where capital goods with different ages are distinguished. To do so a vintage capital stock model is developed. We use Haurie, Sethi and Hartl [5] as basic departure point (see also Appendix 5 of Feichtinger and Hartl [6]).

In order to show what influence ageing has on the age distribution of the capital stock we consider a situation where there is no technological progress and there is constant returns to scale. Productivity only depends on its age. This means that capital stocks of the same age have the same productivity independent of the year in which they are operating. Thus each capital good of the same age produces a fixed amount.

The vintage capital model has become increasingly popular among economists, especially because it provides an appealing framework for the analysis of investment volatility. However, Barucci and Gozzi [7] state that, apart from their paper, in the literature the vintage differentiation of the capital goods has not been analyzed in a complete dynamic optimization framework; often capital goods are not durable, they cannot be accumulated and therefore the capital accumulation problem either becomes a simple intertemporal budget allocation problem (e.g., Grossman and Helpman [8]) or capital is completely absent as an explicit input factor (e.g., Chari and Hopenhayn [9]). Xepapadeas and De Zeeuw [10] limit their analysis to the OSSP (Optimal Steady State Problem). Jovanovic [11] argues that full dynamics are notoriously difficult in such models. Our paper offers a complete dynamic optimization framework, but contrary to Barucci and Gozzi [7] who concentrate on technological progress, we focus on the effects of ageing on the dynamic investment rates and on the age distribution of capital goods in the steady state. Like Xepapadeas and De Zeeuw [10], our analysis thus mainly considers the steady state, but additionally we show that it is in fact optimal for the firm to reach this steady state as soon as possible. The steady state does not exist in Barucci and Gozzi's model due to the technological progress considered there.

By analyzing this model we are able to determine the firm's optimal investment decisions in capital goods of different ages. It turns out that the firm always invests in such a way that the net present value of marginal investment equals zero, so that the discounted extra revenue stream caused by the addition of a capital good exactly balances the marginal investment costs. Investments in younger machines have the advantage that due to ageing they are more productive than older ones, but the disadvantage is that older machines are cheaper and the costs of depreciation and discounting are less. The presence of the latter effects may explain why, according to Chari and Hopenhayn [9], it is undeniable that new technologies are often adopted on a large scale only after a prolonged period of time (see Mansfield [12] for empirical evidence). For the steady state it turns out that, provided that the discount rate is sufficiently low, the firm should invest mostly in new capital goods. Disinvestment only occurs if acquisition costs are high and machines are sufficiently old.

The paper is organized as follows. The model is formulated in Section 2. In Section 3 the optimality conditions are formulated and expressions for the investment rate in capital stocks of different age are derived and economically analyzed. Moreover, Section 3.3 considers the firm in steady state in order to see what the age distribution of capital goods then looks like.

# 2 The Model

In a recent paper Xepapadeas and De Zeeuw [10] studied the ideal age composition of the capital stock subject to environmental regulation. Here we consider a related version of the model of Xepapadeas and De Zeeuw [10]: where they concentrate on environmental regulation by specifying pollution output, we leave this out. Instead, we extend their framework by adding discounting and depreciation, so that this paper is a natural extension to the capital accumulation literature mentioned in the first paragraph of the Introduction. As in their paper, here it also holds that the age of the machine is denoted by $\tau \in [0, h]$, so that the maximum age of machines is $h$.

$v(\tau)$ is the output produced by a machine of age $\tau$, with $v'(\tau) \leq 0$. That is, a newer machine cannot produce less output than an older one. Since $v$ is independent of time $t$ no technological progress is included.[1]

The stock of capital goods of age $\tau$ at time $t$ is denoted by $K(t, \tau)$. Then total output produced in year $t$ is defined as

$$Q(t) = \int_0^h v(\tau)K(t, \tau)d\tau \, .$$

It is assumed that markets exist for machines of any age from 0 to $h$. Let $b(\tau)$ be the cost of buying a machine of age $\tau$, with $b'(\tau) \leq 0$ (older machines cannot be more expensive than newer machines) and $b(h) = 0$ (a machine at the maximum age is not worth anything).

Let $I(t, \tau)$ be the number of machines of age $\tau$ bought (if $I(t, \tau) > 0$) or sold (if $I(t, \tau) < 0$) in year $t$. The total cost or revenue to the firm from transactions in the machine market is defined as $b(\tau)I(t, \tau) + \frac{1}{2}c\left[I(t, \tau)\right]^2$, with the second term reflecting the adjustment costs in buying or selling machines. These costs are, for example, adaptation costs or search costs. The quadratic form of this cost

---

[1]This model feature is taken from Xepapadeas and De Zeeuw [10] (see also Barucci and Gozzi [13]) who argue that this implies that new machines are more productive because they embody superior technology. However, this argument seems to be wrong. To see this, note that $v(\tau)$ is the same for different $t$. Now consider two points of time: $t_1$ and $t_2$ so that $t_2 > t_1$. Then a machine constructed at time $t_2$, say $m_2$, has the same productivity at the same age as a machine constructed at time $t_1$ ($m_1$), i.e., $m_2$ produces at $t_2 + \tau : v(\tau)$, which is also the amount that $m_1$ produces at $t_1 + \tau$. Hence there is no superior technology embedded in $m_2$. Therefore, in order to include technological progress, output should be modelled by $v(t, \tau)$ with, at least, $v_t > 0$.

term leads to a simple expression for optimal purchases. It is further imposed that machines of age $\tau$ depreciate with rate $\delta(\tau)$, which is the same for every vintage.

The firm chooses to buy or sell machines of different ages in order to maximize profits, with $p$ the price of output. That is, the firm chooses at each point in time an age distribution of machines to maximize profits. In addition to Xepapadeas and De Zeeuw [10], our model also includes discounting, where $r$ is the discount rate. The dynamic model of the firm is now given by

$$\max_{I(t,\tau),I_0(t)} \int_0^\infty e^{-rt} \int_0^h \left[ pv(\tau)K(t,\tau) - b(\tau)I(t,\tau) - \frac{c}{2}\left[I(t,\tau)\right]^2 \right] d\tau dt -$$

$$- \int_0^\infty e^{-rt} \left[ b_0 I_0(t) + \frac{c_0}{2}\left[I_0(t)\right]^2 \right] dt, \tag{1}$$

subject to

$$\frac{\partial K(t,\tau)}{\partial t} + \frac{\partial K(t,\tau)}{\partial \tau} = I(t,\tau) - \delta(\tau)K(t,\tau), \tag{2}$$

$$K(t,0) = I_0(t), \qquad K(0,\tau) = K_0(\tau). \tag{3}$$

This is an infinite horizon optimal control problem with transition dynamics described by a linear partial differential equation (Carlson *et al.* [14]). The transition equation indicates that the rate of change in the number of machines of a given age, $\tau$, at a given time, $t$, is determined by two factors. These are the reduction or increase in the number of machines brought about by the sale or acquisition of machines of the given age $\tau$ (the first term of the transition equation), and the reduction due to depreciation at rate $\delta(\tau)$. The initial condition on the number of machines implies that the firm starts with given amount $K_0(\tau)$ of machines of age $\tau$. At each time $t$ it is possible to buy new machines. This purchase rate of new machines is denoted by the boundary control $I_0$.

# 3   Analysis of the Model

First, by using the maximum principle analytical expressions are obtained for investment and capital stock in Section 3.1. It is shown that after $h$ years the steady state will be reached. In Section 3.2 the expressions for investment and capital stock are economically analyzed. Section 3.3 focuses entirely on the steady state to see what the optimal age distribution in the steady state looks like.

## 3.1   Maximum principle

The current value Hamiltonian $H$ for this problem is given by (see, e.g., Feichtinger and Hartl [6]):

$$H = pv(\tau)K(t,\tau) - b(\tau)I(t,\tau) - \frac{c}{2}\left[I(t,\tau)\right]^2 + \lambda(t,\tau)\left[I(t,\tau) - \delta(\tau)K(t,\tau)\right], \tag{4}$$

while the boundary Hamiltonian is

$$H_0 = -b_0 I_0(t) - \frac{c_0}{2} [I_0(t)]^2 + \lambda(t,0) I_0(t) .$$

Consequently, the first-order conditions for optimality are

$$\frac{\partial H}{\partial I} = 0, \quad \text{or} \quad cI(t,\tau) = \lambda(t,\tau) - b(\tau), \tag{5}$$

$$\frac{\partial H_0}{\partial I_0} = 0, \quad \text{or} \quad c_0 I_0(t) = \lambda(t,0) - b_0, \tag{6}$$

$$\frac{\partial \lambda(t,\tau)}{\partial t} + \frac{\partial \lambda(t,\tau)}{\partial \tau} = r\lambda - \frac{\partial H}{\partial K} = (r + \delta(\tau))\lambda(t,\tau) - pv(\tau), \tag{7}$$

$$\lambda(t,h) = 0. \tag{8}$$

Solving the partial differential equation (7), while taking into account the boundary condition (8) yields:

$$\lambda(t,\tau) = \int_\tau^h \exp\left(-\int_\tau^s (r + \delta(\rho))\, d\rho\right) pv(s)ds. \tag{9}$$

From (5) and (9) the optimal investment rate is obtained:

$$I(t,\tau) = \frac{1}{c}\left[\int_\tau^h \exp\left(-\int_\tau^s (r + \delta(\rho))\, d\rho\right) pv(s)ds - b(\tau)\right]. \tag{10}$$

By (6) and (9) it can be concluded that a similar expression holds for the investment in new capital goods:

$$I_0(t) = \frac{1}{c_0}\left[\int_0^h \exp\left(-\int_0^s (r + \delta(\rho))\, d\rho\right) pv(s)ds - b_0\right]. \tag{11}$$

An expression for the stock of capital goods can be derived from (2), assuming for the moment that $\tau \leq t$:

$$K(t,\tau) = \left(\int_0^\tau \exp\left(\int_0^\sigma \delta(\rho)\, d\rho\right) I(t + \sigma - \tau, \sigma)\, d\sigma + A_2\right) \exp\left(-\int_0^\tau \delta(\rho)\, d\rho\right). \tag{12}$$

Note that the initial stock is $A_2 = K(t - \tau, 0) = I_0(t - \tau)$ (see (3)). Combining the last three expressions, we obtain

$$K(t,\tau) = \left\{\int_0^\tau \exp\left(\int_0^\sigma \delta(\rho)d\rho\right) \frac{1}{c}\left[\int_\sigma^h \exp\left(-\int_\sigma^s (r + \delta(\rho))d\rho\right) pv(s)ds - b(\sigma)\right] d\sigma + \right.$$

$$+ \frac{1}{c_0} \left( \int_0^h \exp \left( - \int_0^s (r + \delta(\rho)) \, d\rho \right) pv(s) ds - b_0 \right) \right\} \exp \left( - \int_0^\tau \delta(\rho) \, d\rho \right). \quad (13)$$

Note that this formula is only valid for $\tau \leq t$. In case $\tau > t$, i.e., the vintage already exists at the initial time, it is easily obtained via the second boundary condition in (3) that

$$K(t, \tau) = K_0(\tau - t) \exp \left( - \int_{\tau-t}^\tau \delta(\rho) \, d\rho \right) +$$

$$+ \int_{\tau-t}^\tau \exp \left( - \int_\sigma^\tau \delta(\rho) \, d\rho \right) \frac{1}{c} \left[ \int_\sigma^h \exp \left( - \int_\sigma^s (r + \delta(\rho)) \, d\rho \right) pv(s) ds - b(\sigma) \right] d\sigma.$$

$$(14)$$

An important observation is that (9), (10) and (11) are time invariant. Moreover, $K(t, \tau)$ depends on $t$ only in case $t < \tau$. This means that after $h$ years everything becomes time invariant, that is, the steady state with respect to calendar time is reached.

## 3.2   Economic analysis

Let us analyze by what characteristics the investment rate in machines of different years is influenced. The amount of investment is given by

$$I(t, \tau) = \frac{1}{c} \left[ \int_\tau^h \exp \left( - \int_\tau^s (r + \delta(\rho)) \, d\rho \right) pv(s) ds - b(\tau) \right]$$

for older machines, and

$$I_0(t) = \frac{1}{c_0} \left[ \int_0^h \exp \left( - \int_0^s (r + \delta(\rho)) \, d\rho \right) pv(s) ds - b_0 \right] \quad (15)$$

for new machines. It follows that the net present value of marginal investment equals zero: the term with the integral equals the revenue stream (corrected for discounting and depreciation) generated by an extra unit of capital stock of age $\tau$ (or 0) bought at time $t$, and this extra revenue equals total marginal investment costs $b + cI$.

It is clear that no investment will take place in a machine of age $h$, so that

$$I(t, h) = 0.$$

At a given point of time $t$ the investment rate is influenced by its age as follows:

$$c \frac{\partial I(t, \tau)}{\partial \tau} = (r + \delta(\tau)) \left( \int_\tau^h \exp \left( - \int_\tau^s (r + \delta(\rho)) \, d\rho \right) pv(s) ds \right) - pv(\tau) - b'(\tau).$$

$$(16)$$

Expression (16) shows how investment is affected when the firm compares investing in a machine of age $\tau$ with investing in a machine of a marginally older age. According to the RHS of (16), three effects arise. The *first* effect is positive and consists of a discounting and a depreciation effect. The depreciation effect results from the fact that by buying a machine of older age the machine is depreciated less at the moment that its age is $s$, thus when its productivity equals $v(s)$. The discounting effect is also positive, because the revenue obtained at the moment that the machine is of age $s$ is obtained earlier so that the discounted revenue is higher. The *second* effect is negative and arises from the fact that when buying the machine of a marginally older age than $\tau$, it will not collect the revenue when the machine operates at age $\tau$. The *last* effect is positive which is due to the fact that the acquisition costs of older machines are cheaper.

These effects may help to explain why firms often invest in older technologies even when apparently superior technologies may be available (Chari and Hopenhayn [9]). According to (16) reasons may be that (i) effects of discounting and depreciation are substantial, and (ii) an older machine has a lower acquisition price.

Expression (16) also helps to explain the observation that new technologies are often adopted so slowly, as recognized by, e.g., Chari and Hopenhayn [9]. Reasons for such behavior can thus be that effects of discounting and depreciation (especially during the first years that a new capital good operates) are large and/or that the reduction of the acquisition price when the capital good gets older is substantial.

In case $v' \leq 0$ and $\delta' \geq 0$ it can be easily shown that the first effect is always dominated by the second effect, i.e., the discounting and depreciation effects are more than outweighed by the effect that revenues are earned during a shorter time. We illustrate this by taking $\delta$ and $v$ constant, after which expression (16) becomes

$$\frac{\partial I(t,\tau)}{\partial \tau} = \frac{1}{c}\left[\left(pv\left(1 - e^{-(h-\tau)(r+\delta)}\right)\right) - pv - b'(\tau)\right] =$$

$$= \frac{1}{c}\left[-pv\,e^{-(h-\tau)(r+\delta)} - b'(\tau)\right].$$

Now there are only two contrary effects of age on the investment rate. The advantage of investing in a machine of older age is that investments are cheaper as reflected by the term $-b'(\tau)$. However, the disadvantage is that the planning period during which the firm enjoys revenue from this investment becomes shorter, which is presented by the first term.

Consider now the evolution of the capital stock, where we concentrate on those capital goods for which $\tau < t$, thus at the initial point of time this stock was not present yet. From (12) and $A_2 = I_0(t - \tau)$, it can be obtained that

$$\frac{\partial K(t,\tau)}{\partial \tau} = I(t,\tau) - \delta(\tau)K(t,\tau). \tag{17}$$

Hence, to find out how capital stocks of different age relate to each other at a given point of time, would require substitution of (13) and (10) into (17), and this becomes too messy for drawing clear economic conclusions.

## 3.3   The steady state

As remarked at the end of Section 3.1, from time $h$ onwards the firm is in steady state with respect to calendar time. First we consider the optimal age distribution in general, after which we consider a specific example.

### 3.3.1   The optimal age distribution

From (9) it follows that $\lambda(\tau)$ is given by

$$\lambda(\tau) = \int_{\tau}^{h} \exp\left(-\int_{\tau}^{s} (r + \delta(\rho))\, d\rho\right) pv(s)ds\,. \tag{18}$$

The value of $\lambda$ as given by (18) reflects the benefits from installing one machine of age $\tau$ and keeping it until it becomes of maximum age. From (5) the optimal sales or acquisitions of machines of age $\tau$ is given by

$$cI(\tau) = \lambda(\tau) - b(\tau) = \int_{\tau}^{h} \exp\left(-\int_{\tau}^{s} (r + \delta(\rho))\, d\rho\right) pv(s)ds - b(\tau)\,. \tag{19}$$

Note that

$$I(\tau) \begin{pmatrix} > \\ = \\ < \end{pmatrix} 0\,, \qquad \text{as} \qquad \lambda(\tau) \begin{pmatrix} > \\ = \\ < \end{pmatrix} b(\tau)\,,$$

which is intuitively clear because $\lambda$ denotes the benefits, and $b$ denotes the price of new machines.

The stock of machines of age $\tau$ is partly determined by sales and acquisitions of machines of that age and partly inherited from sales and acquisitions in the past. The set of stocks of all ages is the optimal age distribution of machines and from (13) it is obtained that

$$K(\tau) = \left\{ \frac{1}{c} \int_{0}^{\tau} \exp\left(\int_{0}^{\sigma} \delta(\rho)d\rho\right) \left[\int_{\sigma}^{h} \exp\left(-\int_{\sigma}^{s} (r + \delta(\rho))d\rho\right) pv(s)ds - b(\sigma)\right] d\sigma + \right.$$

$$\left. + \frac{1}{c_0} \left(\int_{0}^{h} \exp\left(-\int_{0}^{s} (r + \delta(\rho))\, d\rho\right) pv(s)ds - b_0\right) \right\} \exp\left(-\int_{0}^{\tau} \delta(\rho)\, d\rho\right).$$

### 3.3.2   Example

In case there is no depreciation $\delta(\tau) = 0$ and no initial investment $I_0 = 0$, as in Xepapadeas–De Zeeuw the solution simplifies to:

$$\lambda(\tau) = \int_{\tau}^{h} e^{-r(s-\tau)} pv(s)ds\,. \tag{20}$$

$$I(\tau) = \frac{\lambda(\tau) - b(\tau)}{c} = \frac{1}{c}\left(\int_\tau^h e^{-r(s-\tau)}pv(s)ds - b(\tau)\right). \tag{21}$$

$$K(\tau) = \int_0^\tau I(\sigma)\,d\sigma = \frac{1}{c}\int_0^\tau \left[\int_\sigma^h e^{-r(s-\sigma)}pv(s)ds - b(\sigma)\right]d\sigma. \tag{22}$$

To see what (21) and (22) look like, consider the following example:

$$v(\tau) = a_0 + a_1(h - \tau), \tag{23}$$

$$b(\tau) = b(h - \tau), \tag{24}$$

where all parameters are nonnegative, and at least $a_1$ is strictly positive. This implies that acquisition cost $b$ declines linearly with age $\tau$ of the machines, and output $v$ is linearly decreasing with age $\tau$.

Substitution of these functions into (21) gives

$$cI(\tau) = \int_\tau^h e^{-r(\rho-\tau)}p\left(a_0 + a_1(h-\rho)\right)d\rho - b(h-\tau) =$$

$$= \int_\tau^h p\left(a_0 + a_1 h\right)e^{-r\rho}e^{r\tau}d\rho - \int_\tau^h pa_1\rho e^{-r\rho}e^{r\tau}d\rho - b(h-\tau) =$$

$$= \left[-\frac{p}{r}(a_0 + a_1 h)e^{-r(\rho-\tau)}\right]_\tau^h + \left[\frac{pa_1}{r}e^{-r(\rho-\tau)}\left(\rho + \frac{1}{r}\right)\right]_\tau^h - b(h-\tau) =$$

$$= \frac{p}{r}e^{-r(h-\tau)}\left[-a_0 + \frac{a_1}{r}\right] + \frac{p}{r}\left[a_0 - \frac{a_1}{r} + a_1(h-\tau)\right] - b(h-\tau) =$$

$$= \left[a_0 - \frac{a_1}{r}\right]\frac{p}{r}\left[1 - e^{-r(h-\tau)}\right] + \left[\frac{pa_1}{r} - b\right](h-\tau). \tag{25}$$

from which it is obtained that

$$c\frac{\partial I(\tau)}{\partial \tau} = p\left(-a_0 + \frac{a_1}{r}\right)e^{-r(h-\tau)} - \frac{a_1 p}{r} + b, \tag{26}$$

so that

$$c\frac{\partial^2 I(\tau)}{\partial \tau^2} = (-ra_0 + a_1)\,p\,e^{-r(h-\tau)}. \tag{27}$$

This yields the following result:

**Proposition 1** *Under the specifications given by (23) and (24) it holds that*

$$I(h) = 0.$$

*Furthermore, for different cases the following results are obtained:*

*1. Low discount rate: $r < \dfrac{a_1}{a_0}$:*

$$\frac{\partial^2 I}{\partial \tau^2} > 0,$$

*1.1. Low acquisition cost: $b < pa_0$:*

$$\frac{\partial I(\tau)}{\partial \tau} < 0 \quad \forall \tau, \qquad I(\tau) > 0 \quad for \quad \tau \in [0, h),$$

*1.2. High acquisition cost: $b \geq pa_0$:*

$$\frac{\partial I(\tau)}{\partial \tau} \begin{pmatrix} < \\ = \\ > \end{pmatrix} 0 \quad iff \quad \tau \begin{pmatrix} < \\ = \\ > \end{pmatrix} h - \frac{1}{r} \ln \left( \frac{a_1 p - r a_0 p}{a_1 p - rb} \right).$$

*And*

*2. High discount rate: $r > \dfrac{a_1}{a_0}$:*

$$\frac{\partial^2 I}{\partial \tau^2} < 0,$$

*2.1. High acquisition cost: $b > pa_0$:*

$$\frac{\partial I(\tau)}{\partial \tau} > 0 \quad \forall \tau, \qquad I(\tau) < 0 \quad for \quad \tau \in [0, h),$$

*2.2. Low acquisition cost: $b \leq pa_0$:*

$$\frac{\partial I(\tau)}{\partial \tau} \begin{pmatrix} > \\ = \\ < \end{pmatrix} 0 \quad iff \quad \tau \begin{pmatrix} < \\ = \\ > \end{pmatrix} h - \frac{1}{r} \ln \left( \frac{r a_0 p - a_1 p}{rb - a_1 p} \right).$$

We note that the most reasonable cases are probably 1.1 and 1.2. In case 2.1 the solution makes no sense, since $I_0$ being equal to zero and investments being negative for each age imply that $K$ will become negative too.

Next, let us concentrate on the capital stock rather than investment. To do so, we combine (22) and (25) to obtain:

$$cK(\tau) = \frac{p}{r} \left( -a_0 + \frac{a_1}{r} \right) \int_0^\tau e^{-r(h-z)} dz + \int_0^\tau \left[ \frac{p}{r} \left( a_0 - \frac{a_1}{r} + a_1 h \right) - bh \right] dz +$$

$$+ \int_0^\tau \left[ -\frac{pa_1}{r} + b \right] z\, dz = \frac{p}{r^2} \left( -a_0 + \frac{a_1}{r} \right) \left( e^{-r(h-\tau)} - e^{-rh} \right) +$$

$$+ \frac{p}{r} \tau \left[ a_0 - \frac{a_1}{r} + a_1 h \right] - bh\tau + + \frac{1}{2} \tau^2 \left( \frac{-pa_1}{r} + b \right),$$

from which it can be derived that (cf. (25))

$$c\frac{\partial K(\tau)}{\partial \tau} = \frac{p}{r}\left(-a_0 + \frac{a_1}{r}\right)e^{-r(h-\tau)} + \frac{p}{r}\left[a_0 - \frac{a_1}{r} + a_1 h\right] - bh + \tau\left(\frac{-pa_1}{r} + b\right) =$$

$$= cI(\tau).$$

Due to the last two equations and Proposition 1 we can conclude the following proposition:

**Proposition 2** *Consider the problem with the specifications presented in (23) and (24). Then it holds that capital stock is age dependent in the following way:*

$$\left.\frac{\partial K}{\partial \tau}\right|_{\tau=h^-} = 0.$$

*Furthermore, for different cases the following results are obtained:*

*1.1. Low discount rate:* $\left(r < \dfrac{a_1}{a_0}\right)$ *and low acquisition cost:* $(b < pa_0)$:

$$\frac{\partial^2 K(\tau)}{\partial \tau^2} < 0 \quad \forall \tau, \qquad \frac{\partial K}{\partial \tau} > 0 \quad for \quad \tau \in [0, h) \,,$$

*1.2. Low discount rate:* $(r < \dfrac{a_1}{a_0})$ *and high acquisition cost:* $(b \geq pa_0)$:

$$\frac{\partial^2 K(\tau)}{\partial \tau^2} \begin{pmatrix} < \\ = \\ > \end{pmatrix} 0 \quad iff \quad \tau \begin{pmatrix} < \\ = \\ > \end{pmatrix} h - \frac{1}{r}\ln\left(\frac{a_1 p - ra_0 p}{a_1 p - rb}\right),$$

*2.1. High discount rate:* $\left(r > \dfrac{a_1}{a_0}\right)$ *and high acquisition cost:* $(b > pa_0)$:

$$\frac{\partial^2 K(\tau)}{\partial \tau^2} > 0 \quad \forall \tau, \qquad \frac{\partial K}{\partial \tau} < 0 \quad for \quad \tau \in [0, h) \,,$$

*2.2. High discount rate:* $\left(r > \dfrac{a_1}{a_0}\right)$ *and low acquisition cost:* $(b \leq pa_0)$:

$$\frac{\partial^2 K(\tau)}{\partial \tau^2} \begin{pmatrix} > \\ = \\ < \end{pmatrix} 0 \quad iff \quad \tau \begin{pmatrix} < \\ = \\ > \end{pmatrix} h - \frac{1}{r}\ln\left(\frac{ra_0 p - a_1 p}{rb - a_1 p}\right).$$

### 3.3.3 Economic Interpretation

To understand the age-dependent investment level, let us rewrite (26) as follows:

$$c\frac{\partial I(\tau)}{\partial \tau} = b - pa_0 e^{-r(h-\tau)} - \int_{\tau}^{h} pa_1 e^{-r(h-z)} dz. \tag{28}$$

The first term of the right-hand side of (28) reflects that investing in an older machine is advantageous from the point of view that fewer investment costs are incurred. The second term indicates that investing in an older machine implies that the lifetime of this machine is shorter which reduces the revenue stream. The third term of the right-hand side of (28) resembles the fact that production with an older machine leads to a lower revenue flow per time unit.

Explaining Proposition 1 is now an easy job. (28) (cf. (27)) implies that, in case of a low discount rate, $\frac{\partial I(\tau)}{\partial \tau}$ increases with $\tau$ (according to the third term of the right-hand side of (28) the revenue flow reduction takes place during a shorter time interval when $\tau$ increases), implying that $\frac{\partial I(\tau)}{\partial \tau}$ reaches its maximum for $\tau = h$. If $pa_0 > b$, $\frac{\partial I(\tau)}{\partial \tau}$ is negative for $\tau = h$, which implies that it will be negative for all possible ages. Since $I(h) = 0$, this in turn implies that the investment rate is positive for all ages of the capital stock, except of course for $\tau = h$. In case acquisition costs are high ($b > pa_0$), it holds that $\frac{\partial I}{\partial \tau} > 0$ for $\tau$ sufficiently large, which together with $I(h) = 0$ implies that the firm sells machines (only sufficiently young machines may be bought, because for these machines a large lifetime with positive revenues may counterbalance the high acquisition costs).

The fact that machines are sold in the case of large acquisition costs also holds when the discount rate is large. When acquisition costs are low, the firm again makes use of this by keeping the investment rate positive for all ages (except the maximal age). For high discount rate it further holds that $\frac{\partial I}{\partial \tau}$ decreases with $\tau$. This is due to the fact that future revenues are heavily discounted, so that the effect of the shorter lifetime of the machine (given by the second term on the right-hand side of (28)) is less.

The results concerning the levels of the capital stocks presented in Proposition 2 follow directly from the investment levels, but additionally it must be taken into account that older machines have a longer investment history. It holds that capital stock increases in a concave way with age if investment is positive but decreasing, capital stock decreases in a concave way with age if investment is negative (machines are sold) and decreasing, while capital stock decreases in a convex way if investment is negative but increasing.

# Acknowledgment

The authors would like to thank Christian Almeder, who provided us with valuable comments.

# References

1. Eisner, R. and Strotz, R. (1963) The Determinants of Business Investment. In *Impacts of Monetary Policy*. Prentice Hall, Englewood Cliffs, NJ.

2. Lucas, R.E. (1967) Adjustment Costs and the Theory of Supply. *Journal of Political Economy*, **75**, 321–334.

3. Davidson, R. and Harris, R. (1981) Non-Convexities in Continuous Time Investment Theory. *Review of Economic Studies*, **48**, 235–253.

4. Barucci, E. (1998) Optimal Investments with Increasing Returns to Scale. *International Economic Review*, **39**, 789–808.

5. Haurie, A., Sethi, S. and Hartl, R.F. (1984) Optimal Control of an Age-Structured Population Model with Applications to Social Services Planning. *Large Scale Systems*, **6**, 133–158.

6. Feichtinger, G. and Hartl, R.F. (1986) *Optimale Kontrolle Oekonomischer Prozesse: Anwendungen des Maximumprinzips in den Wirtschaftswissenschaften*. de Gruyter, Berlin.

7. Barucci, E. and Gozzi, F. (2001) Technology Adoption and Accumulation in a Vintage Capital Model. *Journal of Economics*, **74** (1), 1–38.

8. Grossman, G. and Helpman, E. (1991) Quality Ladders in the Theory of Growth. *Review of Economic Studies*, **58**, 43–61.

9. Chari, V.V. and Hopenhayn, H. (1991) Vintage Human Capital, Growth, and the Diffusion of New Technology. *Journal of Political Economy*, **99**, 1142–1165.

10. Xepapadeas, A. and De Zeeuw, A. (1999) Environmental Policy and Competitiveness: the Porter Hypothesis and the Composition of Capital, *Journal of Environmental Economics and Management*, **37**, 165–182.

11. Jovanovic, B. (1998) Vintage Capital and Inequality. *Review of Economic Dynamics*, **1**, 497–530.

12. Mansfield, E. (1968) *Industrial Research and Technological Innovation: An Econometric Analysis*. Longmans, London.

13. Barucci, E. and Gozzi, F. (1998) Investment in a Vintage Capital Model. *Research in Economics*, **52**, 159–188.

14. Carlson, D., Haurie, A. and Leizarowitz, A. (1991) *Infinite Horizon Optimal Control*, Springer-Verlag, Berlin.

# A Mathematical Approach towards the Issue of Synchronization in Neocortical Neural Networks

## R. Stoop and D. Blank

*Institut für Neuroinformatik, ETHZ/UNIZH*
*Winterthurerstraße 190, CH-8057 Zürich*
*Tel/Fax: 0041-52411153/1153, E-mail: ruedi@ini.phys.ethz.ch*

When interaction among regularly spiking neurons is simulated, using measured cortical response profiles as experimental input, besides complex network-effects dominated behavior, embedded periodic behavior is observed. This is the starting point for our theoretical analysis of possible emergence of synchronized neocortical neuronal firing, where we start from the model that complex behavior, as observed in natural neural firing, is generated from such periodic behavior, lumped together in time. We address the question of how, during periods of quasistatic activity, different local centers of such behaviors could synchronize, as has been postulated, e.g., by binding theory. It is shown that for synchronization, methods of self-organization are insufficient: additional structure is needed. As a candidate for this task, thalamic input into layer IV is proposed, which, due to the layer's recurrent architecture, may trigger macroscopically synchronized bursting among intrinsically non-bursting neurons, leading in this way to a robust neocortical synchronization paradigm. This collective behavior in layer IV is hyperchaotic and corresponds well with the characterizations obtained from *in vivo* time series measurements of cortical response to visual stimuli.

## 1   Introduction

In mankind's struggle to understand its own intellectual capacity, one question has attracted particular attention. It is the puzzle of how the human cortex can be so variable and efficient, in cognitive tasks and in storage, although the cycle time of cortical computation – if a spike is taken as the basic unit of clock time – is of the order of ten milliseconds, a time that is far slower than what is currently (easily) achieved by computers. This observation leads to the expectation that there may be hidden, still undiscovered, computational principles within the cortex that, if combined with the speed of modern computers, could lead to a jump in the computational power of artificial computation, from the hardware and software point of view. In the explanation of the computational properties of the human brain, an important issue of current interest is the feature-binding problem, which relates to the cortical task of associating one single object with its different features [1–2]. As a solution to this problem – in opposition to the concept of so-called grandmother cells – synchronization among neuron firing has been proposed.

In order to address what ingredients are needed to obtain synchronized ensembles of firing neurons, we proceed as follows: first, we investigate networks of neocortical circuits of pyramidal neurons, that are endowed with excitatory and inhibitory connections, where we restrict ourselves to quasistatic dynamical conditions. Under these conditions, we obtain a picture of insulated sites that mostly are engaged in locked states, which may be expressing computational results [3]. It has been shown that recurrent connections on these computational circuits can be interpreted as controllers of the periodicity of the locking. In other words, they modify the computational results returned by the circuit [4]. The natural step then is to add second order perturbations among these sites. For this refined case, we find strong indications that self-organized synchronization, needed to support the binding by synchronization hypothesis, is virtually impossible. However, when we turn our attention to layer IV, the picture changes. As is known, this layer's task is more centered on amplification and coordination, than on computation. When we perform biophysically detailed simulations of this layer (measurements comparable to those made for the previous layers are difficult to obtain), we find a strong tendency to generate synchronized activity among the participating neurons. As a conclusion, we find that synchronization in neocortical networks – if present at all – will have its origin in layer IV, since synchronization cannot emerge in a self-organized way from the pyramidal neuron circuits alone.

## 2   Absence of Self-Organized Synchronization

To prove that self-organized synchronization is virtually impossible, we approximate the cortical network by weakly coupled centers, consisting of more strongly coupled neurons. In the latter, we focus on binary interaction, although the extension to n-ary interaction, or even to interaction among synchronized ensembles, is straightforward. The description of this interaction is by means of maps of the circle

$$f \; : \; \phi_{i+1} = \phi_i + \Omega - T(\phi_i)/T_0 \;\; (modulo \; 1) \,. \tag{1}$$

This formula describes the response of a previously regularly spiking neuron upon a perturbation by a (in the biological sense strongly connected) neuron (note, however, that in the mathematical literature, this type of interaction is mostly referred to as weak interaction of oscillators [5]). In this formula, $\Omega$ is the ratio of the self-oscillation frequency over the perturbation frequency, and $T(\phi_i)/T_0$ measures the lengthening/shortening of the unperturbed interspike interval due to the perturbation, as a function of the phase at which the perturbation arrives. Both quantities can be measured in experiments; this is how we base our derivation upon experimental data. From experiments of increased perturbation strengths, we found that the effect of increased perturbation strength can be parametrized as

$$g(\phi, K) := T(\phi_i, K)/T_0 = (T(\phi_i, K_0)/T_0 - 1)K + 1 \,, \tag{2}$$

where $K_0$ is a normalization, chosen such that at $K = 1$, 75% of the maximal experimentally applicable perturbation strength is obtained. The perturbation response experiments are performed for excitatory, as well as for inhibitory, perturbations. The first experimental finding is that, in biology, chaotic response may be attained from pair interaction, but only if the interaction is of inhibitory nature [6]. This is essentially a consequence of the greater efficacy of inhibitory synapses, a fact that is well known in physiology. Note also that the biology-motivated normalization we are using differs from the usual mathematical one, which sets the value $K = 1$ as the critical value of the map, i.e., when the map $f$ loses invertibility. The second finding is that, as is predicted by the theory of interacting limit cycles, locking into periodic states is abundant, and that the measure of a quasiperiodic firing relation between the neurons quickly vanishes as a function of the perturbation strength $K$. A last finding is that when going from the static to the quasistatic case, lockings into subsequent periodicities are observed, exactly of the type that is predicted by the associated Farey-tree. In fact, our results can be interpreted as the first experimental proof of the limit cycle nature of regularly firing cortical neurons.

While, consequently, the activity within the centers of stronger interacting neurons is described by locking on Arnold tongues (see, e.g., [3]), beyond the bi- or n-ary strong interaction, there is also weaker exchange of activity. This weaker exchange can be modeled as diffusive coupling-mediated interaction, among the more strongly coupled centers. In this way, we arrive at a coupled map lattice model, which we base on measured binary interaction profiles at physiological conditions (including all kinds of variability, e.g., interaction, coupling strengths)

$$\phi_{i,j}(t_{n+1}) := (1 - k_2 k_{i,j}) f_{K\Omega}(\phi_{i,j}(t_n)) + \frac{k_2}{nn} k_{i,j} \sum_{nn} \phi_{k,l}(t_n) \,, \tag{3}$$

where $\phi$ is the phase of the phase-return map, at the indexed site, and $nn$ again denotes the cardinality of the set of all next-neighbors of site $i, j$. $k_2$ describes the overall coupling among the site maps. This global coupling strength is locally modified by realizations $k_{i,j}$, taken from some distribution, which may or may not have a first moment (in the first case, $k_2$ can be normalized to be the global average over the local coupling strengths). In Eq. 3, the first term reflects the degree of self-determination of the phase at site $\{i, j\}$, the second term reflects the influence by next-neighbor centers, which are again understood in the sense of strongest interaction.

The corresponding statement of synchronized behavior, as we understand it, would be observable emergence of non-local structures within the firing behavior of the neurons in the network. In the case of initially independent behavior, we may expect that due to the coupling, a simpler macroscopic behavior will be attained, which could be taken as the expression of corresponding perceptional state.

Extended simulations, however, yield the result that, for biologically reasonable parameters, the response of the network is, based on this understanding, essentially unsynchronized, despite the coupling. Extrapolations from simpler models, for which exact results are available [7], provide us with the explanation why. Generically, from weakly coupled regular systems, regular behavior can be expected. If only two systems are coupled, generally a simpler period than the maximum of the involved periodicities emerges. If, however, more partners are involved, a competition sets in, and high periodicities most often are the result. Typically, synchronized chaotic behavior, results from coupling chaotic and regular systems, if the chaotic contribution is strong enough. Otherwise, the response will be regular. When chaotic systems are coupled, however, synchronized chaotic behavior as well as macroscopically synchronized regular behavior, may be the result (e.g., [7]). For obtaining fully synchronized networks, the last option is the one to focus on. The evolution of cyclic eigenstates deserves particular attention, as it shows how novel collective behavior may emerge.

We performed simulations using 2-d networks, diffusive coupling between $20 \times 20$ to $100 \times 100$ local maps of excitatory/inhibitory interaction. In agreement with the above expectations, we found no signs of macroscopic, self-organized synchronization, using physiologically motivated variability on the parameters (type of site maps, excitability expressed by means of $K$, locally varying diffusive coupling strength, etc.). To understand this in more detail, we compared it with an idealized model that should be a better candidate for collective synchronization. This model is a diffusively coupled model with tent maps as sites. It corresponds to a situation where all site maps are identical (a situation that also can be implemented in our numerical simulations). In this comparison, it first may be objected that in distinction from the maps derived from the experiments, the model is hyperbolic, which is a non-generic situation. Through simulations, however, it can be shown that the corresponding model with nonhyperbolic site maps (parabola, e.g.), share the primary properties of the tent-map model, i.e., the phenomenology is due to the coupled map model. The advantage of the model of coupled tent maps is that it can be solved analytically. In our case we want to derive the largest network Lyapunov exponent [8]. This can be achieved by using the approach of thermodynamic formalism as follows. First, it must be realized that the coupled map lattice can be mapped onto a matrix representation of the form:

$$M(a, k_2) =$$

$$= \begin{pmatrix} |(1-k_2)a| & \dfrac{k_2}{4}a & 0 & \dots & \dfrac{k_2}{4}a & 0 & \dfrac{k_2}{4}a & 0 & \dfrac{k_2}{4}a \\ \dfrac{k_2}{4}a & |(1-k_2)a| & \dfrac{k_2}{4}a & 0 & \dfrac{k_2}{4}a & \dots & 0 & \dfrac{k_2}{4}a & 0 \\ \vdots & & & \ddots & & & & & \vdots \\ \dfrac{k_2}{4}a & 0 & \dfrac{k_2}{4}a & 0 & \dfrac{k_2}{4}a & \dots & 0 & \dfrac{k_2}{4}a & |(1-k_2)a| \end{pmatrix},$$

$$(4)$$

where $a$ is the slope of the local tent maps, and $k$ is the diffusive coupling strength. The thermodynamic formalism formally proceeds by raising the (matrix) entries to the (inverse) temperature $\beta$, and focusing, as the dominating effect, on the largest

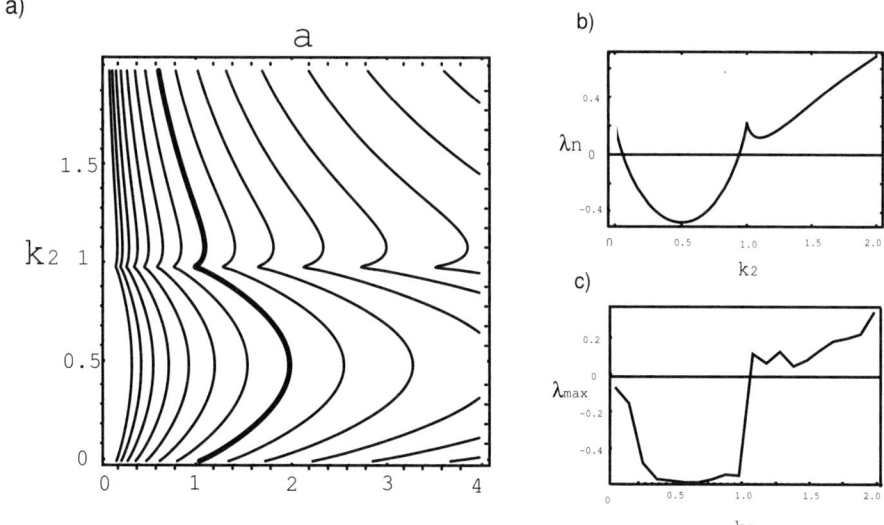

Figure 1 (a) Network Lyapunov exponent $\lambda_n$ describing stability of patterns of a network of coupled tent maps, as a function of the (identical) site map slopes $a$ and coupling $k_2$. Contour lines of distance 0.5 are drawn light, where stable network patterns evolve ($\lambda_n < 0$), bold where unstable patterns evolve; (b) Maximal site-Lyapunov exponent $\lambda_{max}$ of a network of locked inhibitory site maps, as a function of the coupling $k_2$. For the network, the local excitability is $K = 0.5$ for all sites and $\Omega$ is from the interval $[0.8, 0.85]$. The behavior of this network closely follows the behavior predicted by the tent-map model; (c) Cut through the contour plot of (a), slightly above $a = 1$.

eigenvalue as a function of the inverse temperature. For large network sizes, the latter converges towards

$$\mu(\beta, k) = (|(1 - k_2)a|)^{\beta} + (ak_2)^{\beta} . \tag{5}$$

This expression explicitly shows the contributions to the unstable/stable behavior from the two sources: the coupling ($k_2$) and the local instability at the site ($a$). Using this expression of the largest eigenvalue, we obtain the free energy of our model as $F(\beta) = \log((|a(1 - k_2)|)^{\beta} + (ak_2)^{\beta}))$. From the free energy, the largest network Lyapunov exponent is derived as a function of the diffusive coupling strength $k_2$ and the slope of the local maps $a$, according to the formula

$$\lambda = -\left. \frac{d}{d\beta} F(\beta, k_2) \right|_{\beta=1} , \tag{6}$$

which yields the final result

$$\lambda(a, k_2) = \frac{a(1 - k_2) \log(|a(1 - k_2)|) + ak_2 \log(ak_2)}{a|1 - k_2| + ak_2} . \tag{7}$$

Fig. 1a shows a contour plot of $\lambda(a, k_2)$, for identically coupled identical tent maps, over a range of $\{a, k_2\}$-values. In Fig. 1c, a cut through this contour plot is shown, at parameters that correspond to results of numerical simulations of the biologically motivated, variable coupled map lattice, displayed in Fig. 1b. The qualitative equivalence of the two approaches is easily seen. Numerical simulations of coupled parabola show furthermore that the behavior is preserved even in the presence of non-hyperbolicities. As a function of the slope $a$ of the local tent map (which corresponds to the local excitability $K$) and of the coupling strength $k_2$, contour lines indicate the instability of the network patterns. As can be seen, due to the coupling, even for locally chaotic maps ($|a| > 1$), stable network patterns may evolve (often in the form of statistical cycling, see [7]). Upon further increasing the local instability, finally chaotic network behavior of turbulent characteristics emerges. The stable patterns, however, are unlikely to correspond to emergent macroscopic behavior, comparable to synchronized behavior. Therefore, in order to estimate the potential for synchronization, we need to concentrate on the parameter region where macroscopic patterns evolve, that is, on the statistical cycling regime. However, the parameter space that corresponds to this behavior is very small, even in the tent map model. When we compare the model situation with our simulations from biologically motivated variable networks, we again observe that the overall picture provided by the tent map model of identical maps still applies. To show this in a qualitative manner, we compare the contour plot of the tent map model with the numerically calculated Lyapunov exponent of the biological network, which shows the identical qualitative behavior. Based on our insight into the tent-map model behavior, we conclude that in the biologically motivated network, a notable degree of global synchronization would require, at least, all binary inhibitory connections to be in the chaotic regime of interaction (excitatory connections are unable to reach this state [3]). Unfortunately, in case of measured neuronal phase return maps, this possibility only exists for the inhibitory connections. Furthermore, the part of the phase space on which the maps would need to dwell is rather small (although of nonzero measure, see [6]). It is then reasonable to expect that for the network including biological variability, statistical cycling is of vanishing measure, and therefore cannot provide a means of synchronizing neuron firing on a macroscopic scale. To phrase it more formally: this implies that by methods of self-organization, the network cannot achieve states of macroscopic synchronization. In addition, we also investigated whether Hebbian [9] learning rules on the weak connections between centers of stronger coupling could be a remedy for this lack of coherent behavior. Even when using this additional mechanism, it does not result in macroscopic synchronization.

# 3   Synchronization via Thalamic Input

Assuming that synchronization – understood as an emergent, not a feed-forward property – is needed for computational and cognitive tasks, the question remains as to what this property may result from. In simulations of biophysically detailed models of layer IV cortical architecture [10], we discovered a strong tendency of this layer to produce coarse-grained synchronization. This synchronization is based on intrinsically non-bursting neurons that develop the bursting property, as a consequence of the recurrent network architecture and the feed-forward thalamic input.

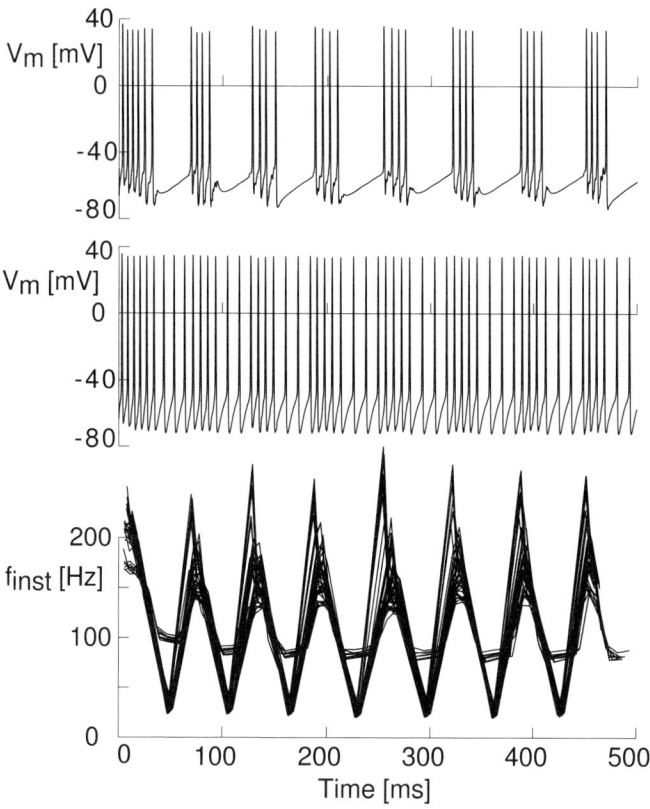

Figure 2  Coarse-grained synchronized activity of layer IV dynamics. Excitatory (upper trace) and inhibitory (lower trace) neuron firing is superimposed (bottom), from several neurons.

Detailed numerical simulations yield the result that, in the major part of the accessible parameter space, collective bursting emerges. That is, all individual neurons are collectivized, in the sense that, in spite of their individual characteristics, they all give rise to dynamics of very similar, synchronized on a coarse-grained scale, characteristics (see Fig. 2). In fact, using methods of noise cleaning (noise, in this sense, is small variations due to the individual neuron characteristics), we find that the collective behavior can be represented in a four-dimensional model, having a strong positive, a small positive, a zero and a very strong negative Lyapunov exponent. This is tantamount saying that the basic behavior of the neuron types involved are identical and hyperchaotic [11]. The validity of the latter characterization has been checked by comparing the Lyapunov dimension ($d_{KY} \sim 3.5$) with the correlation dimension ($d \sim 3.5$). Moreover, different statistical tests have been performed to assess that noise-cleaning did not modify the statistical behavior of the system in an inappropriate way. As a function of the feed-forward input current, we observed an astonishing ability of the layer IV network to generate well-separated characteristic interspike interval lengths.

# 4 Comparison with *in Vivo* Data

When we compared the model data with *in vivo* anesthetized cat measurements (17 time series from 4 neurons of unspecified type from unspecified layers), we found corresponding behavior. Not only were the measured dimensions in the range predicted by the model; specific characteristic patterns found *in vivo* could also be reproduced by our simulation model. Of particular interest are step-wise structures found in the log-log-plots used for the evaluation of the dimensions (see Fig. 3). These steps have previously erroneously been attributed to low dimensions themselves [12] but can be proven to be related to the firing in terms of patterns (the remaining results, although obtained on much smaller data bases, however, agree very well with our findings). These coincidences of modeling and experimental aspects of visual cortex firing lead us to believe that this ability of the network, to fire in well-separated characteristic time scales or in whole patterns, is not accidental, but serves to evoke corresponding responses by means of resonant cortical circuits. Not every neuron, of course, is part of such firing in patterns. In our recent studies of *in vivo* anesthetized cat data, we found essentially three different neuron firing behavior classes upon evoked or spontaneous neuron firing (where the distinction of the stimulation paradigms allowed for no further discrimination of the classes). The first class shows no patterns in their firing at all. The second classes' firing is compatible with the stimulation pattern, whereas the third's firing is incompatible. In the unaffected case, long-tail behavior of the interspike distribution is found. In the compatible case, a clean separation between patterns and individual firing is found, whereas the characteristics of the last class are more associated with the mixture of two behaviors. In all cases, however, the behavior at long interspike interval times is governed by a linear part, i.e., is long-tail.

# 5 Control of Chaotic Network Behavior

Chaotic spiking emerges from my model, as well as from the *in vivo* data that we compared it with. Moreover, nearly identical characterizations in terms of Lyapunov exponents, and of fractal dimensions, emerged. The agreement between Kaplan–Yorke and correlation dimensions [8] corroborates the consistency of the results obtained.

The question then arises of what functional, possibly computational, relevance this phenomenon could be associated with? Cortical chaos essentially reflects the ability of the system to express its internal states (e.g., a result of computation) by choosing among different interspike intervals (as in the last example above) or, more generally, among distinct spiking patterns. This mechanism can be viewed in a broader context. Chaotic dynamics is generated through the interplay of distinct unstable periodic orbits, where the system follows a particular orbit until, due to the instability of the orbit, the orbit is lost and the system follows another orbit, and so on. As it is composed of unstable periodic orbits, chaos therefore is not amorphous. It is then natural to exploit this wealth of structures hidden within chaos, especially for technical applications. The task that needs to be solved to do so is the so-called targeting, and chaos control, problem: The chaotic dynamics first

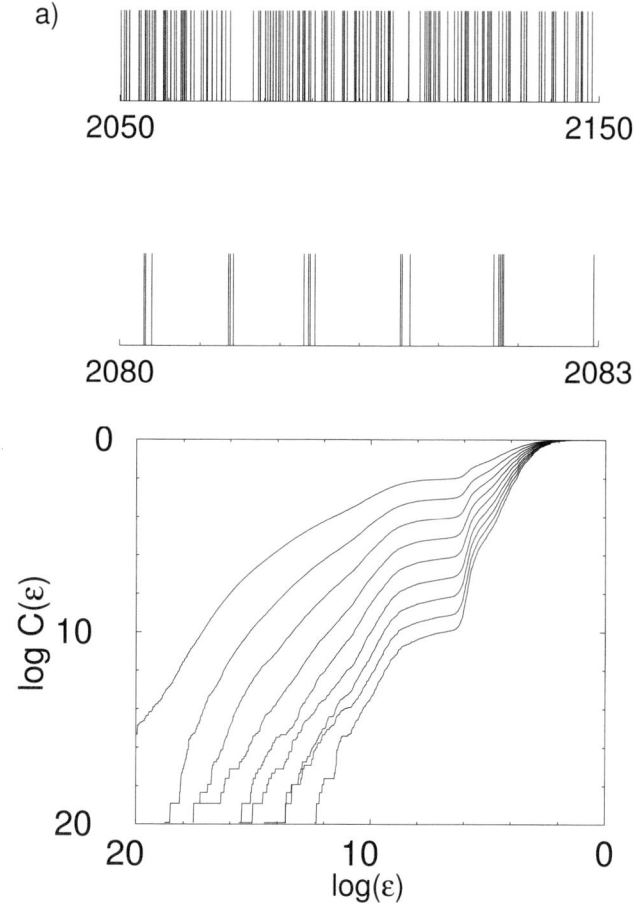

Figure 3 Step-like behavior indicating firing in patterns is observed *in vivo* (picture) and can be reproduced in biologically detailed simulations of layer IV by the interaction of several feed-forward currents to layer IV: (a),(b) from experiments, (c),(d) from corresponding simulations.

needs to be directed onto a desired orbit, on which it then needs to be stabilized, until another choice of orbit is submitted. From an information-theoretic point of view, information content can be associated with the different periodic orbits. This view is related to recent beliefs that information is essentially contained in pattern structures. In the case of the particular *in vivo* measurements discussed above, the different, well-separated interspike interval lengths, can directly be mapped onto symbols (of the same number as there are classes of distinguishable interspike interval lengths). A suitable transition matrix then specifies the allowed, and forbidden, succession of interspike intervals; i.e., this transition matrix provides an approximation to the (in this particular case: almost trivial) grammar of the natural system. In the case of collective bursting, it may be more useful to associate information

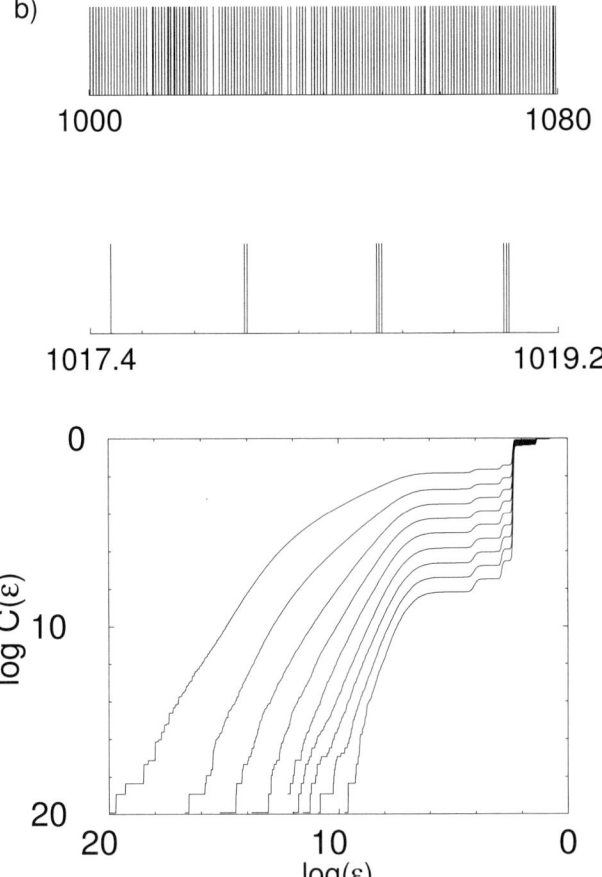

Figure 3  *(continued)*.

content with spiking patterns consisting of characteristic successions of spikes. Such an approach has been shown to be optimally tailored to the description of intermittent systems. In a broader context, the two approaches can be interpreted as realizations of a statistical mechanics description by means of different types of ensembles [13–14].

In the case of artificial systems or technical applications, strategies on how to use chaos to transmit messages, and more general information, are well developed. One basic principle used is that small perturbations applied to a chaotic trajectory are sufficient to make the system follow a desired symbol sequence, containing the transmitted message [15]. This control strategy is based upon the property of chaotic systems known as "sensitive dependence on initial conditions." Another approach, which is currently the focus of applications in areas of telecommunications, is the addition of hard limiters to the system's evolution [16–18]. This very simple and robust control mechanism can, due to its simplicity, even be applied to

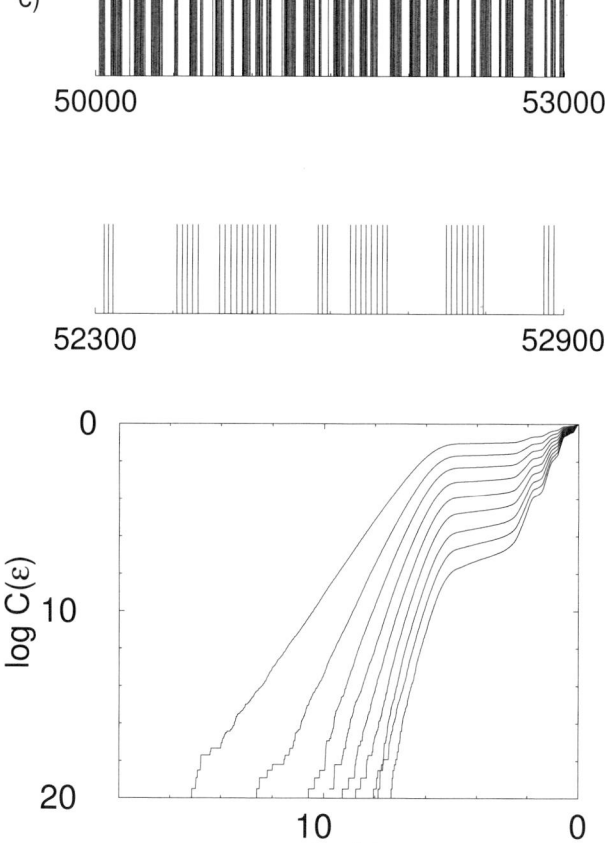

Figure 3 *(continued)*.

systems running at Giga-Hertz frequencies. It recently has been shown [18] that optimal hard limiter control leads to convergence onto periodic orbits in less than exponential time.

In spite of these insights into the nature of chaos control, which kind of control measures should be associated with cortical chaos, however, is unclear. In the collective bursting case of layer IV, one possible biophysical mechanism would be a small excitatory post-synaptic current. When the membrane of an excitatory neuron is perturbed at the end of a collective burst with an excitatory pulse, the cell may fire additional spikes. Alternatively, at this stage inhibitory input may prevent the appearance of spikes and terminate bursts abruptly. In a similar way, the spiking of inhibitory neurons also can be controlled. Another possibility is the use of local recurrent loops to establish delay-feedback control [4]. In fact, such control loops could be one explanation for the abundantly occurring recurrent connections among neurons. The relevant parameters in this approach are the time delay of the

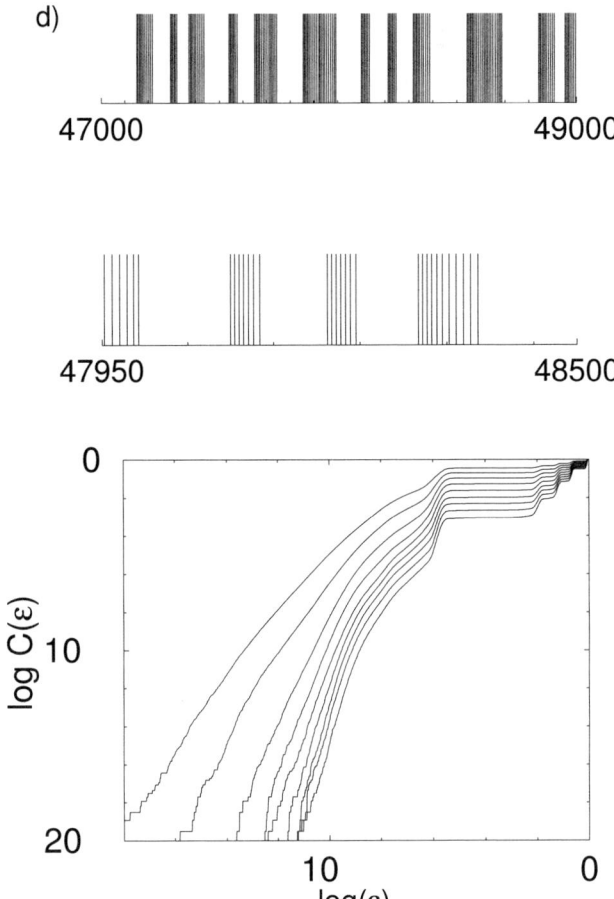

Figure 3  *(continued)*.

re-fed signal, and the synaptic efficacy, where especially the latter seems biologically well accessible.

In addition to the encoding of information, one also needs read-out mechanisms able to decode the signal at the receiver's side. Thinking in terms of encoding strategies, as outlined above, this would amount to the implementation of spike-pattern detection mechanisms. Besides simple straightforward implementations based on decay times, more sophisticated approaches, such as the recently discovered activity-dependent synapses [19–21] seem natural candidates for this task. Also the interactions of synapses, with varying degrees of short-term depression and facilitation, could provide the selectivity for certain spike patterns. Yet another possible mechanism is small populations of neurons, where varying axonal delays, and delays in the propagation time of the synaptic potentials, lead to supra-threshold summation, only for some sequences of input spike intervals.

In conclusion, as we have seen, biological systems provide an abundance of possible information encoding/decoding mechanisms. To explore the contextual dependence in which these alternative strategies are applied in the neocortex will require detailed experimental work in the future. We find that the most appealing explanation of synchronized firing in the cortex originates in layer IV, and is heavily based on recurrent connections and simultaneous LGN feed-forward input. We expect that firing in patterns, in this layer, is able to trigger specific resonant circuits in other layers, where then the actual computation is done (which we propose to be based on the symbol set of an infinity of locked states [22]). In future investigations we will focus on the mathematical properties of the interaction between the two types of networks, and on the relationship of this two-fold structure with the computational task the brain performs. One thing that is easy to predict is that, as a by-product of the network structure, long-range network interactions should emerge. The majority of the interspike interval distributions from *in vivo* cat visual cortex neurons and simple statistical models of neuron interaction, where the emergent long-tail behavior can be traced back to the influence of the inhibitory inputs, support this claim [23]. Long-tail interspike interval distributions are in full contrast to the current assumption of a Poissonian behavior that originates from the assumption of random spike coding. Our conclusion here is that in the cases that can be approximated by Poissonian spike trains, layer IV explicitly shuts down the long-range interactions via inhibitory connections or by pumping energy into new temporal scales that no longer sustain the ongoing activity.

# References

1. Von der Malsburg, C. (1994) The Correlation Theory of Brain Function. In *Models of Neural Networks II*, E. Domany, J. van Hemmen, K. Schulten (eds), Springer-Verlag, Berlin, pp. 95–119.
2. Singer, W. (1994) Putative Functions of Temporal Correlations in Neocortical Processing. In *Large-Scale Neuronal Theories of the Brain*, C. Koch, J. Davis (eds), Bradford Books, Cambridge, MA, pp. 201–237.
3. Stoop, R., Schindler, K. and Bunimovich, L.A. (2000) Noise-Driven Neocortical Interaction: A Simple Generation Mechanism for Complex Neuron Spiking. *Acta Biotheoretica*, **48** (2), 149–171.
4. Stoop, R. (2000) Efficient Coding and Control in Canonical Neocortical Microcircuits. In *Nonlinear Dynamics of Electronic Systems*, G. Setti, R. Rovatti, G. Mazzini (eds), World Scientific, Singapore, pp. 278-282.
5. Hoppensteadt, F.C. and Izhiekevich, E.M. (1997) *Weakly Connected Neural Networks*. Springer, New York.
6. Stoop, R., Schindler, K. and Bunimovich, L.A. (2000) Neocortical Networks of Pyramidal Neurons. *Nonlinearity*, **13**, 1515–1529.
7. Losson, J. and Mackey, M. (1994) Coupling-Induced Statistical Cycling in Diffusively Coupled Maps. *Phys. Rev. E*, **50**, 843–856.
8. Stoop, R. and Meier, P.F. (1988) Evaluation of Lyapunov Exponents and Scaling Functions from Time Series. *Journ. Opt. Soc. Am. B*, **5**, 1037–1045; Peinke, J., Parisi, J., Roessler, O.E. and Stoop, R. (1992) *Encounter with Chaos*. Springer-Verlag, Berlin.

9. Hebb, D. (1949) *The Organization of Behavior*. Wiley & Sons, New York.

10. Blank, D. (2001) PhD thesis. Swiss Federal Institute of Technology ETHZ.

11. Roessler, O.E. (1979) A Hyperchaotic Attractor. *Phys. Lett. A*, **71**, 155–159.

12. Celletti, A. and Villa, A. (1996) Low-Dimensional Chaotic Attractors in the Rat Brain. *Biol. Cybern.*, **74**, 387–393.

13. Stoop, R., Parisi, J. and Brauchli, H. (1991) Convergence Properties for the Evaluation of Invariants from Finite Symbolic Substrings. *Z. Naturforsch. A*, **46**, 642–646.

14. Beck, C. and Schloegel, F. (1993) *Thermodynamics of Chaotics Systems: an Introduction*. Cambridge University Press.

15. Hayes, S., Grebogi, C., Ott, E. and Mark, A. (1994) Experimental Control of Chaos for Communication. *Phys. Rev. Lett.*, **73** (13), 1781–1784.

16. Corron, N., Pethel, S. and Hopper, B. (2000) Control by Hard Limiters. *Phys. Rev. Lett.*, **84**, 3835.

17. Wagner, C. and Stoop, R. (2001) Optimized Chaos Control with Simple Limiters. *Phys. Rev. E*, **63**, 017201.

18. Wagner, C. and Stoop, R. (2002) Renormalization Approach to Optimal Limiter Control. *J. Stat. Phys.*, **10**, 97–107.

19. Abbott, L., Varela, J., Sen, K. and Nelson, S.B. (1997) Synaptic Depression and Cortical Gain Control. *Science*, **275**, 220–224.

20. Tsodyks, M.V. and Markram, H. (1997) The Neural Code between Neocortical Pyramidal Neurons depends on Neurotransmitter Release Probability. In *Proc. Natl. Acad. Sc. USA 94*, pp. 719-723.

21. Thomson, A.M. (1997) Activity Dependent Properties of Synaptic Transmission at Two Classes of Connections Made by Rat Neocortical Pyramidal Neurons in Vitro. *Journ. Physiol.*, **502**, 131–147.

22. Stoop, R., Bunimovich, L.A. and Steeb, W.-H. (2000) Generic Origins of Irregular Spiking in Neocortical Networks. *Biol. Cybern.*, **83**, 481–489.

23. Stoop, R. and Kern, A., unpublished manuscript.

# Optimal Control of Human Posture Using Algorithms Based on Consistent Approximations Theory

Luciano Luporini Menegaldo,[1,2]
Agenor de Toledo Fleury [1,2] and Hans Ingo Weber [3]

[1] *São Paulo State Institute for Technological Research*
*Control System Group / Mechanical and Electrical Engineering Division*
*P.O. Box 0141, CEP 01604-970, São Paulo-SP, Brazil*
*Tel: 55-11-3767-4504, 55-11-3767-4940, Fax: 55-11-3768-5996*
*E-mail: lmeneg@ipt.br*
[2] *University of São Paulo, Politechnic School*
*Department Mechanical Engineering*
*Av. Prof. Mello Moraes 2231, CEP 05508-900, São Paulo-SP, Brasil*
*Tel: 55-11-3818-5491 ext 214, E-mail: agfleury@ipt.br*
[3] *Pontifical Catholic University of Rio de Janeiro*
*Deptartment of Mechanical Engineering*
*Rua Marquês de Sãao Vicente 225/301L Sala 117-L*
*Gáavea CEP 22453-900, Rio de Janeiro, RJ, Brazil*
*Tel: 55-21-529-9532, Fax: 55-21-294-9148, E-mail: hans@mec.puc-rio.br*

This work deals with some analytical and numerical problems for the open-loop optimal control of a dynamical non-linear model of human posture [1]. The dynamical model analyzed is a three-link inverted pendulum representing the human body segments of the shank, thigh and HAT (head, arms and trunk). This model is driven by 10 muscles, expressed by adimensional, 2-nd order, non-linear equations of activation and contraction dynamics. Musculo-tendon geometry was found with the aid of a musculoskeletal modeling software [2]. The optimal control algorithms were developed and implemented on Polak's consistent approximation theory [3] framework by Schwartz [3, 5]. Due to the highly non-linear and unstable nature of the problem, several numerical difficulties were found and the numerical experiments performed suggests the need for a careful estimation of the initial control guess, the gradual imposition of constraints and the appropriate choice of optimization tolerances.

# 1   Introduction

One of the most fascinating problems in the field of motor control theory is the understanding of how the brain controls voluntary movements. The physiological structures and strategies involved in the emergence and execution of such tasks have been intensively studied in recent decades. The nervous impulse paths inside the central nervous system (CNS), as well as the force modulation mechanisms, are well known [6, 7]. However, reliable computational models able to reproduce quantitatively complex motor tasks are still a challenge.

Some authors believe that the brain computation results, in the planning and execution of motor tasks, are similar to the solution of an optimization [8–10] or an optimal control problem [11–16]; nevertheless, such computation does not involve variational calculus or line searches. By this reasoning, when the CNS plans a limb movement, it tries to reach a certain target in the environment space, i.e., a terminal constraint, at the same time that qualitative measure of the "goodness" of the movement is maximized, or an undesirable cost is minimized are along the path. In addition, it could be supposed that the different natures of motor tasks require different measures: maximize smoothness, accuracy, power or swiftness.

The minimization of this performance index is carried over a specific hypersurface that respects the dynamics of the problem, the so-called dynamical constraints. In the physiological version of the problem, this is the very biomechanics of the body interacting with the environment, but in the computational representation, a mechanical description expressed through as ordinary differential equations (ODEs) is required. It means that accuracy of the control depends directly on the hypotheses and the accuracy of the biomechanical modeling.

This work addresses the problem of reproducing human postural movements through a biomechanical representation of body and muscular dynamics, using iterative integration optimal control algorithms based on the consistent approximations theory. Simulation studies were carried out with increasing complexity of biomechanical models of posture developed by the authors [1, 18, 19]. Due to the high nonlinearity and dynamical instability of the problems, several numerical problems were found. Some strategies to overcome them were addressed and some suggestions about the algorithm parameter specifications were carried out.

# 2   Optimal Control Problem

The problem of finding the neuro-muscular activations that drives the human body from a squatting position to an upright posture, while minimizing some physiological effort, can be formulated, in the context of the consistent approximations optimal control theory, as the problem of finding the pairs (control, initial conditions) $(u, \xi) \in L^m_{\infty,2}[a, b] \times \Re^n$ that minimize the cost function

$$\min_{(u,\xi)\in L^m_{\infty,2}[a,b]\times\Re^n} \left\{ \max_{\nu\in\mathbf{q}_o} \left\{ f^\nu(u,\xi) = g^\nu_o(\xi, x(b)) + \int_a^b l^\nu_o(t, x, u)\, dt \right\} \right\} \quad (1)$$

and respect the dynamical constraints $\dot{x} = h(t, x, u)$ with $x(a) = \xi$ (known) initial conditions, $t \in [a, b]$, the control bounds $u^j_{\min}(t) \leq u^j(t) \leq u^j_{\max}(t)$, $j = 1, \ldots, m$

and trajectory endpoint constraints given by

$$g_{ei}^{\nu}(\xi, x(b)) \leq 0, \quad \nu \in \mathbf{q}_{ei} \qquad \text{Inequality constraints,}$$

$$g_{ee}^{\nu}(\xi, x(b)) = 0, \quad \nu \in \mathbf{q}_{ee} \qquad \text{Equality constraints.}$$

In these equations, $x(t) \in \Re^n$, $u(t) \in \Re^m$, $g : \Re^n \times \Re^n \to \Re$, $l : \Re \times \Re^n \to \Re$, $h : \Re \times \Re^n \times \Re^m \to \Re^n$. $f^{\nu}$ is the $\nu$-th function $f^{\nu} : \mathbf{B} \to \Re$, $f^{\nu}(\eta) = \zeta^{\nu}(\xi, x^{\eta}(1))$, where $\zeta : \Re^n \times \Re^n \to \Re$, $\mathbf{q}_0 = \{1, 2, \ldots, q_0\}$, $\mathbf{q}_c = \{1, 2, \ldots, q_c\}$, $q_0$ and $q_c$ positive integers, (expressing the number of objective functions and constraints, respectively). Given the set $\mathbf{q} = \{1, 2, \ldots, q\}$, $q = q_0 + q_c$, $\mathbf{q}_c + q_0 = \{1 + q_0, \ldots, q_c + q_0\}$, $x^{\eta}(1)$ is the solution at the final time of the system, given the initial condition $\eta$. The elements of $\mathbf{q}_{ei}$ and $\mathbf{q}_{ee}$ refer to the number of endpoint equality and inequality constraints, respectively. $L_2^m[0, 1]$ is a square-integrable functional space in the time interval $[0,1]$ in $\Re^m$. $L_{\infty,2}^m[0, 1]$ is a pre-Hilbert space with inner product and norm from $L_2^m[0, 1]$. The elements $\eta = (\xi, u)$ belongs to another pre-Hilbert space $H_{\infty,2} = \Re^n \times L_{\infty,2}^m[0, 1]$, a dense subspace in the Hilbert space $H_2 = \Re^n \times L_2^m[0, 1]$. The function $h$ is limited and continuously differentiable and limited, according to Assumption 3.1 and Theorem 3.2 of [3]. The functions $\zeta^{\nu}$ and their derivatives in $x$ and $\xi$ are Lipschitz continuous in limited sets.

The control elements are contained in the set of available controls $U = \{u \in L_{\infty,2}^m[0, 1] \mid u(t) \in U, \ t \in [0, 1]\}$, supposing that $U$ is a compact and convex set contained in a closed ball of radius $\rho_{\max} : U \subset B(0, \rho_{\max}) = (u \in \Re^m \mid |u| \leq \rho_{\max})$.

Once the optimal control is defined, its numerical solution is forwarded through a reformulation in terms of a non-linear programming problem. This is the core of the iterative integration methods; Canon and Polak [20, 21] shows equivalence between the both problems.

The concept of consistent approximation can be formulated for an optimization problem considering the problem P, where $H$ is a normed linear space and $\mathbf{B} \subset H$ a convex set

$$\text{P} \qquad \min_{\eta \in \mathbf{F}} \psi(\eta) \qquad (2)$$

where $\psi : \mathbf{B} \to \Re$ is at least lower semi-continuous and $\mathbf{F} \subset \mathbf{B}$ a feasible set.

Let $\mathbf{N} = \{1, 2, 3, \ldots\}$ and $\{H_N\}_{N \in \mathbf{N}}$ a family of finite dimensional subspaces of $H$, the approximation problem $\mathbf{P}_N$ is defined as

$$\text{P}_N \qquad \min_{\eta \in \mathbf{F}_N} \psi_N(\eta) \qquad (3)$$

where $\psi_N : H_N \to \Re$ is at least lower semi-continuous and $\mathbf{F}_N \subset H_N \cap \mathbf{B}$.

If the problem $\mathbf{P}_N$ converges epigraphicaly (according to definition 2.1 of [22]) to P, the minimizing accumulation points of $\mathbf{P}_N$ are also minimizing points of P.

Also, if the optimality functions (according to definition 2.5 of [22]) of the approximation problems agree with the relation $\overline{\lim} \theta_N(\eta_N) \leq \theta(\eta)$, it can be stated that problem $\mathbf{P}_N$ is a *consistent approximation* of problem P.

Schwartz and Polak [3, 5] formulated the optimal control problem as an approximation problem, using Runge–Kutta formulas to discretize system dynamics through numerical integration of the ODEs, considering known initial conditions and an actual estimation of the control vector, at the same time that a specific structure for expressing the control samples at the integration knots was introduced. A subspace of the control discretized space was used to express the control samples as spline coefficients, at the same time that specific formulas were introduced by the authors to certify the mathematical equivalence of the internal product and norm operations between the defined functional and the Euclidean spaces. By these transformations, the available optimization algorithms can be used in a safe and efficient way to solve the associated non-linear optimal control problem.

The algorithms briefly described above were implemented by the authors in the Matlab Toolbox RIOTS [23] (Recursive Integration Optimal Trajectory Solver). This software solves the OCP as stated above, allowing the use of 1-st, 2-nd, 3-rd and 4-th order fixed step-size Runge–Kutta integrator and 1-st to 4-th order splines. The LSODA [24] variable step-size integrator can also be used. The optimization problem is solved with a class of conjugate-gradients [25] or with SQP solver NPSOL [26, 27]. User-defined cost and constraint functions, as well as their symbolic derivatives, are written in ANSI C code and dynamically linked to RIOTS.

# 3   Biomechanical Model

The dynamical system equations $h$ of the optimal control problem defined above were formulated as a mathematical representation of human postural mechanics. This biomechanical model has been studied by the authors in recent years [1, 18, 19]. Human rigid-body system behavior is described as a three-link inverted pendulum in the saggital plane [18, 19] (Figure 1). The six resulting equations of motion form the non-linear dynamical system

$$[M(x)]\,\ddot{x} + [C(x)]\,\dot{x}^2 + \{g(x)\} = [D]\,\{\tau\}\ ,  \tag{3}$$

where $[D]$ is a matrix relating joint to rigid-body torques, $[M(x)]$, $[C(x)]$ are mass and centripetal terms matrices, $\{g(x)\}$ represents gravity terms and $\{\tau\}$ the external torques at the joints. Muscle forces and joint torques are related by a moment-arm matrix $[r]$, where $[r]\{F\} = \{\tau\}$. Inertial parameters were estimated for an adult male from regression equations. Muscle mechanics is modeled trough a visco-elastic representation of human skeletal muscle [31], (see Figure 2) based on Zajac's adimentional formulation of contraction dynamics [32, 33].

Each muscle is represented mathematically by two 1-st order non-linear differential equations. The first is a bilinear equation that models the activation dynamics, with different activation and deactivation times, depending on whether the muscle is contracting or relaxing.

$$\dot{a} = (u - a)(k_1 u + k_2)\ ,  \tag{5}$$

where $T_{\text{act}} = \dfrac{1}{k_1 + k_2}$ and $T_{\text{deac}} = \dfrac{1}{k_2} \cdot a(t)$ is the muscle activation and $u(t)$ represents the neuromuscular excitation [32], and is used as the control variable of the model. The values assumed by these functions are restricted to [0,1], where 0 represents complete inactivity and 1 maximum contraction.

The second equation expresses the contraction dynamics:

$$\dot{\tilde{F}}^T = \tilde{k}^T \left( \tilde{v}^{MT} - \tilde{v}^M \cos \alpha \right)  \tag{6}$$

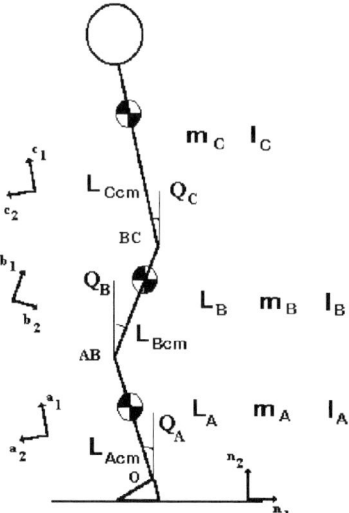

Figure 1 Three-link inverted pendulum representing human postural dynamics. The joints O, AB and BC are the ankle, knee and hip and the limbs A, B and C the shank, thigh and HAT (Head, arms and trunk). $m$ is the mass of the limb, $L$ is the length and $I$ the moment of inertia. $Q_A$, $Q_B$ and $Q_C$ are the generalized coordinates, i.e., the joint angles. $L_{XCM}$ is the distance from the joint to the center of mass.

where $\dot{\tilde{F}}$ is the adimensional time derivative of musculotendon force, $\tilde{v}^{MT}$ is the musculotendon velocity and $\tilde{v}^M$ is the muscle contraction unit velocity, which is in found by inverting the force–velocity relation of the muscle [31]. The sign ˜ means that the variable or the parameters are adimensional [32].

A public domain lower limb model developed by Scott Delp [2, 34, 35] was used as the choice for musculotendon parameters. Table 1 shows the mechanical and geometrical parameters for the 40 muscles of this model, obtained from several cadaver measurements [36–39].

Some simplifying hypothesis were assumed:

- Force–length relationship is always optimal (equal to unity).
- Musculotendon length are fixed at anatomical position (isometric contraction).
- Muscle moment arms are fixed at anatomical position.
- The same activation dynamics is used for all muscles.

In order to decrease the number of equations in the model and save computational effort, 10 muscular groups were selected, according to the following criterion. An estimator for the mechanical relevance of each muscle was created, by multiplying the moment arm by the maximum force (Table 2). Some muscles were eliminated, when this product was too small. Muscles with the same biomechanical function were grouped, and the parameters assumed as a weighted mean according to the value of $\mathbf{r}\,F_0^M$ for each generalized coordinate.

The resulting overall dynamical equations of the biomechanical system are:

$$
\begin{pmatrix} \dot{x}_1 \\ \dot{x}_2 \\ \dot{x}_3 \\ \dot{x}_4 \\ \dot{x}_5 \\ \dot{x}_6 \\ \dot{x}_7 \\ \dot{x}_8 \\ \vdots \\ \dot{x}_{7+2n} \end{pmatrix} = \begin{pmatrix} x_4 \\ x_5 \\ x_6 \\ M(x_1,x_2,x_3)^{-1}\left[ [D]\begin{pmatrix} r_{11} & r_{12} & \cdots & r_{1n} \\ r_{21} & r_{22} & \cdots & r_{2n} \\ r_{31} & r_{32} & \cdots & r_{3n} \end{pmatrix}\begin{pmatrix} x_{7+n} \\ x_{7+n+1} \\ \vdots \\ x_{7+2n} \end{pmatrix} \right. \\ \left. -[C(x_1,x_2,x_3)]\begin{pmatrix} x_4^2 \\ x_5^2 \\ x_6^2 \end{pmatrix} - g(x_1,x_2,x_3) \right) \\ \dot{a}_1 = f_1(u_1,a_1) \\ \dot{a}_2 = f_2(u_2,a_2) \\ \vdots \\ \dot{a}_n = f_n(u_n,a_n) \\ \dot{F}_1^T = g_1(a_1,\tilde{L}^{MT},\tilde{F}^T,\tilde{k}^T,\tilde{L}_s^T,F_0^M,\ldots) \\ \dot{F}_2^T = g_2(a_2,\tilde{L}^{MT},\tilde{F}^T,\tilde{k}^T,\tilde{L}_s^T,F_0^M,\ldots) \\ \vdots \\ \dot{F}_n^T = g_n(a_n,\tilde{L}^{MT},\tilde{F}^T,\tilde{k}^T,\tilde{L}_s^T,F_0^M,\ldots) \end{pmatrix}
$$

$$(7)$$

# 4  Optimal Control Results with Simplified Models of Posture

Preliminary implementations of the model described in the previous section were carried out with simpler models of human posture for testing the methodology. The first model studied was a simple inverted pendulum, shown in Figure 3.

This model was controlled using an initial estimation of the control based on an LQR regulator, with final time fixed and free. For a fixed final time of 2 seconds, the optimal control was found after 256 iterations. Using an open final time, keeping the same tolerances and other parameters, the optimal control did not converge after 4200 iterations, and the cost function increased by 8%. This effect was not studied exhaustively, but the results indicate that an open final time introduces non-trivial numerical difficulties for this problem.

Table 1 Muscular parameters at anatomical positions. HF, KA and AA are the three gerneralized coordinates active in the saggital plane: Hip Flexion, Knee Angle and Ankle Angle. $r_1$, $r_2$ and $r_3$ are the moment arms, with relation to each axis, for the muscles. In the last column the selected groups are shown, according to functional similarity of the composing muscles. $r_1$, $r_2$ and $r_3$ are the moment arms with relation to ankle, hip and knee, respectively. $F_0^M$ is the maximum force and $L^{MT}$ is the musculotendon length.

| Muscle | $r_1$ (HF) | $r_2$ (KA) | $r_3$ (AA) | $F_0^M$ | $\mathbf{r}_1 F_0^M$ | $\mathbf{r}_2 F_0^M$ | $\mathbf{r}_3 F_0^M$ | $L^{MT}$ | groups |
|---|---|---|---|---|---|---|---|---|---|
| gmed1 | −0.0125 | | | 546 | −6.825 | | | 0.12 | 1 |
| gmed2 | −0.0225 | | | 382 | −8.595 | | | 0.13 | 1 |
| gmed3 | −0.0250 | | | 435 | −10.875 | | | 0.11 | 1 |
| gmin1 | −0.002 | | | 180 | −0.36 | | | 0.08 | |
| gmin2 | −0.005 | | | 190 | −0.95 | | | 0.08 | |
| gmin3 | −0.010 | | | 215 | −2.15 | | | 0.09 | |
| semimem | −0.055 | −0.040 | | 1030 | −56.65 | −41.20 | | 0.42 | 2 |
| semiten | −0.065 | −0.042 | | 328 | −21.32 | −13.78 | | 0.47 | 2 |
| bifemlh | −0.065 | −0.055 | | 717 | −46.605 | −39.435 | | 0.45 | 2 |
| bifemsh | | −0.050 | | 402 | | −20.10 | | 0.21 | 3 |
| sar | 0.05 | −0.020 | | 104 | 5.20 | −2.08 | | 0.56 | |
| addlong | 0.04 | | | 418 | 16.72 | | | 0.20 | |
| addbrev | 0.005 | | | 286 | 1.43 | | | 0.13 | |
| amag1 | −0.01 | | | 346 | −3.46 | | | 0.12 | |
| amag2 | −0.02 | | | 312 | −6.24 | | | 0.20 | |
| amag3 | −0.015 | | | 444 | −6.66 | | | 0.33 | |
| tfl | 0.04 | | | 155 | 6.20 | | | 0.52 | |
| pect | 0.02 | | | 117 | 2.34 | | | 0.10 | |
| gra | 0.015 | −0.035 | | 108 | 1.62 | −3.78 | | 0.41 | |
| gmax1 | −0.04 | | | 382 | −15.28 | | | 0.20 | 4 |
| gmax2 | −0.05 | | | 546 | −27.30 | | | 0.21 | 4 |
| gmax3 | −0.07 | | | 368 | −25.76 | | | 0.24 | 4 |
| iliacus | 0.036 | | | 429 | 15.44 | | | 0.20 | 5 |
| psoas | 0.034 | | | 371 | 12.61 | | | 0.26 | 5 |
| rf | 0.047 | 0.0295 | | 779 | 36.61 | 22.98 | | 0.45 | 6 |
| vasmed | | 0.0316 | | 1294 | | 40.89 | | 0.23 | 7 |
| vasint | | 0.0300 | | 1365 | | 40.95 | | 0.25 | 7 |
| vaslat | | 0.0305 | | 1871 | | 57.06 | | 0.26 | 7 |
| medgas | | −0.0150 | −0.0395 | 1113 | | −16.69 | −43.96 | 0.41 | 8 |
| latgas | | −0.0125 | −0.0405 | 488 | | −6.10 | −19.76 | 0.41 | 8 |
| sol | | | −0.0384 | 2839 | | | −109.02 | 0.30 | 9 |
| tibpost | | | −0.010 | 1270 | | | −12.70 | 0.35 | 9 |
| flexdig | | | −0.010 | 310 | | | −3.10 | 0.44 | |
| flexhal | | | −0.014 | 322 | | | −4.51 | 0.42 | |
| tibant | | | 0.0426 | 603 | | | 25.69 | 0.30 | 10 |
| perbrev | | | −0.004 | 348 | | | 1.39 | 0.21 | |
| perlong | | | −0.008 | 754 | | | 6.03 | 0.40 | 10 |
| pertert | | | 0.0274 | 90 | | | 2.47 | 0.17 | |
| extdig | | | 0.040 | 341 | | | 13.64 | 0.44 | 10 |
| exthal | | | 0.042 | 108 | | | 4.54 | 0.40 | 10 |

Other simplified models tested were the triple pendulum with non-dynamic (torque and muscular) actuators and with linear dynamic (torque and muscular) actuators. When non-dynamic torque-actuators were used, the system converged using a few minutes of CPU time in a Pentium 600 MHz and LQR-like cost function. However, when the final time was greater than 0.6 seconds, there was a fatal

Table 2  Muscular groups (gm) selected and their parameters. Optimal length, pennation angles and tendon slack length were calculated as a weighted mean of the product $r\,F_0^M$ in each generalized coordinate.

| Muscle | groups | $L_0^M$ | $\alpha$ | $L^{ST}$ | $F_0^M$ | $r_1 F_0^M$ | $r_2 F_0^M$ | $r_3 F_0^M$ | $L^{MT}$ |
|---|---|---|---|---|---|---|---|---|---|
| gmed1 | 1 | 0.0535 | 8.0 | 0.0780 | 546 | −6.825 | | | 0.12 |
| gmed2 | 1 | 0.0845 | 0.0 | 0.0530 | 382 | −8.595 | | | 0.13 |
| gmed3 | 1 | 0.0643 | 19.0 | 0.0530 | 435 | −10.875 | | | 0.11 |
| gm1 | | 0.0681 | 9.9344 | 0.0595 | 1363 | −26.295 | | | 0.1191 |
| semimem | 2 | 0.0800 | 15.0 | 0.3590 | 1030 | −56.65 | −41.20 | | 0.42 |
| semiten | 2 | 0.2010 | 5.0 | 0.2620 | 328 | −21.32 | −13.78 | | 0.47 |
| bifemlh | 2 | 0.1090 | 0.0 | 0.3410 | 717 | −46.605 | −39.435 | | 0.45 |
| gm2 | | 0.1108 | 7.5027 | 0.3363 | 2075 | −124.58 | −94.415 | | 0.4398 |
| bifemsh | 3 | 0.1730 | 23.0 | 0.1000 | 402 | | −20.10 | | 0.21 |
| gm3 | | 0.1730 | 23.0 | 0.1000 | 402 | | −20.10 | | 0.21 |
| gmax1 | 4 | 0.1420 | 5.0 | 0.1250 | 382 | −15.28 | | | 0.20 |
| gmax2 | 4 | 0.1470 | 0.0 | 0.1270 | 546 | −27.30 | | | 0.21 |
| gmax3 | 4 | 0.1440 | 5.0 | 0.1450 | 368 | −25.76 | | | 0.24 |
| gm4 | | 0.1448 | 3.0026 | 0.1333 | 1296 | −68.34 | | | 0.2191 |
| iliacus | 5 | 0.1000 | 7.0 | 0.0900 | 429 | 15.44 | | | 0.20 |
| psoas | 5 | 0.1040 | 8.0 | 0.1300 | 371 | 12.61 | | | 0.26 |
| gm5 | | 0.1018 | 7.4496 | 0.1080 | 800 | 28.05 | | | 0.2270 |
| rf | 6 | 0.0840 | 5.0 | 0.3460 | 779 | 36.61 | 22.98 | | 0.45 |
| gm6 | | 0.0840 | 5.0 | 0.3460 | 779 | 36.61 | 22.98 | | 0.45 |
| vasmed | 7 | 0.0890 | 5.0 | 0.1260 | 1294 | | 40.89 | | 0.23 |
| vasint | 7 | 0.0870 | 3.0 | 0.1360 | 1365 | | 40.95 | | 0.25 |
| vaslat | 7 | 0.0840 | 5.0 | 0.1570 | 1871 | | 57.06 | | 0.26 |
| gm7 | | 0.0857 | 4.3788 | 0.1407 | 4530 | | 139.90 | | 0.2464 |
| medgas | 8 | 0.0450 | 17.0 | 0.4080 | 1113 | | −16.69 | −43.96 | 0.41 |
| latgas | 8 | 0.0640 | 8.0 | 0.3850 | 488 | | −6.10 | −19.76 | 0.41 |
| gm8 | | 0.0507 | 14.3097 | 0.4011 | 1601 | | −22.79 | −63.72 | 0.41 |
| sol | 9 | 0.0300 | 25.0 | 0.2860 | 2839 | | | −109.02 | 0.30 |
| tibpost | 9 | 0.0310 | 12.0 | 0.3100 | 1270 | | | −12.70 | 0.35 |
| gm9 | | 0.0301 | 23.6436 | 0.2885 | 4109 | | | −121.72 | 0.3052 |
| tibant | 10 | 0.0980 | 5.0 | 0.2230 | 603 | | | 25.69 | 0.30 |
| perlong | 10 | 0.0490 | 10.0 | 0.3450 | 754 | | | 6.03 | 0.40 |
| extdig | 10 | 0.1020 | 8.0 | 0.3450 | 341 | | | 13.64 | 0.44 |
| exthal | 10 | 0.1110 | 6.0 | 0.3050 | 108 | | | 4.54 | 0.40 |
| gm10 | | 0.0944 | 6.5152 | 0.2786 | 1806 | | | 49.90 | 0.3595 |

memory crash and the simulation stopped. This problem will be discussed later. With a linear dynamic torque actuator, tests performed with the LQR-like cost function did not converge, but promising results were achieved with the following cost minimized, similar to the function suggested by Pandy [12]:

$$f(u, \xi) = \int_0^{t_f} \left( \dot{\tau}_1^2 + \dot{\tau}_2^2 + \dot{\tau}_3^2 \right) dt \,. \tag{8}$$

This index represents the sum of the torques derivatives over all movement. Minimizing this quantity is equivalent to maximizing the smoothness of the driving torque (for one isolated muscle attached to a mass, Pandy [12] shows that minimizing this cost function is the equivalent of minimizing jerk).

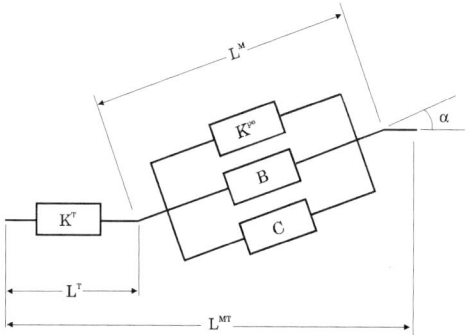

**Figure 2** Muscle model elements: $K^{pe}$ is the parallel elastic element, B is the damping element and C represents the contractile one. $k^{T}$ is the tendon stiffness, where $L^{ST}$ is the tendon slack length. $\alpha$ represents the pennation angle.

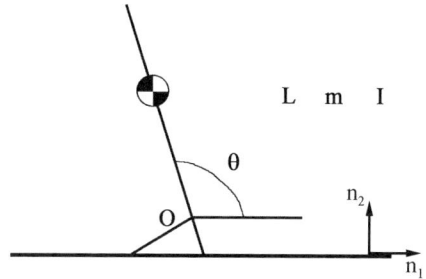

**Figure 3** One degree-of-freedom pendulum representing the entire body rotating about the ankle joint.

A model comprising eight linear muscular actuators was developed [1], to test the ability of the methodology to solve the actuator redundancy problem [8], which arises due to the existence of a greater number of actuators than degrees of freedom. The actuators here were able to respond for positive or negative values of excitation, and the maximum force was chosen to be about 9000 N for all actuators, strong enough to avoid control saturation. Two of the actuators were bi-articular, and the other six mono-articular, i.e., one flexor and one extensor for each joint. The cost function used was:

$$f(u, x(t_0)) = \int_0^2 \sum_{i=1}^8 \left( \dot{F}_i^2 + ku_i^2 \right) dt \qquad (9)$$

with $k = 0.2$.

Several tests were performed using all the options of spline interpolation and Runge–Kutta integrators and discretization was refined until the machine memory limit was reached, and no convergence was achieved. The reason for this failure may be due to the numerical degradation of the LQR estimated initial guess of the controls. Runge–Kutta methods of order greater than two makes evaluations of the function $h$ at intermediary points in the mesh, at $t_k + \Delta/2$. When the gain matrix is incorporated in the system, the value of the controls ($\{u\} = -[K]\{x\}$) is evaluated correctly at the integration step. In the numerical experiments carried out, the control vector was estimated off-line, by multiplying the values of $x$ in the mesh knots by the gain matrix. When the simulation of the first iteration is done, the value of the control out of the knot is approximated by the limit at the left [5]:

$$\text{if} \quad \tau_k = t_k, \quad u[\tau_{k,i}] = u(\tau_{k,i}),$$

$$\text{else if} \quad \tau_k = t_k + \Delta/2, \quad u[\tau_{k,i}] = \lim_{t \to \tau_{k,i}} u(t).$$

This approximation leads to an error that is accumulated over the integration, and corrupts the stability properties of the initial guess obtained with the LQR. To avoid this effect, 1-st order Runge–Kutta may be used, but only with linear muscle models, as this low-order integrator is not able to handle the numerical difficulties introduced by muscle non-linear behavior.

Another strategy used was to replace terminal equality constraints by inequality constraints with a wide value of $\pi/5$, imposing the lower control bound of 0, as the linear LQR does not deal with this constraint. Convergent results were obtained using the last result as initial control guess, terminal equality constraints, 1-st order Runge–Kutta.

# 5   Optimal Control Results with 10 Non-Linear Muscular Actuators

Optimal control trajectories were found for the complete 10 non-linear 2-nd order muscular actuator model. Several tests were performed, using different values of discretization level, tolerances, control bounds and initial guesses of the control vector. Final time was 0.4 seconds and initial conditions corresponding to a knee flexion of $\pi/6$, both fixed. Spline interpolation order was kept fixed to 4-th. Terminal constraints were all treated as inequality constraints of the kind $x_f^2 \leq K^2$, where $K$ is the value of the terminal constraint. The optimal cost function used in all simulations was

$$f(u, x(t_0)) = \int_0^{t_f} \sum_{i=1}^{10} \dot{F}_i^2 \, dt, \tag{10}$$

where the derivatives of the muscle forces were the same analytical expressions as used in the dynamical system.

One series of tests used 300 points and an LSODA integrator. The simulations obeyed the following sequence, where the control results of one simulation were the initial guess of the next:

- Initial guess found by trial and error and terminal constraints for displacements and velocities set as $\pi/5$.

- Last simulation as initial guess and terminal constraints of $\pi/15$.

- Idem, terminal constraints of $\pi/30$.

Setting the optimization and constraint violation tolerances of NPSOL at 0.1, total CPU time in a Pentium III 600 Mhz was 18h. When the tolerances were decreasing to 0.01, 5.5h were spent, and further narrowing of the tolerances led to a memory crash error. The responses obtained with these conditions showed a great amount of numerical noise and the objective function value grew from 9.2596 to 10.4078 when optimization tolerance decreased from 0.1 to 0.01 and the same constraint violation tolerance was maintained.

Other sequences of tests were performed using 150 points and 4-th order fixed step-size Runge–Kutta, and both tolerances 0.1. Figure 4 indicates the initial guess of the control chosen and the corresponding trajectory. As can be seen, the pendulum makes a divergent trajectory, but does not fall completely. Using this initial control guess to find the optimal control that respects $\pi/5$ terminal constraints, the trajectory and control shown in Figure 5 were obtained. After that, for terminal constraints of $\pi/15$ with the initial guess being the result of the anterior simulation and the same for $\pi/30$, the results displayed in Figures 6 and 7 were reached.

# 6 Critical Final Time Problem

All the biomechanical models studied could only be controlled if the final time $(t_f)$ of the optimal control problem was chosen as shorter than a critical time, about 0.4 to 0.6 seconds, depending on the complexity of the actuator model. The reasons for this behavior are not well known, but are certainly related to the choice of the control initial guess in the optimal control iterative numerical process.

System dynamics is unstable, as no feedback is considered, and the desired equilibrium point to be reached, where the terminal constraints are respected, is an unstable point. As the initial guesses initially used were not guaranteed to lead the system close to the solution, when time was sufficiently large, the system diverged far from the equilibrium point, leading the pendulum to *fall*. When this happened, the system was *trapped* by an undesired basin of attraction. Sometimes, it converged to an absolute minimum of the muscle effort cost function, but not to the augmented Lagrangean minimal solution. Using shorter times, the pendulum had no time to *fall too much*, and the subsequent iterations were able to lead to the solution.

However, the central nervous system is able to drive the musculoskeletal system from a squatting position to an upright posture using virtually any desired time to perform the task, and it is desired that the proposed controller should do the same.

Some techniques were proposed to increase the simulation time. With torque actuators, using an LQR estimated initial control guess and RK1, it was possible to achieve a desired final time of 1 s. When linear muscular actuators with control bounds of [0,1] were used, two time intervals had to be defined: from 0 to 0.5 s and from $0.5 + \Delta$ to 1.0 s. In the first interval, a large terminal constraint of $\pi/4$ for all trajectory states was stated. The achieved terminal states were used as the initial values for the 2-nd interval, whose terminal constraints were fixed to be small, around $\pi/30$. If a 1-st order Runge–Kutta was used, and a corresponding small step-size, the pendulum should be controlled using the controls of both the intervals bonded. Using a greater integrator order, the system diverged from the stable solution in the 2-nd interval. This effect must be due to an inconsistency

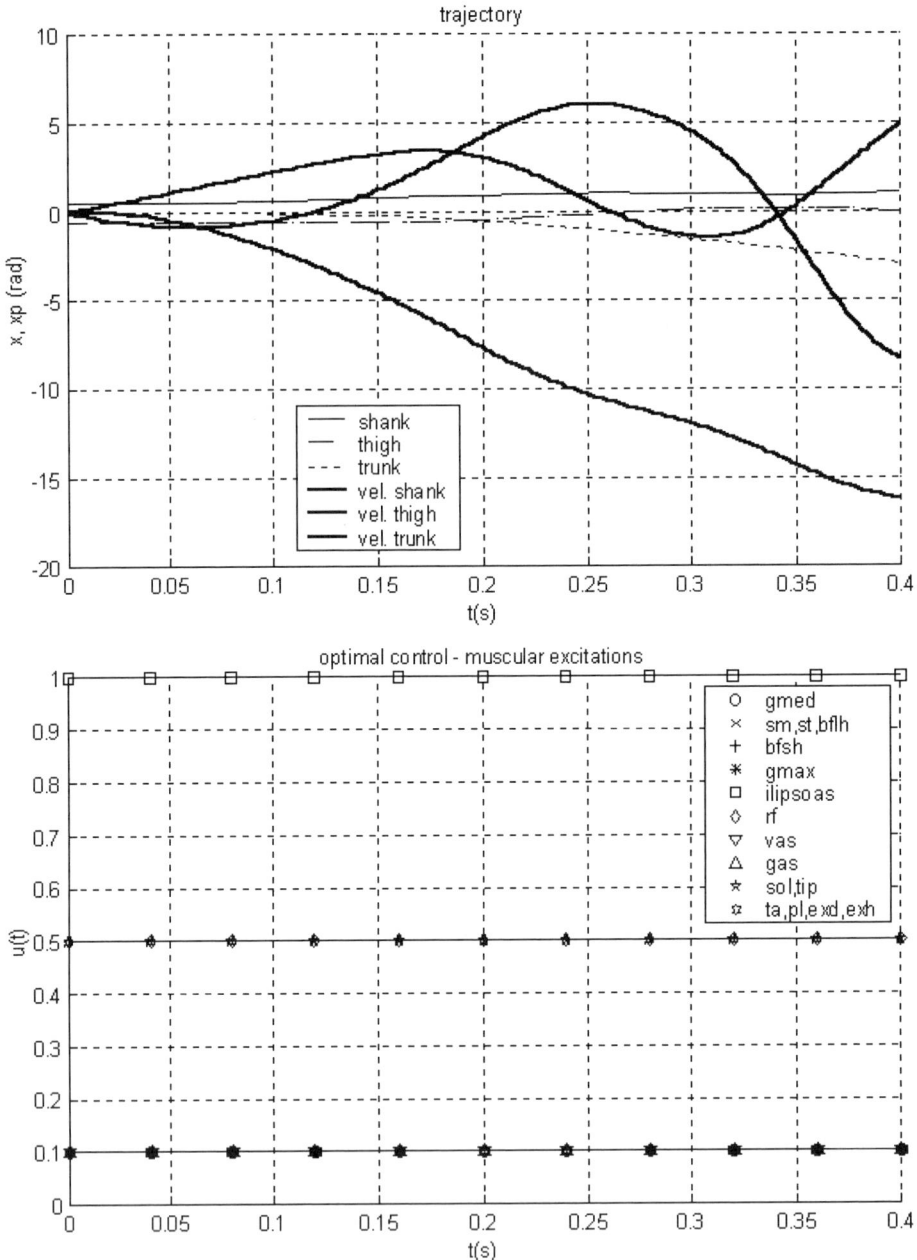

Figure 4  Trajectory and initial guess of control vector.  4-th order Runge–Kutta, 150 points.

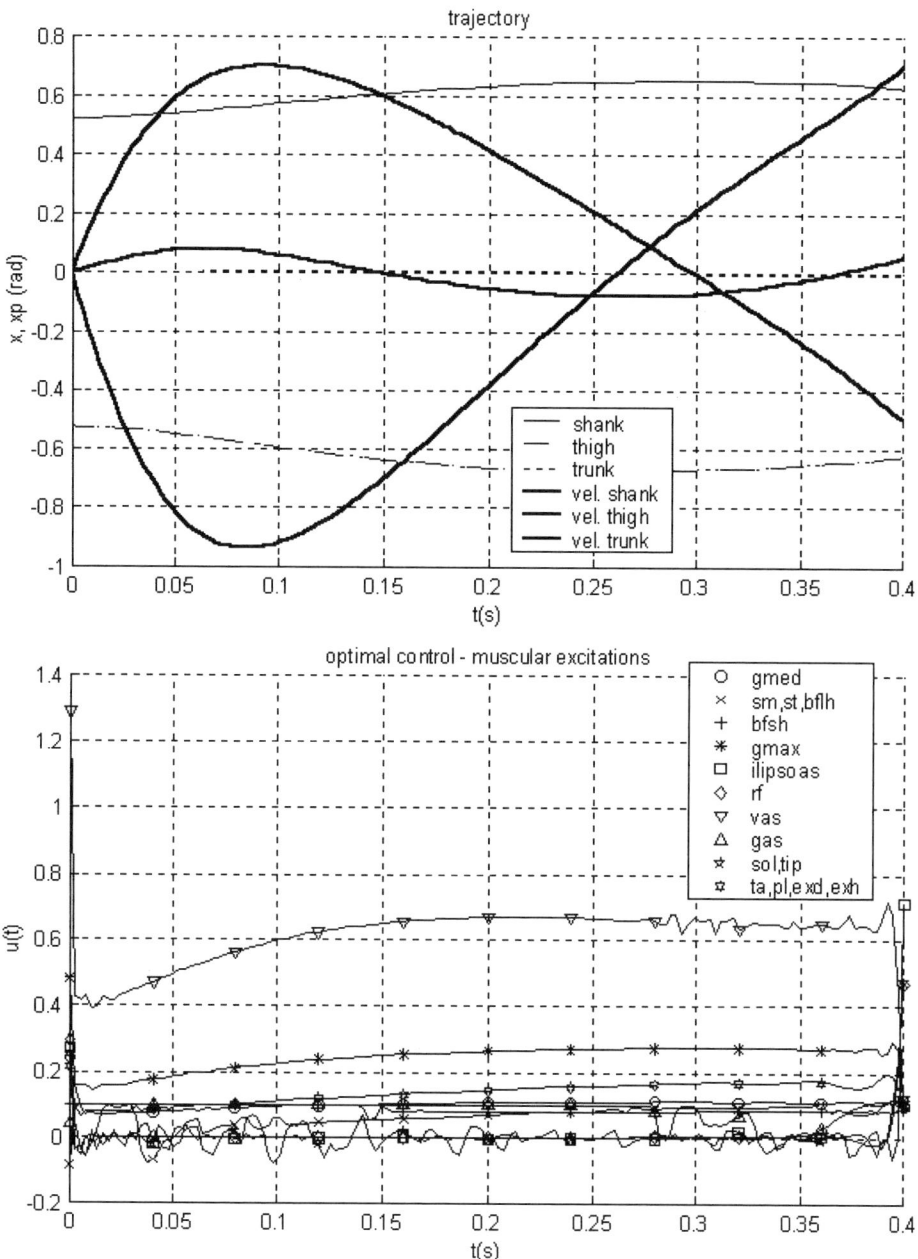

Figure 5  Trajectory and optimal control for $\pi/5$ terminal constraints. 4-th order Runge–Kutta, 150 points, control bounds [0,10]. CPU time Pentium III 600 MHz: 96 min. Cost function $f = 3.4314$.

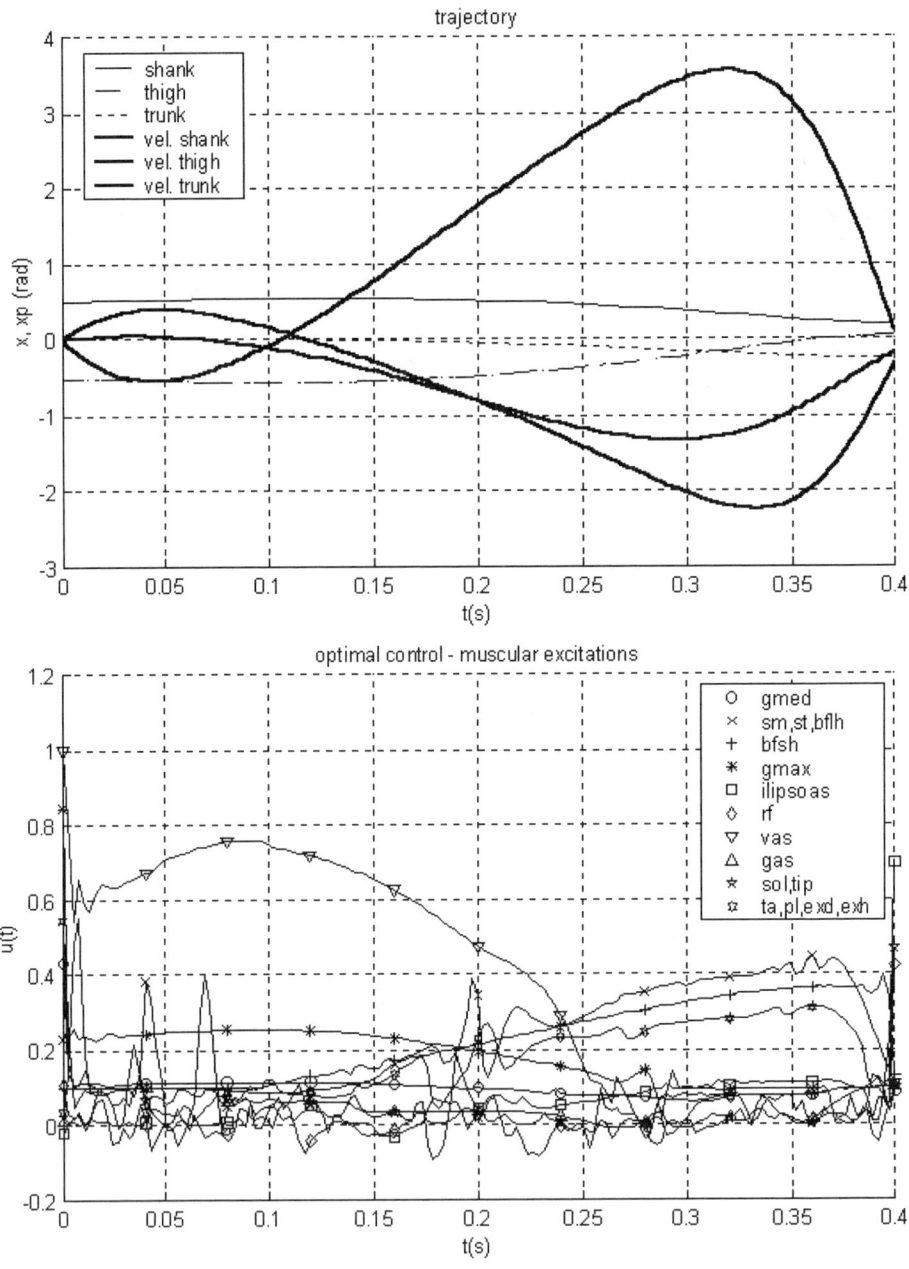

Figure 6  Trajectory and optimal control for $\pi/15$ terminal constraints. 4-th order Runge–Kutta, 150 points, control bounds [0,10]. CPU time Pentium III 600 MHz: 43 min. Cost function $f = 9.2629$.

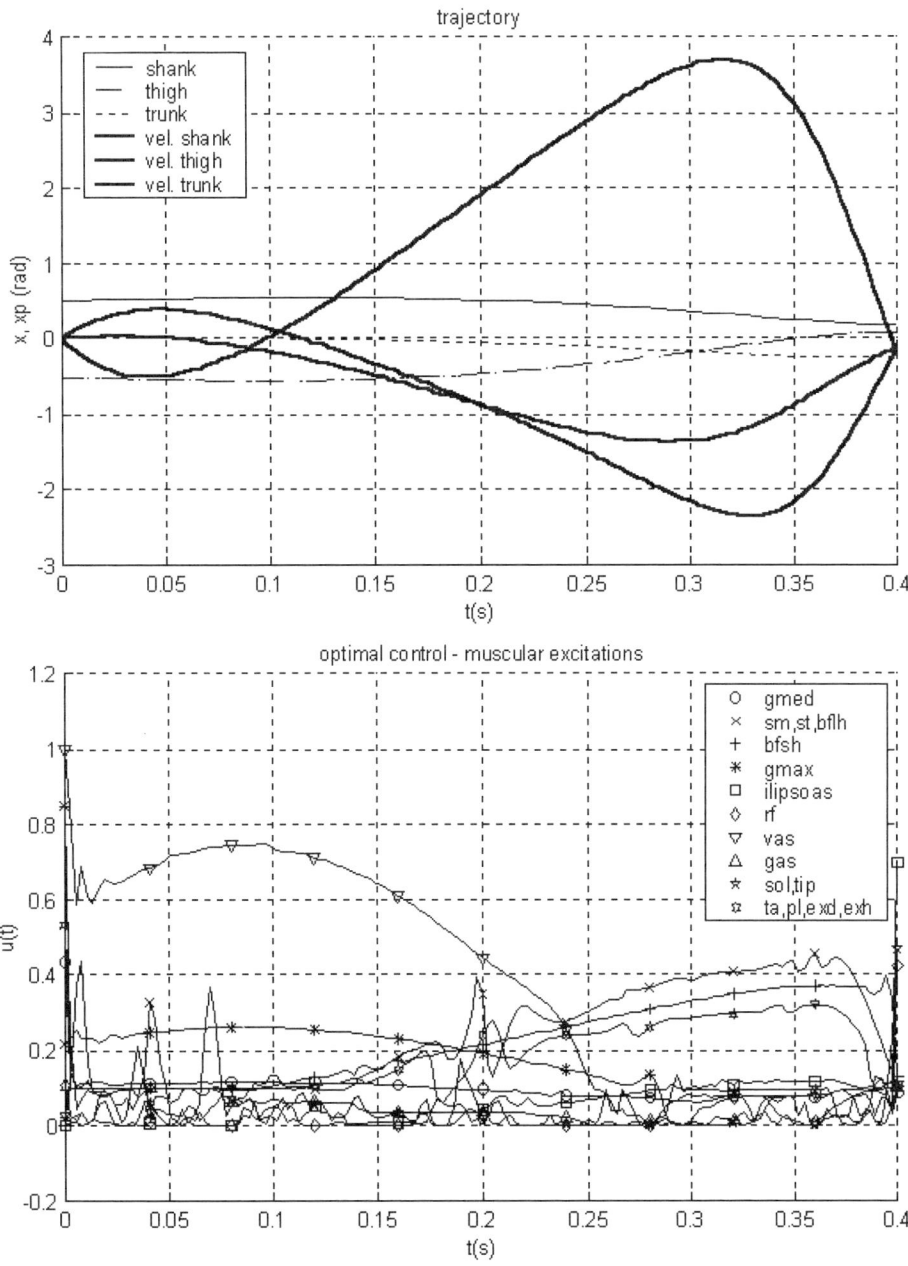

Figure 7 Trajectory and optimal control for $\pi/30$ terminal constraints. 4-th order Runge–Kutta, 150 points, control bounds [0,10]. CPU time Pentium III 600 MHz: 28 min. Cost function $f = 9.3786$.

in the calculus of the first point after the initial condition in the mesh of the 2-nd interval that is not evaluated with the information of the $i-1$ knot. The control functions obtained showed a sharp discontinuity in the interface between the intervals. In the sequence, these results were used as an initial control guess for the OCP, and smooth controls were obtained.

For the biomechanical model containing 10 non-linear muscles the strategies described above are not adequate, due to the low order of the integrator, and other methods were tested to increase the simulation time. The first considered a complete solution for 0.4 seconds and added a new set of points with constant control values, using this augmented control vector as an initial guess. After a large number of numerical tests, convergent results were found for, in the maximum, 50 added points, as shown in Figure 8.

The other strategy, tested for 1-second final time, was decreasing the value of gravity, finding an optimal solution with some initial guess and applying the solution to another simulation with a greater value of gravity, until the real 9.81 m/s$^2$ acceleration was reached. Initially, $g = 2$ m/s$^2$ was tested for an initial control guess of constant values of control corresponding to the vector [0.04 0.04 0.04 0.04 0.10 0.08 0.04 0.04 0.08 0.04], where the sequence of muscles is the same shown in the legend of the figures. The results for terminal inequality constraints of $\pi/5$ are shown in Figure 9. Subsequent tests were performed with 5, 7.5 and finally 9.81 m/s$^2$ (see details in [1]). After six subsequent simulations, spending a total of 58.5 hours of CPU time, the results indicated in Figure 10 were achieved.

# 7   Discussion

The results obtained allow us to point out some of numerical problems that arise when the optimal trajectories for a musculoskeletal model are determined by an iterative integration method, like the ones based on the consistent approximation theory. The strong dependency of the initial guess of the control vector on the convergence of the results, for every degree of complexity of the musculoskeletal model, should be noted. This effect may be assigned to the intrinsically unstable nature of the problem, which in the real biological system is changed by the feedback circuits related to the spinal cord and brain stem. As in this work such feedback mechanisms are not considered, any initial guess for the iterative optimal control procedure must lie in some region of the phase-space where a feasible solution can be reached when the augmented Lagrangean is minimized. This equivalent to saying that the initial control guess does not fall out of the optimization solution attraction basin.

Analysis of the strategies used to generate initial control vectors indicates that the most appropriate, for every specified final time, is varying gravity. This procedure is able to find a stable first solution from the "not too bad" guess derived from some heuristics consisting of activating the main extensors, for example. Multiple bonded simulations are restricted to an Euler integrator, and the LQR controller leads to negative values of the control variable, besides the numerical corruption addressed in item 4. Augmenting a solution with more points can lead to at most 30% of increase in the final time.

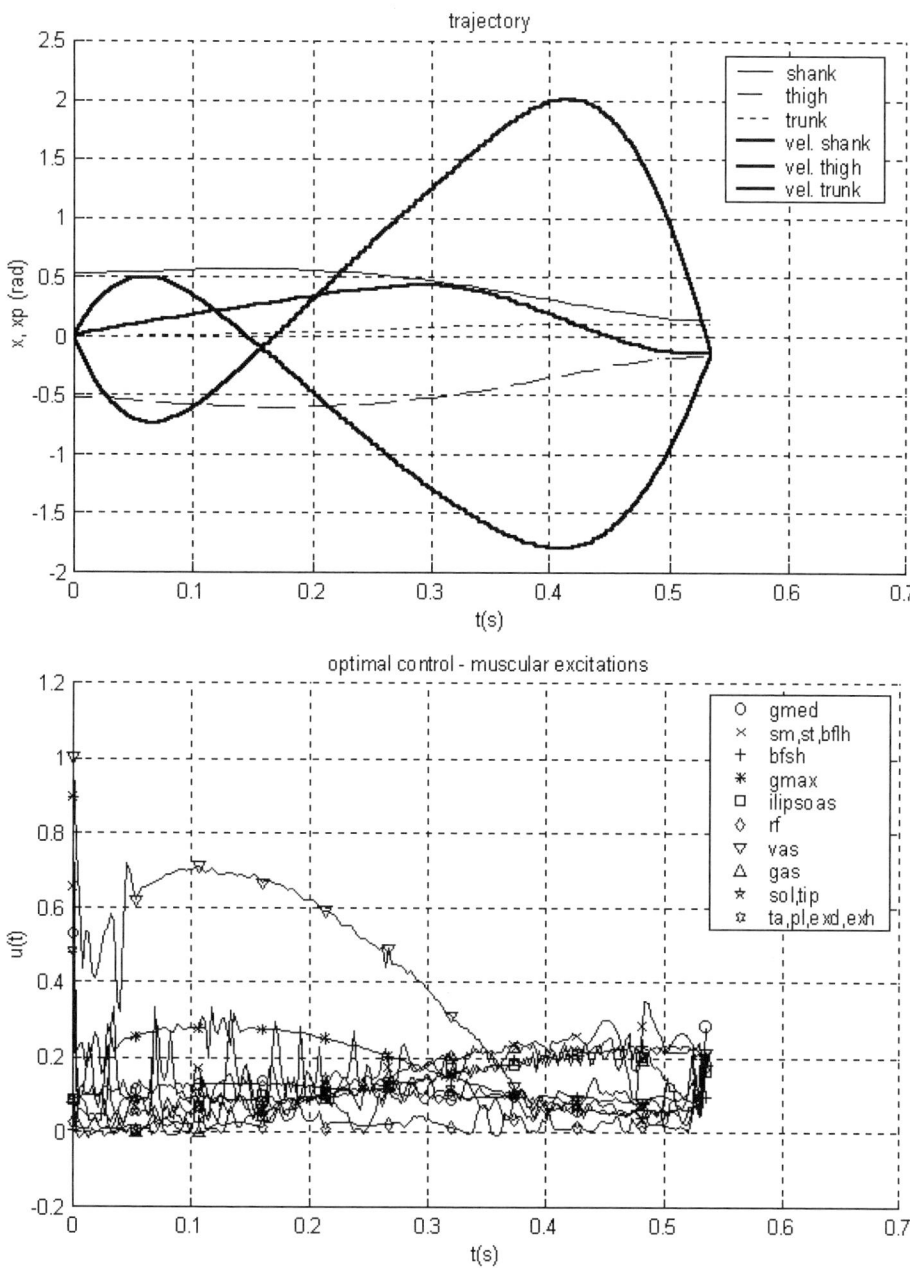

Figure 8 Trajectory and optimal control for $\pi/30$ terminal constraints. LSODA integrator, 150+50 points, control bounds [0,10]. CPU time Pentium III 600 MHz (4 simulation sequence, using 0.4 seconds result): 3.7 hours. $f = 6.8569$.

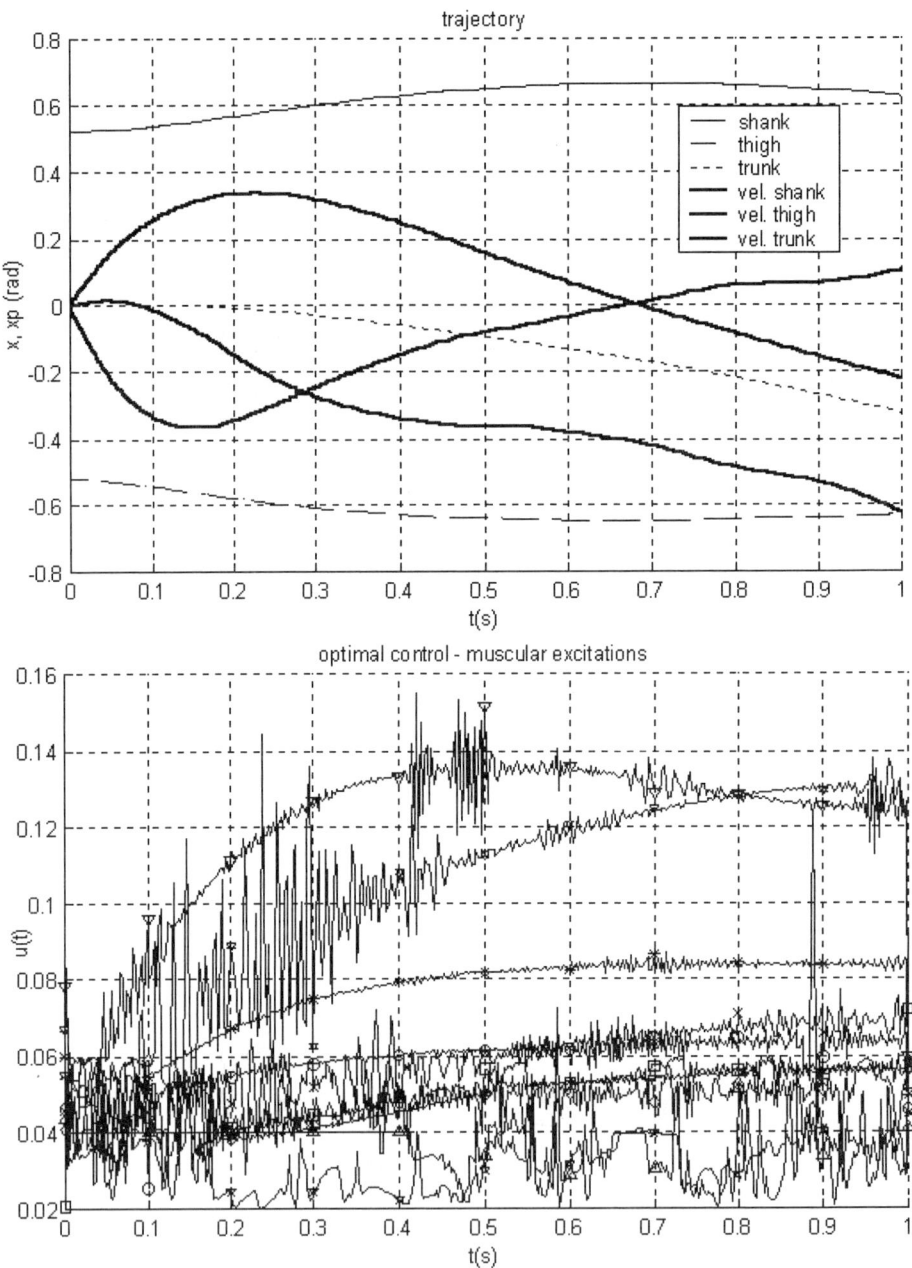

Figure 9 Trajectory and optimal control for $\pi/5$ terminal constraints. $g = 2\,\mathrm{m/s}^2$. Runge–Kutta 4-th order integrator, 400 points, control bounds [0.04,10], optim. and constr. tol. of 0.05 and 0.02. CPU time Pentium III 600 MHz: 22.6 hours. $f = 0.2201$.

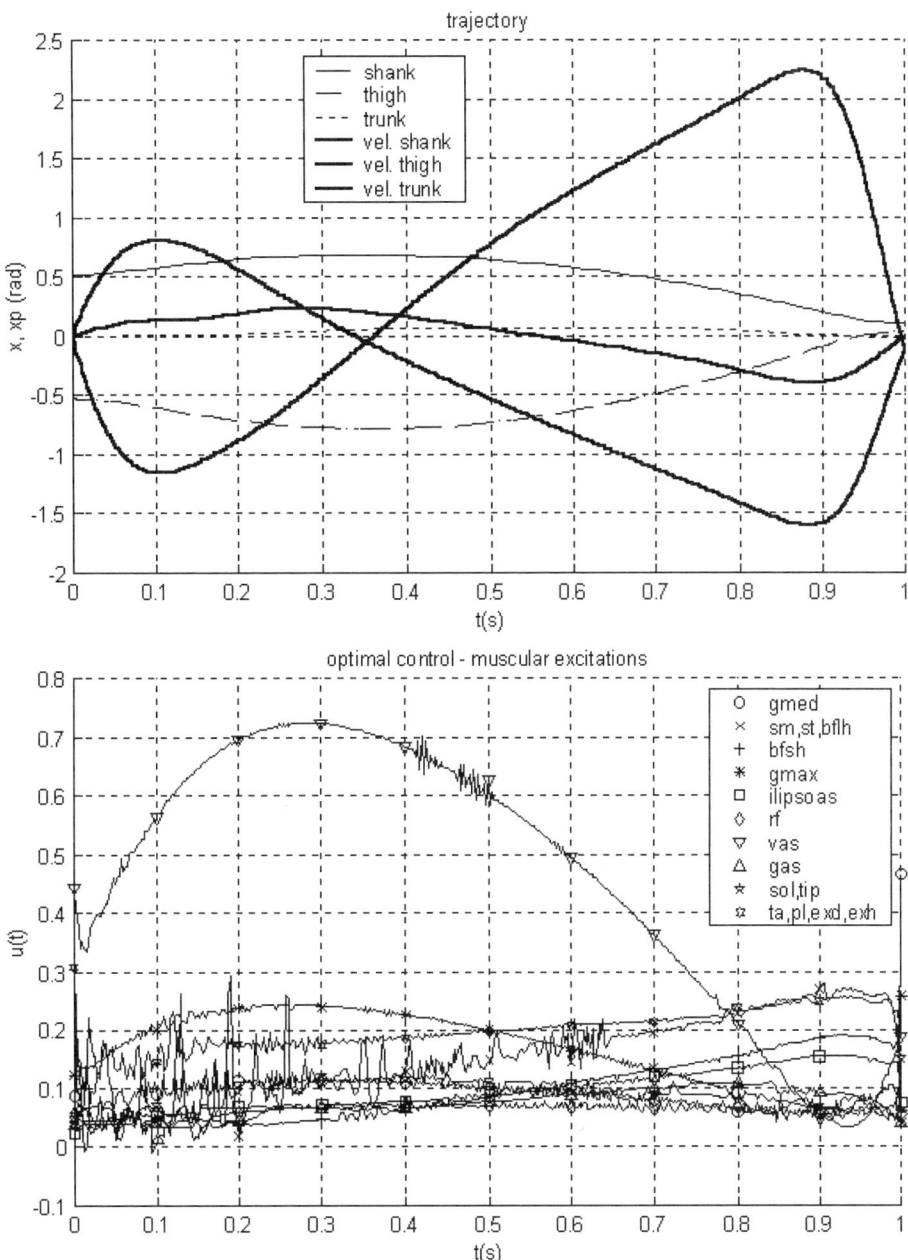

Figure 10  Trajectory and optimal control for $\pi/30$ terminal constraints. $g = 9.81 \text{ m/s}^2$. Runge–Kutta 4-th order integrator, 400 points, control bounds [0,10], optim. and constr. tol. of 0.1 and 0.12. Total CPU time Pentium III 600 MHz (6 simulations): 58.5 hours. $f = 4.0576$.

It should be noted that no convergence was reached for Runge–Kutta and spline interpolation order lower than 4-th, although LSODA can be used. This last choice provided very noisy responses (see Figure 8), and was slower than Runge–Kutta.

Terminal inequality constraints should be used, instead of equality, for the system proposed here. This choice allows a greater flexibility in imposing the constraints progressively, and the results are almost equivalent to imposing equality constraints. In this case, the final error of the states obtained at the end of the simulation is around the value of the prescribed violation tolerance, when in other cases it is equal to the value of the inequality plus the tolerance. These constraints should be imposed progressively, and convergence could not be obtained if a very narrow value was used in the first iteration. The same can be said about control constraints. In the postural task proposed here, maximum values of muscle excitation exceeded only slightly the upper control bounds. However, in the results presented in Figure 8, the previous simulation was done with the same conditions, but the upper bound was set to a very high value. In these results, two muscles only in the first point of the mesh presented values of excitation close to 1.5. The cutting-off from 1.5 to 1.0 took 2.4 hours of CPU time, and the cost increased from 6.0173 to 6.8569.

Care must be taken in the choice of optimization and constraint violation tolerances of the NPSOL routine. Optimization tolerance around 0.1 was used in the most of the convergent results, and the minimum value with convergence was 0.01. Constraint violation was set to the same value in most cases. Reduced values should be used, and terminal constraints would be reached more closely, but must be set smaller than the lower control bound, or a division-by-zero error occurs.

The methodology proposed here for the model with fixed moment-arm and musculotendon length should be extended to variable values of these parameters, using the regression equations obtained in [1]. Other initial conditions and postural tasks should be simulated, such as rising from a chair or gait. The optimal cost function used seems to produce physiologically plausible responses, but the muscular excitation patterns obtained should be compared with filtered EMG in experimental studies; this may suggest the use of other optimal cost functions.

The results obtained could be used for finding feedback matrices, simulating some of the spinal cord and brain stem control mechanisms. As the open-loop optimal trajectory is known, extended linearization might be applied for finding such variable control gains.

More reliable models of the human musculoskeletal complex should consider individual specific parameters. When the methodologies for finding these parameters non-invasively is available, the optimal control technique may become a functional tool for simulating the neuro-mechanical effect of orthopedic simulations and the planning of rehabilitation therapies.

# Acknowledgments

The authors gratefully acknowledge Fundação de Amparo à Pesquisa do Esatdo de São Paulo (FAPESP) for funding the research project and CNPq, for the Ph.D. scholarship to one of the authors, L.L. Menegaldo. The authors are also grateful to Dr. Raul Gonzales and José Jaime Cruz for critical discussion of the results.

# References

1. Menegaldo, L.L. (2001) Biomechanical Modeling and Optimal Control of Human Posture using Consistent Approximations Theory Algorithms (in Portuguese). Ph.D. Thesis, University of São Paulo.

2. MusculoGraphics Inc. (1997) SIMM Software for Interactive Musculoskeletal Modeling. User Manual, Version 1.2.5, May.

3. Polak, E. (1997) Optimization: Algorithms, and Consistent Approximations, In *Applied Mathematical Series*, J.E. Marsden, L. Sirovich (eds), Springer-Verlag, New York.

4. Schwartz, A.L. (1996) Theory and Implementation of Numerical Methods Based on Runge–Kutta Integration for Solving Optimal Control Problems. Ph.D. dissertation, University of California at Berkley.

5. Schwartz, A.L. and Polak, E. (1996) Consistent Approximations for Optimal Control Problems based on Runge–Kutta Integration. *SIAM Journal of Control and Optimization*, **34** (4) 1235–1296.

6. Kandel, E.K., Shwartz, J.H. and Jessel, T.M. (1991) *Principles of Neural Science*, 3-d Edition. Elsevier Science Publishing, New York.

7. Enoka, R.M. (1994) *Neuromechanical basis of Kinesiology*, 2-nd Edition, Human Kinetics, Champain, IL.

8. An, K.N., Kaufman, K.R. and Chao, E.Y.-S. (1995) Estimation of Muscle and Joint Forces. In *Three-Dimensional Analysis of Human Movement*, P. Allard, I.A.F. Stokes, J.-P. Blanchi (eds), 1st Edition, Human Kinetics.

9. Tsirakos, D., Baltzopoulos, V. and Bartlett, R. (1997) Inverse Optimization: Functional and Physiological Considerations Related to the Force-Sharing Problem. *Critical Reviews in Biomedical Engineering*, **25** (4-5) 371–407.

10. Yamaguchi, G.T., Moran, D.W. and Si, J. (1995) A Computationally Efficient Method for Solving the Redundant Problem in Biomechanics. *Journal of Biomechanics*, **28**, 999–1005.

11. Anderson, F.C., Pandy, M.G. and Hull, D.G. (1992) A Parameter Optimization Approach for the Optimal Control of Large-Scale Musculoskeletal Systems. *ASME Journal of Mechanical Engineering*, **114**, 450–460.

12. Pandy, M.G., Garner, B.A. and Anderson, F.C. (1995) Optimal Control of Non-Ballistic Muscular Movements: A Constraint-Based Performance Criterion for Rising from a Chair. *ASME Jounal of Biomechanical Engineering*, **117**, 15–26.

13. Anderson, F.C. and Pandy, M.G. (1999) A Dynamic Optimization Solution for Vertical Jumping in Three Dimensions. *Computer Methods in Biomechanics and Biomedical Engineering*, **2**, 201–231.

14. Anderson, F.C., Ziegler, J.M., Pandy, M.G. and Whalen, R.T. (1996) Solving Large-Scale Optimal Control Problems for Human Movement using Supercomputers. In *Building a Man in the Machine: Computational Medicine, Public Health, and Biotechnology*, J. Witten, D.J. Vincent (eds), Part 2, World Scientific, pp. 1088–1118.

15. Kaplan, M.L. and Heegaard, J.H. (1999) An Efficient Optimal Control Algorithm for Human Locomotion. In *Applied Mechanics in the Americas*, vol. 6, B. Gonçalves, I. Jasiuk, D. Pamplona, C.R. Steele, H.I. Weber, C. Bevilacqua (eds), American Academy of Mechanics and Brazilian Society of Mechanical Sciences, Rio de Janeiro, pp. 31–34.

16. Kuo, A. (1995) An Optimal Control Model for Analyzing Human Postural Balance. *IEEE Transactions on Biomedical Engineering*, **42** (1), 97–101.

17. Spägele, T., Kistner, A. and Gollhofer, A. (1999) Modelling, Simulation and Optimization of a Human Vertical Jump. *Journal of Biomechanics*, **32**, 521–530.

18. Menegaldo, L.L. (1997) Mathematical Modelling, Simulation and Artificial Control of Human Posture (in Portuguese). Master Thesis, Faculdade de Engenharia Mecânica - UNICAMP.

19. Menegaldo, L.L. and Weber, H.I. (1998) Biomechanics of Upright Standing in Humans: a Simulation Study. In *Computer Methods in Biomechanics and Biomedical Engineering – 2*, J. Middleton, M.L. Jones, G.N. Pande (eds), Gordon and Breach Science Publishers, Amsterdam, The Netherlands, pp. 775–782.

20. Canon, M.D., Cullum Jr., C.D. and Polak, E. (1970) *Theory of Optimal Control and Mathematical Programming*. McGraw-Hill Series in Systems Science, McGraw-Hill, Inc., New York.

21. Polak, E. (1971) *Computational Methods in Optimization - A Unified Approach*. Mathematics in Science and Engineering, vol. 77, Academic Press, Inc., New York.

22. Polak, E. (1993) On the Use of Consistent Approximations in the Solution of Semi-Infinite Optimization and Optimal Control Problems. *Math. Progr.*, **62**, 385–415.

23. Schwartz, A.L., Polak, E. and Chen, Y. (1997) RIOTS - Recursive Integration Optimal Control Trajectory Solver. A MATLAB toolbox for solving optimal control problems. Version 1.0, Dept. of Electrical Engineering and Computer Science, University of California at Berkley.

24. Byrne, G.D., and Hindsmarsh, A.C. (1987) Stiff ODE Solvers: A Review of Current and Coming Attractions. *Journal of Computational Physics*, **70**, 1–62.

25. Schwartz, A.L. and Polak, E. (1997) Family of Projected Descent Methods for Optimization Problems with Simple Bounds. *Journal of Optimization Theory and Applications*, **92**, 1–31.

26. Gill, P.E., Murray, W., Saunders, M.A. and Wright, M. (1998) User's Guide for NPSOL 5.0: a Fortran Package for Nonlinear Programming, Systems Optimization Laboratory, Stanford University, Technical Report SOL 86-1, Revised.

27. Gill, P.E., Murray, W. and Wright, M. (1981) *Practical Optimization*. Academic Press Limited, London.

28. Vaughan, C.L., Davis, B.L. and O'Connor, J.C. (1992) *Dynamics of Human Gait*, 1st Edition. Human Kinetics Publishers, Chicago, IL.

29. Yeadon, M.R. and Morlock, M. (1989) The Appropriate Use of Regression Equations for the Estimation of Segmental Inertial Parameters. *Journal of Biomechanics*, **22**, 683–689.

30. Drills, R., Contini, R. and Bluestein, W. (1972) Body Segment Parameters. *Artificial Limbs*, **16** (1).

31. Menegaldo, L.L., Cestari, I.A., Carvalho Jr., A.C.S., Fleury, A.T. and Weber, H.I. (1999) A Model of Skeletal Muscular Dynamics: Analytical Development and Experimental Testing in Canine Lastissumus Dorsi. In *Applied Mechanics in the Americas*, vol. 6, B. Gonçalves, I. Jasiuk, D. Pamplona, C.R. Steele, H.I. Weber, C. Bevilacqua (eds), American Academy of Mechanics and Brazilian Society of Mechanical Sciences, Rio de Janeiro, pp. 11–14.

32. Zajac, F.E. (1989) Muscle and Tendon: Properties, Models, Scaling and Application to Biomechanics and Motor Control. *CRC Critical Reviews in Biomedical Engineering*, **17** (4), 359–411.

33. Zajac, F.E. and Gordon, M.E. (1989) Determining Muscle's Force and Action in Multi-Articular Movement. In *Exercise and Sport Science Reviews*, K. Pandolf (ed.), vol. 17, Williams & Wilkins, Baltimore, pp. 187–230.

34. Delp, S.L. and Loan, J.P. (1995) A Software System to Develop and Analyze Models of Musculoskeletal Structures. *Computers in Biology and Medicine*, **25**, 21–34.

35. Delp, S.L., Loan, J.P., Hoy, M.G., Zajac, F.E., Topp, E.L. and Rosen, J.M. (1990) An Interactive Graphics-Based Model of the Lower Extremity to Study Orthopaedic Surgical Procedures. *IEEE Transactions on Biomedical Engineering*, **37**, 757–767.

36. Brand, R.A., Crowninshield, R.D., Wittstock, C.E., Pederson, D.R., Clark, C.R. and van Krieken, F.M. (1982) A Model of Lower Extremity Muscular Anatomy. *ASME Journal of Biomechanical Engineering*, **104**, 304–310.

37. Brand, R.A., Pedersen, D.R. and Friederich, J.A. (1986) The Sensitivity of Muscle Force Predictions to Changes in the Physiologic Cross-Sectional Area. *Journal of Biomechanics*, **19**, 589–596.

38. Wickewicz, T.L., Roy, R.R., Powell, P.L. and Edgerton, V.R. (1983) Muscle Architecture of the Human Lower Limb. *Clinical Orthopaedics and Related Research*, **179**, 275–283.

39. Friederich, J.A. (1990) Muscle Fiber Architecture in the Human Lower Limb. *Journal of Biomechanics*, **23**, 91–95.

# Subject Index